Distributions

To Laurent Schwartz,

For his Theory of Distributions, *obviously,*
without which this book could not have existed,
but also, and above all, for his kindness and courage.

The clarity of Schwartz's analysis classes at the École
Polytechnique in 1968 made the dunce that I was there happy.
Even if I arrived late, even if I had skipped a few sessions,
everything was clear, lively and easy to understand.
His soft voice, benevolent smile, mischievous eye —
especially when, with an air of nothing, he was
watching for reactions to one of his veiled jokes,
"a tore, from the Greek toro, the tyre"—
he made people love analysis.

When master's students at the university demanded
"a grade average for all" in 1969, most professors
either complied or slunk away. Not Schwartz.
When the results were posted—
I was there, to make up easily for the
calamitous grades I had earned at Polytechnique—
he came alone, frail, in front of a fairly excited horde.
He explained, in substance:

"An examination given to all, without any value, would
no longer allow one to rise in society through knowledge.
Removing selection on the basis of merit would leave
the field open to selection by money or social origin".

Premonitory, alas.

Analysis for PDEs Set

coordinated by
Jacques Blum

Volume 3

Distributions

Jacques Simon

iSTE

WILEY

First published 2022 in Great Britain and the United States by ISTE Ltd and John Wiley & Sons, Inc.

ISTE Ltd
27-37 St George's Road
London SW19 4EU
UK

www.iste.co.uk

John Wiley & Sons, Inc.
111 River Street
Hoboken, NJ 07030
USA

www.wiley.com

Library of Congress Control Number: 2022936452

British Library Cataloguing-in-Publication Data
A CIP record for this book is available from the British Library
ISBN 978-1-78630-525-1

Contents

Introduction

Objective. This book is the third of seven volumes dedicated to solving partial differential equations in physics:

Volume 1: *Banach, Frechet, Hilbert and Neumann Spaces*
Volume 2: *Continuous Functions*
Volume 3: *Distributions*
Volume 4: *Integration*
Volume 5: *Sobolev Spaces*
Volume 6: *Traces*
Volume 7: *Partial Differential equations*

This third volume aims to construct the space of distributions with real or vectorial values and to provide the main properties that are useful in studying partial differential equations.

Intended audience. We [1] have looked for simple methods that require a minimal level of knowledge to make this tool accessible to as wide an audience as possible — doctoral students, university students, engineers — without loosing generality and even generalizing certain results, which may be of interest to some researchers.

This has led us to choose an unconventional approach that prioritizes semi-norms and sequential properties, whether related to completeness, compactness or continuity.

1. **We?** We, it's just "me"! There's no intention of using the Royal We, dear reader, but this *modest (?)* "we" is commonly used in scientific texts when an author wishes to speak of themselves. *It is out of modesty that the writers of Port-Royal made this the trend so that they could avoid, they say, the vanity of "me"* [Louis-Nicolas BESCHERELLE, *Dictionnaire universel de la langue française*, 1845].

Utility of distributions. The main advantage of distributions is that they provide derivatives of all continuous or integrable functions, even those which are not differentiable, and thus broaden the scope of application of differential calculus. This is especially useful for solving partial differential equations.

To this end, a family of objects, the distributions, is defined, with the following properties.

— Any continuous function is a distribution.
— Any distribution has partial derivatives, which are distributions.
— For a differentiable function, we find the conventional derivatives.
— Any limit of distributions is a distribution.
— Any Cauchy sequence of distributions has a limit.

These properties may be roughly summarized by saying that the space $\mathcal{D}'$ of distributions is the *completion with respect to derivation* of the space $\mathcal{C}$ of continuous functions. This construction, due to Laurent SCHWARTZ, [69] and [72], is completed here for distributions on an open subset Ω of $\mathbb{R}^d$ with values in a Neumann space E, i.e. a sequentially complete separable semi-normed space. This includes values in a Banach or Fréchet space.

Originality. The quest for simple methods [2] giving general properties led us to proceed as follows.

— Directly consider *vectorial values*, i.e. constructing $\mathcal{D}'(\Omega; E)$ without any prior study of real distributions.
— Assume that E *is sequentially complete*, i.e. a Neumann space.
— Use *semi-norms* to construct the topologies of E, $\mathcal{D}(\Omega)$, $\mathcal{D}'(\Omega; E)$, *etc.*
— Equip $\mathcal{D}'(\Omega; E)$ with the *simple topology*.
— Introduce *weighting* to generalize the convolution to open domains.
— Explicitly construct the *primitives*.
— *Separate the variables* using a "basic" method.
— Only use *integration* for *continuous functions*.

Let us take a closer look at these points that lie off the beaten track.

Vector values. We consider distributions with values in a general Neumann space E even though the partial differential equations in physics generally have real values. This is useful in evolution equations to separate the time t from the variable of space x. A distribution over t, x with real values is then identified with a distribution over t with values in a space E of distributions on x, for example, with an element of

2. **Focus and simplicity.** This was one of Steve JOBS' favourite mantras: "Simple can be harder than complex: You have to work hard to get your thinking clean to make it simple. But it's worth it in the end because once you get there, you can move mountains." [*BusinessWeek*, 1998].

$\mathcal{D}'((0,T);E)$ where $E = \mathcal{D}'(\Omega)$, which is itself a Neumann space. This identification is made possible by the fundamental *kernel theorem*, p. 312.

A list of the most useful Neumann spaces is given on page 43.

For stationary equations, the real distributions (that is, the case where $E = \mathbb{R}$) are sufficient. We will directly work on the case where E is a Neumann space in order to avoid repetitions, the generalization often consisting of replacing $\mathbb{R}$ with E and the absolute value $|\ |$ with a semi-norm of E in the statements and proofs, when using appropriate methods.

Particular features in the case of vector values. The main differences as compared to distributions with real values are as follows, for a general space E.

— The space $\mathcal{D}'(\Omega;E)$ is not reflexive and its topology of pointwise convergence on $\mathcal{D}(\Omega)$ does not coincide with its weak topology.

— The bounded subsets of $\mathcal{D}'(\Omega;E)$ are not relatively compact.

— The distributions over Ω are not of a finite order over its compact parts: they cannot always be expressed as finite order derivatives of continuous functions.

— Variables may be separated by constructing a bijection from $\mathcal{D}'(\Omega_1 \times \Omega_2;E)$ onto $\mathcal{D}'(\Omega_1;\mathcal{D}'(\Omega_2;E))$ (even for a real distribution, i.e. for $E = \mathbb{R}$, this brings in vector values, in this case in $\mathcal{D}'(\Omega_2)$).

Sequential completeness. We assume that E is a Neumann space, i.e. that all its Cauchy series converge, since this is an essential condition for continuous functions to be distributions. That is, for $\mathcal{C}(\Omega;E) \subset \mathcal{D}'(\Omega;E)$, see section 3.4, *The case where E is not a Neumann space*, p. 53.

This property is simpler than the completeness, i.e. the convergence of all the Cauchy filters, and is especially more general: for example, if H is a Hilbert space with infinite dimensions, H-weak is sequentially complete but is not complete [Vol. 1, Property (4.11), p. 63].

It is also simpler and more general than quasi-completeness, i.e. the completeness of bounded subsets, used by Laurent SCHWARTZ [72, p. 2, 50 and 52].

Semi-norms. We use families of semi-norms rather than locally convex topologies, which are equivalent, in order to be able to define $L^p(\Omega;E)$ in Volume 4. Indeed, it is possible to raise a semi-norm to a power p, but not a convex neighborhood!

The handling of semi-normed spaces is simple, although it is less familiar than that of topological spaces: it follows the handling of normed spaces, the main difference being that there are several semi-norms or norms instead of a single norm. For example, we bring in the topology of $\mathcal{D}(\Omega)$ through the family of semi-norms

$\|\varphi\|_{\mathcal{D}(\Omega);p} = \sup_{x\in\Omega,\, |\beta|\le p(x)} p(x)|\partial^\beta\varphi(x)|$ indexed by $p \in \mathcal{C}^+(\Omega)$, which is much simpler than its (equivalent) construction as the inductive limit of the $\mathcal{D}_K(\Omega)$.

Simply topology. We equip the space $\mathcal{D}'(\Omega; E)$ with the family of semi-norms $\|f\|_{\mathcal{D}'(\Omega;E);\varphi,\nu} = \|\langle f, \varphi\rangle\|_{E;\nu}$ indexed by $\varphi \in \mathcal{D}(\Omega)$ and $\nu \in \mathcal{N}_E$ (set indexing the semi-norms of E), i.e. with the topology of simple convergence on $\mathcal{D}(\Omega)$, as it is well-suited to our study ... and is simple. This simplicity is achieved without restricting ourselves to a *pseudo-topology* as is done in several texts.

In addition, this topology has the same convergent sequences and the same bounded sets as the topology of uniform convergence on the bounded subsets of $\mathcal{D}(\Omega)$ used by Laurent SCHWARTZ. The reasons for our choices are detailed on p. 45.

Open domain and weighting. We consider distributions defined on an open subset Ω of $\mathbb{R}^d$. As these do not necessarily have an extension to all of $\mathbb{R}^d$, we introduce an operation, we call it weighting, which plays a role for Ω that is similar to the role played by convolution for $\mathbb{R}^d$ and which we constantly use.

The weighted distribution $f \diamond \mu$ of a distribution f, defined on an open set Ω, by a weight μ, which is a real distribution on $\mathbb{R}^d$ with a compact support D, is a distribution defined on the open set $\Omega_D = \{x \in \mathbb{R}^d : x + D \subset \Omega\}$. When f and μ are functions, it is given by $(f \diamond \mu)(x) = \int_{\mathring{D}} f(x+y)\mu(y)\,dy$. When $\Omega = \mathbb{R}^d$, the convolution is recovered up to a symmetry on μ, and all its properties are recovered up to a possible sign.

Primitives. We show that a field of distributions $q = (q_1, \dots, q_d)$ has a primitive f, that is $\nabla f = q$, if and only if it satisfies $\langle q, \psi\rangle = 0_E$ for all the test fields $\psi = (\psi_1, \dots, \psi_d)$ such that $\nabla \cdot \psi = 0$. It is the *orthogonality theorem.* We explicitly determine all the primitives and among these determine one which depends continuously on q.

We also demonstrate that when Ω is simply connected it is necessary and sufficient that $\partial_i q_j = \partial_j q_i$ for all i and j. It is the *Poincaré's generalized theorem.*

Separation of variables. We show that the separation of variables is bijective from $\mathcal{D}'(\Omega_1 \times \Omega_2; E)$ onto $\mathcal{D}'(\Omega_1; \mathcal{D}'(\Omega_2; E))$ by means of inequalities. These are certainly laborious to establish, but they avoid the difficult topological properties used by Laurent SCHWARTZ in his diabolical proof of this *kernel theorem.*

The advantage of this method is laid out in the commentary *Originality...*, p. 317.

Integration. The integration of continuous functions is essential to identify them with distributions through the equality $\langle f, \varphi\rangle = \int f\varphi$, for all test functions φ. Since the theory of integration was not developed for values in a Neumann space, in Volume 2 we established results relative to uniformly continuous functions that meet our requirements. We remind them before using them in this volume.

The general theory of integration with values in a Neumann space will be done in a later volume in the context of *integrable distributions*, which play the role of the usual *classes of almost equal integrable functions*. Indeed, it has seemed simpler to thus construct general integration.

Prerequisite. The proofs in the main body of the text only use the definitions and results already established in Volumes 1 and 2, recalled either in the Appendix or in the text, with references to their proofs.

This book has been written such that it can be read in an out-of-order fashion by a non-specialist: the proofs are detailed and include arguments that may be trivial for an expert and the numbers of the theorems being used are systematically recalled. These details are even more necessary [3] since the majority of the results are generalizations, that are new, to functions and distributions with values in a Neumann space of properties that are classic for values in a Banach space.

I request the reader to be lenient with how heavy this may make the text.

Comments. Unlike the main body of the text, the comments, appearing in smaller font, may refer to external results or those not yet established. The appendix 'Reminders' is also written in smaller font as it is assumed that the content is familiar.

Historical overview. Wherever possible, the origin of the concepts and results is specified in footnotes [4].

3. **Necessary details.** As Laurent SCHWARTZ explained in the preamble to one of his articles [71, p. 88]: "*Although many proofs are relatively easy, we find it useful to write them in extenso, because whenever topological vector spaces come into play there are so many 'traps' that great rigor is needed*".
Given that the great and rigorous Augustin CAUCHY has himself arrived at an erroneous result, we have not treated any detail too lightly. Let us recall that, in 1821, in his remarkable *Cours d'Analyse de l'École Royale Polytechnique*, he declared that he had 'easily' [18, p. 46] proven that if a real function with two real variables is continuous with respect to both the variables, it is continuous with respect to their couple. It was not till 1870 that Carl Johannes THOMAE [89, p. 15] demonstrated that this was inexact.

4. **Historical overview. Objective.** This is first of all to honour the mathematicians whose work has made this book possible and inspire it. Although some may be missing, either due to limited space or knowledge. The other objective is to show that the world of mathematics is an ancient human construction, not a "revealed truth", and that behind each theorem there are one or more humans, our contemporaries or distant ancestors who — including the Greeks — reasoned just as well as us, without internet, computers or even printing and paper.

The forgotten. The French are probably over-represented here, as they are in all french libraries and teaching and, often, in french hearts. Among the French, I am over-represented, because this book is the result of thirty years of work I have carried out to simplify and generalize distributions with vectorial values.

Navigating this book.
— The **table of contents**, at the beginning of this book, lists the topics discussed.
— The **index**, p. 371, provides another thematic access.
— The **table of notations**, p. xv, specifies the meaning of the symbols used.
— The hypotheses are all stated within the theorems themselves.
— The numbering is common to all the statements, so that they can be easily found in numerical order (for instance, Theorem 2.2 is found between statements 2.1 and 2.3, which are definitions).

Acknowledgements. I am particularly grateful to Enrique FERNÁNDEZ-CARA who has proofread countless versions of this text, indefatigable, and who has given me various friendly suggestions for equally countless improvements in each version.

Fulbert MIGNOT suggested (among other things) that each chapter should be preceded by a brief introduction. This was very helpful: in order to reveal the guiding principle, I had to highlight it and re-write several sections.

The meticulous and knowledgeable readings carried out by Olivier BESSON, Didier BRESCH and Pierre DREYFUSS contributed substantial improvements to the text.

Jérôme LEMOINE, my disciple — a stigmata and a cross he'll bear for life! —, had the task of proofreading the demonstrations: he is, thus, entirely responsible for any errors there may be ... except, perhaps, those I may have added since.

Jacques BLUM was able to convince me that it was time to publish this. Indeed.

Thank you, my friends, for all your help and warm support.

Jacques SIMON
Chapdes-Beaufort, March 2021

Laurent SCHWARTZ, my primary source of inspiration and admiration, is also perhaps over-represented, since, his treatises having no historical notes due to his great modesty, I have attributed to him the totality of their contents. On the other hand, the Russians and Eastern Europeans are probably particularly under-represented, due to the language barrier, aggravated by the mutual ignorance of the West and East during the Cold War period.

Novelties. At the risk of sounding immodest (well, *nobody is perfect*) I have marked out a large number of results that I believe are new, both to arouse the reader's vigilance — it's not impossible that this book may contain some careless mistake — and to draw their attention to the new tools available to them.

Appeal to the reader. A number of important results lack historical notes because I am not familiar with their origin. I beg the reader's indulgence for these lacunae and, above all, the injustices that may result. And I call upon the erudite among you to flag any improvements to me for future editions.

Notations

SPACES OF DISTRIBUTION

$\mathcal{D}(\Omega)$	space of test functions (infinitely differentiable with compact support)	21
$\mathcal{D}_K(\Omega)$	*id.* with support in the compact set $K \subset \Omega$ (another notation for $\mathcal{C}_K^\infty(\Omega)$)	24
$\mathcal{D}(\Omega; \mathbb{R}^d)$	space of test fields	271
$\mathcal{D}'(\Omega)$	space of real distributions	42
$\mathcal{D}'(\Omega; E)$	space of distributions with values in E	42
$\mathcal{D}'_K(\Omega; E)$	*id.* with support in the closed set $K \subset \Omega$	137
$\mathcal{D}'(\Omega; E^d)$	space of distribution fields	82
$\mathcal{D}'_\nabla(\Omega; E^d)$	space of gradients	253
$\mathcal{D}'(\Omega)$-weak	$\mathcal{D}'(\Omega)$ equipped with its weak topology	80
$\mathcal{D}'(\Omega; E\text{-weak})$	space of distributions with values in E-weak	78
$\mathcal{D}'(\Omega; E)$-unif	$\mathcal{D}'(\Omega; E)$ with topology of uniform convergence on bounded subsets of $\mathcal{D}(\Omega)$	46
$\mathcal{D}'(\Omega_1; \mathcal{D}'(\Omega_2; E)))$	space of distributions of distributions	285

SET OF DISTRIBUTIONS

$\mathcal{D}'^+(\Omega)$	set of real positive distributions	111

SPACE OF FUNCTIONS

$\mathcal{B}(\Omega; E)$	space of uniformly continuous functions with bounded support	14
$\mathcal{C}(\Omega; E)$	space of continuous functions	12
$\mathcal{C}_b(\Omega; E)$	*id.* bounded	12
$\mathcal{C}_K(\Omega; E)$	*id.* with support in the compact set $K \subset \Omega$	18
$\mathcal{C}^+(\Omega)$	set of real positive continuous functions	19
$\mathcal{C}^m(\Omega; E)$	space of m times continuously differentiable functions; case of $m = \infty$	12, 13
$\mathcal{C}_b^m(\Omega; E)$	*id.* with bounded derivatives, case of $m = \infty$	12, 13
$\mathcal{C}_K^m(\Omega; E)$	*id.* with support in the compact set $K \subset \Omega$	18
$\mathbf{C}(\Omega; E)$	space of uniformly continuous functions	342
$\mathbf{C}_b(\Omega; E)$	*id.* bounded	13
$\mathbf{C}_D(\Omega; E)$	*id.* with support in the compact set $D \subset \mathbb{R}^d$	18
$\mathbf{C}_b^m(\Omega; E)$	space $\mathcal{C}^m$ with uniformly continuous and bounded derivatives	13
$\mathcal{K}(\Omega)$	space of real continuous functions with compact support	57
$\mathcal{K}^\infty(\Omega)$	*id.* infinitely differentiable	24
$\mathcal{K}(\Omega; E)$	space of continuous functions with compact support	186
$\mathcal{K}^\infty(\Omega; E)$	space of infinitely differentiable functions with compact support	178

SEPARATED SEMI-NORMED SPACES

SUBSETS AND MAPPINGS OF SEMI-NORMED SPACES

POINTS AND SUBSETS OF $\mathbb{R}^d$

OTHER SETS [1]

AND ALSO

$\emptyset$	empty set
$\subset$	algebraic inclusion
$\setminus$	set difference: $U \setminus V = \{u \in U : u \notin V\}$
$\times$	product: $U \times V = \{(u, v) : u \in U,\ v \in V\}$

SPECIFIC FUNCTIONS AND DISTRIBUTIONS

TYPOGRAPHY

▮	end of statement
□	end of proof or comment

FIGURES

1. **Notation of natural numbers.** The notation of $\mathbb{N}$ and $\mathbb{N}^*$ follows the ISO 80000-2 standard, see p. 349.

Chapter 1

Semi-Normed Spaces and Function Spaces

In this chapter, we provide definitions for the following essential notions.
— Semi-normed spaces and, in particular, Neumann, Fréchet and Banach spaces (§ 1.1).
— Topological equality and inclusion of semi-normed spaces (§ 1.2).
— Continuous mappings (§ 1.3) and differentiable functions (§ 1.4).
— Spaces of continuously differentiable functions and their semi-norms (§ 1.5).
— The integral of a uniformly continuous function with values in a Neumann space (§ 1.6).

We will make extensive use of their properties established in Volumes 1 and 2, and refer, as to make this book self-contained, to their precise statements in the course of the text or in the Appendix with references to their proofs.

1.1. Semi-normed spaces

Let us define separated semi-normed spaces [1] (the definitions of vector spaces and of semi-norms are recalled in the Appendix, § A.2).

Definition 1.1.– *A* ***semi-normed space*** *is a vector space E endowed with a family of semi-norms $\{\| \, \|_{E;\nu} : \nu \in \mathcal{N}_E\}$.*

Any such space is said to be ***separated*** *(or* ***Hausdorff****) if $u = 0_E$ is the only element such that $\|u\|_{E;\nu} = 0$ for all $\nu \in \mathcal{N}_E$.*

---

1. **History of the notion of semi-normed space.** John von NEUMANN introduced semi-normed spaces in 1935 [56] (with an unnecessary countability condition). He also showed [56, Theorem 26, p. 19] that they coincide with the locally convex topological vector spaces that Andrey KOLMOGOROV previously introduced in 1934 [46, p. 29].

History of the notion of Hausdorff space. Felix HAUSDORFF had included the separation condition in its original definition of a topological space in 1914 [40].

*A **normed space** is a vector space E endowed with a norm $\| \ \|_E$.* ■

Caution. Definition 1.1 of a separated semi-normed space is general but not universal. For Laurent SCHWARTZ [73, p. 240], a semi-normed space is a space endowed with a *filtering* family of semi-norms (Definition 1.8). This definition is equivalent, since every family is equivalent to a filtering family [Vol. 1, Theorem 3.15].

For Nicolas BOURBAKI [10, editions published after 1981, Chap. III, p. III.1] and Robert EDWARDS [30, p. 80], a semi-normed space is a space endowed with a *single* semi-norm, which drastically changes its meaning. □

Let us define bounded subsets [2] of a separated semi-normed space.

Definition 1.2.– *Let U be a subset of a separated semi-normed space E, whose family of semi-norms is denoted by $\{\| \ \|_{E;\nu} : \nu \in \mathcal{N}_E\}$. We say that U is **bounded** if, for every $\nu \in \mathcal{N}_E$,*

$$\sup_{u\in U} \|u\|_{E;\nu} < \infty. \ ■$$

Let us define convergent and Cauchy sequences [3] in a semi-normed space.

Definition 1.3.– *Let $(u_n)_{n\in\mathbb{N}}$ be a sequence in a separated semi-normed space E, whose family of semi-norms is denoted by $\{\| \ \|_{E;\nu} : \nu \in \mathcal{N}_E\}$.*

(a) *We say that $(u_n)_{n\in\mathbb{N}}$ **converges to a limit** $u \in E$, and we denote $u_n \to u$, if, for every $\nu \in \mathcal{N}_E$,*

$$\|u_n - u\|_{E;\nu} \to 0 \ \text{when } n \to \infty.$$

(b) *We say that $(u_n)_{n\in\mathbb{N}}$ is a **Cauchy sequence** if, for every $\nu \in \mathcal{N}_E$,*

$$\sup_{m\geq n} \|u_m - u_n\|_{E;\nu} \to 0 \ \text{when } n \to \infty. \ ■$$

CAUTION. We denote $\mathbb{N} \stackrel{\text{def}}{=} \{0, 1, 2, \ldots\}$ and $\mathbb{N}^* \stackrel{\text{def}}{=} \{1, 2, \ldots\}$, conforming to the ISO 80000-2 standard for mathematical and physics notation (edited in 2009).

Any possible confusion will be of no consequence, apart from surprising the reader used to the opposite notation when seeing a term u_0 of a series indexed by $\mathbb{N}$, or the inverse $1/n$ of a number n in $\mathbb{N}^*$. □

---

2. **History of the notion of bounded set.** Bounded sets in a semi-normed space were introduced in 1935 by John von NEUMANN [56]. Andrey KOLMOGOROV had introduced them in 1934 [46] for topological vector spaces.

3. **History of the notion of convergent sequence.** Baron Augustin CAUCHY gave Definition 1.3 for convergence in $\mathbb{R}$, in 1821 [18, p. 19]. Niels ABEL contributed to the emergence of this notion.

History of the notion of Cauchy sequence. Augustin CAUCHY introduced the convergence criterion of Definition 1.3 for real series, in 1821 [18, p. 115-116], admitting it (i.e. by implicitly considering $\mathbb{R}$ as the completion of $\mathbb{Q}$). Bernard BOLZANO previously stated this criterion in 1817 in [6], trying unsuccessfully to justify it due to the lack of a coherent definition for $\mathbb{R}$.

Let us define several types of sequentially complete spaces.

Definition 1.4.– *A separated semi-normed space is* ***sequentially complete*** *if all its Cauchy sequences converge.*

A ***Neumann space*** *is a sequentially complete separated semi-normed space.*

A ***Fréchet space*** *is a sequentially complete metrizable semi-normed space.*

A ***Banach space*** *is a sequentially complete normed space.* ▮

Neumann spaces. We named these spaces in Volume 1 in homage to John VON NEUMANN, who introduced sequentially complete separated semi-normed spaces in 1935 [56]. Thus, readers should recall the definition before using it elsewhere. Examples of such spaces are given in the commentary *Examples of Neumann spaces*, p. 43. □

Completeness and sequential completeness. A semi-normed space is **complete** if every Cauchy filter converges [SCHWARTZ, 73, Chap. XVIII, § 8, Definition 1, p. 251]. We shall not use this notion since sequential completeness is much simpler and more general (complete implies sequentially complete [SCHWARTZ, 73, p. 251]) and especially since certain useful spaces are sequentially complete but not complete. For example, it is the case of any reflexive Hilbert or Banach space of infinite dimension endowed with its weak topology [Vol. 1, Property (4.11), p. 63]. □

Completeness and Metrizability. Recall that, for a metrizable space, *completeness* is equivalent to *sequential completeness* [SCHWARTZ, 73, Theorem XVIII, 8; 1, p. 251]. This is why some authors speak of completeness for Banach or Fréchet spaces, while in reality they only use sequential completeness. □

Let us give a definition of metrizability.

Definition 1.5.– *A semi-normed space is* ***metrizable*** *if it is separated and if its family of semi-norms is countable or is equivalent to a countable family of semi-norms.* ▮

Definitions of the equivalence of families of semi-norms and of their countability are recalled on pages 4 and 350 (Definitions 1.6 and A.1).

Justification for the name "metrizable". We refer here to a *metrizable* space since every countable family or, what leads to the same thing, any sequence $(\| \|_k)_{k\in\mathbb{N}}$ of semi-norms can be associated with a **distance**, or *metric*, d which generates the same topology, for instance

$$d(u,v) = \sum_{k\in\mathbb{N}} 2^{-k} \frac{\|u-v\|_k}{1+\|u-v\|_k}.$$

Definition 1.5 is, in fact, that of a **separated countably semi-normable space**. We shall abuse terminology and speak of *metrizable* spaces since this equivalent notion is more familiar. To be precise, a **metrizable** space is a space that is "*topologically equal to a metric space*". □

Superiority of a sequence of semi-norms over a distance. The semi-norms of a metrizable space E allow us to characterize its bounded subsets U by (Definition 1.2)

$$\sup_{u\in U} \|u\|_{E;k} < \infty, \text{ for all } k \in \mathbb{N}.$$

On the contrary, *if E is not normable* and if d is a distance that generates its topology, its bounded subsets **are not characterized by**

$$\sup_{u\in U} d(u, 0_E) < \infty.$$

What is worse, is that no "ball" $\{u \in E : d(u, z) < r\}$, for $r > 0$, is bounded. Indeed, the existence of a non-empty bounded open subset is equivalent to normability, due to **Kolmogorov's Theorem** [4]. □

1.2. Comparison of semi-normed spaces

First, let us compare families of semi-norms on the same vector space.

Definition 1.6.– *Let $\{\| \|_{1;\nu} : \nu \in \mathcal{N}_1\}$ and $\{\| \|_{2;\mu} : \mu \in \mathcal{N}_2\}$ be two families of semi-norms on the same vector space E.*

The first family ***dominates*** *the second if, for every $\mu \in \mathcal{N}_2$, there exist a finite subset N_1 of $\mathcal{N}_1$ and $c_1 \in \mathbb{R}$ such that for every $u \in E$,*

$$\|u\|_{2;\mu} \le c_1 \sup_{\nu\in N_1} \|u\|_{1;\nu}.$$

Both families are ***equivalent*** *if each one dominates the other. We also say that they* ***generate the same topology****.* ▮

Terminology. The **topology** of E is the family of its open subsets. We can say that two families of semi-norms *generate the same topology* instead of saying that they *are equivalent*, since the equivalence of the families of semi-norms implies the equality of the families of open subsets [Vol. 1, Theorem 3.4], and reciprocally [Vol. 1, Theorem 7.14 (a) and 8.2 (a), with $L = T =$ Identity]. □

Let us see how we can compare two semi-normed spaces (the definition of a vector subspace is recalled in the Appendix, § A.2).

Definition 1.7.– *Let E and F be two semi-normed spaces, whose families of semi-norms are denoted by $\{\| \|_{E;\nu} : \nu \in \mathcal{N}_E\}$ and $\{\| \|_{F;\mu} : \mu \in \mathcal{N}_F\}$.*

(a) *We denote $E \underset{\leftrightarrows}{\equiv} F$ if $E = F$ and if their additions, multiplications and families of semi-norms coincide. That is to say, if they have the same vector space structure and the same semi-norms.*

4. **History of Kolmogorov's Theorem.** Andrey KOLMOGOROV showed in 1934 [46, p. 33] that a topological vector space is normable if and only if there exists a bounded convex neighborhood of the origin, which is equivalent here to the existence of a bounded open set.

(b) *We say that E is* ***topologically equal*** *to F and we denote $E \underset{\leftrightarrows}{=} F$ if $E = F$, if their additions and multiplications coincide and if their families of semi-norms are equivalent.*

(c) *We say that E is* ***topologically included*** *in F and we denote $E \underset{\rightarrow}{\subseteq} F$ if E is a* ***vector subspace*** *of F and if the family of semi-norms of E dominates the family of restrictions to E of the semi-norms of F.*

That is to say if, for every semi-norm $\mu \in \mathcal{N}_F$, there exist a finite subset N of $\mathcal{N}_E$ and $c \in \mathbb{R}$ such that, for every semi-norm $u \in E$,

$$\|u\|_{F;\mu} \leq c \sup_{\nu \in N} \|u\|_{E;\nu}.$$

(d) *We say that E is a* ***topological subspace*** *of F if it is a vector subspace of F and if it is endowed with the restrictions to E of the semi-norms of F or, more generally, with a family equivalent to the family of these restrictions.* ▮

Caution. Suppose that $E = F$ and that, for every $\mu \in \mathcal{N}_F$, there exists $\nu \in \mathcal{N}_E$ such that, for every $u \in E$,

$$\|u\|_{F;\mu} = \|u\|_{E;\nu}.$$

This equality **does not imply topological equality** $E \underset{\leftrightarrows}{=} F$: it only implies $E \underset{\rightarrow}{\subseteq} F$. Indeed, it does not ensure the existence, for every $\nu \in \mathcal{N}_E$, of a μ satisfying this equality or of a finite family of μ such that $\|u\|_{E;\nu} \leq c \sup_{\mu \in M} \|u\|_{F;\mu}$, which is necessary for the converse inclusion $F \underset{\rightarrow}{\subseteq} E$.

Such an equality of semi-norms occurs for example in step 3, p. 28, of the proof of Theorem 2.12, where we thus prove a converse inequality to get topological equality. □

Let us finally define filtering families of semi-norms.

Definition 1.8.– *A family $\{\|\ \|_\nu : \nu \in \mathcal{N}\}$ of semi-norms on a vector space E is* ***filtering*** *if, for every finite subset N of $\mathcal{N}$, there exist $\mu \in \mathcal{N}$ such that, for every $u \in E$,*

$$\sup_{\nu \in N} \|u\|_\nu \leq \|u\|_\mu. \ ▮$$

Utility of filtering families. The use of filtering families simplifies some statements, by substituting a single semi-norm to the upper envelope of a finite number of semi-norms. This is for example the case with the characterization of continuous linear mappings from Theorem 1.12 where we consider both cases. Definition 1.9 of continuous mappings could similarly be simplified with a filtering family.

Any family of semi-norms is equivalent to a filtering family [Vol. 2, Theorem 3.15], but this one is not necessarily pleasant to use. □

Spaces endowed with filtering families. The "natural" family of some spaces is filtering. For example, $\mathcal{D}(\Omega)$ is endowed (Definition 2.5) with the family, which is filtering (Theorem 2.7), of the semi-norms, indexed by $p \in \mathcal{C}^+(\Omega)$,

$$\|\varphi\|_{\mathcal{D}(\Omega);p} = \sup_{x \in \Omega} \sup_{|\beta| \leq p(x)} p(x) |\partial^\beta \varphi(x)|.$$

A single norm also constitutes, on its own, a filtering family of semi-norms. □

Utility of non-filtering families. The "natural" family of some spaces is not filtering. For example, $\mathcal{D}'(\Omega)$ is endowed (Definition 3.1) with the non-filtering family of the semi-norms, indexed by $\varphi \in \mathcal{D}(\Omega)$,

$$\|f\|_{\mathcal{D}'(\Omega);\varphi} = |\langle f, \varphi \rangle|.$$

If the aim was to consider filtering families only, then $\mathcal{D}'(\Omega)$ should be endowed with the semi-norms

$$|||f|||_{\mathcal{D}'(\Omega);N} = \sup_{\varphi \in N} |\langle f, \varphi \rangle|$$

indexed by the finite subsets N of $\mathcal{D}(\Omega)$. The maximum over finite subsets would of course disappear from Definition 1.9 of continuous functions but it would reappear here. □

1.3. Continuous mappings

We now define various notions of continuity[5] of a mapping from a semi-normed space into another.

Definition 1.9.– *Let T be a mapping from a subset X of a separated semi-normed space E into another separated semi-normed space F, and let $\{\| \|_{E;\nu} : \nu \in \mathcal{N}_E\}$ and $\{\| \|_{F;\mu} : \mu \in \mathcal{N}_F\}$ be the families of semi-norms of E and F.*

(a) *We say that T is **continuous at the point** u of X if, for every $\nu \in \mathcal{N}_E$ and $\epsilon > 0$, there exist a finite subset M of $\mathcal{N}_F$ and $\eta > 0$ such that:*

$$v \in X, \ \sup_{\mu \in M} \|v - u\|_{F;\mu} \leq \eta \quad \Rightarrow \quad \|T(v) - T(u)\|_{E;\nu} \leq \epsilon.$$

*We say that T is **continuous** if it is so at every point of X.*

(b) *We say that T is **uniformly continuous** if, for every $\nu \in \mathcal{N}_E$ and $\epsilon > 0$, there exist a finite subset M of $\mathcal{N}_F$ and $\eta > 0$ such that:*

$$u \in X, \ v \in X, \ \sup_{\mu \in M} \|v - u\|_{F;\mu} \leq \eta \quad \Rightarrow \quad \|T(v) - T(u)\|_{E;\nu} \leq \epsilon.$$

(c) *We say that T is **sequentially continuous at the point** u of X if, for every sequence $(u_n)_{n\in\mathbb{N}}$ of X:*

$$u_n \to u \text{ in } E \quad \Rightarrow \quad T(u_n) \to T(u) \text{ in } F.$$

*We say that T is **sequentially continuous** if it is so at every point of X.*

---

5. **History of the notions of continuity.** Augustin CAUCHY defined sequential continuity for a real function on a line segment in 1821, in [18]. Bernard Placidus Johann Nepomuk BOLZANO also contributed to the emergence of this notion.
Eduard HEINE defined the uniform continuity of a function defined on a part of $\mathbb{R}^d$ in 1870, in [42]. It had already been used implicitly by Augustin CAUCHY in 1823 to define the integral of a real function [19, p. 122-126], and then explicitly by Peter DIRICHLET.

(d) *We say that T is **bounded** if its image $T(X) = \{T(u) : u \in X\}$ is bounded in F. That is to say if, for every $\mu \in \mathcal{N}_F$,*

$$\sup_{u \in X} \|T(u)\|_{F;\mu} < \infty. \blacksquare$$

Recall that continuity always implies sequential continuity [Vol. 1, Theorem 7.2] [6].

Theorem 1.10.– *Any continuous mapping from a subset of a separated semi-normed space into a separated semi-normed space is sequentially continuous.* ▮

The converse is true if the initial space is metrizable [Vol. 1, Theorem 9.1].

Theorem 1.11.– *A mapping from a subset of a metrizable separated semi-normed space into a separated semi-normed space is continuous if and only if it is sequentially continuous.* ▮

For linear mappings, Definition 1.9 gives [Vol. 1, Theorem 7.14]:

Theorem 1.12.– *Let L be a linear mapping from a separated semi-normed space E into a separated semi-normed space F, and let $\{\| \|_{E;\nu} : \nu \in \mathcal{N}_E\}$ and $\{\| \|_{F;\mu} : \mu \in \mathcal{N}_F\}$ be the families of semi-norms of E and F. Then:*

(a) *L is continuous if and only if, for every $\mu \in \mathcal{N}_F$, there exist a finite subset N of $\mathcal{N}_E$ and $c \geq 0$ such that: for every $u \in E$,*

$$\|Lu\|_{F;\mu} \leq c \sup_{\nu \in N} \|u\|_{E;\nu}.$$

This is also equivalent to the statement: L is uniformly continuous.

(b) *If the family of semi-norms of E is filtering, then L is continuous if and only if, for every $\mu \in \mathcal{N}_F$, there exist $\nu \in \mathcal{N}_E$ and $c \geq 0$ such that: for every $u \in E$,*

$$\|Lu\|_{F;\mu} \leq c \|u\|_{E;\nu}. \blacksquare$$

Observe that topological inclusion is equivalent to the continuity of identity.

---

6. **Numbering of statements.** The numbering is common to all the statements — **Definition 1.1**, ..., **Definition 1.9**, **Theorem 1.10**, **Theorem 1.11**, etc. — , to make it easier to find a given result by following the order of the numbers. It is not worthwhile therefore to look for Theorems 1.1 to 1.9, as these numbers have been assigned to definitions. There is also no need to look for Definitions 1.10, 1.11, etc.

Theorem 1.13.– *The topological inclusion $E \underset{\to}{\subsetneq} F$ of a separated semi-normed space into another is equivalent to the continuity of the identity mapping from E into F.* ▌

Proof. Definition 1.7 (c) of topological inclusion coincides with the characterization of continuity from Theorem 1.12 (a) applied to the identity, i.e. for $Lu = u$. □

For semi-norms, Definition 1.9 gives [Vol. 1, Theorem 7.11]:

Theorem 1.14.– *A semi-norm p on a separated semi-normed space E with a filtering family $\{\| \, \|_{E;\nu} : \nu \in \mathcal{N}_E\}$ of semi-norms is continuous if and only if there exist $\nu \in \mathcal{N}_E$ and $c \geq 0$ such that: for every $u \in E$,*

$$p(u) \leq c\|u\|_{E;\nu}. \ ▌$$

Addition and multiplication by a real number t for mappings with values in a vector space are defined by

$$(T + S)(u) \overset{\text{def}}{=} T(u) + S(u), \quad (tT)(u) \overset{\text{def}}{=} tT(u). \tag{1.1}$$

1.4. Differentiable functions

We reserve the term **function** for a mapping defined on a subset of $\mathbb{R}^d$, which in general is denoted by Ω. Throughout the book, the dimension d is an integer ≥ 1.

Recall that a function is continuous if and only if it is sequentially continuous (Theorem 1.11, since $\mathbb{R}^d$ is normed).

Gradient being defined in $E^d \overset{\text{def}}{=} \{(u_1, \ldots, u_d) : u_i \in E, \ \forall i\}$, recall that this product space is endowed with the semi-norms, indexed by $\nu \in \mathcal{N}_E$,

$$\|u\|_{E^d;\nu} \overset{\text{def}}{=} (\|u_1\|^2_{E;\nu} + \cdots + \|u_d\|^2_{E;\nu})^{1/2}, \tag{1.2}$$

which makes it a separated semi-normed space [Vol. 1, Theorem 6.11].

We denote by $|z| \overset{\text{def}}{=} (z_1^2 + \cdots + z_d^2)^{1/2}$ the Euclidean norm of $z \in \mathbb{R}^d$ and, for $u \in E^d$,

$$z \cdot u \overset{\text{def}}{=} z_1 u_1 + \cdots + z_d u_d.$$

Observe that

$$\|z \cdot u\|_{E;\nu} \leq |z| \|u\|_{E^d;\nu}. \tag{1.3}$$

Indeed, from the Cauchy-Schwarz inequality in $\mathbb{R}^d$, see (A.1), p. 354,

$$\|z \cdot u\|_{E;\nu} \leq \sum_{i=1}^{d} |z_i| \|u_i\|_{E;\nu} \leq \Big(\sum_{i=1}^{d} z_i^2\Big)^{1/2} \Big(\sum_{i=1}^{d} \|u_i\|_{E;\nu}^2\Big)^{1/2} = |z| \|u\|_{E^d;\nu}.$$

Let us define various levels of differentiability for a function with values in a separated semi-normed space.

Definition 1.15.– *Let f be a function from an open subset Ω of $\mathbb{R}^d$ into a separated semi-normed space E, whose family of semi-norms is denoted by $\{\| \ \|_{E;\nu} : \nu \in \mathcal{N}_E\}$.*

We say that f is ***differentiable at the point*** *x of Ω, if there exists an element of E^d, denoted by $\nabla f(x)$ and called the* ***gradient*** *of f at the point x, such that, for every $\nu \in \mathcal{N}_E$ and $\epsilon > 0$, there exists $\eta > 0$ such that, if $z \in \mathbb{R}^d$, $|z| \leq \eta$ and $x + z \in \Omega$, then*

$$\|f(x+z) - f(x) - z \cdot \nabla f(x)\|_{E;\nu} \leq \epsilon |z|.$$

We say that f is ***differentiable*** *if it is differentiable at every point of Ω, and that it is* ***continuous differentiable*** *if, moreover, ∇f is continuous from Ω into E^d.*

We say that f is m ***times differentiable****, where $m \in \mathbb{N}^*$, if it has successive gradients $\nabla f, \nabla^2 f, \ldots, \nabla^m f$ (these are elements of E^d, $E^{d^2}, \ldots, E^{d^m}$ respectively).*

We say that f is m ***times continuously differentiable*** *if, moreover, its successive gradients are continuous. We extend this notion and the previous one to $m = 0$ by denoting*

$$\nabla^0 f \stackrel{\text{def}}{=} f.$$

We say that f is ***infinitely differentiable*** *if it is m times differentiable for every $m \in \mathbb{N}^*$.* ▌

When $d = 1$, the differentiability of f at the point x reduces to the existence of an element of E, denoted by $f'(x)$ and called the **derivative**[7] at the point x, such

7. **History of the notion of the derivative of a real function.** EUCLIDE, in his *Elements* [31, Book III, p. 16] was already looking for the tangent to a curve. Pierre de FERMAT, in 1636, found the tangent for the curve of equation $y = x^m$ with a calculation prefiguring that of the derivative. Isaac NEWTON introduced in 1671 the *fluxion* of a function $y = f(x)$ which he denoted by $\dot{y}$ [57, p. 76]. Gottfried von LEIBNIZ developed *infinitesimal calculus* in 1675 [51]. The notion of derivative was made rigorous in 1821 by Augustin CAUCHY [18, p. 22].
History of the notation. The notation $\mathrm{d}y/\mathrm{d}x$ was introduced by Gottfried von LEIBNIZ in 1675 [51]. This was what Joseph Louis LAGRANGE denoted by $f'x$ in 1772 [49].
The symbol ∇ was introduced by Sir William Rowan HAMILTON in 1847, by inverting the Greek letter Δ, which had already been used in an analogous context (to designate the *Laplacian*); the name *nabla* was given to him by Peter Guthrie TAIT on the advice of William Robertson SMITH, in 1870, in analogy to the form of a Greek harp which in Antiquity bore this name ($\nu\acute{\alpha}\beta\lambda\alpha$).

that, for every $\nu \in \mathcal{N}_E$ and $\epsilon > 0$, there exists $\eta > 0$ such that, if $t \in \mathbb{R}$, $|t| \leq \eta$ and $x + t \in \Omega$, then

$$\|f(x+t) - f(x) - tf'(x)\|_{E;\nu} \leq \epsilon\,|t|. \tag{1.4}$$

The gradient here has only one component: $\nabla f(x) = f'(x)$. The derivative is often denoted as $\mathrm{d}f/\mathrm{d}x$ instead of f', in particular when we wish to specify the variable with respect to which we are differentiating.

Utility of assuming that Ω is open. This hypothesis guarantees the uniqueness of the gradient at every point where it exists [Vol. 2, Theorem 2.2]. If not, for example if Ω was just a point, then every function would be differentiable and would admit every element of E as its gradient. However, the notion of differentiability can be extended to the closure of an open set while preserving the uniqueness of the gradient [Vol. 2, Definition 2.26]. □

Let us define partial derivatives [8], denoting by $\mathbf{e}_i$ the i-th basis vector of $\mathbb{R}^d$, i.e.

$$(\mathbf{e}_i)_i = 1, \quad (\mathbf{e}_i)_j = 0 \text{ if } j \neq i,$$

and denoting by $[\![m, n]\!]$ the interval of integers $\{i \in \mathbb{N} : m \leq i \leq n\}$.

Definition 1.16.– *Let f be a function from an open subset Ω of $\mathbb{R}^d$ into a separated semi-normed space E, whose family of semi-norms is denoted $\{\|\ \|_{E;\nu} : \nu \in \mathcal{N}_E\}$.*

We say that f has a ***partial derivative*** *$\partial_i f(x) \in E$ at the point x of Ω, where $i \in [\![1, d]\!]$, if the function $x_i \mapsto f(x)$ is differentiable at the point x_i with derivative $\partial_i f(x)$.*

That is to say, if, for every $\nu \in \mathcal{N}_E$ and all $\epsilon > 0$, there exists $\eta > 0$ such that, if $t \in \mathbb{R}$, $|t| \leq \eta$ and $x + t\mathbf{e}_i \in \Omega$, then

$$\|f(x+t\mathbf{e}_i) - f(x) - t\partial_i f(x)\|_{E;\nu} \leq \epsilon\,|t|. \tag{1.5}$$

■

Clarification. More precisely, f has a partial derivative $\partial_i f(x)$ at the point x if the function

$$s \mapsto f(x_1, \ldots, x_{i-1}, s, x_{i+1}, \ldots, x_d),$$

which is defined on the open subset $\{s \in \mathbb{R} : (x_1, \ldots, x_{i-1}, s, x_{i+1}, \ldots, x_d) \in \Omega\}$ of $\mathbb{R}$, has the derivative $\partial_i f(x)$ at the point x_i. □

8. **History of partial derivatives.** Partial derivatives appeared in 1747 with Alexis Claude CLAIRAUT and Jean le Rond D'ALEMBERT [24], and in 1755 with Leonhard EULER [32].
The symbol ∂ was introduced by Nicolas de Caritat, Marquis of CONDORCET, in 1773 [22].

CAUTION. A function which has partial derivatives is not necessarily differentiable nor even continuous. For example, the function defined on $\mathbb{R}^2$ by

$$f(x_1,x_2)=\frac{x_1x_2}{(x_1^2+x_2^2)} \text{ if } x\neq(0,0) \text{ and } f(0,0)=0$$

has partial derivatives $\partial_1 f$ and $\partial_2 f$ at every point, but it is neither differentiable nor continuous at $(0,0)$.

On the other hand, a function whose partial derivatives are continuous is continuously differentiable and continuous (Theorem A.55). □

We often use the notation $\partial f/\partial x_i$ instead of $\partial_i f$, in particular when we wish to emphasize the variable with respect to which we are differentiating.

Let us now recall **Schwarz's Theorem** [9] [Vol. 2, Theorem 2.12]:

Theorem 1.17.– *Let f be a twice differentiable function from an open subset of $\mathbb{R}^d$ into a separated semi-normed space. Then, for every i and j in $[\![1,d]\!]$,*

$$\partial_j\partial_i f=\partial_i\partial_j f. \blacksquare$$

Since they commute, we can reorder successive partial derivatives to express them as follows. For any $\beta\in\mathbb{N}^d$, we denote

$$\partial^\beta f \overset{\text{def}}{=} \partial_1^{\beta_1}\cdots\partial_d^{\beta_d} f,\quad \partial^0 f\overset{\text{def}}{=} f.$$

And we denote $|\beta|=\beta_1+\cdots+\beta_d$.

1.5. Spaces $\mathcal{C}^m(\Omega;E)$, $\mathcal{C}_{\rm b}^m(\Omega;E)$ and $\mathrm{C}_{\rm b}^m(\Omega;E)$

Let us define spaces [10] of functions that are m times differentiable with values in a semi-normed space.

Definition 1.18.– *Let Ω be an open subset of $\mathbb{R}^d$, E a separated semi-normed space whose family of semi-norms is denoted by $\{\|\ \|_{E;\nu} : \nu\in\mathcal{N}_E\}$, and $m\in\mathbb{N}$.*

---

9. **History of Schwarz's Theorem.** Karl WEIERSTRASS proved in his unpublished teachings of 1861 that the second partial derivatives of a function f commute at every point x where they are continuous, namely $\partial^2 f/\partial x_1\partial x_2(x_1,x_2)=\partial^2 f/\partial x_2\partial x_1(x_1,x_2)$.
Hermann Amandus SCHWARZ proved in 1873 that this result is valid as soon as one of the two members is continuous.

10. **History of the notion of function space.** Bernhard RIEMANN introduced the concept of function space (of infinite dimension) in 1892 in his inaugural lecture *On the Hypotheses Which Lie at the Bases of Geometry* [64, p. 276], see the extract in french cited in [12, p. 176].

(a) *We denote by* $\mathcal{C}^m(\Omega; E)$ *the vector space of* m*-times continuously differentiable functions from* Ω *into* E *endowed with the semi-norms, indexed by the compact subsets* K *of* Ω *and by* $\nu \in \mathcal{N}_E$,

$$\|f\|_{\mathcal{C}^m(\Omega;E);K,\nu} \stackrel{\text{def}}{=} \sup_{0\le|\beta|\le m} \sup_{x\in K} \|\partial^\beta f(x)\|_{E;\nu}.$$

We denote this space $\mathcal{C}(\Omega; E)$ *when* $m = 0$*, and* $\mathcal{C}^m(\Omega)$ *when* $E = \mathbb{R}$.

(b) *We denote*

$$\mathcal{C}_{\mathrm{b}}^m(\Omega; E) \stackrel{\text{def}}{=} \{f \in \mathcal{C}^m(\Omega; E) : \partial^\beta f \text{ is bounded, } \forall\beta,\ 0 \le |\beta| \le m\}$$

endowed with the semi-norms, indexed by $\nu \in \mathcal{N}_E$,

$$\|f\|_{\mathcal{C}_{\mathrm{b}}^m(\Omega;E);\nu} \stackrel{\text{def}}{=} \sup_{0\le|\beta|\le m} \sup_{x\in \Omega} \|\partial^\beta f(x)\|_{E;\nu}.$$

We denote this space $\mathcal{C}_{\mathrm{b}}(\Omega; E)$ *when* $m = 0$*, and* $\mathcal{C}_{\mathrm{b}}^m(\Omega)$ *when* $E = \mathbb{R}$. ▮

A shortcut. We write "the compact subsets K of Ω", for "the compact subsets K of $\mathbb{R}^d$ included in Ω". However, this is not ambiguous. □

Justification of Definition 1.18. **1.** *Vector structures.* Addition and scalar multiplication of functions defined in (1.1), p. 8, make $\mathcal{C}^m(\Omega; E)$ and $\mathcal{C}_{\mathrm{b}}^m(\Omega; E)$ into vector spaces.

2. *Semi-norms of* $\mathcal{C}^m(\Omega; E)$. In (a), the mapping

$$f \mapsto \sup_{0\le|\beta|\le m} \sup_{x\in K} \|\partial^\beta f(x)\|_{E;\nu}$$

is indeed a semi-norm on $\mathcal{C}^m(\Omega; E)$, because the upper envelope of semi-norms is a semi-norm if it is everywhere finite (Theorem A.4)[11]. This is the case here, since:

— for each $x \in \Omega$ and $\beta \in \mathbb{N}^d$ such that $|\beta| \le m$, the mapping $f \mapsto \|\partial^\beta f(x)\|_{E;\nu}$ is a semi-norm on $\mathcal{C}^m(\Omega; E)$;

— for each $f \in \mathcal{C}^m(\Omega; E)$,

$$\sup_{0\le|\beta|\le m} \sup_{x\in K} \|\partial^\beta f(x)\|_{E;\nu} < \infty,$$

since continuous functions are bounded on compact sets (Theorem A.32).

3. *Semi-norms of* $\mathcal{C}_{\mathrm{b}}^m(\Omega; E)$. The mapping $f \mapsto \sup_{|\beta|\le m} \sup_{x\in\Omega} \|\partial^\beta f(x)\|_{E;\nu}$, in (b), is similarly a semi-norm on $\mathcal{C}_{\mathrm{b}}^m(\Omega; E)$. □

11. **Theorem A.4.** The theorems numbered as A.n can be found in the **Appendix**.

Let us now define spaces of infinitely differentiable functions.

Definition 1.19.– *Let Ω be an open subset of $\mathbb{R}^d$ and E a separated semi-normed space, whose family of semi-norms is denoted by $\{\| \|_{E;\nu} : \nu \in \mathcal{N}_E\}$.*

(a) *We denote by $\mathcal{C}^\infty(\Omega; E)$ the vector space of infinitely differentiable functions from Ω into E endowed with the semi-norms, indexed by $m \in \mathbb{N}$, by the compact subsets K of Ω and by $\nu \in \mathcal{N}_E$,*

$$\|f\|_{\mathcal{C}^\infty(\Omega;E);m,K,\nu} \stackrel{\text{def}}{=} \sup_{0\le|\beta|\le m} \sup_{x\in K} \|\partial^\beta f(x)\|_{E;\nu}.$$

(b) *We denote*

$$\mathcal{C}_{\mathrm{b}}^\infty(\Omega; E) \stackrel{\text{def}}{=} \{f \in \mathcal{C}^\infty(\Omega; E) : \partial^\beta f \text{ is bounded, } \forall \beta \in \mathbb{N}^d\}$$

endowed with the semi-norms, indexed by $m \in \mathbb{N}$ and $\nu \in \mathcal{N}_E$,

$$\|f\|_{\mathcal{C}_{\mathrm{b}}^\infty(\Omega;E);m,\nu} \stackrel{\text{def}}{=} \sup_{0\le|\beta|\le m} \sup_{x\in\Omega} \|\partial^\beta f(x)\|_{E;\nu}. \blacksquare$$

Justification. One shows that these semi-norms are well defined as in the justification of Definition 1.18. For (a), it should be remarked that the $\partial^\beta f$ are continuous, as is every differentiable function (Theorem A.54), and hence bounded on compact sets. □

Observe that Definitions 1.19 (b) and 1.18 (b) lead to

$$\|f\|_{\mathcal{C}_{\mathrm{b}}^\infty(\Omega;E);m,\nu} = \|f\|_{\mathcal{C}_{\mathrm{b}}^m(\Omega;E);\nu}.$$

For real values, this yields in particular, for every function $f \in \mathcal{C}^\infty(\Omega)$,

$$\|f\|_{\mathcal{C}_{\mathrm{b}}^\infty(\Omega);m} = \|f\|_{\mathcal{C}_{\mathrm{b}}^m(\Omega)}. \tag{1.6}$$

We will use $\|f\|_{\mathcal{C}_{\mathrm{b}}^m(\Omega)}$ when we wish to remain in familiar territory, and we will use $\|f\|_{\mathcal{C}_{\mathrm{b}}^\infty(\Omega);m}$ when we wish to highlight that f belongs to $\mathcal{C}_{\mathrm{b}}^\infty(\Omega)$.

Let us define the space of functions that are uniformly continuous along with their derivatives.

Definition 1.20.– *Let Ω be an open subset of $\mathbb{R}^d$, E a separated semi-normed space and $m \in [\![0, \infty]\!]$. We denote*

$$\mathbf{C}_{\mathrm{b}}^m(\Omega; E) \stackrel{\text{def}}{=} \{f \in \mathcal{C}_{\mathrm{b}}^m(\Omega; E) : \partial^\beta f \text{ is uniformly continuous, } 0 \le |\beta| \le m\}$$

endowed with the semi-norms of $\mathcal{C}_{\mathrm{b}}^m(\Omega; E)$.

We denote this space $\mathbf{C}_{\mathrm{b}}(\Omega; E)$ when $m = 0$, and $\mathbf{C}_{\mathrm{b}}^m(\Omega)$ when $E = \mathbb{R}$. ■

Another shortcut. We write that $\mathbf{C}_{\mathrm{b}}^m(\Omega; E)$ is "endowed with the semi-norms of $\mathcal{C}_{\mathrm{b}}^m(\Omega; E)$", when in reality it is endowed with their restrictions to $\mathbf{C}_{\mathrm{b}}^m(\Omega; E)$. Here, there is also no possible ambiguity. □

1.6. Integral of a uniformly continuous function

Integration of continuous functions is essential for their identification with distributions by the equality $\langle f, \varphi \rangle = \int f\varphi$, for any test function φ. The theory of integration not having been performed yet for values into a Neumann space, we did it in the most basic way possible in Volume 2, for uniformly continuous functions which were sufficient for our needs. Let us recall this construction.

Let us first define the vector space $\mathcal{B}(\Omega; E)$ of functions for which the Cauchy integral will be defined.

Definition 1.21.– *Let Ω be an open subset of $\mathbb{R}^d$ and E a separated semi-normed space. We denote by $\mathcal{B}(\Omega; E)$ the set of uniformly continuous functions from Ω into E that are zero ouside of a bounded set.* ▮

Utility of the space $\mathcal{B}(\Omega; E)$. The integral of continuous functions is often introduced on the space $\mathcal{K}(\Omega; E)$ of functions with compact support, see for example [BOURBAKI, 9, Chap. III, § 3, Definition 1, p. 75]. We prefer to define it on $\mathcal{B}(\Omega; E)$ so that we are able to apply it to functions, such as constant functions, that are not zero in a neighborhood of the boundary.

This space is not commonly used, and its **notation** $\mathcal{B}(\Omega; E)$ is **not standard**. We do not endow it with semi-norms, because we not need them. □

Now we define the **Cauchy integral** [12] of a uniformly continuous function with bounded support, with values in a Neumann space.

12. **History of the integral of a continuous function. Real integral.** Augustin CAUCHY defined the integral of a real continuous function on a real interval in 1823 in his *Leçons a l'Ecole Royale Polytechnique sur le calcul infinitesimal* [19, p. 122-126], by the method that we use in Definition 1.22. He showed [19, p. 125] that the approximate sums $\mathbb{S}^n_\omega f$ converge to a limit using the uniform continuity of the function (which was only shown later by Eduard HEINE [44] in 1872!). He also introduced the integral of a function defined on $\mathbb{R}^2$ in 1827 [20].

Precursors. Johann BERNOULLI wrote the first treatise on differential and integral calculus in 1691–1692 [3], whose part concerning the integral calculus was not published until 1742. In 1768, Leonhard EULER published his important treatise *Institutionum calculi integralis*, in which he used the sums $\mathbb{S}^n f$ [33, p. 178] with the assumption that they converge. He calculated integrals on $\mathbb{R}^2$ in 1770 [34]. For more details, see *Histoire de l'intégration* by Jean-Paul PIER [60, p. 79–86].

Vector-valued integral. The integral of an integrable function with values in a Banach space E was defined in 1933 by Salomon BOCHNER [5] using a different method.
Nicolas BOURBAKI defined in 1966 [9, Chap. III, § 3, Definition 1, p. 75] the integral of a function in $\mathcal{K}(\Omega; E\text{-weak})$, where E is a *Hausdorff locally convex topological vector space*, a notion that is equivalent to a *separated semi-normed space*. If E-weak is sequentially complete, this integral coincides with the Cauchy integral with values in E-weak; if not, it takes values in the completion $\widehat{E}$ of E, which coincides with the completion of E-weak.

Originality. Extending the Cauchy integral to values in a Neumann space E seems new to us, although it is straightforward.

Definition 1.22.– *Let $f \in \mathcal{B}(\Omega; E)$, i.e. a uniformly continuous function from Ω into E that is zero ouside of a bounded set, where Ω is an open subset of $\mathbb{R}^d$ and E is a Neumann space, and let ω be an open subset of Ω.*

The ***approximate integral*** [13] *of f over ω is the element of E defined, for $n \in \mathbb{N}$, by*

$$\mathbb{S}_\omega^n f \stackrel{\text{def}}{=} \sum_{s \in \mathbb{Z}^d : \Delta_{s,n} \subset \omega} 2^{-nd} f(2^{-n} s),$$

where $\Delta_{s,n}$ is the closed cube $\mathbb{R}^d$ of side length 2^{-n} centered at $2^{-n}s$.

The ***integral*** *over ω of f is the element of E defined by* [14]

$$\int_\omega f \stackrel{\text{def}}{=} \lim_{n \to \infty} \mathbb{S}_\omega^n f. \blacksquare$$

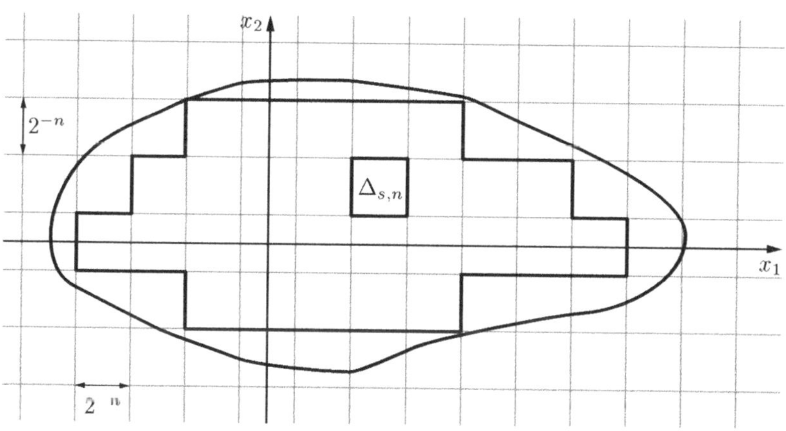

Figure 1.1. *Paving of ω by the cubes $\Delta_{s,n}$*

Definition 1.22 is justified since, as was established in Volume 2 [Lemma 4.10, p. 90], $\mathbb{S}_\omega^n f$ has a limit when $n \to \infty$.

13. **Paternity of the approximate integral.** Approximate integrals $\mathbb{S}_\omega^n f$ are sometimes called *Riemann sums* with reference to a later generalization of Bernhard RIEMANN, in his habilitation thesis of 1854 published in 1867, after his death, [65, p. 239]. In the case of continuous functions, according to BOURBAKI [12, p. 248], it would be better to name them *Archimedes sums* or *Eudoxe sums*.

14. **History of the notation of the integral.** The symbol $\int$, "*a long* s *being the first letter of the word* summa", was introduced by Siegfried von LEIBNIZ in a letter in 1675 to Henry OLDENBURG, reproduced by Florian CAJORI [14, Vol. 2, p. 243]. The notation $\int_a^b f$ was introduced by Joseph FOURIER [35, p. 252] in 1822. The term *integral* was introduced in 1690 by Jacob BERNOULLI [2], see the facsimile in [60, p. 78].

Convention. The definition of the approximate integral does not make sense if none of the cubes $\Delta_{s,n}$ is included in ω, since the sum of zero element is not defined! We avoid this dilemma by agreeing that $\mathbb{S}^n_\omega f$ is zero in this case. In particular, $\mathbb{S}^n_\emptyset f = 0_E$ for all n and thus

$$\int_\emptyset f = 0_E. \quad \square$$

Notation. We use the notation $\int_\omega f(x)\,\mathrm{d}x$ when we wish to emphasize the variable with respect to which we integrate, for example in the expression $\int_\omega f(x,y)\,\mathrm{d}x$, for a function f of two variables. $\square$

The classical properties of the Cauchy integral with values in a Banach space (or in a complete semi-normed space) extend to values in a Neumann space. This is the case with all the properties that we use in this book. These were proven in Volume 2 and their precise statements are recalled in the Appendix or in the text, along with references for their proofs.

The semi-norms of E play the role held by the norm in the case of a Banach space. In particular, we have the following inequalities established respectively in Theorems 4.11 and 4.15 of Volume 2 [82], that we will use very often.

Theorem 1.23.– *Let $f \in \mathcal{B}(\Omega; E)$, where Ω is an open subset of $\mathbb{R}^d$ and E is a Neumann space, and let ω be an open subset of Ω and $\| \ \|_{E;\nu}$ a semi-norm of E. Then:*

(a) $\|f\|_{E;\nu} \in \mathcal{B}(\Omega)$ *and*

$$\Big\| \int_\omega f \Big\|_{E;\nu} \leq \int_\omega \|f\|_{E;\nu} \leq \int_\Omega \|f\|_{E;\nu}.$$

(b)
$$\Big\| \int_\omega f \Big\|_{E;\nu} \leq |\omega| \sup_{x\in\omega} \|f(x)\|_{E;\nu}. \quad \blacksquare$$

The definition of the measure $|\omega|$ of an open subset ω of $\mathbb{R}^d$ is recalled in the Appendix (Definition A.83).

Chapter 2

Space of Test Functions

In order to be able to build in the next chapter the space $\mathcal{D}'(\Omega; E)$ of distributions, i.e. of continuous linear mappings from $\mathcal{D}(\Omega)$ into E, we construct here the space $\mathcal{D}(\Omega)$ of test functions.

We endow it with the semi-norms $\|\varphi\|_{\mathcal{D}(\Omega);p} = \sup_{x\in\Omega} \sup_{0\leq|\beta|\leq p(x)} p(x)|\partial^\beta \varphi(x)|$ indexed by $p \in \mathcal{C}^+(\Omega)$, and give the following properties.

— It is a Neumann space (Theorem 2.8) which is sequentially separable (Theorem 2.16) and in which every bounded sequence has a convergent subsequence (Theorem 2.14).

— We give characterizations of its convergent sequences, including a new one (Theorem 2.13 (c)) which will serve later (Theorem 4.14) to show that $\mathcal{D}'(\Omega; E\text{-weak}) = \mathcal{D}'(\Omega; E)$.

— We establish an inequality (Theorem 2.22), also new, that allows us to "control" a series of norms of the $\mathcal{C}^m_K(\Omega)$ by one semi-norm of the $\mathcal{D}(\Omega)$, which gives rise to properties usually obtained by resorting to the delicate *inductive limit* topology of the $\mathcal{D}_K(\Omega)$.

We begin with a condition for all functions of a family to have their support included in a same compact set (Theorem 2.4).

2.1. Functions with compact support

Let us first define the support of a function.

Definition 2.1.– *The* ***support*** *of a function f from a subset Ω of $\mathbb{R}^d$ into a separated semi-normed space E is the set*

$$\operatorname{supp} f \stackrel{\text{def}}{=} \overline{\{x \in \Omega : f(x) \neq 0_E\}} \cap \Omega. \blacksquare$$

We denote by $\overline{X}$ the closure of a set X, and $\mathring{X}$ its interior, see Definition A.8.

Let us give elementary conditions for the support of a function to be compact (recall that, in $\mathbb{R}^d$, a compact set is a closed and bounded set, due to the Borel–Lebesgue Theorem (Theorem A.26 (b)).

Theorem 2.2.– *For every function f from a subset Ω of $\mathbb{R}^d$ into a separated semi-normed space, each of the following properties are equivalent to:*

The support of f is compact.

(a) *The support of f is included in a compact subset of Ω.*

(b) *f is zero ouside of a compact subset of Ω.* ▮

Proof. The equivalences result from the three following implications:

1. supp f *is compact* ⇒ *(a).* Indeed, if the support of f is compact, then it is included in a compact subset of Ω (itself).

2. *(a)* ⇒ *(b).* Indeed, by Definition 2.1, f is zero ouside of its support so, if this is included in a compact subset of Ω, f is zero ouside of the said compact subset.

3. *(b)* ⇒ supp f *is compact.* Suppose that f is zero ouside of a compact subset K of Ω. Then $\{x \in \Omega : f(x) \neq 0_E\} \subset K$, thus

$$\overline{\{x \in \Omega : f(x) \neq 0_E\}} \subset \overline{K} = K \subset \Omega.$$

Definition 2.1 then gives

$$\operatorname{supp} f = \overline{\{x \in \Omega : f(x) \neq 0_E\}},$$

which is closed (by Definition) and bounded (as K is) in $\mathbb{R}^d$ and thus is compact by the Borel–Lebesgue Theorem (Theorem A.26 (b)). □

Let us now define spaces of differentiable functions with support in a given compact.

Definition 2.3.– *Let Ω be an open subset of $\mathbb{R}^d$, E a separated semi-normed space and $m \in \mathbb{N}$.*

(a) *Given a compact subset K of Ω, we denote*

$$\mathcal{C}_K^m(\Omega; E) \stackrel{\text{def}}{=} \{f \in \mathcal{C}^m(\Omega; E) : \operatorname{supp} f \subset K\}$$

endowed with the semi-norms of $\mathcal{C}_\text{b}^m(\Omega; E)$.

(b) *Given a compact subset Q of $\mathbb{R}^d$ (not necessarily included in Ω), we denote*

$$\mathbf{C}_Q^m(\Omega; E) \stackrel{\text{def}}{=} \{f \in \mathbf{C}_\text{b}^m(\Omega; E) : \operatorname{supp} f \subset Q\}$$

endowed with the semi-norms of $\mathcal{C}_\text{b}^m(\Omega; E)$. ▮

Justification of (a). The semi-norms of $\mathcal{C}^m_{\mathrm{b}}(\Omega; E)$ can be used for every function f of $\mathcal{C}^m_K(\Omega; E)$ since f is bounded as well as its derivatives $\partial^\beta f$ of order $|\beta| \leq m$, as is every continuous function which has a compact support (Theorem A.33). □

We denote respectively these spaces by $\mathcal{C}_K(\Omega; E)$ and $\mathbf{C}_Q(\Omega; E)$ when $m = 0$, and by $\mathcal{C}^m_K(\Omega)$ and $\mathbf{C}^m_Q(\Omega)$ when $E = \mathbb{R}$.

Caution about the behavior at the boundary. When Ω is open, every function in $\mathcal{C}^m_K(\Omega; E)$ is zero in a neighborhood of the boundary $\partial\Omega$, since K is a compact subset of Ω. Indeed, there exists $r > 0$ such that $K + B(0, r) \subset \Omega$, according to the strong inclusion theorem (Theorem A.22).

In contrast, the functions of $\mathbf{C}^m_Q(\Omega; E)$ are not necessarily zero in a neighborhood of $\partial\Omega$, when Q is not included in Ω.

We denote the compact subsets Q of $\mathbb{R}^d$ differently to those, K, of Ω, to emphasize this difference. □

Remark. Every function of $\mathcal{C}^m(\Omega; E)$ with support included in Q and whose derivatives (of order $|\beta| \leq m$) are uniformly continuous belongs to $\mathbf{C}^m_Q(\Omega; E)$. Indeed, its derivatives are bounded since every uniformly continuous function with bounded support is itself bounded. □

2.2. Compactness in their whole of support of functions

Let us give a condition [1] for a family of real functions to have all of their support in a single compact set. For this, denote

$$\mathcal{C}^+(\Omega) \stackrel{\text{def}}{=} \{p \in \mathcal{C}(\Omega) : \forall x \in \Omega,\ p(x) \geq 0\}.$$

Theorem 2.4.– *Let $\mathcal{F} \subset \mathcal{C}(\Omega)$, where $\Omega \subset \mathbb{R}^d$, be such that, for every function p of $\mathcal{C}^+(\Omega)$,*

$$\sup_{f \in \mathcal{F}} \sup_{x \in \Omega} p(x)|f(x)| < \infty. \tag{2.1}$$

Then all the functions of $\mathcal{F}$ have their support in a same compact subset of Ω. ∎

Proof. Let

$$K \stackrel{\text{def}}{=} \overline{\bigcup_{f \in \mathcal{F}} \operatorname{supp} f}.$$

1. **History of the condition of compactness of the supports.** Theorem 2.4, a new property, is equivalent to the result of BOURBAKI [9, Chap. III, § 1, Proposition 2, (ii), p. 42] which states that: all the functions of a bounded subset of $\mathcal{K}(\Omega)$ endowed with the topology of inductive limit of the $\mathcal{C}_K(\Omega)$ have their support in the same compact subset K of Ω.
The proof of BOURBAKI, which is based on topological arguments (infinite topological direct sums and strict inductive limits), is quite different to ours.

This set contains the support of every function in $\mathcal{F}$, so it is sufficient to verify that it is compact and included in Ω.

1. *Inclusion in* Ω. Let us show by contradiction that

$$K \subset \Omega. \tag{2.2}$$

If not, then since closures in $\mathbb{R}^d$ coincide with the set of limits of convergent sequences (Theorem A.23 (b)), there would exist z and a sequence $(y_m)_{m\in\mathbb{N}}$ such that

$$z \in K \setminus \Omega, \quad y_m \in \bigcup_{f\in\mathcal{F}} \operatorname{supp} f, \quad |y_m - z| \leq \frac{1}{m}.$$

For every $m \in \mathbb{N}$, there would exist $f_m \in \mathcal{F}$ such that y_m belongs to the support of f_m. By Definition 2.1 of the support,

$$y_m \in \overline{\{x \in \Omega : f_m(x) \neq 0\}}.$$

According again to the characterization of closure from Theorem A.23 (b), there would exist z_m such that

$$z_m \in \Omega, \quad f_m(z_m) \neq 0, \quad |z_m - y_m| \leq \frac{1}{m}.$$

Thus, when $m \to \infty$,

$$z_m \to z.$$

By passing to a subsequence if necessary, we could (because $z_m \neq z$ since $z \notin \Omega$), for each $m \in \mathbb{N}$, choose $r_m > 0$ such that, by denoting

$$B_m \stackrel{\text{def}}{=} \{x : |x - z_m| \leq r_m\},$$

the ball B_m does not contain z and is disjoint from B_i for all $i \leq m-1$.

We would then define a function p on all of $\mathbb{R}^d$ by

$$p(x) = \begin{cases} \dfrac{m}{|f_m(z_m)|}\left(1 - \dfrac{|x - z_m|}{r_m}\right) & \text{on } B_m, \\ 0 & \text{ouside of } B_m. \end{cases}$$

It would be everywhere continuous except at z, thus its restriction to Ω would be continuous. For every $m \in \mathbb{N}$, it would satisfy

$$p(z_m)|f_m(z_m)| = m.$$

This would contradict our hypothesis (2.1). Hence, (2.2) is satisfied, that is, K is included in Ω.

2. *Compactness.* Let us now show, using contradiction again, that

$$K \text{ is bounded.}$$

If not, we could choose y_m, z_m and r_m, and then p, as before with $|y_m| \to \infty$ in the place of $y_m \to z$. Thus, $|z_m| \to \infty$ takes the place of $z_m \to z$, so p would be continuous everywhere, and again

$$p(z_m)|f_m(z_m)| = m.$$

This would contradict once more our hypothesis (2.1). Thus, K is bounded.

Furthermore, by construction K is closed in $\mathbb{R}^d$ and is therefore compact by the Borel–Lebesgue theorem (Theorem A.26 (b)). □

2.3. The space $\mathcal{D}(\Omega)$

Let us define the space $\mathcal{D}(\Omega)$ of **test functions**[2].

Definition 2.5.– *Let Ω be an open subset of $\mathbb{R}^d$. We denote by $\mathcal{D}(\Omega)$ the vector space of infinitely differentiable real functions on Ω with compact support, endowed with the semi-norms, indexed by $p \in \mathcal{C}^+(\Omega)$,*

$$\|\varphi\|_{\mathcal{D}(\Omega);p} \stackrel{\text{def}}{=} \sup_{x\in\Omega} \sup_{0\le|\beta|\le p(x)} p(x)|\partial^\beta \varphi(x)|. \blacksquare$$

We need to justify that this indeed defines a vector space and semi-norms. To this end, we will use the following property.

Theorem 2.6.– *Let $\varphi \in \mathcal{D}(\Omega)$, where Ω is an open subset of $\mathbb{R}^d$, and $\beta \in \mathbb{N}^d$. Then,*

$$\partial^\beta \varphi \in \mathcal{D}(\Omega),$$

$\partial^\beta \varphi$ is continuous and is zero ouside of $\operatorname{supp} \varphi$, *and* $\operatorname{supp} \partial^\beta \varphi \subset \operatorname{supp} \varphi$. ■

---

2. **History of the space $\mathcal{D}(\Omega)$.** Laurent SCHWARTZ gave in 1945 [68, Definition 1, p. 60] Definition 2.5 of the space of test functions on $\mathbb{R}^d$, which he denoted by Φ at that time, without a topology but instead with convergent sequences in the sense of Theorem 2.13. In 1950 [69], he endowed it with the *inductive limit topology of the* $\mathcal{D}_K(\mathbb{R}^d)$ and denoted it by $(\mathcal{D})$.

Semi-norms on $\mathcal{D}(\Omega)$. The semi-norms in Definition 2.5 were introduced by Jacques SIMON in 1996 [80], see also [LEMOINE & SIMON, 52, p. 32].

In 1950, Laurent SCHWARTZ introduced [69, Chap. III, § 1, p. 65] another family of semi-norms on $\mathcal{D}(\mathbb{R}^d)$, equivalent to this one and therefore generating the same topology. It is less simple, since it is indexed by two increasing sequences, one of bounded open sets covering $\mathbb{R}^d$ and the other of integers.

Proof. **1.** *Cancellation.* Let K be the support of φ. By Definition 2.1 of the support,

$$\Omega \setminus K = \Omega \setminus \left(\overline{\{x \in \Omega : \varphi(x) \neq 0\}} \cap \Omega \right) = \Omega \cap \left(\mathbb{R}^d \setminus \overline{\{x \in \Omega : \varphi(x) \neq 0\}} \right).$$

Thus, $\Omega \setminus K$ is an open set, since every finite intersection of open sets (Theorem A.11) is, and φ is zero on it. Definition 1.16 of the partial derivatives $\partial_i \varphi$ then shows that they are zero there too. By successive partial differentiation, the same is also true of the $\partial^\beta \varphi$, that is,

$$\partial^\beta \varphi = 0 \ \text{ on } \Omega \setminus \operatorname{supp} \varphi. \tag{2.3}$$

2. *Support.* From (2.3), and again from Definition 2.1 of the support K of φ,

$$\{x \in \Omega : \partial^\beta \varphi(x) \neq 0\} \subset K \subset \overline{\{x \in \Omega : \varphi(x) \neq 0\}},$$

thus

$$\operatorname{supp} \partial^\beta \varphi = \overline{\{x \in \Omega : \partial^\beta \varphi(x) \neq 0_E\}} \cap \Omega \subset \overline{\{x \in \Omega : \varphi(x) \neq 0\}} \cap \Omega = \operatorname{supp} \varphi.$$

3. *Test function.* The function $\partial^\beta \varphi$ belongs to $\mathcal{D}(\Omega)$ since it is infinitely differentiable (because φ is) and has a compact support (from Theorem 2.2 (b), because it is included in that of φ which is compact and included in Ω).

4. *Continuity.* The function $\partial^\beta \varphi$ is continuous, as is every differentiable function (Theorem A.54). □

Let us now justify the definition of $\mathcal{D}(\Omega)$.

Justification of Definition 2.5. **1.** *Vector structure.* The addition and multiplication of functions defined by (1.1), p. 8, makes $\mathcal{D}(\Omega)$ into a vector space; in particular, because the support of a sum of two functions with compact support is also compact.

Indeed, if f_1 and f_2 both have compact support, then each is zero ouside of its support, so $f_1 + f_2$ is zero ouside of the union of their supports, which is compact (from the Borel–Lebesgue Theorem A.26 (b), since it is closed and bounded). The support of $f_1 + f_2$ is thus compact by Theorem 2.2 (b).

2. *Semi-norms.* The mapping

$$\varphi \mapsto \sup_{x \in \Omega} \sup_{|\beta| \leq p(x)} p(x) |\partial^\beta \varphi(x)|$$

is indeed a semi-norm on $\mathcal{D}(\Omega)$, since the upper envelope of semi-norms is also a semi-norm if it is everywhere finite (Theorem A.4). This is the case here, since:

— on one hand, for every $x \in \Omega$ and $\beta \in \mathbb{N}^d$, the mapping $\varphi \mapsto p(x)|\partial^\beta \varphi(x)|$ is a semi-norm on $\mathcal{D}(\Omega)$;

— on the other hand, for every $\varphi \in \mathcal{D}(\Omega)$,

$$\sup_{x\in\Omega}\ \sup_{|\beta|\le p(x)} p(x)|\partial^\beta\varphi(x)| \le m \sup_{|\beta|\le m}\ \sup_{x\in K} |\partial^\beta\varphi(x)| < \infty,$$

where K is the support of φ and m is an upper bound of p over K. Indeed, as we have just seen in Theorem 2.6, the $\partial^\beta\varphi$ are zero ouside of K, and are continuous and thus bounded on K as well as p, as is any continuous function on a compact set (Theorem A.32). □

Support of test functions. Due to Theorem 2.2 (a), Definition 2.5 of the set $\mathcal{D}(\Omega)$ is equivalent to:

$$\mathcal{D}(\Omega) \stackrel{\text{def}}{=} \{\varphi \in \mathcal{C}^\infty(\Omega) : \operatorname{supp}\varphi \textit{ is a compact subset of } \mathbb{R}^d \textbf{\textit{ included in }} \Omega\}.$$

The condition ***included in*** Ω added here is redundant with Definition 2.1 of support. But it needs to be specified when we use a function φ defined on all of $\mathbb{R}^d$ as a test function on Ω, and we still denote it by φ in the place of $\varphi|_\Omega$. Indeed, in such a case, the compactness of $\operatorname{supp}\varphi$ is not sufficient. □

Existence of test functions. Let us give an example [3] of an infinitely differentiable function (this is established in Volume 2 [82, proof of Theorem 3.19]) in $\mathbb{R}^d$ with compact support:

$$\rho(x) \stackrel{\text{def}}{=} \exp\Big(\frac{-1}{1-|x|^2}\Big) \text{ if } |x| \le 1, \quad \rho(x) \stackrel{\text{def}}{=} 0 \text{ if } |x| \ge 1. \tag{2.4}$$

More generally, $\mathcal{D}(\Omega)$ does not reduce to the zero function if Ω is non-empty, as then the latter contains a ball $B(a, r)$ and the function $x \mapsto \rho(2(x-a)/r)$ belongs to $\mathcal{D}(\Omega)$. □

Let us prove that $\mathcal{D}(\Omega)$ is separated.

Theorem 2.7.– *The space $\mathcal{D}(\Omega)$, where Ω is an open subset of $\mathbb{R}^d$, is semi-normed and separated, and its family of semi-norms is filtering.* ▮

Proof. **1.** *Separated.* The space $\mathcal{D}(\Omega)$ is semi-normed by Definition 2.5. It is separated (Definition 1.1) since, if all the semi-norms of φ are zero, then $\varphi = 0$, because its semi-norm indexed by the constant function $p_{1/2}$ of value $1/2$ equals

$$\|\varphi\|_{\mathcal{D}(\Omega);p_{1/2}} = \tfrac{1}{2}\sup_{x\in\Omega}|\varphi(x)| = 0.$$

2. *Filtering.* The family of semi-norms of $\mathcal{D}(\Omega)$ is filtering (Definition 1.8) since, given functions $p_1, p_2, \ldots$, and p_n in $\mathcal{C}^+(\Omega)$, their sum $p = p_1 + \cdots + p_n$ belongs to $\mathcal{C}^+(\Omega)$ and is greater than all the p_i, thus

$$\sup_{1\le i\le n} \|\varphi\|_{\mathcal{D}(\Omega);p_i} \le \|\varphi\|_{\mathcal{D}(\Omega);p}. \ \square$$

3. **History of the construction of a function $\mathcal{C}^\infty$ with support in the unit ball.** Jean LERAY introduced the function ρ defined by (2.4) and proved that it is infinitely differentiable in 1934 [53, Note 1, p. 206].

Interest of the semi-norms of $\mathcal{D}(\Omega)$. The family of semi-norms in Definition 2.5 avoids appealing to the delicate **topology of inductive limit of the** $\mathcal{D}_K(\Omega)$ introduced by Laurent SCHWARTZ [69, Chap. III, § 1, p. 64]. Indeed, this family generates the said topology [SIMON, 84], since it is equivalent to the family of semi-norms that SCHWARTZ considers in [69, p. 65]. □

Spaces $\mathcal{D}_K(\Omega)$. The spaces $\mathcal{D}_K(\mathbb{R}^d)$ introduced by Laurent SCHWARTZ coincide algebraically and topologically with the space $\mathcal{C}_K^\infty(\mathbb{R}^d)$ in our Definition 2.3 (a). He denotes it by $\mathcal{D}_K$, since he used it as a milestone in the construction of $\mathcal{D}$ [69, Chap. I, § 2, p. 24, or Chap. III, § 1, p. 64]. We prefer to denote it by $\mathcal{C}_K^\infty(\Omega)$, as we introduce it as a topological subspace of $\mathcal{C}_{\mathrm{b}}^\infty(\Omega)$, and not of $\mathcal{D}(\Omega)$. □

Another topology on $\mathcal{D}(\Omega)$. In Volume 2 [82, Definition 2.15 (e)], we introduced the following family of semi-norms on $\mathcal{D}(\Omega)$, then denoted by $\mathcal{K}^\infty(\Omega)$, indexed by $m \in \mathbb{N}$ and $p \in \mathcal{C}^+(\Omega)$,

$$\|f\|_{\mathcal{K}^\infty(\Omega);m,p} = \sup_{x\in\Omega} \sup_{0\leq|\beta|\leq m} p(x)|\partial^\beta f(x)|.$$

(Here, the weight $p(x)$ controls only the value $|\partial^\beta f(x)|$ of the derivatives whereas, in Definition 2.5 of the semi-norms of $\mathcal{D}(\Omega)$, it also controls their order by the inequality $|\beta| \leq p(x)$.)

These semi-norms define a topology on $\mathcal{D}(\Omega)$ intermediate between those of $\mathcal{D}(\Omega)$ and $\mathcal{C}_{\mathrm{b}}^\infty(\Omega)$ since, for every $\varphi \in \mathcal{D}(\Omega)$, $m \in \mathbb{N}$ and $p \in \mathcal{C}^+(\Omega)$, denoting by m_Ω the function of constant value m, we have

$$\|\varphi\|_{\mathcal{C}_{\mathrm{b}}^\infty(\Omega);m} \leq \frac{1}{m}\|\varphi\|_{\mathcal{K}^\infty(\Omega);m,m_\Omega} \quad\text{and}\quad \|\varphi\|_{\mathcal{K}^\infty(\Omega);m,p} \leq \|\varphi\|_{\mathcal{D}(\Omega);p+m_\Omega}.$$

These semi-norms give rise to the same convergent sequences as the semi-norms of $\mathcal{D}(\Omega)$. Indeed, on this space, the semi-norms of $\mathcal{C}_{\mathrm{b}}^\infty(\Omega)$ give rise to them as too by Theorem 2.13 (b).

We could therefore replace one family of semi-norms by the other one almost everywhere. However, the family of $\mathcal{D}(\Omega)$ facilitates certain proofs. Moreover, it generates the topology of inductive limit of the $\mathcal{D}_K(\Omega)$ used by Laurent SCHWARTZ, and it makes $\mathcal{D}(\Omega)$ reflexive, which is not the case of $\mathcal{K}^\infty(\Omega)$. □

2.4. Sequential completeness of $\mathcal{D}(\Omega)$

Let us show that $\mathcal{D}(\Omega)$ is sequentially complete [4].

Theorem 2.8.– *The space $\mathcal{D}(\Omega)$, where Ω is an open subset of $\mathbb{R}^d$, is a Neumann space.* ▮

The proof will use the following property.

Theorem 2.9.– *The functions of a bounded subset of $\mathcal{D}(\Omega)$, where Ω is an open subset of $\mathbb{R}^d$, all have their support in the same compact subset of Ω.* ▮

4. **History of Theorem 2.8.** Laurent SCHWARTZ proved in 1950 [69, chap. III, § 1, Theorem I, p. 66] that $\mathcal{D}(\mathbb{R}^d)$ is complete, which results in sequentially complete.

Proof. Let $\mathcal{B}$ be a bounded subset of $\mathcal{D}(\Omega)$, $\varphi \in \mathcal{B}$ and $p \in \mathcal{C}^+(\Omega)$. Since $\partial^0\varphi = \varphi$, Definition 2.5 of the semi-norms of $\mathcal{D}(\Omega)$ and Definition 1.2 of a bounded set give

$$\sup_{\varphi\in\mathcal{B}}\sup_{x\in\Omega} p(x)|\varphi(x)| \leq \sup_{\varphi\in\mathcal{B}}\sup_{x\in\Omega}\sup_{0\leq|\beta|\leq p(x)} p(x)|\partial^\beta\varphi(x)| = \sup_{\varphi\in\mathcal{B}}\|\varphi\|_{\mathcal{D}(\Omega);p} < \infty.$$

This holds for every function p of $\mathcal{C}^+(\Omega)$, so that, by Theorem 2.4, all the functions φ of $\mathcal{B}$ have their support included in a same compact subset K of Ω. □

We can now prove that $\mathcal{D}(\Omega)$ is sequentially complete.

Proof of Theorem 2.8. **1.** *Preliminary.* Given $m \in \mathbb{N}$, Definition 1.19 (b) of the semi-norms of $\mathcal{C}_b^\infty(\Omega)$ and Definition 2.5 of the semi-norms of $\mathcal{D}(\Omega)$ give, for every $\varphi \in \mathcal{D}(\Omega)$,

$$\|\varphi\|_{\mathcal{C}_b^\infty(\Omega);m} = \sup_{|\beta|\leq m}\sup_{x\in\Omega}|\partial^\beta\varphi(x)| =$$
$$= \frac{1}{m}\sup_{x\in\Omega}\sup_{|\beta|\leq m_\Omega(x)} m_\Omega(x)|\partial^\beta\varphi(x)| = \frac{1}{m}\|\varphi\|_{\mathcal{D}(\Omega);m_\Omega}, \quad (2.5)$$

where m_Ω is the constant function with value m.

2. *Sequential completeness.* Let

$$(\varphi_n)_{n\in\mathbb{N}} \text{ be a Cauchy sequence in } \mathcal{D}(\Omega).$$

By Definition 1.3 (b) of a Cauchy sequence, equality (2.5) implies that $(\varphi_n)_{n\in\mathbb{N}}$ is Cauchy in $\mathcal{C}_b^\infty(\Omega)$. Since this one is sequentially complete (Theorem A.53), there is a limit, say

$$\varphi_n \to \varphi \text{ in } \mathcal{C}_b^\infty(\Omega).$$

The sequence $(\varphi_n)_{n\in\mathbb{N}}$ is bounded in $\mathcal{D}(\Omega)$, as is every Cauchy sequence (Theorem A.5), and thus, from Theorem 2.9,

$$\text{all the } \varphi_n \text{ have their support in a same compact subset } K \text{ of } \Omega.$$

By Definition 2.1 of the support, the φ_n are zero in $\Omega \setminus K$. Their limit φ is zero too on this set. Its support is thus compact, as is the case for every function that is zero ouside of a compact subset of Ω (Theorem 2.2 (b)). Hence,

$$\varphi \in \mathcal{D}(\Omega).$$

For every $x \in \Omega$ and $\beta \in \mathbb{N}^d$, $\partial^\beta\varphi_n(x) \to \partial^\beta\varphi(x)$, thus

$$|\partial^\beta\varphi_n(x) - \partial^\beta\varphi(x)| \leq \sup_{\ell\geq n}|\partial^\beta\varphi_n(x) - \partial^\beta\varphi_\ell(x)|.$$

Whence, again by Definition 2.5 of the semi-norms of $\mathcal{D}(\Omega)$, for every $p \in \mathcal{C}^+(\Omega)$,

$$\|\varphi_n - \varphi\|_{\mathcal{D}(\Omega);p} = \sup_{x\in\Omega} \sup_{|\beta|\le p(x)} p(x)|\partial^\beta \varphi_n(x) - \partial^\beta \varphi(x)| \le$$

$$\le \sup_{\ell\ge n} \sup_{x\in\Omega} \sup_{|\beta|\le p(x)} p(x)|\partial^\beta \varphi_n(x) - \partial^\beta \varphi_\ell(x)| = \sup_{\ell \ge n} \|\varphi_n - \varphi_\ell\|_{\mathcal{D}(\Omega);p}.$$

The last term tends to 0 when $n \to \infty$, again by Definition 1.3 (b) of a Cauchy sequence. Hence,

$$\varphi_n \to \varphi \text{ in } \mathcal{D}(\Omega).$$

This proves that $\mathcal{D}(\Omega)$ is sequentially complete, that is to say Neumann. □

2.5. Comparison of $\mathcal{D}(\Omega)$ to various spaces

Let us compare $\mathcal{D}(\Omega)$ to the spaces $\mathcal{C}^\infty(\Omega)$, $\mathcal{C}^\infty_b(\Omega)$ and $\mathcal{C}^\infty_K(\Omega)$.

Theorem 2.10.– *For every open subset Ω of $\mathbb{R}^d$ and every compact subset K of Ω,*

$$\mathcal{C}^\infty_K(\Omega) \subsetneq \mathcal{D}(\Omega) \subsetneq \mathcal{C}^\infty_b(\Omega) \subsetneq \mathcal{C}^\infty(\Omega). \ ▮$$

The proof will use the following inequality, which will also be used later.

Theorem 2.11.– *Let $p \in \mathcal{C}^+(\Omega)$, where Ω is an open subset of $\mathbb{R}^d$, and K be a compact subset of Ω. Then, for every $\varphi \in \mathcal{C}^\infty_K(\Omega)$,*

$$\|\varphi\|_{\mathcal{D}(\Omega);p} \le m\|\varphi\|_{\mathcal{C}^m_b(\Omega)},$$

where m is any integer upper bound of p over K. ▮

Proof of Theorem 2.11. **1.** *Existence of m.* The existence of an upper bound m of p over K results from the fact that any continuous function is bounded on any compact set due to Heine's theorem (Theorem A.32).

2. *Inequality.* Let $\varphi \in \mathcal{C}^\infty_K(\Omega)$. Since the $\partial^\beta\varphi$ are zero ouside of K according to Theorem 2.6, Definition 2.5 of the semi-norms of $\mathcal{D}(\Omega)$ gives

$$\|\varphi\|_{\mathcal{D}(\Omega);p} = \sup_{x\in\Omega} \sup_{|\beta|\le p(x)} p(x)|\partial^\beta\varphi(x)| \le m \sup_{|\beta|\le m} \sup_{x\in K} |\partial^\beta\varphi(x)|,$$

which is bounded from above, by Definition 1.18 (b) of the semi-norms of $\mathcal{C}^m_b(\Omega)$, by

$$\le m \sup_{|\beta|\le m} \sup_{x\in\Omega} |\partial^\beta\varphi(x)| = m\,\|\varphi\|_{\mathcal{C}^m_b(\Omega)}. \ \square$$

Now we prove the topological inclusions.

Proof of Theorem 2.10. **1.** *Inclusion* $\mathcal{C}_K^\infty(\Omega) \subsetneqq \mathcal{D}(\Omega)$. According to Theorem 2.11 and equality (1.6), p. 13, for every function $p \in \mathcal{C}^+(\Omega)$, there exists $m \in \mathbb{N}$ such that, for every $\varphi \in \mathcal{C}_K^\infty(\Omega)$,

$$\|\varphi\|_{\mathcal{D}(\Omega);p} \leq m\|\varphi\|_{\mathcal{C}_b^m(\Omega)} = m\|\varphi\|_{\mathcal{C}_b^\infty(\Omega);m}.$$

Since $\mathcal{C}_K^\infty(\Omega)$ is, by Definition 2.3 (a), endowed with the semi-norms of $\mathcal{C}_b^\infty(\Omega)$, this inequality gives, by Definition 1.7 (c) of topological inclusion,

$$\mathcal{C}_K^\infty(\Omega) \subsetneqq \mathcal{D}(\Omega).$$

2. *Inclusion* $\mathcal{D}(\Omega) \subsetneqq \mathcal{C}_b^\infty(\Omega)$. This results from the fact that, as established in (2.5), p. 25, for every $m \in \mathbb{N}$, we have, for every $\varphi \in \mathcal{D}(\Omega)$,

$$\|\varphi\|_{\mathcal{C}_b^\infty(\Omega);m} = \frac{1}{m}\|\varphi\|_{\mathcal{D}(\Omega);m_\Omega},$$

where m_Ω is the constant function of value m.

3. *Inclusion* $\mathcal{C}_b^\infty(\Omega) \subsetneqq \mathcal{C}^\infty(\Omega)$. This results from the fact that, by Definition 1.19 of the semi-norms of $\mathcal{C}^\infty(\Omega)$ and $\mathcal{C}_b^\infty(\Omega)$, for every $m \in \mathbb{N}$ and every compact subset K of Ω, we have, for every $\varphi \in \mathcal{C}_b^\infty(\Omega)$,

$$\|\varphi\|_{\mathcal{C}^\infty(\Omega);m,K} = \sup_{|\beta|\leq m}\sup_{x\in K}|\partial^\beta\varphi(x)| \leq \sup_{|\beta|\leq m}\sup_{x\in\Omega}|\partial^\beta\varphi(x)| = \|\varphi\|_{\mathcal{C}_b^\infty(\Omega);m}. \quad \square$$

Let us show that the topologies of these spaces coincide on $\mathcal{C}_K^\infty(\Omega)$.

Theorem 2.12.– *For every open subset Ω of $\mathbb{R}^d$ and every compact subset K of Ω, the topologies of $\mathcal{C}_K^\infty(\Omega)$, $\mathcal{D}(\Omega)$, $\mathcal{C}_b^\infty(\Omega)$ and $\mathcal{C}^\infty(\Omega)$ coincide on $\mathcal{C}_K^\infty(\Omega)$.*

That is, their families of semi-norms are equivalent on $\mathcal{C}_K^\infty(\Omega)$. ▮

A shortcut. We write "their families of semi-norms are equivalent", when actually it is the families of their restrictions to $\mathcal{C}_K^\infty(\Omega)$ that are. This is not ambiguous, however. □

Proof of Theorem 2.12. On $\mathcal{C}_K^\infty(\Omega)$, let us compare these topologies to that of $\mathcal{C}_b^\infty(\Omega)$.

1. *Topology of* $\mathcal{C}_K^\infty(\Omega)$. The topology of $\mathcal{C}_K^\infty(\Omega)$ is that of $\mathcal{C}_b^\infty(\Omega)$, according to Definition 2.3 (a).

2. Topology of $\mathcal{D}(\Omega)$. The space $\mathcal{D}(\Omega)$ topologically lies between $C_K^\infty(\Omega)$ and $C_b^\infty(\Omega)$ due to Theorem 2.10, thus on $C_K^\infty(\Omega)$ its topology coincides with that of $C_b^\infty(\Omega)$ from step 1.

3. Topology of $C^\infty(\Omega)$. On $C_K^\infty(\Omega)$, the topology of $C^\infty(\Omega)$ coincides with that of $C_b^\infty(\Omega)$ since, by Definition 1.19 of their semi-norms:

— on one hand, for every $\varphi \in C_K^\infty(\Omega)$ and $m \in \mathbb{N}$, the $\partial^\beta\varphi$ are zero ouside of K due to Theorem 2.6, and thus

$$\|\varphi\|_{C_b^\infty(\Omega);m} = \sup_{|\beta|\le m} \sup_{x\in\Omega} |\partial^\beta\varphi(x)| = \sup_{|\beta|\le m} \sup_{x\in K} |\partial^\beta\varphi(x)| = \|\varphi\|_{C^\infty(\Omega);m,K};$$

— conversely, for every $\varphi \in C_K^\infty(\Omega)$, every $m \in \mathbb{N}$ and every compact subset Q of Ω,

$$\|\varphi\|_{C^\infty(\Omega);m,Q} = \sup_{|\beta|\le m} \sup_{x\in Q} |\partial^\beta\varphi(x)| \le \sup_{|\beta|\le m} \sup_{x\in\Omega} |\partial^\beta\varphi(x)| = \|\varphi\|_{C_b^\infty(\Omega);m}. \quad \square$$

Remark. In step 3 of the proof, the equality of semi-norms $\|\varphi\|_{C_b^\infty(\Omega);m} = \|\varphi\|_{C^\infty(\Omega);m,K}$ for every $m \in \mathbb{N}$ is not sufficient for the equality of the topologies of $C_b^\infty(\Omega)$ and $C^\infty(\Omega)$ on $C_K^\infty(\Omega)$. The necessity for an inverse inequality is explained in the comment *Caution*, p. 5. □

2.6. Convergent sequences in $\mathcal{D}(\Omega)$

Let us characterize the convergent sequences[5] in $\mathcal{D}(\Omega)$.

Theorem 2.13.– *Let* $(\varphi_n)_{n\in\mathbb{N}}$ *be a sequence in* $\mathcal{D}(\Omega)$ *and* $\varphi \in \mathcal{D}(\Omega)$, *where* Ω *is an open subset of* $\mathbb{R}^d$. *Then each of the following properties is equivalent to*

$$\varphi_n \to \varphi \ \text{ in } \mathcal{D}(\Omega).$$

(a) *There exists a compact subset* K *of* Ω *such that all the* φ_n *and* φ *have their support in* K *and*

$$\varphi_n \to \varphi \ \text{ in } C^\infty(\Omega).$$

(b) *There exists a compact subset* K *of* Ω *such that*

$$\varphi_n \to \varphi \ \text{ in } C_K^\infty(\Omega).$$

5. **History of Theorem 2.13 (a).** Laurent SCHWARTZ proved in 1950 [69, Chap. III, § 1, proof of Theorem I, p. 66] that the functions of a convergent sequence in $\mathcal{D}(\mathbb{R}^d)$ and their limit all have their support in the same compact set.

History of Theorem 2.13 (c). This characterization, which is new including for real values, seems to be related to the fact that $\mathcal{D}(\Omega)$ is bornological [SCHWARTZ, 69, Chap. III, Theorem II, § 1, p. 66], see note 2, p. 46.

(c) *There exists a compact subset K of Ω, a real sequence $(t_n)_{n\in\mathbb{N}}$ and a sequence $(\phi_n)_{n\in\mathbb{N}}$ of $\mathcal{C}_K^\infty(\Omega)$ such that*

$$\varphi_n - \varphi = t_n\phi_n \textit{ for each } n, \quad t_n \to 0, \quad (\phi_n)_{n\in\mathbb{N}} \textit{ is bounded in } \mathcal{C}_K^\infty(\Omega). \blacksquare$$

Proofs of characterizations (a) and (b). These properties are equivalent to the convergence in $\mathcal{D}(\Omega)$ due to the three following implications.

1. *Convergence in $\mathcal{D}(\Omega) \Rightarrow$ convergence in $\mathcal{C}^\infty(\Omega)$ with support in K.* Suppose that

$$\varphi_n \to \varphi \text{ in } \mathcal{D}(\Omega).$$

Since $\mathcal{D}(\Omega) \underset{\rightarrow}{\subsetneq} \mathcal{C}^\infty(\Omega)$ (Theorem 2.10), it follows (Theorem A.6) that

$$\varphi_n \to \varphi \text{ in } \mathcal{C}^\infty(\Omega).$$

Moreover, $(\varphi_n)_{n\in\mathbb{N}}$ is bounded in $\mathcal{D}(\Omega)$, as is every convergent sequence (Theorem A.5), so the set composed of the φ_n and φ is bounded too, which implies by Theorem 2.9 that

all of the φ_n and φ have their support in a same compact subset K of Ω.

2. *Convergence in $\mathcal{C}^\infty(\Omega)$ with support in $K \Leftrightarrow$ convergence in $\mathcal{C}_K^\infty(\Omega)$.* Indeed, the topologies of $\mathcal{C}^\infty(\Omega)$ and $\mathcal{C}_K^\infty(\Omega)$ coincide on $\mathcal{C}_K^\infty(\Omega)$ (Theorem 2.12).

3. *Convergence in $\mathcal{C}_K^\infty(\Omega) \Rightarrow$ convergence in $\mathcal{D}(\Omega)$.* This results from the topological inclusion $\mathcal{C}_K^\infty(\Omega) \underset{\rightarrow}{\subsetneq} \mathcal{D}(\Omega)$ (Theorem 2.10), again due to Theorem A.6. □

Proof of characterization (c) of Theorem 2.13. It is sufficient to establish the equivalence of (c) with (b), namely that, for a given compact subset K of Ω, the convergence $\varphi_n \to \varphi$ in $\mathcal{C}_K^\infty(\Omega)$ is equivalent to the existence of t_n and ϕ_n such that

$$\varphi_n - \varphi = t_n\phi_n \text{ for each } n, \quad t_n \to 0, \quad (\phi_n)_{n\in\mathbb{N}} \text{ is bounded in } \mathcal{C}_K^\infty(\Omega). \qquad (2.6)$$

Recall that $\mathcal{C}_K^\infty(\Omega)$ is (by Definition 2.3 (a)) endowed with the semi-norms of $\mathcal{C}_{\mathrm{b}}^\infty(\Omega)$ which are (by Definition 1.19 (b)) indexed by $m \in \mathbb{N}$ and are equal to

$$\|\varphi\|_{\mathcal{C}_{\mathrm{b}}^\infty(\Omega);m} = \sup_{0\le|\beta|\le m}\ \sup_{x\in\Omega} |\partial^\beta\varphi(x)|.$$

The properties (2.6) imply the convergence $\varphi_n \to \varphi$ in $\mathcal{C}_K^\infty(\Omega)$ (since, for every $m \in \mathbb{N}$, they give $\|\varphi_n - \varphi\|_{\mathcal{C}_{\mathrm{b}}^\infty(\Omega);m} = |t_n|\|\phi_n\|_{\mathcal{C}_{\mathrm{b}}^\infty(\Omega);m} \to 0$), therefore it remains to prove the converse.

So suppose that

$$\varphi_n \to \varphi \text{ in } \mathcal{C}_K^\infty(\Omega), \tag{2.7}$$

and let us construct suitable t_n and ϕ_n, in four steps.

1. *The integer ℓ_m.* Given $m \in \mathbb{N}$, let

$$X_m \stackrel{\text{def}}{=} \{k \in \mathbb{N} : m\|\varphi_k - \varphi\|_{\mathcal{C}_b^\infty(\Omega);m} \geq 1\}$$

and

$$\ell_m \stackrel{\text{def}}{=} \sup X_m + m + 1 \tag{2.8}$$

(if X_m is empty, its upper bound in $\mathbb{N}$ is 0 and so $\ell_m \stackrel{\text{def}}{=} m + 1$). Observe that ℓ_m is finite since, from (2.7), k does not belong to X_m if it is large enough.

2. *Choice of t_n.* Observe also that

$$\ell_{m+1} \geq \ell_m + 1, \tag{2.9}$$

because $X_{m+1} \supset X_m$, since $\|\varphi_k - \varphi\|_{\mathcal{C}_b^\infty(\Omega);m+1} \geq \|\varphi_k - \varphi\|_{\mathcal{C}_b^\infty(\Omega);m}$.

Given $n \in \mathbb{N}$, there exists then a unique $i_n \in \mathbb{N}$ such that

$$\ell_{i_n} \leq n < \ell_{i_n+1}. \tag{2.10}$$

We choose

$$t_n \stackrel{\text{def}}{=} \frac{1}{i_n}.$$

3. *Convergence of t_n.* Since ℓ_m grows with m,

$$i_n \text{ grows with } n. \tag{2.11}$$

When $n \to \infty$, we have $i_n \to \infty$ (if not, i_n would be bounded by an integer N, so we would have $\ell_{i_n+1} \leq \ell_{N+1}$ from (2.8), and then n would be bounded due to (2.10)). Thus,

$$t_n \to 0 \text{ when } n \to \infty.$$

4. *Bounding $(\phi_n)_{n\in\mathbb{N}}$.* From our choice of t_n,

$$\phi_n = \frac{1}{t_n}(\varphi_n - \varphi) = i_n(\varphi_n - \varphi). \tag{2.12}$$

From (2.10) and (2.8), $n \geq \ell_{i_n} > \sup X_{i_n}$. Therefore $n \notin X_{i_n}$, i.e.

$$i_n\|\varphi_n - \varphi\|_{\mathcal{C}_b^\infty(\Omega);i_n} < 1. \tag{2.13}$$

Given $m \in \mathbb{N}$, observe that

$$i_{\ell_m} = m. \tag{2.14}$$

Indeed, inequality (2.10) with $n = \ell_m$ is written as $\ell_{i_{\ell_m}} \leq \ell_m < \ell_{i_{\ell_m}+1}$, from which follows $i_{\ell_m} \leq m < i_{\ell_m} + 1$ since ℓ_m grows strictly with m due to (2.9).

If $n \geq \ell_m$, then $i_n \geq i_{\ell_m}$ (from (2.11)) and therefore (with (2.14)) $i_n \geq m$, from which (since the semi-norm $\| \ \|_{\mathcal{C}_b^\infty(\Omega);m}$ grows with m by its Definition 1.19 (b)), with (2.12) and (2.13),

$$\|\phi_n\|_{\mathcal{C}_b^\infty(\Omega);m} \leq \|\phi_n\|_{\mathcal{C}_b^\infty(\Omega);i_n} = i_n \|\varphi_n - \varphi\|_{\mathcal{C}_b^\infty(\Omega);i_n} \leq 1.$$

Thus,

$$\sup_{n\in\mathbb{N}} \|\phi_n\|_{\mathcal{C}_b^\infty(\Omega);m} \leq \sup\{\|\phi_1\|_{\mathcal{C}_b^\infty(\Omega);m}, \ldots, \|\phi_{\ell_m}\|_{\mathcal{C}_b^\infty(\Omega);m}, 1\} < \infty.$$

Therefore, since ϕ_n belongs to $\mathcal{C}_K^\infty(\Omega)$ (as φ_n and φ belong to it too, since they converge there by hypothesis)

$$(\phi_n)_{n\in\mathbb{N}} \text{ is bounded in } \mathcal{C}_K^\infty(\Omega). \quad \square$$

Utility of the characterization of Theorem 2.13 (c). This new characterization of convergent sequences of test functions serves to prove, via the characterization of distributions from Theorem 3.4 (e), the equality $\mathcal{D}'(\Omega; E\text{-weak}) = \mathcal{D}'(\Omega; E)$ of Theorem 4.14. □

Let us show that every bounded sequence in $\mathcal{D}(\Omega)$ has a convergent subsequence [6].

Theorem 2.14.– *Every bounded sequence in $\mathcal{D}(\Omega)$, where Ω is an open subset of $\mathbb{R}^d$, has a convergent subsequence.* ▌

The proof will use the following property, whose proof is given in Volume 2 [82, Theorem 2.25].

Theorem 2.15.– *Every bounded sequence in $\mathcal{C}_K^{m+1}(\Omega)$, where Ω is an open subset of $\mathbb{R}^d$, K is a compact subset of Ω and $m \in \mathbb{N}$, has a subsequence that converges in $\mathcal{C}_K^m(\Omega)$.* ▌

---

6. **History of Theorem 2.14.** Laurent SCHWARTZ proved in 1950 [69, Chap. III, § 2, Theorem VII, p. 70] that every bounded subset of $\mathcal{D}(\mathbb{R}^d)$ is relatively compact.
A priori, this does not involve Theorem 2.14, as there are semi-normed spaces in which *relatively compact* **does not imply** *relatively sequentially compact* [Vol. 1, Properties (2.6) and (2.7), p. 27]. But the argument used in [69], that we restate here, shows that, in $\mathcal{D}(\Omega)$, *relatively compact* coincides with *relatively sequentially compact* and also with *bounded*.

Proof of Theorem 2.14. Let

$$(\varphi_n)_{n\in\mathbb{N}} \text{ be a bounded sequence in } \mathcal{D}(\Omega).$$

1. *Boundedness in $\mathcal{C}_K^m(\Omega)$ for every m.* All these functions have their support in a same compact subset K of Ω from Theorem 2.9, so

$$\varphi_n \in \mathcal{C}_K^\infty(\Omega).$$

For every $m \in \mathbb{N}$, $\mathcal{D}(\Omega)$ is topologically included in $\mathcal{C}_b^\infty(\Omega)$ (by Theorem 2.10) and thus in $\mathcal{C}_b^{m+1}(\Omega)$. Therefore, since $\mathcal{C}_K^{m+1}(\Omega)$ is (by Definition 2.3 (a)) endowed with the semi-norms of $\mathcal{C}_b^{m+1}(\Omega)$,

$$(\varphi_n)_{n\in\mathbb{N}} \text{ is bounded in } \mathcal{C}_K^{m+1}(\Omega).$$

2. *Passing to subsequences.* According to Theorem 2.15, the sequence $(\varphi_n)_{n\in\mathbb{N}}$ therefore has a subsequence $(\varphi_{\sigma_m(n)})_{n\in\mathbb{N}}$ which converges in $\mathcal{C}_K^m(\Omega)$. That is, when $n \to \infty$,

$$\varphi_{\sigma_m(n)} \to \ell_m \text{ in } \mathcal{C}_K^m(\Omega).$$

By successively passing to subsequences, we can choose $(\varphi_{\sigma_{m+1}(n)})_{n\in\mathbb{N}}$ which is a subsequence of $(\varphi_{\sigma_m(n)})_{n\in\mathbb{N}}$. So, $\ell_{m+1} = \ell_m$, since every subsequence of a convergent sequence converges to the same limit (Theorem A.5). Hence,

$$\ell_m = \ell_0.$$

3. *Diagonal subsequence.* The diagonal subsequence $(\varphi_{\sigma_n(n)})_{n\in\mathbb{N}}$ also converges to ℓ_0 in $\mathcal{C}_K^m(\Omega)$, since it is (for the $n \geq m$) a subsequence of $(\varphi_{\sigma_m(n)})_{n\in\mathbb{N}}$. That is,

$$\varphi_{\sigma_n(n)} \to \ell_0 \text{ in } \mathcal{C}_K^m(\Omega). \tag{2.15}$$

This holds for each m, so

$$\ell_0 \in \mathcal{C}_K^\infty(\Omega).$$

The spaces $\mathcal{C}_K^\infty(\Omega)$ and $\mathcal{C}_K^m(\Omega)$ are (by Definition 2.3 (a)) endowed respectively with the semi-norms of $\mathcal{C}_b^\infty(\Omega)$ and $\mathcal{C}_b^m(\Omega)$, and, from equality (1.6), p. 13,

$$\|\varphi_{\sigma_n(n)} - \ell_0\|_{\mathcal{C}_b^\infty(\Omega);m} = \|\varphi_{\sigma_n(n)} - \ell_0\|_{\mathcal{C}_b^m(\Omega)}.$$

Since the right-hand side tends to 0 according to (2.15),

$$\varphi_{\sigma_n(n)} \to \ell_0 \text{ in } \mathcal{C}_K^\infty(\Omega).$$

4. *Convergence in $\mathcal{D}(\Omega)$.* Since (Theorem 2.10) $\mathcal{C}_K^\infty(\Omega) \subsetneq \mathcal{D}(\Omega)$, convergence in the first space leads to (Theorem A.6) convergence in the second, thus

$$\varphi_{\sigma_n(n)} \to \ell_0 \text{ in } \mathcal{D}(\Omega). \ \square$$

Observe now that $\mathcal{D}(\Omega)$ is sequentially separable, and so is separable.

Theorem 2.16.– *Given an open subset Ω of $\mathbb{R}^d$, there exists a sequence $(\phi_k)_{k\in\mathbb{N}}$ which is sequentially dense, and hence dense, in $\mathcal{D}(\Omega)$.* ∎

The proof will use the following property of density, which is proven in Volume 2 [82, Theorem 7.21] since the space $\mathcal{K}^\infty(\Omega)$ used there is algebraically equal to $\mathcal{D}(\Omega)$.

Theorem 2.17.– *Given an open subset Ω of $\mathbb{R}^d$, there exists a sequence $(\phi_k)_{k\in\mathbb{N}}$ of $\mathcal{D}(\Omega)$ such that, for every $\varphi \in \mathcal{D}(\Omega)$, there exist a compact subset K of Ω and a subsequence $(\phi_{\sigma(n)})_{n\in\mathbb{N}}$ converging to φ in $\mathcal{C}_K^\infty(\Omega)$.* ∎

Proof of Theorem 2.16. The sequence $(\phi_k)_{k\in\mathbb{N}}$ given by Theorem 2.17 is sequentially dense (Definition A.12) in $\mathcal{D}(\Omega)$, since the convergence in all the $\mathcal{C}_K^\infty(\Omega)$ is equivalent to the convergence in $\mathcal{D}(\Omega)$ (according to the characterization of convergent sequences in $\mathcal{D}(\Omega)$ from Theorem 2.13 (b)).

This sequence is equally dense in $\mathcal{D}(\Omega)$, since density always follows from sequential density (Theorem A.13). □

2.7. Covering by crown-shaped sets and partitions of unity

We define a cover by crown-shaped sets of an open subset of $\mathbb{R}^d$, which we will use in the following section to "control" the norms of the $\mathcal{C}_K^m(\Omega)$.

Definition 2.18.– *The **cover by crown-shaped sets** of an open subset Ω of $\mathbb{R}^d$ is the sequence $(\kappa_n)_{n\in\mathbb{N}}$ of crown-shaped sets defined, for $n \geq 1$, by*

$$\kappa_n \stackrel{\text{def}}{=} \Omega^{n+2}_{1/(n+2)} \setminus \overline{\Omega^n_{1/n}},$$

and by $\kappa_0 \stackrel{\text{def}}{=} \Omega^2_{1/2}$, where

$$\Omega^n_r \stackrel{\text{def}}{=} \{x \in \Omega : |x| < n,\ B(x,r) \subset \Omega\}$$

and $B(x,r) \stackrel{\text{def}}{=} \{y \in \mathbb{R}^d : |y - x| \leq r\}$. ∎

Let us recall some elementary properties of crown-shaped sets which are proven in Volume 2 [82, Theorem 7.17].

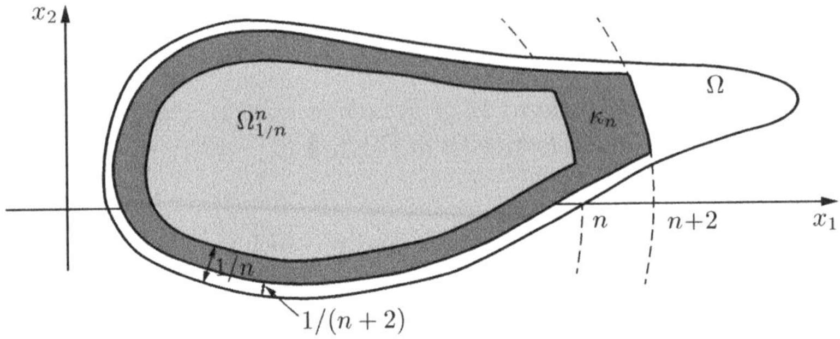

Figure 2.1. *Crown-shaped set κ_n and potato-shaped set $\Omega^n_{1/n}$. κ_n is dark gray, $\Omega^n_{1/n}$ is light gray*

Theorem 2.19.– *Let $(\kappa_n)_{n\in\mathbb{N}}$ be the cover by crown-shaped sets of an open subset Ω of $\mathbb{R}^d$. Then:*

(a) *The union of the κ_n is Ω.*

(b) *Each κ_n is open, bounded, only intersects κ_{n-1} and κ_{n+1}, and its closure $\overline{\kappa_n}$ is a compact subset of Ω.*

(c) *Each compact subset K of Ω intersects only finitely many κ_n.* ▮

Let us introduce partitions of unity [7].

Definition 2.20.– *A* ***partition of unity*** *subordinate to the cover of an open subset Ω of $\mathbb{R}^d$ by a family $\{\omega_i\}_{i\in\mathcal{I}}$ of open sets is a family $\{\alpha_i\}_{i\in\mathcal{I}}$ of functions such that: for every $i \in \mathcal{I}$,*

$$\alpha_i \in \mathcal{C}^\infty(\Omega),\ \ 0 \le \alpha_i \le 1,\ \ \operatorname{supp}\alpha_i \subset \omega_i;$$

$$\textstyle\sum_{i\in\mathcal{I}} \alpha_i = 1 \ \text{ on } \Omega;$$

for every compact subset K of Ω, the set of the α_i that are not identically zero on K is finite. ▮

Justification. The sum $\sum_{i\in\mathcal{I}} \alpha_i(x)$ is well-defined at each point x of Ω, since only finitely many $\alpha_i(x)$ are not zero (from Theorem 2.19 (c) for the compact set $K = \{x\}$). □

7. **History of partition of unity.** Jean DIEUDONNÉ introduced the notion of partition of unity in 1937 [25]. Salomon BOCHNER used it simultaneously but independently. Laurent SCHWARTZ spoke about it earlier.

Recall [Vol. 2, Theorem 7.18]:

Theorem 2.21.– *Every cover of an open subset of $\mathbb{R}^d$ by open sets possesses a subordinate partition of unity.* ▌

Support of α_i. The condition $\operatorname{supp} \alpha_i \subset \omega_i$ in Definition 2.20 of a partition of unity requires, by Definition 2.1 of the support, that α_i be zero in $\Omega \setminus \omega_i$, and thus on $\partial\omega_i \setminus \partial\Omega$, i.e. on the part of the boundary $\partial\omega_i$ of ω_i which is not included in $\partial\Omega$.

In contrast, α_i is not necessarily close to 0 in the neighborhood of the part of $\partial\omega_i$ which is included in $\partial\Omega$. For example, this is the case if the cover of Ω is reduced to $\omega_1 = \Omega$, since then α_1 equals 1 on Ω. □

Partition whose functions are not uniformly continuous. There are covers of open sets for which every function of every subordinate partition of unity satisfies

$$\alpha_i \notin \mathbf{C}_{\mathrm{b}}(\Omega).$$

This may occur even if Ω is bounded. For example, it is the case if Ω is the union of two disjoint open sets ω_1 and ω_2 whose boundaries are not disjoint, since then α_1 equals 1 on ω_1 and 0 on ω_2, so it *jumps* in the neighborhood of every common point in their boundaries. □

2.8. Control of the $\mathcal{C}_K^m(\Omega)$-norms by the semi-norms of $\mathcal{D}(\Omega)$

We saw in (2.5), p. 25, that for every m there exists a function p (in fact, p is constant of value m) such that

$$\|\varphi\|_{\mathcal{D}(\Omega);p} \geq m \|\varphi\|_{\mathcal{C}_{\mathrm{b}}^m(\Omega)}, \quad \forall \varphi \in \mathcal{D}(\Omega).$$

We propose to establish a stronger inequality which allows m to become arbitrarily large when the support of φ concentrates in the neighbourhood of the boundary of Ω.

To do this, we will decompose φ into a sum of functions $\alpha_k\varphi$ where the support of α_k concentrates in the neighbourhood of the boundary when $k \to \infty$ and build, for given m_k, a function p such that $\|\varphi\|_{\mathcal{D}(\Omega);p}$ is larger than a sum of the $\|\alpha_k\varphi\|_{\mathcal{C}_{\mathrm{b}}^{m_k}(\Omega)}$.

It is the **first theorem of control of the $\mathcal{C}_K^m(\Omega)$-norms**, that is here [8] (the second is Theorem 2.25).

Theorem 2.22.– *Let Ω be an open subset of $\mathbb{R}^d$, $(\kappa_k)_{k\in\mathbb{N}}$ its cover by crown-shaped sets (Definition 2.18), $(\alpha_k)_{k\in\mathbb{N}}$ a subordinate partition of unity (Definition 2.20), and $(m_k)_{k\in\mathbb{N}}$ a sequence of integers.*

8. **History of Theorem 2.22.** This inequality which "controls" series of $\mathcal{C}_K^m(\Omega)$-norms by a semi-norm of $\mathcal{D}(\Omega)$ is new, as is everything related to the semi-norms $\| \ \|_{\mathcal{D}(\Omega);p}$. It expresses that the topology they generate is finer than (and thus equal to) that of inductive limit of the $\mathcal{C}_K^\infty(\Omega)$.

Then there exists a function $p \in \mathcal{C}^+(\Omega)$ such that, for every $\varphi \in \mathcal{D}(\Omega)$,

$$\sum_{k \in \mathbb{N}} m_k \|\alpha_k \varphi\|_{\mathcal{C}_b^{m_k}(\Omega)} \leq \|\varphi\|_{\mathcal{D}(\Omega);p}.$$

Only a finite number of terms in the left-hand sum are not zero. ▌

Utility of Theorem 2.22. This result is used to prove, via the *second theorem of control of the $\mathcal{C}_K^m(\Omega)$-norms* (Theorem 2.25), the following.

— Every semi-norm that is continuous on all the $\mathcal{C}_K^\infty(\Omega)$ is continuous on $\mathcal{D}(\Omega)$ (Theorem 2.26).

— Every linear mapping that is continuous on all the $\mathcal{C}_K^\infty(\Omega)$ is a distribution (Theorem 3.4 (d)).

— Bounded subsets of $\mathcal{D}'(\Omega; E)$ are equicontinuous on $\mathcal{D}(\Omega)$ (Theorem 4.2 (b)), a property which is fundamental to proving the sequential completeness of $\mathcal{D}'(\Omega; E)$ (Theorem 4.5) and the continuity of $\langle \, , \, \rangle$ (Theorem 4.4 (c)).

This avoids us resorting to the difficult topology of inductive limit of the $\mathcal{D}_K(\Omega)$, that is of the $\mathcal{C}_K^\infty(\Omega)$, with which Laurent SCHWARTZ endows $\mathcal{D}(\Omega)$ [69, Chap. III, § 1, Theorems II and III, p. 66]. □

The proof will use the following bound for the $\mathcal{C}^m(\Omega)$-norm of a product of real functions, which is established in Volume 2 [82, Theorem 3.10].

Theorem 2.23.– *Let f_1 and f_2 be two functions in $\mathcal{C}_b^m(\Omega)$, where Ω is an open subset of $\mathbb{R}^d$ and $m \in \mathbb{N}$. Then, $f_1 f_2 \in \mathcal{C}_b^m(\Omega)$ and*

$$\|f_1 f_2\|_{\mathcal{C}_b^m(\Omega)} \leq 2^m \|f_1\|_{\mathcal{C}_b^m(\Omega)} \|f_2\|_{\mathcal{C}_b^m(\Omega)}.$$ ▌

We will also make use of the following upper bound of the sum of a series (which extends to infinite series) by a weighted supremum.

Lemma 2.24.– *Let $(r_k)_{k \in N}$ be a finite sequence (that is, N is a finite subset of $\mathbb{N}$) of non-negative numbers. Then,*

$$\sum_{k \in N} r_k \leq \sup_{k \in N} 2^{k+1} r_k.$$ ▌

Proof of Lemma 2.24. With the bound of the sum of the geometric series given by Theorem A.68, we have

$$\sum_{k \in N} r_k = \sum_{k \in N} 2^{-k} 2^k r_k \leq \Big(\sum_{k \in N} 2^{-k}\Big) \sup_{k \in N} 2^k r_k \leq 2 \sup_{k \in N} 2^k r_k.$$ □

We now have the tools to establish the inequality which controls the $C^m_K(\Omega)$-norms by a semi-norm of $\mathcal{D}(\Omega)$.

Proof of Theorem 2.22. **1.** *Aim.* By Definition 2.5 of the semi-norms of $\mathcal{D}(\Omega)$, we have to construct a function $p \in \mathcal{C}^+(\Omega)$ such that, for each $\varphi \in \mathcal{D}(\Omega)$,

$$\sum_{k \in \mathbb{N}} m_k \|\alpha_k \varphi\|_{\mathcal{C}_b^{m_k}(\Omega)} \leq \sup_{x \in \Omega} \sup_{|\beta| \leq p(x)} p(x) |\partial^\beta \varphi(x)|. \tag{2.16}$$

2. *Reduction to a finite set of indices.* If

$$\kappa_k \cap \operatorname{supp} \varphi = \emptyset,$$

then $\alpha_k \varphi = 0$, since α_k and φ are zero ouside of their respective supports (by Definition 2.1 of support), and since the support of α_k is included in κ_k (by Definition 2.20 of a partition of unity).

We can therefore limit ourselves to the set N_φ of the k such that

$$\kappa_k \cap \operatorname{supp} \varphi \neq \emptyset.$$

This collection N_φ is finite since the support of φ, as any compact subset of Ω, only intersects finitely many crown-shaped sets (Theorem 2.19 (c)).

3. *A preliminary inequality.* Observe first that

$$\|\alpha_k \varphi\|_{\mathcal{C}_b^{m_k}(\Omega)} = \|\alpha_k \varphi\|_{\mathcal{C}_b^{m_k}(\kappa_k)}.$$

Indeed, the support of $\alpha_k \varphi$ is included in $\overline{\kappa_k}$ (by Definition 2.1 of support, since $\alpha_k \varphi = 0$ ouside of κ_k, by Definition 2.20 of the α_k), which is (Theorem 2.19 (b)) a compact subset of Ω, therefore it is compact (Theorem 2.2 (a)); and so the $\partial^\beta(\alpha_k \varphi)$ are zero ouside of this support (Theorem 2.6), and *a fortiori* ouside of κ_k.

On the other hand, the inequality from Theorem 2.23 gives

$$\|\alpha_k \varphi\|_{\mathcal{C}_b^{m_k}(\kappa_k)} \leq 2^{m_k} \|\alpha_k\|_{\mathcal{C}_b^{m_k}(\kappa_k)} \|\varphi\|_{\mathcal{C}_b^{m_k}(\kappa_k)}.$$

With the inequality from Lemma 2.24, this results in

$$\sum_{k \in N_\varphi} m_k \|\alpha_k \varphi\|_{\mathcal{C}_b^{m_k}(\Omega)} \leq \sup_{k \in N_\varphi} m_k 2^{k+1+m_k} \|\alpha_k\|_{\mathcal{C}_b^{m_k}(\kappa_k)} \|\varphi\|_{\mathcal{C}_b^{m_k}(\kappa_k)}.$$

Denoting $s_k = m_k 2^{k+1+m_k} \|\alpha_k\|_{\mathcal{C}_b^{m_k}(\kappa_k)}$, we can write using Definition 1.18 (b) of $\|\varphi\|_{\mathcal{C}_b^{m_k}(\kappa_k)}$,

$$\sum_{k \in N_\varphi} m_k \|\alpha_k \varphi\|_{\mathcal{C}_b^{m_k}(\Omega)} \leq \sup_{k \in N_\varphi} \sup_{x \in \kappa_k} \sup_{|\beta| \leq m_k} s_k |\partial^\beta \varphi(x)|. \tag{2.17}$$

4. Construction of p. Define $p \in \mathcal{C}^+(\Omega)$ by

$$p(x) \stackrel{\text{def}}{=} \sum_{k\in\mathbb{N}} \sup\{s_k, m_k\} a_k(x),$$

where, denoting $\alpha_{-1} = 0$,

$$a_k \stackrel{\text{def}}{=} \alpha_{k-1} + \alpha_k + \alpha_{k+1}.$$

This sum makes sense, since only finitely many of these terms are non-zero. Indeed the crown-shaped set κ_k only intersects κ_{k-1} and κ_{k+1} (Theorem 2.19 (b)), thus only α_{k-1}, α_k and α_{k+1} are non-zero in κ_k, and hence only a_{k-2}, a_{k-1}, a_k, a_{k+1} and a_{k+2} are non-zero here. So, if $x \in \kappa_k$,

$$p(x) = \sum_{i=k-2}^{k+2} \sup\{s_i, m_i\} a_i(x). \tag{2.18}$$

Each of these five functions a_i is continuous and positive, since the α_k are (by Definition 2.20), therefore p is continuous on each κ_k, and thus on their union Ω.

5. Inequality. Again by definition of a partition of unity, we have, always for $x \in \kappa_k$,

$$a_k(x) = \alpha_{k-1}(x) + \alpha_k(x) + \alpha_{k+1}(x) = \sum_{i\in\mathbb{N}} \alpha_i(x) = 1.$$

So (2.18) gives, with the positivity of the a_i,

$$p(x) \geq \sup\{s_k, m_k\}.$$

Inequality (2.17) therefore leads to

$$\sum_{k\in N_\varphi} m_k \|\alpha_k\varphi\|_{\mathcal{C}_b^{m_k}(\Omega)} \leq \sup_{k\in\mathbb{N}} \sup_{x\in\kappa_k} \sup_{|\beta|\leq p(x)} p(x)|\partial^\beta\varphi(x)|.$$

Inequality (2.16) follows since (Theorem 2.19 (a)) Ω is the union of the κ_k. □

2.9. Semi-norms that are continuous on all the $\mathcal{C}_K^\infty(\Omega)$

We intend to show that, if a semi-norm π is continuous on all the $\mathcal{C}_K^\infty(\Omega)$, then it is continuous on $\mathcal{D}(\Omega)$. We will deduce this from the **second theorem of control of the $\mathcal{C}_K^m(\Omega)$-norms**, that is here.

Theorem 2.25.– *Let Ω be an open subset of $\mathbb{R}^d$ and, for every compact subset K of Ω, let $c_K \in \mathbb{R}$ and $m_K \in \mathbb{N}$.*

Then there exists a function $p \in \mathcal{C}^+(\Omega)$ such that, if a semi-norm π on $\mathcal{D}(\Omega)$ satisfies, for every compact subset K of Ω and every $\varphi \in \mathcal{C}_K^\infty(\Omega)$,

$$\pi(\varphi) \leq c_K \|\varphi\|_{\mathcal{C}_b^{m_K}(\Omega)}, \tag{2.19}$$

then it satisfies, for every $\varphi \in \mathcal{D}(\Omega)$,

$$\pi(\varphi) \leq \|\varphi\|_{\mathcal{D}(\Omega);p}. \blacksquare$$

Proof. Let $(\kappa_k)_{k\in\mathbb{N}}$ be the cover by crown-shaped sets (Definition 2.18) of Ω, and $(\alpha_k)_{k\in\mathbb{N}}$ be a subordinate partition of unity (Definition 2.20; it exists according to Theorem 2.21).

The support of a function $\varphi \in \mathcal{D}(\Omega)$, like any compact subset of Ω, only intersects a finite number of κ_k (Theorem 2.19 (c)). So, denoting by N_φ the (finite) set of these k,

$$\varphi = \sum_{k \in N_\varphi} \alpha_k \varphi.$$

Moreover, $\alpha_k\varphi \in \mathcal{C}_{K_k}^\infty(\Omega)$ where $K_k = \overline{\kappa_k}$. Indeed, this product is infinitely differentiable since α_k and φ are (Theorem A.60), and its support is included in $\overline{\kappa_k}$ (by Definition 2.1, since $\alpha_k\varphi = 0$ outside of κ_k by Definition 2.20 of the α_k).

So, for each semi-norm π satisfying hypothesis (2.19),

$$\pi(\varphi) \leq \sum_{k \in N_\varphi} \pi(\alpha_k\varphi) \leq \sum_{k\in\mathbb{N}} c_{K_k} \|\alpha_k\varphi\|_{\mathcal{C}_b^{m_{K_k}}(\Omega)} \leq \sum_{k\in\mathbb{N}} n_k \|\alpha_k\varphi\|_{\mathcal{C}_b^{n_k}(\Omega)},$$

where n_k is any integrer greater than c_{K_k} and m_{K_k} (the last inequality results from the fact that $\| \ \|_{\mathcal{C}_b^\infty(\Omega);m}$ grows with m, according to Definition 1.19 (b)).

The first theorem of control of the $\mathcal{C}_K^m(\Omega)$-norms (Theorem 2.22) then shows that there exists $p \in \mathcal{C}^+(\Omega)$ such that $\|\varphi\|_{\mathcal{D}(\Omega);p}$ is greater than the last term. □

Now we show that a continuous semi-norm on all the $\mathcal{C}_K^\infty(\Omega)$ is continuous on $\mathcal{D}(\Omega)$ (the inverse is true since $\mathcal{C}_K^\infty(\Omega) \subsetneq \mathcal{D}(\Omega)$).

Theorem 2.26.– *If a semi-norm on $\mathcal{D}(\Omega)$, where Ω is an open subset of $\mathbb{R}^d$, is continuous on $\mathcal{C}_K^\infty(\Omega)$ for every compact subset K of Ω, then it is continuous on $\mathcal{D}(\Omega)$.* $\blacksquare$

Proof. Let π be a semi-norm on $\mathcal{D}(\Omega)$ which is continuous on $\mathcal{C}_K^\infty(\Omega)$.

The characterization of continuous semi-norms from Theorem 1.14 shows, since $\mathcal{C}_K^\infty(\Omega)$ is (by Definition 2.3 (a)) endowed with the family of semi-norms of $\mathcal{C}_b^\infty(\Omega)$ which is filtering (Theorem A.52), that there exist $c_K \in \mathbb{R}$ and $m_k \in \mathbb{N}$ such that, for each $\varphi \in \mathcal{C}_K^\infty(\Omega)$,

$$\pi(\varphi) \leq c_K \|\varphi\|_{\mathcal{C}_b^\infty(\Omega);m_K}.$$

If this is true for every compact subset K of Ω, the second theorem of control of the $\mathcal{C}_K^m(\Omega)$-norms (Theorem 2.25) shows, since $\|\varphi\|_{\mathcal{C}_b^\infty(\Omega);m_K} = \|\varphi\|_{\mathcal{C}_b^{m_K}(\Omega)}$, that there exists $p \in \mathcal{C}^+(\Omega)$ such that

$$\pi(\varphi) \leq \|\varphi\|_{\mathcal{D}(\Omega);p}.$$

Hence, again according to the characterization from Theorem 1.14, π is continuous on $\mathcal{D}(\Omega)$. □

Chapter 3

Space of Distributions

The goal of this chapter is to construct the space $\mathcal{D}'(\Omega; E)$ of distributions on an open subset Ω of $\mathbb{R}^d$ with values in a Neumann space E. It is the space of continuous linear mappings from $\mathcal{D}(\Omega)$ into E.

We endow it (Definition 3.1) with the semi-norms $\|f\|_{\mathcal{D}'(\Omega;E);\varphi,\nu} = \|\langle f, \varphi\rangle\|_{E;\nu}$ indexed by $\varphi \in \mathcal{D}(\Omega)$ and $\nu \in \mathcal{N}_E$, namely with the topology of simple (pointwise) convergence on $\mathcal{D}(\Omega)$. We then give various characterizations of distributions (Theorem 3.4) and we identify every continuous function with a distribution (Theorem 3.8).

In § 3.4, we show that if E is not a Neumann space, i.e. if it is not sequentially complete, this construction does not give the expected properties, since not every continuous function is (identifiable with) a mapping of $\mathcal{L}(\mathcal{D}(\Omega); E)$, mappings that should not then be called "distributions".

Finally in § 3.5, we construct the space $\mathcal{M}(\Omega; E)$ of measures and we identify it with a subspace of $\mathcal{D}'(\Omega; E)$ that contains $\mathcal{C}(\Omega; E)$ (Theorems 3.18 and 3.21).

3.1. The space $\mathcal{D}'(\Omega; E)$

Let us define the space of distributions [1].

1. **History of the space of distributions. Real distributions on $\mathbb{R}^d$.** Laurent SCHWARTZ defined real distributions on $\mathbb{R}^d$ in 1945 [68, Definition 2, p. 60] as being sequentially continuous linear mappings on $\mathcal{D}(\mathbb{R}^d)$ endowed with convergent sequences in the sense of Theorem 2.13 (b). He endowed the space of distributions with the topology of uniform convergence on the $\mathcal{D}_K(\mathbb{R}^d)$. In 1950, he defined distributions as being continuous linear mappings on $\mathcal{D}(\mathbb{R}^d)$ [69, p. 24], and he verified that this definition coincides with the previous one, denoting their space $(\mathcal{D}')$. He described the genesis of this invention in his autobiography, *A Mathematician Grappling with His Century* (*Un mathématicien aux prises avec le siècle*) [75].

Real distributions on an open subset of $\mathbb{R}^d$. Laurent SCHWARTZ indicated in 1950 [69, p. 26] "that one could study distributions on an open subset Ω of $\mathbb{R}^d$", while limiting himself in giving their definition. This case has been studied by a number of authors, for example John HORVÁTH in 1966 [45] or François TREVES in 1967 [90]. We will not indicate the history of generalizations to such an open set when they are obtained by repeating the proofs of SCHWARTZ, which is often the case, but not always.

Definition 3.1.– *Let Ω be an open subset of $\mathbb{R}^d$ and E a Neumann space, whose family of semi-norms is denoted by $\{\| \, \|_{E;\nu} : \nu \in \mathcal{N}_E\}$.*

*A **distribution** on Ω with values in E is a continuous linear mapping from $\mathcal{D}(\Omega)$ into E.*

We denote by $\mathcal{D}'(\Omega; E)$ the vector space of these distributions endowed with the semi-norms, indexed by $\varphi \in \mathcal{D}(\Omega)$ and $\nu \in \mathcal{N}_E$,

$$\|f\|_{\mathcal{D}'(\Omega;E);\varphi,\nu} \stackrel{\text{def}}{=} \|\langle f, \varphi\rangle\|_{E;\nu}. \ \blacksquare$$

We denote by $\langle f, \varphi\rangle$ the value $f(\varphi)$ of a distribution f applied to φ. We denote it by $\langle f, \varphi\rangle_\Omega$ or $\langle f, \varphi\rangle_{\mathcal{D}'(\Omega;E)\times\mathcal{D}(\Omega)}$ when we want to be precise with the domain or the spaces involved. We denote

$$\mathcal{D}'(\Omega) \stackrel{\text{def}}{=} \mathcal{D}'(\Omega; \mathbb{R}).$$

Examples of distributions. — To any continuous function $f \in \mathcal{C}(\Omega; E)$ we associate (Theorem 3.5, p. 48) the distribution $\bar{f} \in \mathcal{D}'(\Omega; E)$ defined by

$$\langle \bar{f}, \varphi\rangle \stackrel{\text{def}}{=} \int_\Omega f\varphi.$$

— The Dirac measure δ_x at a point x is (Definition 3.14, p. 59) the real distribution defined by

$$\langle \delta_x, \varphi\rangle \stackrel{\text{def}}{=} \varphi(x). \tag{3.1}$$

More generally, to any measure $f \in \mathcal{M}(\Omega; E)$ we associate (Theorem 3.15, p. 60) the distribution $\bar{f} \in \mathcal{D}'(\Omega; E)$ which is the restriction of f to $\mathcal{D}(\Omega)$.

— To the "singular" function $x \mapsto |x|^{-\lambda}$, where $0 < \lambda < d$, which becomes infinite in the neighborhood of 0, we associate (Theorem 9.5, p. 193) the distribution on all of $\mathbb{R}^d$, $\bar{f} \in \mathcal{D}'(\mathbb{R}^d)$, defined by

$$\langle \bar{f}, \varphi\rangle \stackrel{\text{def}}{=} \lim_{\epsilon\to 0} \int_{x\in\mathbb{R}^d : |x|>\epsilon} |x|^{-\lambda}\varphi(x)\, dx.$$

— The incompressible flow $\vec{\delta}_\Gamma \in \mathcal{D}'(\mathbb{R}^d; \mathbb{R}^d)$ "concentrated" along a closed path $\Gamma \in \mathcal{C}^1([0,1]; \mathbb{R}^d)$ is the distribution defined, p. 240, by

$$\langle \vec{\delta}_\Gamma, \varphi\rangle \stackrel{\text{def}}{=} \int_0^1 (\varphi \circ \Gamma)\, \Gamma'. \ \square$$

Vector values. In 1957, Laurent SCHWARTZ introduced [72, I, Definition, p. 49] distributions on $\mathbb{R}^d$ with values in a *locally convex separated topological vector space* (which is equivalent to *separated semi-normed*), which he generally assumed to be quasi-complete [72, p. 50].

Originality of Definition 3.1. Neumann spaces, i.e. *sequentially complete* spaces, considered here are more general than *quasi-complete* spaces.

Precursors. Sergei SOBOLEV was the first to explicitly and rigorously define distributions, and in particular their derivatives [86, p. 62]. Previously, S. BOCHNER used them implicitly [4].
The Prehistory of the Theory of Distributions was written by Jesper LÜTZEN [55], and this topic was extensively explored by Anne-Sandrine PAUMIER [59] ... that I unfortunately did not have time to read.

Examples of Neumann spaces. The following spaces are Neumann, which allows the use (for example in evolution problems) of spaces such as $\mathcal{D}'((0,T); E)$ where E is one of them.

— Lebesgue and Sobolev spaces $L^p(\Omega)$ and $W^{m,p}(\Omega)$ and their *local* variants $L^p_{\text{loc}}(\Omega)$ and $W^{m,p}_{\text{loc}}(\Omega)$, for $1 \le p \le \infty$ and m a positive or negative integer (we will see them in Volumes 4 and 5).

— The spaces $L^p(\Omega)$-weak, $L^p_{\text{loc}}(\Omega)$-weak, $W^{m,p}(\Omega)$-weak and $W^{m,p}_{\text{loc}}(\Omega)$-weak, for $1 < p < \infty$ and m a positive or negative integer [Vol. 4 and 5].

— The space $L^\infty(\Omega)$-$*$weak [Vol. 4].

— The spaces of continuously differentiable functions $\mathcal{C}^m(\Omega)$, $\mathcal{C}^m_{\text{b}}(\Omega)$, $\mathrm{C}^m_{\text{b}}(\Omega)$, $\mathcal{C}^m_K(\Omega)$ and $\mathcal{K}^m(\Omega)$, for $m \in \mathbb{N}$ or $m = \infty$, [Vol. 2, Theorem 2.24].

— The spaces $\mathcal{D}'(\Omega)$ and $\mathcal{D}(\Omega)$ of distributions and test functions (Theorems 4.5 and 2.8).

— The spaces $\mathcal{M}(\Omega)$ of measures and $\mathcal{M}_{\text{b}}(\Omega)$ of bounded measures [Vol. 4].

— The space E-weak, when E is a Hilbert, reflexive, or semi-reflexive space [Vol. 1, Theorems 17.7 and 17.12].

— The duals E', E'', E''', ... of a normed or semi-normed metrizable space E [Vol. 1, Theorem 13.8 (a) and Property (17.3), p. 248].

— The duals E'-weak, E''-weak, E'''-weak, ... of a Hilbert or reflexive space E [Vol. 1, Property (17.6), p. 248].

— The duals E'-$*$weak, E''-$*$weak, E'''-$*$weak, ... of a Hilbert, Banach, Fréchet, or reflexive space E [Vol. 1, Theorem 13.8 (b) and Properties (17.4) and (17.6)].

— Each of their closed or sequentially closed subspaces (Theorem A.24).

CAUTION. The semi-norms in Definition 3.1 generate a **distinct topology from the one introduced and used by** Laurent SCHWARTZ on $\mathcal{D}'(\Omega; E)$, namely the topology of uniform convergence on the bounded subsets of $\mathcal{D}(\Omega)$. When (in comments) we endow $\mathcal{D}'(\Omega; E)$ with the said topology, we denote it by $\mathcal{D}'(\Omega; E)$-unif to avoid confusion. The relation between these topologies is specified in the comment *Schwartz's topology*, p. 45. □

Let us show that $\mathcal{D}'(\Omega; E)$ is separated (we will see in Theorem 4.5 that it is a Neumann space).

Theorem 3.2.– *The space $\mathcal{D}'(\Omega; E)$ is separated and semi-normed, for every open subset Ω of $\mathbb{R}^d$ and every Neumann space E.* ▮

Proof. By Definition 1.1 of a separated space, it is a question of verifying that a distribution f is zero when all its semi-norms are. Thus, let $f \in \mathcal{D}'(\Omega; E)$ such that, for every $\varphi \in \mathcal{D}(\Omega)$ and each semi-norm $\| \ \|_{E;\nu}$ of E,

$$\|f\|_{\mathcal{D}'(\Omega;E);\varphi,\nu} = \|\langle f, \varphi \rangle\|_{E;\nu} = 0.$$

Since E is separated, as is every Neumann space (Definition 1.4),

$$\langle f, \varphi \rangle = 0_E.$$

So,

$$f = 0. \ \square$$

Let us characterize distributions (other characterizations are in Theorem 3.4).

Theorem 3.3.– *Let Ω be an open subset of $\mathbb{R}^d$ and E a Neumann space. Then,*

$$f \in \mathcal{D}'(\Omega; E)$$

if and only if: f is a linear mapping from $\mathcal{D}(\Omega)$ into E and, for every semi-norm $\| \ \|_{E;\nu}$ of E, there exist $p \in \mathcal{C}^+(\Omega)$ and $c \in \mathbb{R}$ such that, for every $\varphi \in \mathcal{D}(\Omega)$,

$$\|\langle f, \varphi \rangle\|_{E;\nu} \leq c \|\varphi\|_{\mathcal{D}(\Omega);p}.$$

That is to say,

$$\|\langle f, \varphi \rangle\|_{E;\nu} \leq c \sup_{x \in \Omega} \sup_{0 \leq |\beta| \leq p(x)} p(x) |\partial^\beta \varphi(x)|. \ \blacksquare$$

Proof. **1.** *First inequality.* This inequality follows from the characterization of continuous linear mappings from Theorem 1.12 (b), since a distribution f is, by Definition 3.1, a continuous linear mapping from $\mathcal{D}(\Omega)$ into E, and since the family of semi-norms of $\mathcal{D}(\Omega)$ is filtering (Theorem 2.7).

2. *Second inequality.* This inequality is a rephrasing of the first with Definition 2.5 of the semi-norms of $\mathcal{D}(\Omega)$. $\square$

Motivation for introducing distributions. The primary interest in distributions is to provide a framework to model numerous physical concepts and to solve the partial differential equations governing them, which we will see in Volume 7.

Function spaces do not suffice for this, on the one hand because concepts like measures cannot be represented by a function, whereas it can by the Dirac measure introduced in (3.1), p. 42, and also because an equation like the Poisson equation $-\Delta u = f$ with $f \in \mathcal{C}(\Omega)$ may be unsatisfied in the sense of differentiable functions, whilst being satisfied in the sense of distributions, see p. 263. $\square$

Motivation to study vector-valued distributions. To solve stationary equations, we will only use real-valued distributions; in contrast, for evolution equations, we will use time-dependent distributions valued in spaces of space-dependent distributions.

These spaces of values will often be Hilbert spaces (for example, for the linearized Navier-Stokes equations, we obtain the velocity u in $L^2(0,T;(L^2(\Omega))^d)$), but sometimes Fréchet (for the same equations, the pressure p belongs to $W^{-1,\infty}(0,T;L^2_{\mathrm{loc}}(\Omega))$, or even Neumann spaces (when transiting through $\mathcal{D}'(0,T;(\mathcal{D}'(\Omega))^d)$ to separate time from space). □

Distributions with values in a separated locally convex topological vector space. The space E can be any sequentially complete separated locally convex topological vector space, since any locally convex topology on a vector space can be generated by a family of semi-norms, as follows from a **von Neumann's theorem** [NEUMANN, 56, th. 26, p. 19]. □

The case where E is complete or quasi-complete. Definition 3.1 encompasses the case where E is a complete semi-normed space since then it is sequentially complete, see for example [SCHWARTZ, 73, p. 251]. It also encompasses the case where E is quasi-complete, i.e. where its bounded subsets are complete, which is the hypothesis used by Laurent SCHWARTZ [72, p. 2, 50 and 52]. □

Interest of assuming E is sequentially complete rather than complete. A semi-normed space is called *complete* if every Cauchy filter converges [SCHWARTZ, 73, Chap. XVIII, § 8, Definition 1, p. 251]. We do not use this notion, since sequential complete is much simpler and more general (complete implies sequentially complete) and, above all, as certain useful spaces are sequentially complete but not complete. For example, this is the case for every Hilbert space or reflexive Banach space which is infinite-dimensional and endowed with the weak topology, as we have mentioned in Volume 1 [81, Property (4.11), p. 63]. □

Interest of assuming E is sequentially complete rather than quasi-complete. It is the lack of completeness for many spaces that we have just mentioned which led Laurent SCHWARTZ to assume [72, p. 55] that the vector-valued space E is quasi-complete, i.e. that its bounded subsets are complete. Indeed, any semi-reflexive space endowed with its weak topology is quasi-complete [SCHAEFER, 67, Result 5.5, p. 144], which allows us to conclude in the case mentioned in the previous comment.

However, again here, sequential complete is much simpler and more general. Indeed, *quasi-complete* implies *sequential complete*, but the converse is false. For example, the vector space of real functions f on $\mathbb{R}$ such that the set $\{x \in \mathbb{R} : f(x) \neq 0\}$ is countable endowed with the simple convergence is sequentially complete but not quasi-complete [SIMON, 83, to appear]. □

Motivation for choosing the simple topology. We endow $\mathcal{D}'(\Omega;E)$ with the **simple topology**, namely with the topology of simple (pointwise) convergence on $\mathcal{D}(\Omega)$ because it is... simple: its semi-norms

$$\|f\|_{\mathcal{D}'(\Omega;E);\varphi,\nu} = \|\langle f,\varphi\rangle\|_{E;\nu}$$

are indexed by $\varphi \in \mathcal{D}(\Omega)$ and $\nu \in \mathcal{N}_E$, that does not involve the topology of $\mathcal{D}(\Omega)$.

On the contrary, the **uniform topology** used by SCHWARTZ does involve the topology of $\mathcal{D}(\Omega)$, via its bounded subsets, in its semi-norms (see (3.2), below) and thus for each operation, convergence, boundedness, and others. This is less simple than the simple topology and does not bring any additional property when studying PDEs, since its convergent sequences, bounded subsets and compact subsets are the same [SIMON, 84, to appear].

Another motive in favor of the simple topology is that it is weaker than the uniform topology, which permits more functional spaces to be topologically embedded in $\mathcal{D}'(\Omega;E)$, and thus enlarges their field of utilization. □

Schwartz's topology. The topology which we endow $\mathcal{D}'(\Omega;E)$ with **is not** the topology introduced and used by Laurent SCHWARTZ. Recall that he considers the space, that we denote by $\mathcal{D}'(\Omega;E)$-unif to avoid any confusion, of distributions endowed with the *topology of uniform convergence on the bounded subsets* of $\mathcal{D}(\Omega)$ [72, Chap. I, § 2 p. 50] (in fact, he considered the case where $\Omega = \mathbb{R}^d$). It is generated by the semi-norms, indexed by the bounded subsets $\mathcal{B}$ of $\mathcal{D}(\Omega)$ and $\nu \in \mathcal{N}_E$,

$$\|f\|_{\mathcal{D}'(\Omega;E)\text{-unif};\mathcal{B},\nu} \stackrel{\text{def}}{=} \sup_{\varphi\in\mathcal{B}} \|\langle f,\varphi\rangle\|_{E;\nu}. \tag{3.2}$$

We will call it the **uniform topology**. It is the topology of $\mathcal{L}(\mathcal{D}(\Omega); E)$ (Definition A.36).

Our simple topology and the uniform topology, and hence the spaces $\mathcal{D}'(\Omega; E)$ and $\mathcal{D}'(\Omega; E)$-unif, have the same convergent sequences (Theorem 8.28) and the same bounded subsets, and they coincide on bounded sets [SIMON, 84], to appear. The spaces $\mathcal{D}'(\Omega; E)$ and $\mathcal{D}'(\Omega; E)$-unif thus also have the same compact subsets and the same sequentially continuous mappings. On the other hand, they do not have the same open subsets nor the same continuous mappings [SIMON, 84]. The uniform topology therefore provides stronger topological results: $\mathcal{D}'(\Omega)$-unif is reflexive, whereas $\mathcal{D}'(\Omega)$ is only semi-reflexive.

In the particular case of real-valued distributions, i.e. for $\mathcal{D}'(\Omega)$, the *simple topology* coincides with the *weak uniform topology*, i.e.

$$\mathcal{D}'(\Omega) \rightleftarrows (\mathcal{D}'(\Omega)\text{-unif})\text{-weak}.$$

Indeed, Schwartz's topology is that of $(\mathcal{D}(\Omega))'$, its weak topology is that of $(\mathcal{D}(\Omega))'$-weak, and our simple topology is that of $(\mathcal{D}(\Omega))'$-$*$weak. These last two topologies coincide, since $\mathcal{D}(\Omega)$ is reflexive [SCHWARTZ, 69, Theorem XIV, p. 75] and since, by [Volume 1, Theorem 17.21], for any reflexive separated semi-normed space E,

$$E'\text{-weak} \rightleftarrows E'\text{-}*\text{weak}.$$

In general, and more precisely when E does not coincide with E-weak, the simple topology of $\mathcal{D}'(\Omega; E)$ does not coincide with the weak uniform topology: it is intermediate between the uniform topology and the weak uniform topology (it is "*weak with respect to* Ω, *but not with respect to* E"), i.e. [SIMON, 84]

$$\mathcal{D}'(\Omega; E)\text{-unif} \subsetneq \mathcal{D}'(\Omega; E) \subsetneq (\mathcal{D}'(\Omega; E)\text{-unif})\text{-weak}. \quad \square$$

3.2. Characterization of distributions

Let us now show that any linear mapping which is sequentially continuous on $\mathcal{D}(\Omega)$, or on all the $\mathcal{C}_K^\infty(\Omega)$, is a distribution [2], and other characterizations.

Theorem 3.4.– *Let Ω be an open subset of $\mathbb{R}^d$, E a Neumann space and*

f a linear mapping from $\mathcal{D}(\Omega)$ into E.

Then, each of the following properties is equivalent to

$$f \in \mathcal{D}'(\Omega; E).$$

(a) *f is continuous from $\mathcal{D}(\Omega)$ into E.*

2. **History of Theorem 3.4 (d).** Laurent SCHWARTZ proved in 1950 [69, Chap. III, § 1, Theorem III, p. 66] that a linear mapping from $\mathcal{D}(\mathbb{R}^d)$ into a locally convex topological vector space is continuous if and only if it is continuous on each $\mathcal{C}_K^\infty(\mathbb{R}^d)$.
The proof given here is new: whereas SCHWARTZ used the fact that $\mathcal{D}(\mathbb{R}^d)$ is bornological and barrelled, since it is an inductive limit of Fréchet spaces, we use the *theorems of control of the $\mathcal{C}_K^m(\Omega)$-norms* (Theorem 2.22, via Theorem 2.25).

History of Theorem 3.4 (e). That a bounded linear mapping on each bounded subset of $\mathcal{C}_K^\infty(\Omega)$ is a distribution seems to be a new statement to us, along with a new proof.

(b) *f is sequentially continuous from $\mathcal{D}(\Omega)$ into E.*

(c) *For every compact subset K of Ω, f is sequentially continuous from $\mathcal{C}_K^\infty(\Omega)$ into E.*

(d) *For every compact subset K of Ω, f is continuous from $\mathcal{C}_K^\infty(\Omega)$ into E.*

(e) *For every compact subset K of Ω, f is bounded on every bounded subset of $\mathcal{C}_K^\infty(\Omega)$.* ∎

Complement to Theorem 3.4. Characterization (b) is equivalent to "f is sequentially continuous from $\mathcal{K}^\infty(\Omega)$ into E", since $\mathcal{K}^\infty(\Omega)$ and $\mathcal{D}(\Omega)$ have the same convergent sequences, see the comment *Another topology on $\mathcal{D}(\Omega)$*, p. 24. □

Bornological character of $\mathcal{D}(\Omega)$. Characterization (e) of Theorem 3.4 is equivalent to "$\mathcal{D}(\Omega)$ is a bornological space", since such a space is characterized by "every linear mapping that sends bounded sets into bounded sets is continuous" [BOURBAKI, 8, Proposition 5, p. 10, and comment, p. 11]. Indeed, a subset is bounded in $\mathcal{D}(\Omega)$ if and only if it is bounded in one of the $\mathcal{C}_K^\infty(\Omega)$. □

Proof of Theorem 3.4. We proceed step-by-step, denoting by $\{\| \ \|_{E;\nu} : \nu \in \mathcal{N}_E\}$ the family of semi-norms of E.

1. *$f \in \mathcal{D}'(\Omega; E) \Leftrightarrow$ continuous in $\mathcal{D}(\Omega)$.* This is Definition 3.1 of a distribution.

2. *Continuous in $\mathcal{D}(\Omega) \Rightarrow$ sequentially continuous in $\mathcal{D}(\Omega)$.* Continuity always leads to sequential continuity (Theorem 1.10).

3. *Sequentially continuous in $\mathcal{D}(\Omega) \Rightarrow$ sequentially continuous in $\mathcal{C}_K^\infty(\Omega)$.* This follows (Theorem A.29) from the inclusion $\mathcal{C}_K^\infty(\Omega) \subsetneq \mathcal{D}(\Omega)$ (Theorem 2.10).

4. *Sequentially continuous in $\mathcal{C}_K^\infty(\Omega) \Rightarrow$ continuous in $\mathcal{C}_K^\infty(\Omega)$.* This follows (Theorem 1.11) from $\mathcal{C}_K^\infty(\Omega)$ being metrizable (Theorem A.53).

5. *Continuous in the $\mathcal{C}_K^\infty(\Omega) \Rightarrow f \in \mathcal{D}'(\Omega; E)$.* Suppose that, for every compact subset K of Ω, f is continuous from $\mathcal{C}_K^\infty(\Omega)$ into E.

Since $\mathcal{C}_K^\infty(\Omega)$ is (by Definition 2.3 (a)) endowed with the family of semi-norms of $\mathcal{C}_{\mathrm{b}}^\infty(\Omega)$ which is filtering (Theorem A.52), the characterization of continuous linear mappings from Theorem 1.12 (b) shows that, for every $\nu \in \mathcal{N}_E$, there exist $m_{K,\nu} \in \mathbb{N}$ and $c_{K,\nu} \in \mathbb{R}$ such that, for every $\varphi \in \mathcal{C}_K^\infty(\Omega)$,

$$\|\langle f, \varphi \rangle\|_{E;\nu} \le c_{K,\nu} \|\varphi\|_{\mathcal{C}_{\mathrm{b}}^\infty(\Omega); m_{K,\nu}}.$$

Since $\|\varphi\|_{\mathcal{C}_{\mathrm{b}}^\infty(\Omega); m_{K,\nu}} = \|\varphi\|_{\mathcal{C}_{\mathrm{b}}^{m_{K,\nu}}(\Omega)}$, the second theorem of control of the $\mathcal{C}_K^m(\Omega)$-norms (Theorem 2.25, for $\pi(\varphi) = \|\langle f, \varphi \rangle\|_{E;\nu}$) then shows that, for every $\nu \in \mathcal{N}_E$, there exist $p_\nu \in \mathcal{C}^+(\Omega)$ such that, for every $\varphi \in \mathcal{D}(\Omega)$,

$$\|\langle f, \varphi \rangle\|_{E;\nu} \le \|\varphi\|_{\mathcal{D}(\Omega); p_\nu}.$$

Hence (Theorem 3.3), $f \in \mathcal{D}'(\Omega; E)$.

These steps 1 to 5 prove the equivalence of $f \in \mathcal{D}'(\Omega; E)$ with (a), (b), (c) and (d).

6. *Sequentially continuous on* $\mathcal{C}_K^\infty(\Omega) \Leftrightarrow$ *bounded on every bounded subset of* $\mathcal{C}_K^\infty(\Omega)$. Every sequentially continuous linear mapping sends bounded sets into bounded sets (Theorem A.37), so it remains to establish the converse statement.

Suppose then that

$$f \text{ sends bounded subsets of } \mathcal{C}_K^\infty(\Omega) \text{ into bounded subsets of } E \qquad (3.3)$$

and let $(\varphi_n)_{n\in\mathbb{N}}$ be a sequence converging to φ in $\mathcal{C}_K^\infty(\Omega)$. Since characterizations (b) and (c) of the convergent sequences in $\mathcal{C}_K^\infty(\Omega)$ stated in Theorem 2.13 are equivalent, such a sequence can be written in the form

$$\varphi_n - \varphi = t_n\phi_n, \text{ where } t_n \to 0 \text{ in } \mathbb{R} \text{ and } \{\phi_n\}_{n\in\mathbb{N}} \text{ is bounded in } \mathcal{C}_K^\infty(\Omega).$$

From (3.3), $\{\langle f, \phi_n\rangle\}_{n\in\mathbb{N}}$ is bounded in E, i.e., for every $\nu \in \mathcal{N}_E$,

$$\sup_{n\in\mathbb{N}} \|\langle f, \phi_n\rangle\|_{E;\nu} = c_\nu \text{ is finite.}$$

Then, $\|\langle f, \varphi_n - \varphi\rangle\|_{E;\nu} \leq t_n c_\nu$, which tends to 0 when $n \to \infty$. Hence,

$$\langle f, \varphi_n\rangle \to \langle f, \varphi\rangle \text{ in } E.$$

Therefore,

$$f \text{ is sequentially continuous from } \mathcal{C}_K^\infty(\Omega) \text{ into } E.$$

So, (e) is equivalent to (c), and thus to $f \in \mathcal{D}'(\Omega; E)$. □

3.3. Inclusion of $\mathcal{C}(\Omega; E)$ into $\mathcal{D}'(\Omega; E)$

Let us associate a distribution $\overline{f}$ with each continuous function f [3].

Theorem 3.5.– *Let $f \in \mathcal{C}(\Omega; E)$, where Ω is an open subset of $\mathbb{R}^d$ and E is a Neumann space.*

We define $\overline{f} \in \mathcal{D}'(\Omega; E)$ by: for every $\varphi \in \mathcal{D}(\Omega)$,

$$\langle \overline{f}, \varphi\rangle \overset{\text{def}}{=} \int_\Omega f\varphi. \blacksquare$$

3. **History of identifying a continuous function with a distribution.** Laurent SCHWARTZ identified in 1945 [68, p. 60] every summable function, and *a fortiori* every continuous function, with a distribution. He proved in 1950 [69, Chap. III, § 3, Theorem XVI, p. 76] that the uniform convergence on compact sets of continuous functions implies their convergence in the sense of distributions, which he had stated in 1945.

Proof. **1.** *Well-definedness of the integral.* The integral of $f\varphi$ makes sense by Definition 1.22 of the Cauchy integral with values in a Neumann space, since $f\varphi$ is uniformly continuous with bounded support, i.e. belongs to $\mathcal{B}(\Omega; E)$ (Definition 1.21).

This follows, due to a corollary (Theorem A.33) of Heine's theorem, from the fact that $f\varphi$ is continuous (as is any product of continuous functions, see Theorem A.60) with compact support (which follows from Theorem 2.2 (a), since it is zero outside of the support of φ, which is a compact subset of Ω).

2. *Obtaining a distribution.* Let us show that $\overline{f}$ satisfies the characterizations of a distribution from Theorem 3.3. Given a semi-norm $\| \ \|_{E;\nu}$ of E, we define $p \in \mathcal{C}^+(\Omega)$ by

$$p(x) = (2 + |x|)^{d+2} \|f(x)\|_{E;\nu}. \tag{3.4}$$

Let $\varphi \in \mathcal{D}(\Omega)$ and $\omega = \Omega \cap B$, where B is any open ball containing the support of φ. Only ω contributes to the integral of $f\varphi$ from Theorem A.77, so the bound of the semi-norms of the integral from Theorem 1.23 (a) gives

$$\Big\| \int_\Omega f\varphi \Big\|_{E;\nu} = \Big\| \int_\omega f\varphi \Big\|_{E;\nu} \le \int_\omega \|f\|_{E;\nu} |\varphi| = \int_\omega \frac{p(x)|\varphi(x)|}{(2+|x|)^{d+2}} \, \mathrm{d}x.$$

Since $\int_\omega 1/(2+|x|)^{d+2} \, \mathrm{d}x \le 2^{d+1}$ (Lemma A.82), the growth of the real integral (Theorem A.76 (a)) and its linearity give

$$\Big\| \int_\Omega f\varphi \Big\|_{E;\nu} \le 2^{d+1} \sup_{x \in \Omega} p(x) |\varphi(x)|. \tag{3.5}$$

From which,

$$\Big\| \int_\Omega f\varphi \Big\|_{E;\nu} \le 2^{d+1} \sup_{x \in \Omega} \sup_{0 \le |\beta| \le p(x)} p(x) |\partial^\beta \varphi(x)|.$$

That is, with Definition 2.5 of the semi-norms of $\mathcal{D}(\Omega)$,

$$\|\langle \overline{f}, \varphi \rangle\|_{E;\nu} \le 2^{d+1} \|\varphi\|_{\mathcal{D}(\Omega);p}.$$

Since the mapping $\overline{f}$ is linear, because the integral is linear (Theorem A.74), this inequality implies, by the characterization of distributions from Theorem 3.3, that

$$\overline{f} \in \mathcal{D}'(\Omega; E). \ \square$$

Let us now show that two distinct continuous functions give rise to two distinct distributions, which will allow us to identify continuous functions with distributions.

Theorem 3.6.– *The mapping* $f \mapsto \overline{f}$ *given by Theorem 3.5 is linear, injective and continuous, and therefore sequentially continuous, from* $\mathcal{C}(\Omega; E)$ *into* $\mathcal{D}'(\Omega; E)$. ▮

Proof. **1.** *Injectivity.* Let $f \in \mathcal{C}(\Omega; E)$ be such that

$$\overline{f} = 0.$$

That is, by definition of $\overline{f}$, for every $\varphi \in \mathcal{D}(\Omega)$,

$$\int_\Omega f\varphi = 0_E.$$

The Du Bois-Reymond lemma for values in a Neumann space (Theorem 3.7, below) then gives

$$f = 0.$$

The mapping $f \mapsto \overline{f}$, being linear, is thus injective.

2. *Continuity.* Denote by $\{\| \ \|_{E;\nu} : \nu \in \mathcal{N}_E\}$ the family of semi-norms of E. Let $\varphi \in \mathcal{D}(\Omega)$, $\nu \in \mathcal{N}_E$ and

$$\omega = \{x \in \Omega : \varphi(x) \neq 0\}.$$

Since only ω contributes to the integral (according to Theorem A.77), Definition 3.1 of the semi-norms of $\mathcal{D}'(\Omega; E)$ and the bound of the semi-norms of the integral from Theorem 1.23 (a) give

$$\|\overline{f}\|_{\mathcal{D}'(\Omega;E);\varphi,\nu} = \|\langle \overline{f}, \varphi\rangle\|_{E;\nu} = \left\|\int_\Omega f\varphi\right\|_{E;\nu} = \left\|\int_\omega f\varphi\right\|_{E;\nu} \leq \int_\omega \|f\|_{E;\nu}|\varphi|.$$

The growth of the real integral (Theorem A.76 (a)) and Definition 1.18 (a) of the semi-norms of $\mathcal{C}(\Omega; E)$ (for $K = \overline{\omega}$, which is compact) then give, denoting $c_\varphi = \int_\Omega |\varphi|$,

$$\|\overline{f}\|_{\mathcal{D}'(\Omega;E);\varphi,\nu} \leq c_\varphi \sup_{x\in\overline{\omega}} \|f(x)\|_{E;\nu} = c_\varphi \|f\|_{\mathcal{C}(\Omega;E);\overline{\omega},\nu}.$$

Due to the characterization of continuous linear mappings from Theorem 1.12 (a), this proves that the mapping $f \mapsto \overline{f}$ is continuous from $\mathcal{C}(\Omega; E)$ into $\mathcal{D}'(\Omega; E)$.

3. *Sequential continuity.* The mapping $f \mapsto \overline{f}$ is sequentially continuous, as is every continuous mapping (Theorem 1.10). □

It remains to prove the **du Bois-Reymond lemma** [4], for values in a Neumann space.

4. **History of the du Bois-Reymond lemma.** Friedrich Ludwig STEGMANN stated Theorem 3.7 for a real function on an interval in 1854 [87], with an incorrect justification. Correct proofs were given by Eduard HEINE, in 1870 [43], then by Paul DU BOIS-REYMOND, in 1879 [28]. This result is also called the **fundamental lemma of calculus of variations**.

Theorem 3.7.– *Let $f \in \mathcal{C}(\Omega; E)$, where Ω is an open subset of $\mathbb{R}^d$ and E is a Neumann space, be such that, for every $\varphi \in \mathcal{D}(\Omega)$,*

$$\int_\Omega f\varphi = 0_E.$$

Then,

$$f = 0. \blacksquare$$

Proof. If f was not zero, there would exist $y \in \Omega$ such that

$$f(y) \neq 0_E.$$

Since E is separated (Definition 1.1), as is every Neumann space (Definition 1.4), there would exist a semi-norm $\| \ \|_{E;\nu}$ of E such that

$$\|f(y)\|_{E;\nu} = c > 0.$$

Since Ω is open and f is continuous, there would exist $r > 0$ such that $|x - y| \leq r$ implies that $x \in \Omega$ and

$$\|f(x) - f(y)\|_{E;\nu} \leq \frac{c}{2}.$$

Let $\varphi \in \mathcal{D}(\Omega)$ whose is support is included in the ball $B = \{x \in \mathbb{R}^d : |y - x| \leq r\}$ be such that

$$\varphi > 0 \text{ on } \mathring{B}.$$

This is for example satisfied by $\varphi(x) = \rho((x - y)/r)$, where the function ρ is defined by (2.4), p. 23. Then (Theorem A.76 (c)),

$$\int_\Omega \varphi = b > 0.$$

Since only $\mathring{B}$ contributes here to the integral (Theorem A.77), the bound of its semi-norms from Theorem 1.23 (a), the growth of the real integral (Theorem A.76 (a)) and its linearity (Theorem A.74), would successively give

$$\Big\| \int_\Omega (f - f(y))\varphi \Big\|_{E;\nu} = \Big\| \int_{\mathring{B}} (f - f(y))\varphi \Big\|_{E;\nu} \leq \int_{\mathring{B}} \|f - f(y)\|_{E;\nu}\, \varphi \leq \frac{c}{2} b$$

and

$$\Big\| \int_\Omega f(y)\varphi \Big\|_{E;\nu} = \Big\| f(y) \int_\Omega \varphi \Big\|_{E;\nu} = cb,$$

hence

$$\Big\| \int_\Omega f\varphi \Big\|_{E;\nu} \geq \frac{cb}{2}.$$

This would contradict the hypothesis $\int_\Omega f\varphi = 0_E$, since E is separated. Therefore,

$$f = 0. \square$$

From now on:

$$\begin{cases} \textit{We identify each continuous function } f \in \mathcal{C}(\Omega; E) \textit{ with the} \\ \textit{distribution } \bar{f} \in \mathcal{D}'(\Omega; E) \textit{ defined in Theorem 3.5.} \end{cases} \tag{3.6}$$

This is permitted since the mapping $f \mapsto \bar{f}$ is injective from Theorem 3.6.

Let us show that, thus, the space of continuous functions is topologically included in that of distributions.

Theorem 3.8.– *Given Ω an open subset of $\mathbb{R}^d$ and E a Neumann space,*

$$\mathcal{C}(\Omega; E) \subsetneq_{\rightarrow} \mathcal{D}'(\Omega; E)$$

and the identity is continuous, and therefore sequentially continuous, from $\mathcal{C}(\Omega; E)$ into $\mathcal{D}'(\Omega; E)$. ▮

Proof. **1.** *Continuity of identity.* With identification (3.6), the mapping $f \mapsto \bar{f}$ constructed in Theorem 3.5 becomes the identity from $\mathcal{C}(\Omega; E)$ into $\mathcal{D}'(\Omega; E)$, so the stated continuity properties are those of Theorem 3.6.

2. *Inclusion.* The topological inclusion follows from the continuity of the identity, according to Theorem 1.13. □

We now have the following equalities.

Theorem 3.9.– *For every $f \in \mathcal{C}(\Omega; E)$, where Ω is an open subset of $\mathbb{R}^d$ and E is a Neumann space, and every $\varphi \in \mathcal{D}(\Omega)$,*

$$\langle f, \varphi \rangle_{\mathcal{D}'(\Omega;E)\times\mathcal{D}(\Omega)} = \int_\Omega f\varphi$$

and, for every semi-norm $\| \ \|_{E;\nu}$ of E,

$$\|f\|_{\mathcal{D}'(\Omega;E);\varphi,\nu} = \left\| \int_\Omega f\varphi \right\|_{E;\nu}. \ ▮$$

Proof. **1.** *First equality.* The expression for $\langle f, \varphi \rangle$ is the rewriting of the equality in Theorem 3.5 with identification (3.6).

2. *Second equality.* The expression for $\|f\|_{\mathcal{D}'(\Omega;E);\varphi,\nu}$ is the rewriting of Definition 3.1 of the semi-norms of $\mathcal{D}'(\Omega; E)$ with the above expression for $\langle f, \varphi \rangle$. □

Compatibility with the operations on functions. To generalize to distributions an operation on functions (differentiation, support, restriction, etc.), it is thus necessary to verify that the generalization is compatible with identification (3.6), namely **to verify that we obtain the usual operations for functions**.

Recall that care should be taken with identifications, as we have seen in the section *Dangerous identifications* of Volume 1 [81, § 14.6, p. 216–219]. □

Convergence of continuous functions. The convergence in $\mathcal{C}(\Omega; E)$, i.e. the uniform convergence on compact sets of continuous functions, implies their convergence in the sense of distributions, due to Theorem 3.6.

In contrast, their pointwise convergence is neither stronger nor weaker that their convergence in the sense of distributions.

For example, the functions on $\mathbb{R}^d$ defined by

$$f_n(x) = n^d \rho(n^2 x),$$

where $\rho \in \mathcal{C}(\mathbb{R}^d)$ has its support in the ball $B(0,1)$ and $\int_{\mathbb{R}^d} \rho = 1$, converge to 0 in $\mathcal{D}'(\mathbb{R}^d)$ when $n \to \infty$, since, for every $\varphi \in \mathcal{D}(\mathbb{R}^d)$,

$$\Big| \int_{\mathbb{R}^d} f_n \varphi \Big| \leq c_\varphi \int_{\mathbb{R}^d} |f_n| = cn^{-d}.$$

But they do not converge pointwise if $\rho(0) \neq 0$, since $f_n(0) = n^d \rho(0)$.

Conversely, the functions defined by

$$g_n(x) = n^{2d} \rho(nx)$$

converge pointwise to 0 if $\rho(0) = 0$, since $g_n(x) = 0$ as soon as $n \geq 1/|x|$ and $g_n(0) = 0$. But they do not converge in $\mathcal{D}'(\mathbb{R}^d)$, since, if $\varphi = 1$ in $B(0,1)$,

$$\int_{\mathbb{R}^d} g_n \varphi = \int_{\mathbb{R}^d} g_n = n^d. \ \Box$$

Identification of integrable functions with distributions. Since integrable functions with values in a Neumann space have not been defined, it is not possible (for the moment) to identify them with distributions. Rather than integrable functions, in Volume 4 we will define *integrable measures* that are equivalent to *classes of integrable functions that are equal almost everywhere* while being more pleasant to manipulate and simpler to define. □

3.4. The case where E is not a Neumann space

If a separated semi-normed space E is not a Neumann space, we can still define the space $\mathcal{L}(\mathcal{D}(\Omega); E)$ of continuous linear mappings from $\mathcal{D}(\Omega)$ into E, but **we should not** denote it by $\mathcal{D}'(\Omega; E)$ nor call it a "*space of distributions*", because it does not have the expected properties. In particular if, in addition, E is metrizable:

$$\left\{ \begin{array}{l} \textit{We can no longer identify each function} \\ \textit{of } \mathcal{C}(\Omega; E) \textit{ with an element of } \mathcal{L}(\mathcal{D}(\Omega); E). \end{array} \right.$$

Firstly, because the integral with values in such a space E is not defined, the definition of $\bar{f}$ in Theorem 3.5 no longer makes sense. Secondly because, even if we extended

Definition 1.22 of an integral to such a space E, the equality $\overline{f}(\varphi) = \int_\Omega f\varphi$ would not always define a mapping from $\mathcal{D}(\Omega)$ into E.

To be able to state this precisely, recall that every separated semi-normed space E possesses [Vol. 1, Theorem 4.24] a **sequential completion** $\widehat{E}$, namely a space such that:

$$\begin{cases} \widehat{E} \text{ is a Neumann space in which } E \text{ is included and dense,} \\ \text{whose semi-norms extend those of } E, \text{ and which has no} \\ \text{sequentially closed proper vector subspace that contains } E. \end{cases} \tag{3.7}$$

(E has infinitely many sequential completions, but we can pass from one to the other by a bijection which is sequentially continuous as well as its inverse; they are included in the classical *completions*. Recall that E is sequentially dense in $\widehat{E}$ if E is metrisable [Vol. 1, Property (4.10), p. 63] but this is not always the case.)

Let us now extend the definition of the integral to a separated semi-normed space E, not necessarily sequentially complete:

$$\begin{cases} \textit{Given } f \in \mathcal{B}(\Omega; E), \textit{ its } \textbf{completed integral } \widehat{\int_\omega} f \textit{ is the limit} \\ \textit{in } \widehat{E} \textit{ of the approximate integrals } \mathbb{S}^n_\omega f \textit{ of Definition 1.22.} \end{cases} \tag{3.8}$$

This definition is allowed, because $(\mathbb{S}^n_\omega f)_{n\in\mathbb{N}}$ is a Cauchy sequence in E (the proof given in Volume 2 [82, Lemma 4.10] does not use the sequential completion of E); it is thus also a Cauchy sequence in $\widehat{E}$, so that it converges there. **We should not** denote it by $\int_\omega f$ nor call it "the integral of f", because it does not belong to E and since $\widehat{E}$ is not unique.

Given now $f \in \mathcal{C}(\Omega; E)$ and $\varphi \in \mathcal{D}(\Omega)$, then $f\varphi \in \mathcal{B}(\Omega; E)$ and we define $\overline{f}(\varphi)$ belonging to $\widehat{E}$ by

$$\overline{f}(\varphi) \stackrel{\text{def}}{=} \widehat{\int_\Omega} f\varphi. \tag{3.9}$$

(It happens that $\overline{f}(\varphi)$ belongs to E for some f and φ, but this is not general.)

We are finally in the position to give a precise statement, as follows.

Theorem 3.10.– *If E is a metrizable separated semi-normed space that is not sequentially complete and Ω is a non-empty open subset of $\mathbb{R}^d$, the mapping $f \mapsto \overline{f}$, defined by (3.8) and (3.9), is not a mapping from $\mathcal{C}(\Omega; E)$ into $\mathcal{L}(\mathcal{D}(\Omega); E)$.*

Indeed, there exist $f \in \mathcal{C}(\Omega; E)$ and $\varphi \in \mathcal{D}(\Omega)$ such that

$$\widehat{\int_\Omega} f\varphi \in \widehat{E} \setminus E. \ \blacksquare$$

Proof. **1.** *Reduction of the problem.* Let $B(z,r)$ be a closed ball included in Ω. It suffices to construct a continuous function f, whose support is in that ball, such that

$$\widehat{\int_\Omega} f \in \widehat{E} \setminus E. \tag{3.10}$$

Indeed, for $\varphi \in \mathcal{D}(\Omega)$ such that $\varphi = 1$ on $B(z,r)$, we will then have

$$\widehat{\int_\Omega} f\varphi = \widehat{\int_\Omega} f \in \widehat{E} \setminus E,$$

and so

$$\bar{f}(\varphi) \notin E.$$

2. *Construction of f.* Since E is not sequentially complete, it possesses a Cauchy sequence $(u_n)_{n\in\mathbb{N}}$ which does not converge. This sequence is also Cauchy in $\widehat{E}$ which is sequentially complete, so here it has a limit u. Therefore,

$$u_n \to u \text{ in } \widehat{E}, \quad u \in \widehat{E} \setminus E. \tag{3.11}$$

Let $(B_n)_{n\in\mathbb{N}}$ be a sequence of balls included in $B(z,r)$, pairwise disjoint and such that $B_n = B(z_n, r_n)$ where

$$r_n > 0, \quad r_n \to 0, \quad z_n \to z.$$

Since E is assumed to be metrizable, it can [Vol. 1, Theorem 4.4] be endowed with an increasing sequence of semi-norms, say $(\| \ \|_m)_{m\in\mathbb{N}}$. Let $(|||\ |||_m)_{m\in\mathbb{N}}$ be the semi-norms on $\widehat{E}$ extending those of E.

After passing to a subsequence of the u_n if necessary, we can suppose that, for $n \geq 1$,

$$\|u_n - u_{n-1}\|_n \leq \frac{1}{n}|r_n|^d. \tag{3.12}$$

Indeed, it suffices to consider a subsequence $(u_{\sigma(n)})_{n\in\mathbb{N}}$ such that, for each n,

$$|||u_{\sigma(n)} - u|||_n \leq \frac{1}{2}\inf\Big\{\frac{1}{n}|r_n|^d, \frac{1}{n+1}|r_{n+1}|^d\Big\}.$$

Finally let $\phi \in \mathcal{C}(\mathbb{R}^d)$ be such that

$$\operatorname{supp}\phi \subset \{x \in \mathbb{R}^d : |x| < 1\} \text{ and } \int_{\mathbb{R}^d} \phi = 1.$$

We define f on Ω by

$$f(x) = \begin{cases} |r_0|^{-d}\phi((x-z_0)/r_0)u_0 & \text{if } x \in B_0, \\ |r_n|^{-d}\phi((x-z_n)/r_n)(u_n - u_{n-1}) & \text{if } x \in B_n,\ n \geq 1, \\ 0_E & \text{if } x \notin \bigcup_{n\geq 0} B_n. \end{cases} \tag{3.13}$$

This function is continuous away from the point z. It is equally so at z, because $f(z) = 0_E$ and, for every $m \in \mathbb{N}$, (3.12) and (3.13) give, when $n \to \infty$ and $n \geq m$,

$$\sup_{x \in B_n} \|f(x)\|_m \leq \sup_{x \in B_n} \|f(x)\|_n \leq \frac{1}{n} \sup_{y \in \mathbb{R}^d} |\phi(y)|$$

which goes to 0 as $n \to \infty$.

So, f is continuous in the whole Ω and its support is included in $B(z, r)$.

3. Verification of (3.10). Let $\widehat{f}$ be the same function whose values $\widehat{f}(x) = f(x)$ are considered as elements of $\widehat{E}$. This one is a Neumann space from (3.7), which allows us to use the properties of an integral with values in such a space.

Firstly, from the homothety effect on the integral (Theorem A.88),

$$\int_{\overset{\circ}{B}_0} \widehat{f} = u_0, \quad \int_{\overset{\circ}{B}_n} \widehat{f} = u_n - u_{n-1}.$$

Then, denoting $\omega_n = \bigcup_{0 \leq i \leq n} \overset{\circ}{B}_i$, the additivity of the integral with respect to disjoint open sets [Vol. 1, Theorem 4.21] gives

$$\int_{\omega_n} \widehat{f} = \sum_{0 \leq i \leq n} \int_{\overset{\circ}{B}_i} \widehat{f} = u_n. \tag{3.14}$$

And, denoting $\omega = \bigcup_{i \in \mathbb{N}} \overset{\circ}{B}_i$, the continuity of the integral with respect to increasing open sets [Vol. 1, Theorem 4.19] gives, when $n \to \infty$,

$$\int_{\omega_n} \widehat{f} \to \int_{\omega} \widehat{f}.$$

From which, by (3.11) and (3.14),

$$\int_{\omega} \widehat{f} = u.$$

Finally, since the domain where the function is zero does not contribute to the integral (Theorem A.77), we can replace ω here with Ω. Definition (3.8) of the completed integral then gives

$$\widehat{\int_{\Omega}} f = \int_{\Omega} \widehat{f} = u.$$

It does not belong to E, since u does not belong to it from (3.11), which proves (3.10), and hence the theorem. □

The case where E is neither sequentially complete nor metrizable. If the space E is not metrizable, we no longer know how to show that, if it is not sequentially complete, $f \mapsto \overline{f}$ is not a mapping from $\mathcal{C}(\Omega; E)$ into $\mathcal{L}(\mathcal{D}(\Omega); E)$.

But we still know how to show that $\mathcal{L}(\mathcal{D}(\Omega); E)$ does not have the expected properties of a space of distributions. More precisely, it is desirable that distributions also generalize certain singular functions, for example the main value, whose definitions involve the finite part of integrals. However, the minimal hypothesis for the finite part to be an element of E, as soon as it exists, is that E be a Neumann space, whether it is metrizable or not [ŠIMON, 84, to appear]. □

To work in the sense of distributions for a function with values in a separated semi-normed space E which is not Neumann, **it suffices** to embed E in its sequential completion $\widehat{E}$ defined by (3.7), p. 54. For example, if

$$f \in \mathcal{C}(\Omega; E),$$

then it equally belongs to $\mathcal{C}(\Omega; \widehat{E})$, so it can be identified with a distribution $\widehat{f}$ in $\mathcal{D}'(\Omega; \widehat{E})$, since $\widehat{E}$ is a Neumann space. Among other things, this allows us to define its partial derivatives

$$\partial^\beta \widehat{f} \in \mathcal{D}'(\Omega; \widehat{E}).$$

(We could write $\widehat{\partial^\beta f}$ instead of $\partial^\beta \widehat{f}$; in contrast, $\partial^\beta f$ should be avoided here.)

3.5. Measures

Let us first define the space $\mathcal{K}(\Omega)$ [5] which serves to construct the space of measures.

Definition 3.11.– *We denote by $\mathcal{K}(\Omega)$, where Ω is an open subset of $\mathbb{R}^d$, the vector space of continuous real functions on Ω having compact support endowed with the semi-norms, indexed by $p \in \mathcal{C}^+(\Omega)$,*

$$\|f\|_{\mathcal{K}(\Omega);p} \overset{\text{def}}{=} \sup_{x \in \Omega} p(x)|f(x)|. \ \blacksquare$$

The notation $\mathcal{K}(\Omega)$. This is the notation used by BOURBAKI [9, Chap. III, § 1, No. 1, p. 40]. □

Interest in the semi-norms of $\mathcal{K}(\Omega)$. The semi-norms in Definition 3.11 provide the properties usually obtained by using the delicate **topology of inductive limit of the $\mathcal{C}_K(\Omega)$**. In particular, all the functions of a bounded subset of $\mathcal{K}(\Omega)$ have their support in a same compact subset K of Ω due to Theorem 2.4. These semi-norms generate the said inductive limit topology [ŠIMON, 84, to appear], while being simpler to use. □

Let us now define the space of measures.

5. **History of the semi-norms of $\mathcal{K}(\Omega)$.** BOURBAKI endowed $\mathcal{K}(\Omega)$ [9, Chap. III, § 1, No. 1, p. 41] with the topology of inductive limit of the $\mathcal{C}_K(\Omega)$. The semi-norms in Definition 3.11 are new to us.

Definition 3.12.– *Let Ω be an open subset of $\mathbb{R}^d$ and E a Neumann space, whose family of semi-norms is denoted by $\{\| \|_{E;\nu} : \nu \in \mathcal{N}_E\}$.*

*A **measure** [6] on Ω with values in E is a continuous linear mapping from $\mathcal{K}(\Omega)$ into E.*

We denote by $\mathcal{M}(\Omega; E)$ the space of these measures endowed with the semi-norms, indexed by $\varphi \in \mathcal{K}(\Omega)$ and $\nu \in \mathcal{N}_E$,

$$\|f\|_{\mathcal{M}(\Omega;E);\varphi,\nu} \stackrel{\text{def}}{=} \|\langle f, \varphi \rangle\|_{E;\nu}. \blacksquare$$

We denote by $\langle f, \varphi \rangle$ the value $f(\varphi)$ of a measure f applied to φ, and

$$\mathcal{M}(\Omega) \stackrel{\text{def}}{=} \mathcal{M}(\Omega; \mathbb{R}).$$

The case where E is not a Neumann space. If E is a separated semi-normed space that is not sequentially complete, we can still define the space $\mathcal{L}(\mathcal{K}(\Omega); E)$ of continuous linear mappings from $\mathcal{K}(\Omega)$ into E, but **we must not** call it a "space of measures", nor denote it by $\mathcal{M}(\Omega; E)$, because it does not have the expected properties.

In particular, we cannot identify every continuous function with an element of $\mathcal{L}(\mathcal{K}(\Omega); E)$ when in addition E is metrizable, since we cannot even identify it with an element of $\mathcal{L}(\mathcal{D}(\Omega); E)$ according to Theorem 3.10. □

Terminology. The measures given by Definition 3.12 are sometimes called **Radon measures**, for example in [EDWARDS, 30, § 4.3, p. 177], although Johann RADON defined them as *completely additive functions of sets* [62]. For our part, we follow the terminology of Nicolas BOURBAKI [9, Chap. III, § 1, No 3, p. 47]. □

Vague topolgy. The topology generated by the semi-norms in Definition 3.12, namely the topology of simple (pointwise) convergence on $\mathcal{K}(\Omega)$, is called the **vague topology** of $\mathcal{M}(\Omega; E)$. □

Lebesgue measure. The mapping $\varphi \mapsto \int_{\mathbb{R}^d} \varphi$ defines a real measure on $\mathbb{R}^d$, called the **Lebesgue measure**, since, due to inequality (3.5), p. 49, for $f \equiv 1$, we have, for every $\varphi \in \mathcal{D}(\mathbb{R}^d)$,

$$\Big| \int_{\mathbb{R}^d} \varphi \Big| \leq 2^{d+1} \sup_{x \in \mathbb{R}^d} p(x)|\varphi(x)| = 2^{d+1} \|\varphi\|_{\mathcal{K}(\mathbb{R}^d);p},$$

where p is given by (3.4), that is here $p(x) = (2 + |x|)^{d+2}$.

Henri LEBESGUE defined it in his thesis [50] as an *additive function of sets*, which in the particular case of an open set ω coincides with its measure $|\omega|$ constructed in Definition A.83. □

Let us characterize measures.

---

6. **History of measures**. The definition of measures as *linear mappings on* $\mathcal{K}(\Omega)$ seems to us to be due to Nicolas BOURBAKI, in 1965 [9, Chap. III, § 1, No. 3, p. 47].
Previously, the measures were introduced as *additive functions of sets*, following the works of Emile BOREL, in 1894 [7], Henri LEBESGUE, in 1902 [50], and Johann RADON, in 1913 [62]. Their genesis is detailed in *Elements of History of Mathematics* of Nicolas BOURBAKI [12, Chap. 22, p. 219–230].

Theorem 3.13.– *Let Ω be an open subset of $\mathbb{R}^d$ and E a Neumann space. Then,*

$$f \in \mathcal{M}(\Omega; E)$$

if and only if: f is a linear mapping from $\mathcal{K}(\Omega)$ into E and, for every semi-norm $\| \ \|_{E;\nu}$ of E, there exist $p \in \mathcal{C}^+(\Omega)$ and $c \in \mathbb{R}$ such that, for every $\varphi \in \mathcal{K}(\Omega)$,

$$\|\langle f, \varphi \rangle\|_{E;\nu} \leq c \|\varphi\|_{\mathcal{K}(\Omega);p}.$$

That is,

$$\|\langle f, \varphi \rangle\|_{E;\nu} \leq c \sup_{x \in \Omega} p(x)|\varphi(x)|. \ \blacksquare$$

Proof. A measure being, by Definition 3.12, a continuous linear mapping from $\mathcal{K}(\Omega)$ into E, the result is given by the characterization of continuous linear mappings from Theorem 1.12 (b), because the family of semi-norms of $\mathcal{K}(\Omega)$ is filtering.

Indeed, this family, which is given by Definition 3.11, satisfies Definition 1.8 of a filtering family since, given functions $p_1, \ldots, p_n$ in $\mathcal{C}^+(\Omega)$, the sum $p = p_1 + \cdots + p_n$ belongs to $\mathcal{C}^+(\Omega)$ and bounds the p_i from above, and thus

$$\sup_{1 \leq i \leq n} \|\varphi\|_{\mathcal{K}(\Omega);p_i} = \sup_{1 \leq i \leq n} \sup_{x \in \Omega} p_i(x)|\varphi(x)| \leq \sup_{x \in \Omega} p(x)|\varphi(x)| = \|\varphi\|_{\mathcal{K}(\Omega);p}. \ \square$$

We now define the **Dirac measure**[7] δ_x, which is a very useful real measure, in particular because δ_0 is the neutral element of weighting (Theorem 7.19).

Definition 3.14.– *Given $x \in \mathbb{R}^d$, we define $\delta_x \in \mathcal{M}(\mathbb{R}^d)$ by: for every $\varphi \in \mathcal{K}(\mathbb{R}^d)$,*

$$\langle \delta_x, \varphi \rangle \stackrel{\text{def}}{=} \varphi(x). \ \blacksquare$$

Justification. The mapping δ_x is indeed a measure, since it is continuous linear from $\mathcal{K}(\mathbb{R}^d)$ into $\mathbb{R}$ because, denoting by $1_{\mathbb{R}^d}$ the constant function of value 1 on $\mathbb{R}^d$,

$$|\langle \delta_x, \varphi \rangle| = |\varphi(x)| \leq \sup_{z \in \mathbb{R}^d} |\varphi(z)| = \|\varphi\|_{\mathcal{K}(\mathbb{R}^d);1_{\mathbb{R}^d}}. \ \square$$

7. **History of the Dirac measure**. The measure that now carries his name was introduced by Paul DIRAC in 1926 [26, p. 625], in the following terms:

> "this function [...] is defined by $\delta(x) = 0$ when $x \neq 0$ and $\int_{-\infty}^{+\infty} \delta(x)\, dx = 1$. Strictly, of course, $\delta(x)$ is not a proper function of x. [...] All the same one use $\delta(x)$ as though it were a proper function for practically all the purposes of quantum mechanics without getting incorrect results."

Let us associate a distribution to each measure.

Theorem 3.15.– *Let $f \in \mathcal{M}(\Omega; E)$, where Ω is an open subset of $\mathbb{R}^d$ and E is a Neumann space.*

We define $\bar{f} \in \mathcal{D}'(\Omega; E)$ by: for every $\varphi \in \mathcal{D}(\Omega)$,

$$\langle \bar{f}, \varphi \rangle \stackrel{\text{def}}{=} \langle f, \varphi \rangle.$$

In other words, $\bar{f}$ is the restriction of f to $\mathcal{D}(\Omega)$. ▮

Proof. According to the characterization of measures from Theorem 3.13 and to Definition 3.11 of the semi-norms of $\mathcal{K}(\Omega)$, for every semi-norm $\| \ \|_{E;\nu}$ of E, there exist $p \in \mathcal{C}^+(\Omega)$ and $c \in \mathbb{R}$ such that, for every $\varphi \in \mathcal{K}(\Omega)$,

$$\|\langle f, \varphi \rangle\|_{E;\nu} \leq c \|\varphi\|_{\mathcal{K}(\Omega);p} = c \sup_{x \in \Omega} p(x) |\varphi(x)|.$$

When $\varphi \in \mathcal{D}(\Omega)$, it follows, with Definition 2.5 of the semi-norms of $\mathcal{D}(\Omega)$,

$$\|\langle \bar{f}, \varphi \rangle\|_{E;\nu} = \|\langle f, \varphi \rangle\|_{E;\nu} \leq c \sup_{x \in \Omega} \sup_{0 \leq |\beta| \leq p(x)} p(x) |\partial^\beta \varphi(x)| = c \|\varphi\|_{\mathcal{D}(\Omega);p}.$$

So, by the characterization of distributions from Theorem 3.3,

$$\bar{f} \in \mathcal{D}'(\Omega; E). \ \square$$

Let us prove that two distinct measures provide two distinct distributions, which will allow us to identify each measure with a distribution.

Theorem 3.16.– *The mapping $f \mapsto \bar{f}$ given by Theorem 3.15 is linear, continuous, and injective from $\mathcal{M}(\Omega; E)$ into $\mathcal{D}'(\Omega; E)$.* ▮

The proof uses the following density property, which is established in Volume 2 [82, Theorem 7.16, since $\mathcal{K}^\infty(\Omega) = \mathcal{D}(\Omega)$].

Theorem 3.17.– *The space $\mathcal{D}(\Omega)$, where Ω is an open subset of $\mathbb{R}^d$, is sequentially dense in $\mathcal{K}(\Omega)$.* ▮

Proof of Theorem 3.16. **1.** *Injectivity.* Let $f \in \mathcal{M}(\Omega; E)$ be such that

$$f \neq 0.$$

That is, there exists $\varphi \in \mathcal{K}(\Omega)$ such that

$$\langle f, \varphi \rangle \neq 0_E.$$

From Theorem 3.17, there exists a sequence $(\varphi_n)_{n\in\mathbb{N}}$ in $\mathcal{D}(\Omega)$ which converges to φ in $\mathcal{K}(\Omega)$. Not all the $\langle f, \varphi_n \rangle$ are zero, otherwise their limit $\langle f, \varphi \rangle$ would too. Thus, for one of the n,

$$\langle \bar{f}, \varphi_n \rangle = \langle f, \varphi_n \rangle \neq 0_E.$$

From which,

$$\bar{f} \neq 0.$$

The mapping $f \mapsto \bar{f}$, being linear, is therefore injective.

2. *Continuity.* Denote by $\{\| \ \|_{E;\nu} : \nu \in \mathcal{N}_E\}$ the family of semi-norms of E. Definitions 3.1 and 3.12 of the semi-norms of $\mathcal{D}'(\Omega; E)$ and $\mathcal{M}(\Omega; E)$ give, for every $\varphi \in \mathcal{D}(\Omega)$ and $\nu \in \mathcal{N}_E$,

$$\|\bar{f}\|_{\mathcal{D}'(\Omega;E);\varphi,\nu} = \|\langle \bar{f}, \varphi \rangle\|_{E;\nu} = \|\langle f, \varphi \rangle\|_{E;\nu} = \|f\|_{\mathcal{M}(\Omega;E);\varphi,\nu}.$$

Due to the characterization of continuous linear mappings from Theorem 1.12 (a), this proves that the mapping $f \mapsto \bar{f}$ is continuous. □

From now on, thanks to the injectivity of the mapping $f \mapsto \bar{f}$ in Theorem 3.16:

$$\begin{cases} \textbf{We identify each measure } f \in \mathcal{M}(\Omega; E) \textbf{ with the} \\ \textbf{distribution } \bar{f} \in \mathcal{D}'(\Omega; E) \textbf{ defined in Theorem 3.15.} \end{cases} \tag{3.15}$$

Let us show that the space of measures is thus topologically included in that of distributions.

Theorem 3.18.– *For every open subset Ω of $\mathbb{R}^d$ and every Neumann space E,*

$$\mathcal{M}(\Omega; E) \subsetneq \mathcal{D}'(\Omega; E). \blacksquare$$

Proof. With identification (3.15), the mapping $f \mapsto \bar{f}$ constructed in Theorem 3.15 becomes the identity from $\mathcal{M}(\Omega; E)$ into $\mathcal{D}'(\Omega; E)$. Theorem 3.16 therefore provides the continuity of the identity. This is equivalent, due to Theorem 1.13, to the stated topological inclusion. □

Properties of measures. Identification (3.15) allows us to extend every property of distributions to measures, and in particular to define their derivatives (§ 5.2), their support (§ 6.6), their weighting (Chapter 7), the separation or the regrouping of their variables (§ 15.4 and 15.6), and obtain some properties and existence conditions for their primitives (Chapters 12 and 13).

Some of these properties of distributions can be improved in the case of measures, since they are particular distributions. In particular, measures are locally derivatives of continuous functions, without being necessary to suppose (as in Theorems 16.6 and 16.8) that E is normed, since they are *locally of order* 0 (Definition 16.1 (b)). □

Let us characterize the distributions that are measures.

Theorem 3.19.– *A distribution $f \in \mathcal{D}'(\Omega; E)$, where Ω is an open subset of $\mathbb{R}^d$ and E is a Neumann space, is a measure of $\mathcal{M}(\Omega; E)$ if, and only if, for every semi-norm $\| \|_{E;\nu}$ of E, there exist $p \in \mathcal{C}^+(\Omega)$ and $c \in \mathbb{R}$ such that, for every $\varphi \in \mathcal{D}(\Omega)$,*

$$\|\langle f, \varphi \rangle\|_{E;\nu} \leq c \sup_{x\in\Omega} p(x)|\varphi(x)|. \tag{3.16}$$

■

Proof. **1.** *Sufficiency of (3.16).* Suppose that $f \in \mathcal{D}'(\Omega; E)$ satisfies (3.16). Then its linearity and Definition 3.11 of the semi-norms of $\mathcal{K}(\Omega)$ give, for φ_1 and φ_2 in $\mathcal{D}(\Omega)$,

$$\|\langle f, \varphi_1 - \varphi_2 \rangle\|_{E;\nu} \leq c \sup_{x\in\Omega} p(x)|(\varphi_1 - \varphi_2)(x)| = c\|\varphi_1 - \varphi_2\|_{\mathcal{K}(\Omega);p}.$$

Therefore, f is uniformly continuous (Definitions 1.9 (b)) from the subset $\mathcal{D}(\Omega)$ of $\mathcal{K}(\Omega)$ into E. Since E is a Neumann space and $\mathcal{D}(\Omega)$ is sequentially dense in $\mathcal{K}(\Omega)$ (Theorem 3.17), the continuous extension theorem (Theorem A.34) shows that there exists a unique mapping g, continuous from $\mathcal{K}(\Omega)$ into E, that extends f, i.e., for every $\varphi \in \mathcal{D}(\Omega)$,

$$\langle f, \varphi \rangle = \langle g, \varphi \rangle.$$

The linearity of f implies, by continuity, that of g, so

$$g \in \mathcal{M}(\Omega; E).$$

Therefore, f is the distribution $\overline{g}$ identified with the measure g by (3.15).

2. *Necessity of (3.16).* Suppose that $f \in \mathcal{D}'(\Omega; E)$ is a measure, namely that it is identified, by (3.15), to a measure

$$\underline{f} \in \mathcal{M}(\Omega; E).$$

According to the characterization of measures from Theorem 3.13, for every semi-norm $\| \|_{E;\nu}$ of E, there exist $p \in \mathcal{C}^+(\Omega)$ and $c \in \mathbb{R}$ such that, for every $\varphi \in \mathcal{K}(\Omega)$,

$$\|\langle \underline{f}, \varphi \rangle\|_{E;\nu} \leq c \sup_{x\in\Omega} p(x)|\varphi(x)|.$$

For $\varphi \in \mathcal{D}(\Omega)$, this gives (3.16) since then $\langle f, \varphi \rangle = \langle \underline{f}, \varphi \rangle$. □

3.6. Continuous functions and measures

We now associate a measure to each continuous function.

Theorem 3.20.– *Let $f \in \mathcal{C}(\Omega; E)$, where Ω is an open subset of $\mathbb{R}^d$ and E is a Neumann space, whose family of semi-norms is denoted by $\{\| \ \|_{E;\nu} : \nu \in \mathcal{N}_E\}$.*

We define $\hat{f} \in \mathcal{M}(\Omega; E)$ by: for every $\varphi \in \mathcal{K}(\Omega)$,

$$\langle \hat{f}, \varphi \rangle \stackrel{\text{def}}{=} \int_\Omega f\varphi.$$

For every $\varphi \in \mathcal{K}(\Omega)$ and $\nu \in \mathcal{N}_E$,

$$\|\hat{f}\|_{\mathcal{M}(\Omega;E);\varphi,\nu} \leq c\|f\|_{\mathcal{C}(\Omega;E);K,\nu},$$

where K is the support of φ and $c = \int_\Omega |\varphi|$. ∎

Proof. **1.** *Well-definedness of the integral.* The integral of $f\varphi$ makes sense, according to Definition 1.22 of the Cauchy integral with values in a Neumann space, since $f\varphi$ is uniformly continuous and has a bounded support, i.e. it belongs to $\mathcal{B}(\Omega; E)$ (Definition 1.21).

This follows, due to a corollary (Theorem A.33) of Heine's theorem, from the fact that $f\varphi$ is continuous (as is every product of continuous functions) and has a compact support (due to Theorem 2.2 (a), since it is zero outside of the support of φ, which is a compact subset of Ω).

2. *Obtaining a measure.* Let us prove that $\hat{f}$ satisfies the characterization of measures from Theorem 3.13. Given $\nu \in \mathcal{N}_E$, we define $p \in \mathcal{C}^+(\Omega)$ by

$$p(x) = (2 + |x|)^{d+2}\|f(x)\|_{E;\nu}.$$

Let $\varphi \in \mathcal{K}(\Omega)$ and $\omega = \Omega \cap B$, where B is any open ball containing the support of φ. Only ω contributes to the integral of $f\varphi$ due to Theorem A.77, hence the bound of the semi-norms of the integral from Theorem 1.23 (a) gives

$$\left\|\int_\Omega f\varphi\right\|_{E;\nu} = \left\|\int_\omega f\varphi\right\|_{E;\nu} \leq \int_\omega \|f\|_{E;\nu}|\varphi| = \int_\omega \frac{p(x)|\varphi(x)|}{(2+|x|)^{d+2}}\, dx.$$

Since $\int_\omega 1/(2 + |x|)^{d+2}\, dx \leq 2^{d+1}$ (Lemma A.82), the growth of the real integral (Theorem A.76 (a)) and its linearity give

$$\left\|\int_\Omega f\varphi\right\|_{E;\nu} \leq 2^{d+1} \sup_{x\in\Omega} p(x)|\varphi(x)|.$$

That is to say, with Definition 3.12 of the semi-norms of $\mathcal{K}(\Omega)$,

$$\|\langle \hat{f}, \varphi \rangle\|_{E;\nu} \leq 2^{d+1} \|\varphi\|_{\mathcal{K}(\Omega);p}.$$

The mapping $\hat{f}$ being linear, since the integral is linear (Theorem A.74), this inequality implies, by the characterization of measures from Theorem 3.13, that

$$\hat{f} \in \mathcal{M}(\Omega; E).$$

3. *Bound of the semi-norms of* $\hat{f}$. Denoting by K the support of φ, we have, for every $y \in \Omega$,

$$\|f(y)\|_{E;\nu} |\varphi(y)| \leq \sup_{x \in K} \|f(x)\|_{E;\nu} |\varphi(y)|.$$

Thus, again using the inequalities from Theorems 1.23 (a) and A.76 (a),

$$\left\| \int_\Omega f\varphi \right\|_{E;\nu} \leq \int_\Omega \|f\|_{E;\nu} |\varphi| \leq \sup_{x \in K} \|f(x)\|_{E;\nu} \int_\Omega |\varphi|.$$

From which, denoting $c = \int_\Omega |\varphi|$,

$$\|\langle \hat{f}, \varphi \rangle\|_{E;\nu} = \left\| \int_\Omega f\varphi \right\|_{E;\nu} \leq c \sup_{x \in K} \|f(x)\|_{E;\nu}.$$

That is, by Definitions 3.12 and 1.18 (a) of the semi-norms of $\mathcal{M}(\Omega; E)$ and $\mathcal{C}(\Omega; E)$,

$$\|\hat{f}\|_{\mathcal{M}(\Omega;E);\varphi,\nu} \leq c \|f\|_{\mathcal{C}(\Omega;E);K,\nu}. \quad \square$$

Finally, we compare the spaces of continuous functions, measures and distributions.

Theorem 3.21.– *For every open subset* Ω *of* $\mathbb{R}^d$ *and every Neumann space* E,

$$\mathcal{C}(\Omega; E) \underset{\to}{\subsetneq} \mathcal{M}(\Omega; E) \underset{\to}{\subsetneq} \mathcal{D}'(\Omega; E). \quad \blacksquare$$

Proof. **1.** *First inclusion.* Given $f \in \mathcal{C}(\Omega; E)$, let $\hat{f} \in \mathcal{M}(\Omega; E)$ be the measure given by Theorem 3.20. The distributions $\overline{f}$ and $\overline{\hat{f}}$ to which they are identified, respectively by (3.6) and (3.15), coincide since both take the value $\int_\Omega f\varphi$ for $\varphi \in \mathcal{D}(\Omega)$. Since $\overline{f} \in \mathcal{M}(\Omega; E)$, this proves that

$$\mathcal{C}(\Omega; E) \subset \mathcal{M}(\Omega; E).$$

The inequality in Theorem 3.20 shows, due to the characterization of continuous linear mappings from Theorem 1.12 (a), that the identity is continuous from $\mathcal{C}(\Omega; E)$ into $\mathcal{M}(\Omega; E)$. Which, according to Theorem 1.13, is equivalent to the topological inclusion

$$\mathcal{C}(\Omega; E) \underset{\to}{\subsetneq} \mathcal{M}(\Omega; E).$$

2. *Second inclusion.* It has been established in Theorem 3.18. $\square$

Chapter 4

Extraction of Convergent Subsequences

This chapter is devoted to the extraction of convergent subsequences of $\mathcal{D}'(\Omega; E)$, which is an important tool for solving PDEs by the *compactness method* that we will use in Volume 7.

We begin by characterizing the bounded subsets of $\mathcal{D}'(\Omega; E)$ (§ 4.1) and its convergent sequences (§ 4.2), then its relatively sequentially compact subsets, i.e. those in which every sequence has a convergent subsequence (Theorem 4.6).

We deduce that:

— $\mathcal{D}'(\Omega; E)$ is a Neumann space, i.e. that its Cauchy sequences converge (Theorem 4.5).

— If every bounded sequence in E has a subsequence converging in a Neumann space F, every bounded sequence in $\mathcal{D}'(\Omega; E)$ has a subsequence converging in $\mathcal{D}'(\Omega; F)$ (Theorem 4.10); in particular, every bounded sequence in $\mathcal{D}'(\Omega)$ has a convergent subsequence (Theorem 4.7).

— If every bounded sequence in E has a subsequence converging in E-weak, this is a Neumann space and every bounded sequence in $\mathcal{D}'(\Omega; E)$ has a subsequence converging in $\mathcal{D}'(\Omega; E\text{-weak})$ (Theorem 4.15).

4.1. Bounded subsets of $\mathcal{D}'(\Omega; E)$

Let us characterize **bounded sets of distributions**, which will be useful since every convergent or Cauchy sequence is bounded.

Theorem 4.1.– *Let $\mathcal{F} \subset \mathcal{D}'(\Omega; E)$, where Ω is an open subset of $\mathbb{R}^d$ and E is a Neumann space.*

Then the following properties are equivalent:

(a) *$\mathcal{F}$ is bounded in $\mathcal{D}'(\Omega; E)$.*

(b) *For every $\varphi \in \mathcal{D}(\Omega)$ and every semi-norm $\| \ \|_{E;\nu}$ of E,*

$$\sup_{f \in \mathcal{F}} \|\langle f, \varphi \rangle\|_{E;\nu} < \infty.$$

(c) *For every* $\varphi \in \mathcal{D}(\Omega)$,

$$\{\langle f, \varphi \rangle : f \in \mathcal{F}\} \text{ is bounded in } E. \blacksquare$$

Proof. By Definitions 1.2 of a bounded set in a semi-normed space and 3.1 of the semi-norms of $\mathcal{D}'(\Omega; E)$, $\mathcal{F}$ is bounded in $\mathcal{D}'(\Omega; E)$ is equivalent to: for every $\varphi \in \mathcal{D}(\Omega)$ and $\nu \in \mathcal{N}_E$,

$$\sup_{f \in \mathcal{F}} \|f\|_{\mathcal{D}'(\Omega;E);\varphi,\nu} = \sup_{f \in \mathcal{F}} \|\langle f, \varphi \rangle\|_{E;\nu} < \infty.$$

That is to say, again from Definition 1.2, $\{\langle f, \varphi \rangle : f \in \mathcal{F}\}$ is bounded in E. □

Let us show that bounded subsets of $\mathcal{D}'(\Omega; E)$ are equicontinuous [1] on $\mathcal{D}(\Omega)$ and on the $\mathcal{C}_K^\infty(\Omega)$.

Theorem 4.2.– *Let*

$$\mathcal{F} \text{ be a bounded subset of } \mathcal{D}'(\Omega; E),$$

where Ω *is an open subset of* $\mathbb{R}^d$ *and* E *is a Neumann space, and let* $\| \ \|_{E;\nu}$ *be a semi-norm of* E*. Then:*

(a) *For every compact subset* K *of* Ω*, there exist* $m_K \in \mathbb{N}$ *and* $c_K \in \mathbb{R}$ *such that, for every* $f \in \mathcal{F}$ *and* $\varphi \in \mathcal{C}_K^\infty(\Omega)$,

$$\|\langle f, \varphi \rangle\|_{E;\nu} \leq c_K \|\varphi\|_{\mathcal{C}_b^{m_K}(\Omega)}.$$

(b) *There exists* $p \in \mathcal{C}^+(\Omega)$ *such that, for every* $f \in \mathcal{F}$ *and* $\varphi \in \mathcal{D}(\Omega)$,

$$\|\langle f, \varphi \rangle\|_{E;\nu} \leq \|\varphi\|_{\mathcal{D}(\Omega);p}. \blacksquare$$

Proof. **1.** Property (a). Let K be a bounded subset of Ω. Let us show that $\mathcal{F}$ satisfies the hypotheses of the Banach–Steinhaus theorem (Theorem A.39) for $F = \mathcal{C}_K^\infty(\Omega)$.
— The space $\mathcal{C}_K^\infty(\Omega)$ is a Fréchet space (Theorem A.53).
— Every distribution $f \in \mathcal{F}$ is continuous from $\mathcal{C}_K^\infty(\Omega)$ into E (Theorem 3.4 (d)).

1. **History of equicontinuity of bounded sets of distributions (Theorem 4.2 (b)).** Laurent SCHWARTZ proved in 1950 [69, Chap. 3, § 3, b) and a), p. 72] that the bounded subsets of $\mathcal{D}'(\mathbb{R}^d)$, i.e. for the simple topology used here, coincide with the bounded subsets of $\mathcal{D}'(\mathbb{R}^d)$-unif, i.e. for the topology of uniform convergence on the bounded subsets of $\mathcal{D}(\mathbb{R}^d)$ that he uses, and that these are equicontinuous on $\mathcal{D}(\mathbb{R}^d)$.

— Due to the characterization of bounded subsets of $\mathcal{D}'(\Omega; E)$ from Theorem 4.1 (b), for every $\varphi \in \mathcal{C}_K^\infty(\Omega)$ and every semi-norm $\| \; \|_{E;\nu}$ of E,

$$\sup_{f \in \mathcal{F}} \|\langle f, \varphi \rangle\|_{E;\nu} < \infty.$$

Theorem A.39 then shows that there exist $c_K \in \mathbb{R}$ and a finite number I of semi-norms of $\mathcal{C}_K^\infty(\Omega)$, i.e. (by Definition 2.3 (a)) of semi-norms of $\mathcal{C}_\mathrm{b}^\infty(\Omega)$ indexed by some $m_i \in \mathbb{N}$, such that, for every $f \in \mathcal{F}$ and $\varphi \in \mathcal{C}_K^\infty(\Omega)$,

$$\|\langle f, \varphi \rangle\|_{E;\nu} \leq c_K \sup_{1 \leq i \leq I} \|\varphi\|_{\mathcal{C}_\mathrm{b}^\infty(\Omega);m_i}.$$

Hence the stated inequality is satisfied with $m_K = \sup_{1 \leq i \leq I} m_i$ since, from Definitions 1.19 (b) and 1.18 (b),

$$\|\varphi\|_{\mathcal{C}_\mathrm{b}^\infty(\Omega);m_i} = \|\varphi\|_{\mathcal{C}_\mathrm{b}^{m_i}(\Omega)} \leq \|\varphi\|_{\mathcal{C}_\mathrm{b}^{m_K}(\Omega)}.$$

2. Property (b). The inequality in (a) implies that in (b) due to the second theorem of control of the $\mathcal{C}_K^m(\Omega)$-norms (Theorem 2.25) for the semi-norm π on $\mathcal{D}(\Omega)$ defined by $\pi(\varphi) = \|\langle f, \varphi \rangle\|_{E;\nu}$. □

4.2. Convergence in $\mathcal{D}'(\Omega; E)$

Let us characterize **convergent sequences of distributions**.

Theorem 4.3.– *Let $(f_n)_{n \in \mathbb{N}}$ be a sequence in $\mathcal{D}'(\Omega; E)$ and $f \in \mathcal{D}'(\Omega; E)$, where Ω is an open subset of $\mathbb{R}^d$ and E is a Neumann space.*

The following properties are equivalent:

(a)
$$f_n \to f \text{ in } \mathcal{D}'(\Omega; E).$$

(b) *For every $\varphi \in \mathcal{D}(\Omega)$ and every semi-norm $\| \; \|_{E;\nu}$ of E,*

$$\|\langle f_n - f, \varphi \rangle\|_{E;\nu} \to 0.$$

(c) *For every $\varphi \in \mathcal{D}(\Omega)$,*

$$\langle f_n, \varphi \rangle \to \langle f, \varphi \rangle \text{ in } E. \blacksquare$$

Proof. Due to Definition 1.3 (a) of a convergent sequence in a semi-normed space and 3.1 of the semi-norms of $\mathcal{D}'(\Omega; E)$, $f_n \to f$ in $\mathcal{D}'(\Omega; E)$ is equivalent to: for every $\varphi \in \mathcal{D}(\Omega)$ and $\nu \in \mathcal{N}_E$,

$$\|f_n - f\|_{\mathcal{D}'(\Omega;E);\varphi,\nu} = \|\langle f_n - f, \varphi \rangle\|_{E;\nu} \to 0.$$

That is, again according to Definition 1.3 (a), to $\langle f_n - f, \varphi \rangle \to 0$ in E. □

Uniform convergence. We show in Theorem 8.28 that the convergence $f_n \to f$ in $\mathcal{D}'(\Omega; E)$ is equivalent to the uniform convergence on every bounded subset $\mathcal{B}$ of $\mathcal{D}(\Omega)$, i.e., for every semi-norm of E,

$$\sup_{\varphi \in \mathcal{B}} \|\langle f_n - f, \varphi\rangle\|_{E;\nu} \to 0.$$

We establish in Theorem 8.27 the equivalence with a stronger property, which gives the uniform convergence on some unbounded subsets of $\mathcal{D}(\Omega)$ (see the comment *Comparison of Theorems 8.27 and 8.28*, p. 190). □

Convergence of continuous functions. Recall that uniform convergence of functions on compact sets implies their convergence in the distribution sense according to the topological inclusion from Theorem 3.8, but that their pointwise convergence is neither stronger nor weaker than their convergence in the distribution sense, see the comment *Convergence of continuous functions*, p. 53. □

We now show that the mapping $(f, \varphi) \mapsto \langle f, \varphi\rangle$ is sequentially continuous [2].

Theorem 4.4.– *Let Ω be an open subset of $\mathbb{R}^d$ and E a Neumann space. Then:*

(a) *For every $\varphi \in \mathcal{D}(\Omega)$, the mapping $f \mapsto \langle f, \varphi\rangle$ is continuous linear from $\mathcal{D}'(\Omega; E)$ into E.*

(b) *For every $f \in \mathcal{D}'(\Omega; E)$, the mapping $\varphi \mapsto \langle f, \varphi\rangle$ is continuous linear from $\mathcal{D}(\Omega)$ into E.*

(c) *The mapping $(f, \varphi) \mapsto \langle f, \varphi\rangle$ is sequentially continuous bilinear from the product $\mathcal{D}'(\Omega; E) \times \mathcal{D}(\Omega)$ into E.* ▮

Proof. Let $\{\| \|_{E;\nu} : \nu \in \mathcal{N}_E\}$ be the family of semi-norms of E.

1. *Property (a).* By Definition 3.1 of the semi-norms of $\mathcal{D}'(\Omega; E)$, for every $\nu \in \mathcal{N}_E$,

$$\|\langle f, \varphi\rangle\|_{E;\nu} = \|f\|_{\mathcal{D}'(\Omega;E);\varphi,\nu}.$$

The mapping $f \mapsto \langle f, \varphi\rangle$ is therefore continuous, due to the characterization of continuous linear mappings from Theorem 1.12 (a).

2. *Property (b).* This is Definition 3.1 of a distribution.

3. *Property (c).* Let $(f_n)_{n\in\mathbb{N}}$ and $(\varphi_n)_{n\in\mathbb{N}}$ be two sequences such that

$$f_n \to f \text{ in } \mathcal{D}'(\Omega; E), \quad \varphi_n \to \varphi \text{ in } \mathcal{D}(\Omega).$$

2. **History of Theorem 4.4.** Laurent SCHWARTZ showed in 1950 [69, Chap. 3, § 3, Theorem XI, p. 73] that $\langle f_n, \varphi_n\rangle \to \langle f, \varphi\rangle$ when $\varphi_n \to \varphi$ in $\mathcal{D}(\mathbb{R}^d)$ and $f_n \to f$ in $\mathcal{D}'(\mathbb{R}^d)$-unif, that is for the topology of uniform convergence on the bounded subsets of $\mathcal{D}(\mathbb{R}^d)$, and thus also in $\mathcal{D}'(\mathbb{R}^d)$, i.e. for the simple convergence used here, since these topologies coincide for convergent sequences according to [69, Theorem XIII, p. 74].

Decompose

$$\langle f_n, \varphi_n \rangle - \langle f, \varphi \rangle = \langle f_n - f, \varphi \rangle + \langle f_n, \varphi_n - \varphi \rangle. \tag{4.1}$$

By the characterization of convergent sequences in $\mathcal{D}'(\Omega; E)$ from Theorem 4.3 (c),

$$\langle f_n - f, \varphi \rangle \to 0_E \text{ in } E. \tag{4.2}$$

On the other hand, the sequence $(f_n)_{n\in\mathbb{N}}$ is bounded, as is every convergent sequence (Theorem A.5). Theorem 4.2 (b) then shows that it is equicontinuous from $\mathcal{D}(\Omega)$ into E; more precisely, it gives, for every $\nu \in \mathcal{N}_E$, the existence of $p \in \mathcal{C}^+(\Omega)$ such that, for every $\phi \in \mathcal{D}(\Omega)$,

$$\sup_{n\in\mathbb{N}} \|\langle f_n, \phi \rangle\|_{E;\nu} \leq \|\phi\|_{\mathcal{D}(\Omega);p}.$$

In particular,

$$\|\langle f_n, \varphi_n - \varphi \rangle\|_{E;\nu} \leq \|\varphi_n - \varphi\|_{\mathcal{D}(\Omega);p},$$

which tends to 0. This holds for every $\nu \in \mathcal{N}_E$, thus

$$\langle f_n, \varphi_n - \varphi \rangle \to 0_E \text{ in } E.$$

Hence, with (4.1) and (4.2),

$$\langle f_n, \varphi_n \rangle \to \langle f, \varphi \rangle \text{ in } E. \ \square$$

Non-continuity of $\langle\ ,\ \rangle$**.** The mapping $(f, \varphi) \mapsto \langle f, \varphi \rangle$ is not continuous from $\mathcal{D}'(\Omega; E) \times \mathcal{D}(\Omega)$ into E, except if Ω is empty or if E reduces to $\{0_E\}$ (in which case $\mathcal{D}'(\Omega; E)$ is zero-dimensional). In contrast, it is continuous on the bounded subsets of $\mathcal{D}'(\Omega; E) \times \mathcal{D}(\Omega)$. These properties are proven in [SIMON, 84].

In the case where $E = \mathbb{R}$, we may observe that the mapping $(f, \varphi) \mapsto \langle f, \varphi \rangle$ is not continuous from $\mathcal{D}'(\Omega) \times \mathcal{D}(\Omega)$ into $\mathbb{R}$, because it is not even continuous when $\mathcal{D}'(\Omega)$ is endowed with the topology of dual of $\mathcal{D}(\Omega)$, which is stronger than the simple topology used here, since the dual bilinear form on $F' \times F$ of a semi-normed space F is only continuous if F is normable [Vol. 1, Theorem 13.22]. $\square$

4.3. Sequential completeness of $\mathcal{D}'(\Omega; E)$

We show that $\mathcal{D}'(\Omega; E)$ is sequentially complete [3], i.e. that its Cauchy sequences converge.

---

3. **History of Theorem 4.5. Real values.** Laurent SCHWARTZ proved in 1950 [69, Chap. 3, § 2, Theorem XIII, p. 74] that, if a sequence in $\mathcal{D}'(\mathbb{R}^d)$ converges simply then its limit is a distribution, which implies that $\mathcal{D}'(\mathbb{R}^d)$ endowed with the simple topology used here is a Neumann space.

Vector values. Laurent SCHWARTZ proved in 1957 [72, Proposition 12, p. 50] that if E is complete or quasi-complete (which is stronger than the sequential completeness used here), then $\mathcal{D}'(\mathbb{R}^d; E)$-unif (that is, endowed with the topology of uniform convergence on the bounded subsets of $\mathcal{D}(\mathbb{R}^d)$, which is stronger than the simple topology used here) has the same property.

This does not imply that $\mathcal{D}'(\Omega; E)$ is sequentially complete if E is, which is the new property expressed by Theorem 4.5.

Theorem 4.5.– *Let Ω be an open subset of $\mathbb{R}^d$ and E a Neumann space. Then,*

$$\mathcal{D}'(\Omega; E) \text{ is a Neumann space.}$$

In particular, $\mathcal{D}'(\Omega)$ is a Neumann space. ▌

Proof. The space $\mathcal{D}'(\Omega; E)$ being semi-normed and separated (Theorem 3.2), it remains, by Definition 1.4 of a Neumann space, to show that it is sequentially complete. Thus, let

$$(f_n)_{n\in\mathbb{N}} \text{ be a Cauchy sequence in } \mathcal{D}'(\Omega; E).$$

1. *Obtaining a limit f.* From Definitions 1.3 (b) of a Cauchy sequence and 3.1 of the semi-norms of $\mathcal{D}'(\Omega; E)$, for every $\varphi \in \mathcal{D}(\Omega)$ and $\nu \in \mathcal{N}_E$, when $n \to \infty$,

$$\sup_{m\geq n} \|f_n - f_m\|_{\mathcal{D}'(\Omega;E);\varphi,\nu} = \sup_{m\geq n} \|\langle f_n - f_m, \varphi\rangle\|_{E;\nu} \to 0.$$

The sequence $(\langle f_n, \varphi\rangle)_{n\in\mathbb{N}}$ is therefore Cauchy in E so, since E is a Neumann space, it has a limit which we denote by $\langle f, \varphi\rangle$, i.e.

$$\langle f_n, \varphi\rangle \to \langle f, \varphi\rangle \text{ in } E.$$

Since this limit is unique (Theorem A.5) as E is separated, this defines a mapping f from $\mathcal{D}(\Omega)$ into E, which is linear since the f_n are.

2. *Property of the limit.* Let us check that f is a distribution. The sequence $(f_n)_{n\in\mathbb{N}}$ is bounded, as is every Cauchy sequence (Theorem A.5). Theorem 4.2 (b) then shows that it is equicontinuous; more precisely, it gives, for every $\nu \in \mathcal{N}_E$, the existence of $p_\nu \in \mathcal{C}^+(\Omega)$ such that, for every $\varphi \in \mathcal{D}(\Omega)$,

$$\sup_{n\in\mathbb{N}} \|\langle f_n, \varphi\rangle\|_{E;\nu} \leq \|\varphi\|_{\mathcal{D}(\Omega);p}.$$

At the limit,

$$\|\langle f, \varphi\rangle\|_{E;\nu} \leq \|\varphi\|_{\mathcal{D}(\Omega);p}.$$

So, due to the characterization of distributions from Theorem 3.3,

$$f \in \mathcal{D}'(\Omega; E).$$

3. *Convergence of f_n.* Since f is a distribution, Definition 3.1, again, gives

$$\|f_n - f\|_{\mathcal{D}'(\Omega;E);\varphi,\nu} = \|\langle f_n, \varphi\rangle - \langle f, \varphi\rangle\|_{E;\nu} \to 0.$$

Thus,

$$f_n \to f \text{ in } \mathcal{D}'(\Omega; E).$$

This proves that $\mathcal{D}'(\Omega; E)$ is indeed sequentially complete.

4. *The real case.* The space $\mathcal{D}'(\Omega) = \mathcal{D}'(\Omega; \mathbb{R})$ is Neumann, since $\mathbb{R}$ is a Banach space (Theorem A.26 (a)) and thus is Neumann by Definition 1.4. □

Non-metrizability. The space $\mathcal{D}'(\Omega; E)$ is not metrizable, and therefore is not a Fréchet space, except if Ω is empty or if E reduces to $\{0_E\}$ (in which case it is zero-dimensional) [SIMON, 84, to appear]. □

Non-completeness. The space $\mathcal{D}'(\Omega; E)$ is not complete, even if E is complete, when it is endowed with the simple topology as in the whole book. But, if E is quasi-complete, i.e. if its bounded subsets are complete, then $\mathcal{D}'(\Omega; E)$ is quasi-complete [SIMON, 84].

In contrast, $\mathcal{D}'(\Omega; E)$-unif, i.e. endowed with the topology of uniform convergence on the bounded subsets of $\mathcal{D}(\Omega)$, is complete if E is complete, see [SCHWARTZ, 72, Proposition 12, p. 50, with $m = \infty$] in the case where $\Omega = \mathbb{R}^d$. □

4.4. Sequential compactness in $\mathcal{D}'(\Omega; E)$

Let us characterize the relatively sequentially compact sets of distributions, i.e. those in which every sequence has a convergent subsequence [4].

Theorem 4.6.– *Let $\mathcal{F} \subset \mathcal{D}'(\Omega; E)$, where Ω is an open subset of $\mathbb{R}^d$ and E is a Neumann space. Then,*

every sequence in $\mathcal{F}$ has a subsequence which converges in $\mathcal{D}'(\Omega; E)$

if, and only if, for every $\varphi \in \mathcal{D}(\Omega)$,

$$\text{every sequence in } \{\langle f, \varphi\rangle : f \in \mathcal{F}\} \text{ has a subsequence which converges in } E. \quad (4.3)$$

■

Proof. Let $\{\| \|_{E;\nu} : \nu \in \mathcal{N}_E\}$ be the family of semi-norms of E.

1. Necessity of (4.3). Suppose that

every sequence in $\mathcal{F}$ has a subsequence which converges in $\mathcal{D}'(\Omega; E)$.

Let $\varphi \in \mathcal{D}(\Omega)$ and $(g_n)_{n\in\mathbb{N}}$ be a sequence in $\{\langle f, \varphi\rangle : f \in \mathcal{F}\}$, i.e. of the form, for every $n \in \mathbb{N}$,

$$g_n = \langle f_n, \varphi\rangle, \text{ where } f_n \in \mathcal{F}.$$

By the hypothesis, $(f_n)_{n\in\mathbb{N}}$ has a subsequence $(f_{\sigma(n)})_{n\in\mathbb{N}}$ which converges in $\mathcal{D}'(\Omega; E)$. Due to the characterization of convergent sequences in $\mathcal{D}'(\Omega; E)$ from Theorem 4.3 (c), $(\langle f_{\sigma(n)}, \varphi\rangle)_{n\in\mathbb{N}}$ converges in E.

That is, $(g_{\sigma(n)})_{n\in\mathbb{N}}$ converges in E, which proves (4.3).

4. **History of the characterization of the relatively sequentially compact sets.** Theorem 4.6 is new, to the knowledge of the author.

2. Sufficiency of (4.3). Suppose now that

every sequence in $\{\langle f, \varphi\rangle : f \in \mathcal{F}\}$ *has a subsequence which converges in* E.

Let $(f_n)_{n\in\mathbb{N}}$ be a sequence in $\mathcal{F}$. We will extract a convergent subsequence in three steps.

2.a. *Equicontinuity of* $\mathcal{F}$. For every $\varphi \in \mathcal{D}(\Omega)$, the set $\langle \mathcal{F}, \varphi\rangle$ is bounded in E, since a set in which every sequence has a convergent subsequence is bounded (see Theorem A.20). So, due to the characterization of bounded sets of distributions from Theorem 4.1 (c),

$$\mathcal{F} \text{ is bounded in } \mathcal{D}'(\Omega; E).$$

Theorem 4.2 (b) also proves that it is equicontinuous; more precisely, it provides, for every $\nu \in \mathcal{N}_E$, the existence of $p_\nu \in \mathcal{C}^+(\Omega)$ such that, for every $f \in \mathcal{F}$ and $\varphi \in \mathcal{D}(\Omega)$,

$$\|\langle f, \varphi\rangle\|_{E;\nu} \leq \|\varphi\|_{\mathcal{D}(\Omega);p_\nu}. \tag{4.4}$$

2.b. *Extraction of a subsequence* $(f_{\sigma_n(n)})_{n\in\mathbb{N}}$. Let

$$(\phi_k)_{k\in\mathbb{N}} \text{ be a sequentially dense sequence in } \mathcal{D}(\Omega),$$

provided by Theorem 2.16.

For every $k \in \mathbb{N}$, the sequence $(\langle f_n, \phi_k\rangle)_{n\in\mathbb{N}}$ possesses, by hypothesis, a convergent subsequence, say $(\langle f_{\sigma_k(n)}, \phi_k\rangle)_{n\in\mathbb{N}}$. By successive extractions, we can choose nested subsequences, i.e. such that $(\sigma_{k+1}(n))_{n\in\mathbb{N}}$ is a subsequence of $(\sigma_k(n))_{n\in\mathbb{N}}$. Then [5],

$$\text{the diagonal sequence } (\langle f_{\sigma_n(n)}, \phi_k\rangle)_{n\in\mathbb{N}} \text{ converges for every } k, \tag{4.5}$$

as does every subsequence of a convergent sequence (Theorem A.5; here, it is a subsequence of $(\langle f_{\sigma_k(n)}, \phi_k\rangle)_{n\geq k}$).

2.c. *Convergence of* $(f_{\sigma_n(n)})_{n\in\mathbb{N}}$. Let us denote $\sigma(n) \stackrel{\text{def}}{=} \sigma_n(n)$. Given $\varphi \in \mathcal{D}(\Omega)$, decompose

$$\langle f_{\sigma(m)} - f_{\sigma(n)}, \varphi\rangle = \langle f_{\sigma(m)} - f_{\sigma(n)}, \phi_k\rangle + \langle f_{\sigma(m)}, \varphi - \phi_k\rangle - \langle f_{\sigma(n)}, \varphi - \phi_k\rangle.$$

For $\nu \in \mathcal{N}_E$ and $\epsilon > 0$, let us estimate the right-hand side by choosing first $k \in \mathbb{N}$ such that

$$\|\varphi - \phi_k\|_{\mathcal{D}(\Omega);p_\nu} \leq \frac{\epsilon}{3}.$$

5. **History of Cantor's diagonal process.** The construction of a diagonal subsequence of nested sequences was introduced by Georg CANTOR, see his works [15].

Then, thanks to (4.5), we choose n_0 large enough so that, for all $m \geq n_0$ and $n \geq n_0$,

$$\|\langle f_{\sigma(m)} - f_{\sigma(n)}, \phi_k \rangle\|_{E;\nu} \leq \frac{\epsilon}{3}.$$

Then, with (4.4),

$$\|\langle f_{\sigma(m)} - f_{\sigma(n)}, \varphi \rangle\|_{E;\nu} \leq \epsilon$$

That is $\|f_{\sigma(m)} - f_{\sigma(n)}\|_{\mathcal{D}'(\Omega;E);\varphi,\nu} \leq \epsilon$, by Definition 3.1 of the semi-norms of $\mathcal{D}'(\Omega; E)$. By Definition 1.3 (b) of a Cauchy sequence, this proves that

$$(f_{\sigma(n)})_{n\in\mathbb{N}} \text{ is a Cauchy sequence in } \mathcal{D}'(\Omega; E).$$

It converges since $\mathcal{D}'(\Omega; E)$ is sequentially complete (Theorem 4.5). Thus,

$$(f_{\sigma(n)})_{n\in\mathbb{N}} \text{ is a convergent subsequence of } (f_n)_{n\in\mathbb{N}}. \ \square$$

Let us show that the bounded subsets of $\mathcal{D}'(\Omega)$ are relatively sequentially compact[6].

Theorem 4.7.– *Every bounded sequence in $\mathcal{D}'(\Omega)$, where Ω is an open subset of $\mathbb{R}^d$, has a subsequence that converges.* ▮

Proof. Let

$$(f_n)_{n\in\mathbb{N}} \text{ be a bounded sequence in } \mathcal{D}'(\Omega).$$

For every $\varphi \in \mathcal{D}(\Omega)$, the set $\{\langle f_n, \varphi\rangle : n \in \mathbb{N}\}$ is bounded in $\mathbb{R}$ according to the characterization of bounded sets of distributions from Theorem 4.1 (c). Therefore, due to the Bolzano–Weierstrass theorem (Theorem A.26 (c)),

$$\text{every sequence in } \{\langle f_n, \varphi\rangle : n \in \mathbb{N}\} \text{ has a convergent subsequence.}$$

So, the characterization of relatively sequentially compact sets of distributions from Theorem 4.6 then shows that

$$(f_n)_{n\in\mathbb{N}} \text{ has a convergent subsequence. } \square$$

6. **History of Theorem 4.7.** Laurent SCHWARTZ proved in 1950 [69, Chap. 3, § 3, Theorem XII, p. 74] that, in $\mathcal{D}'(\mathbb{R}^d)$-unif, every bounded set is relatively compact.
A priori, this does not imply the result of Theorem 4.7, because there are semi-normed spaces in which *relatively compact* **does not imply** *relatively sequentially compact* [Vol. 1, Properties (2.6) and (2.7), p. 27]. Observe however that, in $\mathcal{D}'(\mathbb{R}^d)$, these two notions of compactness coincide due to Property (4.7).

Compact sets of distributions. The relatively compact sets of distributions can be characterized, for every subset Ω of $\mathbb{R}^d$ and every Neumann space E, as follows [SIMON, 84, to appear]:

$$\begin{cases} \textit{A set } \mathcal{F} \textit{ is relatively compact in } \mathcal{D}'(\Omega; E) \textit{ if and only if, for every} \\ \varphi \in \mathcal{D}(\Omega), \textit{ the set } \{\langle f, \varphi \rangle : f \in \mathcal{F}\} \textit{ is relatively compact in } E. \end{cases} \tag{4.6}$$

In $\mathcal{D}'(\Omega; E)$, as in every space, a set is compact if in addition it is closed.

Since *relatively compact* coincides with *relatively sequentially compact* in a metrizable space E (Theorem A.23 (a)), it results, with the characterization from Theorem 4.6, that:

$$\begin{cases} \textit{If } E \textit{ is a Fréchet space, a set is relatively compact in } \mathcal{D}'(\Omega; E) \\ \textit{if and only if it is relatively sequentially compact.} \end{cases} \tag{4.7}$$

□

Compactness versus sequential compactness. There are separated semi-normed spaces in which *compactness* is neither stronger nor weaker than *sequential compactness* [Vol. 1, Properties (2.6) and (2.7), p. 27]. If it is the case for a space E, it is equally the case for $\mathcal{D}'(\Omega; E)$ [SIMON, 84]. □

4.5. Change of the space E of values

We show here that, if the space E of values is included in a "larger" space, the same holds for the corresponding spaces of distributions. Let us begin with the case of a topological inclusion.

Theorem 4.8.– *Let Ω be an open subset of $\mathbb{R}^d$, and E and F two Neumann spaces such that*

$$E \underset{\to}{\subseteq} F.$$

Then,

$$\mathcal{D}'(\Omega; E) \underset{\to}{\subseteq} \mathcal{D}'(\Omega; F). \blacksquare$$

Proof. Denote by $\{\| \ \|_{E;\nu} : \nu \in \mathcal{N}_E\}$ and $\{\| \ \|_{F;\mu} : \mu \in \mathcal{N}_F\}$ the families of semi-norms of E and F.

By Definition 1.7 (c) of topological inclusion, for every $\mu \in \mathcal{N}_F$, there exist a finite subset N of $\mathcal{N}_E$ and $c \in \mathbb{R}$ such that, for every $u \in E$,

$$\|u\|_{F;\mu} \le c \sup_{\nu \in N} \|u\|_{E;\nu}.$$

Then, by Definition 3.1 of the semi-norms of $\mathcal{D}'(\Omega; E)$, for every $f \in \mathcal{D}'(\Omega; F)$ and $\varphi \in \mathcal{D}(\Omega)$,

$$\|f\|_{\mathcal{D}'(\Omega;F);\varphi,\mu} = \|\langle f, \varphi \rangle\|_{F;\mu} \le c \sup_{\nu \in N} \|\langle f, \varphi \rangle\|_{E;\nu} = c \sup_{\nu \in N} \|f\|_{\mathcal{D}'(\Omega;E);\varphi,\nu}.$$

Which, again according to Definition 1.7 (c), gives

$$\mathcal{D}'(\Omega; E) \underset{\to}{\subseteq} \mathcal{D}'(\Omega; F). \ \square$$

Let us come to the case of an inclusion that is only "sequentially continuous".

Theorem 4.9.– *Let Ω be an open subset of $\mathbb{R}^d$, and E and F two Neumann spaces such that E is a vector subspace of F and*

$$\textit{the identity is sequentially continuous from } E \textit{ into } F.$$

Then, $\mathcal{D}'(\Omega; E)$ is a vector subspace of $\mathcal{D}'(\Omega; F)$ and

$$\textit{the identity is sequentially continuous from } \mathcal{D}'(\Omega; E) \textit{ into } \mathcal{D}'(\Omega; F). \blacksquare$$

Proof. **1.** *Inclusion of $\mathcal{D}'(\Omega; E)$ into $\mathcal{D}'(\Omega; F)$.* Let

$$f \in \mathcal{D}'(\Omega; E).$$

Due to the characterization of distributions from Theorem 3.4 (b), f is sequentially continuous from $\mathcal{D}(\Omega)$ into E. That is, $\varphi_n \to \varphi$ in $\mathcal{D}(\Omega)$ implies $\langle f, \varphi_n \rangle \to \langle f, \varphi \rangle$ in E, and so into F by hypothesis. Being linear from $\mathcal{D}(\Omega)$ into E, and thus into F, it follows, again from Theorem 3.4 (b), that

$$f \in \mathcal{D}'(\Omega; F).$$

So, $\mathcal{D}'(\Omega; E)$ is included in $\mathcal{D}'(\Omega; F)$. And it is a vector subspace.

2. *Continuity of identity.* Let

$$f_n \to f \ \text{ in } \mathcal{D}'(\Omega; E).$$

By the characterization of convergent sequences of distributions from Theorem 4.3 (c), for every $\varphi \in \mathcal{D}(\Omega)$, $\langle f_n, \varphi \rangle \to \langle f, \varphi \rangle$ in E, and thus in F by hypothesis. Then, again from Theorem 4.3 (c),

$$f_n \to f \ \text{ in } \mathcal{D}'(\Omega; F).$$

This proves the sequential continuity of the identity from $\mathcal{D}'(\Omega; E)$ into $\mathcal{D}'(\Omega; F)$. □

Other proofs. Theorems 4.8 and 4.9 are particular cases of Properties (b) and (a) of Theorem 5.14, on the continuity in distribution spaces of the mapping $f \mapsto Lf$ associated with a linear mapping L, in this case the identity from E into F. □

Let us finish with the case [7] of a "sequentially compacting" topological inclusion, i.e. such that the bounded subsets of E are relatively sequentially compact in F.

7. **History of Theorem 4.10.** This result is new to our knowledge, albeit simple.

Theorem 4.10.– *Let Ω be an open subset of $\mathbb{R}^d$, and E and F two Neumann spaces such that $E \subsetneq F$ and*

every bounded sequence in E has a subsequence which converges in F.

Then, $\mathcal{D}'(\Omega; E) \subsetneq \mathcal{D}'(\Omega; F)$ and

$$\begin{cases} \text{every bounded sequence in } \mathcal{D}'(\Omega; E) \text{ has a} \\ \text{subsequence which converges in } \mathcal{D}'(\Omega; F). \end{cases}$$ ▮

Proof. The inclusion $\mathcal{D}'(\Omega; E) \subsetneq \mathcal{D}'(\Omega; F)$ being given by Theorem 4.8 since $E \subsetneq F$, it remains to establish the existence of convergent subsequences. Thus, let

$$(f_n)_{n\in\mathbb{N}} \text{ be a bounded sequence in } \mathcal{D}'(\Omega; E).$$

To avoid any ambiguity, denote by $\dot{f}_n$ the mapping from $\mathcal{D}(\Omega)$ into F. According to the characterization of bounded sets of distributions from Theorem 4.1 (c), for every $\varphi \in \mathcal{D}(\Omega)$, the sequence $(\langle f_n, \varphi\rangle)_{n\in\mathbb{N}}$ is bounded in E, therefore, by hypothesis,

$$(\langle \dot{f}_n, \varphi\rangle)_{n\in\mathbb{N}} \text{ has a subsequence which converges in } F.$$

The characterization of relatively sequentially compact sets of distributions from Theorem 4.6 then shows that

$$(\dot{f}_n)_{n\in\mathbb{N}} \text{ has a subsequence which converges in } \mathcal{D}'(\Omega; F).$$ □

Another formulation of Theorem 4.10. If the injection from E into F is **sequentially compacting**, the same is true of the injection from $\mathcal{D}'(\Omega; E)$ into $\mathcal{D}'(\Omega; F)$.

Recall that a mapping is said to be *sequentially compacting* if it transforms bounded sets into relatively sequentially compact sets [Vol. 1, Definition 8.14]. It is a **new definition**, so the reader will need to recall its meaning when using it. For normed spaces, this notion coincides [Vol. 1, Property (8.6), p. 127] with *compact*, namely [SCHWARTZ, 73, Definition 101, p. 419] with the existence of an open set whose image is relatively compact (or relatively sequentially compact). But, in general, the notion of compact mapping is too strong for our needs as, in an infinite-dimensional space, no relatively sequentially compact open sets exist. □

4.6. The space E-weak

Before studying distributions with values in E-weak in the next paragraph, we define here this space and give its properties that we will need.

Let us define the weak topology of a separated semi-normed space E, or more precisely the semi-norms that generate it. They involve the **dual** E' of E, namely the set of continuous linear mappings from E into $\mathbb{R}$. Denote

$$\langle e', e\rangle \overset{\text{def}}{=} e'(e) \text{ when } e' \in E' \text{ and } e \in E.$$

Definition 4.11.– *Let E be a separated semi-normed space. We denote by E-weak the vector space E endowed with the semi-norms, indexed by $e' \in E'$,*

$$\|e\|_{E\text{-weak};e'} \stackrel{\text{def}}{=} |\langle e', e\rangle|. \blacksquare$$

Recall that E-weak is separated [Vol. 1, Theorem 15.2], since, if $\langle e', e\rangle = 0$ for every $e' \in E'$, then $e = 0_E$ according to a corollary [Vol. 1, Theorem 14.4] of the Hahn–Banach theorem. Let us compare its topology with the initial topology.

Theorem 4.12.– *For every separated semi-normed space E,*

$$E \underset{\rightarrow}{\subsetneq} E\text{-weak}. \blacksquare$$

Proof. Let $e' \in E'$. Due to the characterization of continuous linear mappings from Theorem 1.12 (a), there exist a finite subset N of $\mathcal{N}_E$ and $c \in \mathbb{R}$ such that, for every $e \in E$, we have $|\langle e', e\rangle| \leq c \sup_{\nu \in N} \|e\|_{E;\nu}$. That is, by Definition 4.11,

$$\|e\|_{E\text{-weak};e'} \leq c \sup_{\nu \in N} \|e\|_{E;\nu}.$$

This proves that $E \underset{\rightarrow}{\subsetneq} E$-weak, by Definition 1.7 (c) of topological inclusion. □

Let us show that E is sequentially complete if E-weak is [8].

Theorem 4.13.– *Let E be a separated semi-normed space.*

If E-weak is a Neumann space, then E is a Neumann space. ▮

Proof. Let

$$(u_n)_{n \in \mathbb{N}} \text{ be a Cauchy sequence in } E.$$

Since $E \underset{\rightarrow}{\subsetneq} E$-weak (Theorem 4.12), this sequence is Cauchy in E-weak (Theorem A.6). If E-weak is a Neumann space, it converges there to a limit, say

$$u_n \to u \text{ in } E\text{-weak}.$$

Let $\| \ \|_{E;\nu}$ be a semi-norm of E and $\epsilon > 0$. By Definition 1.3 (b) of a Cauchy sequence in E, there exists $k \in \mathbb{N}$ such that, for every $n \geq k$,

$$\|u_n - u_k\|_{E;\nu} \leq \epsilon. \tag{4.8}$$

8. **History of Theorem 4.13.** It was known that, if E-weak is quasi-complete, i.e. if its bounded sets are complete, then so is E, see for example [SCHAEFER, 67, 5.5 (a) and (e) and Corollary 1, p. 144], where authorship is not specified.
This does not imply that E is sequentially complete if E-weak is (i.e. Theorem 4.13, new to our understanding), since quasi-completeness is stronger than sequential completeness.

Consider

$$U \stackrel{\text{def}}{=} \{v \in E : \|v - u_k\|_{E;\nu} \le \epsilon\}.$$

It is a closed convex subset of E. Like any closed convex subset of E, it is closed in E-weak due to Mazur's theorem (Theorem A.42). Since any closed set is sequentially closed (Theorem A.10),

$$U \text{ is sequentially closed in } E\text{-weak.}$$

Since u_n belongs to U for each $n \ge k$, its limit u also belongs to it, i.e.

$$\|u - u_k\|_{E;\nu} \le \epsilon.$$

With (4.8), this gives $\|u - u_n\|_{E;\nu} \le 2\epsilon$, for every $n \ge k$. This holds for every semi-norm of E and every $\epsilon > 0$ (with k depending on ϵ and the semi-norm). Thus,

$$u_n \to u \text{ in } E.$$

This proves that E is sequentially complete, that is Neumann. □

Some cases where E-weak **is a Neumann space.** Recall that E-weak is a Neumann space if E is a Hilbert space, or is reflexive or semi-reflexive [Vol. 1, Theorems 17.7 and 17.12].

Theorem 4.13 shows that E must be Neumann for E-weak to be so. Observe that this is not sufficient: for example, $L^1(\mathbb{R})$ is a Banach space, but $L^1(\mathbb{R})$-weak is not a Neumann space; indeed, the regularizing sequence $(\rho_n)_{n\in\mathbb{N}}$ in Definition 8.1 is Cauchy in $L^1(\mathbb{R})$-weak but does not converge in it, since it converges to δ_0 in $\mathcal{D}'(\mathbb{R})$ (Theorem 8.3). □

4.7. The space $\mathcal{D}'(\Omega; E\text{-weak})$ and extractability

Let us show that every **distribution with values in** E-weak[9] is a distribution with values in E.

Theorem 4.14.– *Let Ω be an open subset of $\mathbb{R}^d$ and E a separated semi-normed space such that*

$$E\text{-weak } \textit{is a Neumann space.}$$

Then, E is a Neumann space,

$$\mathcal{D}'(\Omega; E) = \mathcal{D}'(\Omega; E\text{-weak}),$$

these two spaces have the same bounded sets, and

$$\mathcal{D}'(\Omega; E) \subsetneqq \mathcal{D}'(\Omega; E\text{-weak}). \blacksquare$$

9. **History of Theorem 4.14.** The equality $\mathcal{D}'(\mathbb{R}^d; E) = \mathcal{D}'(\mathbb{R}^d; E\text{-weak})$ is implicit in *Théorie des distributions à valeurs vectorielles* by Laurent SCHWARTZ [72, Property (ε), p. 53, and Examples, p. 55].

Proof. **1.** *Sequential completeness of E.* As the space E-weak is Neumann, E is as well due to Theorem 4.13. This allows us to define $\mathcal{D}'(\Omega; E)$.

2. *Algebraic equality.* A distribution in $\mathcal{D}'(\Omega; E)$ is, from Theorem 3.4 (e), a linear mapping from $\mathcal{D}(\Omega)$ into E which, for every compact subset K of Ω, transforms the bounded subsets of $\mathcal{C}^\infty_K(\Omega)$ into bounded subsets of E.

Therefore, since E and E-weak are algebraically equal (by Definition 4.11 of E-weak) and have the same bounded subsets due to the Banach–Mackey theorem (Theorem A.41),

$$\mathcal{D}'(\Omega; E) = \mathcal{D}'(\Omega; E\text{-weak}).$$

3. *Topological inclusion.* Since $E \underset{\to}{\subseteq} E$-weak (Theorem 4.12), Theorem 4.8 on the change of space of values gives

$$\mathcal{D}'(\Omega; E) \underset{\to}{\subseteq} \mathcal{D}'(\Omega; E\text{-weak}).$$

4. *Equality of the families of bounded sets.* Due to the characterization from Theorem 4.1 (c), a bounded subset of $\mathcal{D}'(\Omega; E)$ is a set $\mathcal{F}$ such that, for every $\varphi \in \mathcal{D}(\Omega)$,

$$\{\langle f, \varphi\rangle : f \in \mathcal{F}\} \text{ is bounded in } E.$$

Here, we can replace E with E-weak since they have the same bounded sets, as we have just recalled, so

$$\mathcal{D}'(\Omega; E) \text{ and } \mathcal{D}'(\Omega; E\text{-weak}) \text{ have the same bounded sets. } \square$$

Let us show that, if every bounded sequence in E has a convergent subsequence in E-weak, then every bounded sequence in $\mathcal{D}'(\Omega; E)$ has a convergent subsequence in $\mathcal{D}'(\Omega; E\text{-weak})$.

Theorem 4.15.– *Let Ω be an open subset of $\mathbb{R}^d$ and E a separated semi-normed space such that*

$$\left\{\begin{array}{l}\textit{every bounded sequence in } E \textit{ has a}\\ \textit{subsequence which converges in } E\text{-weak.}\end{array}\right. \tag{4.9}$$

Then, E and E-weak are Neumann spaces and

$$\left\{\begin{array}{l}\textit{every bounded sequence in } \mathcal{D}'(\Omega; E) \textit{ has a}\\ \textit{subsequence which converges in } \mathcal{D}'(\Omega; E\text{-weak}).\end{array}\right. \blacksquare$$

Proof. Hypothesis (4.9) means that E is extractable (Definition A.43), which implies (Theorem A.44) that E-weak is a Neumann space. Then (Theorem 4.13), E is also a Neumann space.

Since $E \underset{\rightarrow}{\subseteq} E$-weak (Theorem 4.12), this allows us to apply Theorem 4.10 with $F = E$-weak, which gives it the stated property. □

Terminology. Although their property is used abundantly in the literature, extractable spaces were not named and were not studied as such, which we did in Volume 1 [81]. □

For real-valued distributions, we now show that the simple topology on $\mathcal{D}'(\Omega)$ which we use coincides with its weak topology.

Theorem 4.16.– *Let Ω be an open subset of $\mathbb{R}^d$. Then,*

$$\mathcal{D}'(\Omega)\text{-weak} \underset{\leftrightarrows}{=} \mathcal{D}'(\Omega). \ ▮$$

Proof. Given $\varphi \in \mathcal{D}(\Omega)$, we define $T \in (\mathcal{D}'(\Omega))'$ by: for every $f \in \mathcal{D}'(\Omega)$,

$$T(f) \overset{\text{def}}{=} \langle f, \varphi \rangle_{\mathcal{D}'(\Omega) \times \mathcal{D}(\Omega)}.$$

Indeed, T is linear and, by Definition 3.1 of the semi-norms of $\mathcal{D}'(\Omega)$,

$$|T(f)| = \|f\|_{\mathcal{D}'(\Omega);\varphi},$$

which implies that T is continuous from $\mathcal{D}'(\Omega)$ into $\mathbb{R}$ due to the characterization of continuous linear mappings from Theorem 1.12 (a).

Hence, with Definition 4.11 of the semi-norms of the weak topology,

$$\|f\|_{\mathcal{D}'(\Omega);\varphi} = |\langle T, f \rangle_{(\mathcal{D}'(\Omega))' \times \mathcal{D}'(\Omega)}| = \|f\|_{\mathcal{D}'(\Omega)\text{-weak};T}.$$

By Definition 1.7 (c) of topological inclusion, this proves that

$$\mathcal{D}'(\Omega)\text{-weak} \underset{\rightarrow}{\subseteq} \mathcal{D}'(\Omega).$$

The converse inclusion is given by Theorem 4.12. □

Reflexivity. In this book, we do not address the properties of the reflexivity of $\mathcal{D}'(\Omega; E)$ — they are in [SIMON, 84, to appear] —, since they do not serve us when studying PDEs in the following volumes. In contrast, we will use the extractability properties of subsequences from Theorems 4.7 and 4.15.

Semi-reflexivity and extractability are related but distinct properties: the semi-reflexivity of a space E is equivalent to the *relative compactness* in E-weak of its bounded subsets (Banach-Alaoglu–Bourbaki theorem, see Theorem 17.19 in Volume 1), while extractability is equivalent (Definition A.43) to them being *relatively sequentially compact*, again in E-weak.

Observe that $\mathcal{D}'(\Omega)$, endowed with the simple topology as in the whole book, is semi-reflexive but is not reflexive. On the other hand, $\mathcal{D}'(\Omega)$-unif, i.e. endowed with the topology of uniform convergence on the bounded subsets of $\mathcal{D}(\Omega)$ used by Laurent SCHWARTZ, is reflexive. □

Chapter 5

Operations on Distributions

This chapter is dedicated to the derivatives $\partial_i f$ of a distribution $f \in \mathcal{D}'(\Omega; E)$, to its image Lf under a linear mapping L from E into another Neumann space F, to its product αf with a function $\alpha \in \mathcal{C}^\infty(\Omega)$, and to its image $f \circ T$ after a change of variable $T \in \mathcal{C}^\infty(\Lambda; \Omega)$. For each of these four operations:

— we verify that, when $f \in \mathcal{C}(\Omega; E)$, we obtain the usual operation;

— we study its continuity and its "interactions" with the previous operations; for example, the interaction of the derivative with the product is the Leibniz formula $\partial_i(\alpha f) = \alpha \partial_i f + \partial_i \alpha\, f$.

We equally study positive real distributions, and prove that these are measures (Theorem 5.34).

We begin by introducing *distributions fields*, namely distributions with values in E^d, since gradient is such a field, and their separation into components (§ 5.1). In § 5.8, we more generally study distributions with values in a product space $E_1 \times \cdots \times E_I$.

The results of this chapter are "natural" enough and their proofs are simple.

5.1. Distributions fields

We call a **field**, or a **distribution field** if there is a risk of ambiguity, a distribution with values in the Euclidean product space E^d. In the case of functions, we will call it a **vector field** if necessary.

Recall that, if E is a Neumann space, E^d is also a Neumann space as we have established in Volume 1 [81, Theorem 6.12], and therefore $\mathcal{D}'(\Omega; E^d)$ is well-defined. Also recall that E^d is endowed with the semi-norms, indexed by $\nu \in \mathcal{N}_E$,

$$\|u\|_{E^d;\nu} \stackrel{\text{def}}{=} (\|u_1\|^2_{E;\nu} + \cdots + \|u_d\|^2_{E;\nu})^{1/2}. \tag{5.1}$$

Let us define the **components** $(q_1, \ldots, q_d)$ of a distribution field q.

Definition 5.1.– *Let*

$$q \in \mathcal{D}'(\Omega; E^d),$$

where Ω is an open subset of $\mathbb{R}^d$ and E is a Neumann space. We define

$$(q_1, \dots, q_d) \in (\mathcal{D}'(\Omega; E))^d,$$

by: for every $i \in [\![1, d]\!]$ and $\varphi \in \mathcal{D}(\Omega)$,

$$\langle q_i, \varphi \rangle \stackrel{\text{def}}{=} (\langle q, \varphi \rangle)_i. \blacksquare$$

Justification. Due to the characterization of distributions from Theorem 3.3, for every semi-norm $\| \ \|_{E^d;\nu}$ of E^d, there exist $p \in \mathcal{C}^+(\Omega)$ and $c \in \mathbb{R}$ such that, for every $\varphi \in \mathcal{D}(\Omega)$,

$$\|\langle q, \varphi \rangle\|_{E^d;\nu} \leq c\|\varphi\|_{\mathcal{D}(\Omega);p}.$$

From Definition (5.1) of the semi-norms of E^d, the left-hand side is greater than $\|(\langle q, \varphi \rangle)_i\|_{E;\nu}$, which is $\|\langle q_i, \varphi \rangle\|_{E;\nu}$ by definition of q_i, thus the characterization from Theorem 3.3, again, shows that we indeed have

$$q_i \in \mathcal{D}'(\Omega; E). \ \square$$

Utility of separating a field into its components. Fields serve principally to represent the gradient, in particular in the search for primitives. Their separation into components is useful, for example, to write

$$\nabla f = (\partial_1 f, \dots, \partial_d f).$$

These two expressions are equivalent, since the separation into components is an isomorphism from $\mathcal{D}'(\Omega; E^d)$ onto $(\mathcal{D}'(\Omega; E))^d$. Depending on the context, one or the other of the above expressions, and therefore one or the other of the spaces, is easier to use. We prefer to use $\mathcal{D}'(\Omega; E^d)$, if possible, since:
— it is a distribution space, so the properties of distributions directly apply to it, while $(\mathcal{D}'(\Omega; E))^d$, which is a product of distribution spaces, moreover requires properties of product of semi-normed spaces;
— it is more legible than $(\mathcal{D}'(\Omega; E))^d$, which suffers from having nested parentheses. $\square$

Let us specify the families of semi-norms of $\mathcal{D}'(\Omega; E^d)$ and $(\mathcal{D}'(\Omega; E))^d$.

Definition 5.2.– *Let Ω be an open subset of $\mathbb{R}^d$ and E a Neumann space whose family of semi-norms is denoted by $\{\| \ \|_{E;\nu} : \nu \in \mathcal{N}_E\}$. Then:*

(a) *$\mathcal{D}'(\Omega; E^d)$ is endowed with the semi-norms, indexed by $\varphi \in \mathcal{D}(\Omega)$ and $\nu \in \mathcal{N}_E$,*

$$\|q\|_{\mathcal{D}'(\Omega;E^d);\varphi,\nu} \stackrel{\text{def}}{=} \|\langle q, \varphi \rangle\|_{E^d;\nu} \stackrel{\text{def}}{=} \stackrel{\text{def}}{=} (\|(\langle q, \varphi \rangle)_1\|^2_{E;\nu} + \dots + \|(\langle q, \varphi \rangle)_d\|^2_{E;\nu})^{1/2}.$$

(b) *$(\mathcal{D}'(\Omega; E))^d$ is endowed with the semi-norms, indexed by $\varphi \in \mathcal{D}(\Omega)$ and $\nu \in \mathcal{N}_E$,*

$$\|(q_1, \dots, q_d)\|_{(\mathcal{D}'(\Omega;E))^d;\varphi,\nu} \stackrel{\text{def}}{=} (\|q_1\|^2_{\mathcal{D}'(\Omega;E);\varphi,\nu} + \dots + \|q_d\|^2_{\mathcal{D}'(\Omega;E);\varphi,\nu})^{1/2} \stackrel{\text{def}}{=} \stackrel{\text{def}}{=} (\|\langle q_1, \varphi \rangle\|^2_{E;\nu} + \dots + \|\langle q_d, \varphi \rangle\|^2_{E;\nu})^{1/2}. \blacksquare$$

Justification. These expressions follow from Definitions 3.1 of the semi-norms of $\mathcal{D}'(\Omega; E)$ and (5.1) of the semi-norms of E^d. □

Let us show that the separation into components is an isomorphism.

Theorem 5.3.– *Let Ω be an open subset of $\mathbb{R}^d$ and E a Neumann space.*

Then, the separation into components is an isomorphism from $\mathcal{D}'(\Omega; E^d)$ onto $(\mathcal{D}'(\Omega; E))^d$. ▮

Proof. Let $\{\| \ \|_{E;\nu} : \nu \in \mathcal{N}_E\}$ be the family of semi-norms of E.

1. *Injectivity of $q \mapsto (q_1, \ldots, q_d)$.* Let

$$q \in \mathcal{D}'(\Omega; E^d)$$

such that

$$(q_1, \ldots, q_d) = 0.$$

For every $i \in [\![1, d]\!]$ and $\varphi \in \mathcal{D}(\Omega)$, we have $\langle q_i, \varphi \rangle = 0_E$, i.e. $(\langle q, \varphi \rangle)_i = 0_E$ by Definition 5.1 of q_i. So, $\langle q, \varphi \rangle = 0_{E^d}$, from which

$$q = 0.$$

This proves that the mapping $q \mapsto (q_1, \ldots, q_d)$ is injective.

2. *Surjectivity.* Now let

$$(q_1, \ldots, q_d) \in (\mathcal{D}'(\Omega; E))^d.$$

We define a mapping q from $\mathcal{D}(\Omega)$ into E^d by: for every $\varphi \in \mathcal{D}(\Omega)$,

$$\langle q, \varphi \rangle \stackrel{\text{def}}{=} (\langle q_1, \varphi \rangle, \ldots, \langle q_d, \varphi \rangle).$$

Due to the characterization of distributions from Theorem 3.3, for every $i \in [\![1, d]\!]$ and $\nu \in \mathcal{N}_E$, there exist $p_{i,\nu} \in \mathcal{C}^+(\Omega)$ and $c_{i,\nu} \in \mathbb{R}$ such that, for every $\varphi \in \mathcal{D}(\Omega)$,

$$\|\langle q_i, \varphi \rangle\|_{E;\nu} \leq c_{i,\nu} \|\varphi\|_{\mathcal{D}(\Omega);p_{i,\nu}}.$$

Then,

$$\begin{aligned}\|\langle q, \varphi \rangle\|_{E^d;\nu} &= (\|\langle q_1, \varphi \rangle\|_{E;\nu}^2 + \cdots + \|\langle q_d, \varphi \rangle\|_{E;\nu}^2)^{1/2} \leq \\ &\leq (c_{1,\nu} + \cdots + c_{d,\nu})(\|\varphi\|_{\mathcal{D}(\Omega);p_{1,\nu}} + \cdots + \|\varphi\|_{\mathcal{D}(\Omega);p_{d,\nu}}).\end{aligned}$$

This implies, according to the characterization of continuous linear mappings from Theorem 1.12 (a), that q, which is linear, is continuous from $\mathcal{D}(\Omega)$ into E^d, i.e.

$$q \in \mathcal{D}'(\Omega; E^d).$$

This proves that the mapping $q \mapsto (q_1, \ldots, q_d)$ is surjective. It is therefore bijective, due to step 1.

3. *Continuity.* Given $q \in \mathcal{D}'(\Omega; E^d)$, Definitions 5.2 (b) of the semi-norms of $(\mathcal{D}'(\Omega; E))^d$, 5.1 of the components, and 5.2 (a) of the semi-norms of $\mathcal{D}'(\Omega; E^d)$ successively give, for every $\varphi \in \mathcal{D}(\Omega)$ and $\nu \in \mathcal{N}_E$,

$$\|(q_1, \ldots, q_d)\|_{(\mathcal{D}'(\Omega;E))^d;\varphi,\nu} = \Big(\sum_{i=1}^{d} \|\langle q_i, \varphi\rangle\|_{E;\nu}^2\Big)^{1/2} =$$
$$= \Big(\sum_{i=1}^{d} \|(\langle q, \varphi\rangle)_i\|_{E;\nu}^2\Big)^{1/2} = \|q\|_{\mathcal{D}'(\Omega;E)^d;\varphi,\nu}.$$

Therefore, again according to the characterization of continuous linear mappings from Theorem 1.12 (a), the mapping $q \mapsto (q_1, \ldots, q_d)$, which is linear, is continuous from $\mathcal{D}'(\Omega; E^d)$ into $(\mathcal{D}'(\Omega; E))^d$.

4. *Isomorphism.* The above equality also shows that the inverse mapping $(q_1, \ldots, q_d) \mapsto q$ is continuous, since the families of semi-norms of these two spaces are, from their Definition 5.2, indexed by the same set $\mathcal{D}(\Omega) \times \mathcal{N}_E$.

The mapping $q \mapsto (q_1, \ldots, q_d)$ is therefore an isomorphism. □

5.2. Derivatives of a distribution

Now let us define the partial derivatives [1] of a distribution.

Definition 5.4.– *Let $f \in \mathcal{D}'(\Omega; E)$, where Ω is an open subset of $\mathbb{R}^d$ and E is a Neumann space.*

1. **History of the partial derivatives of a distribution.** Laurent SCHWARTZ gave Definition 5.4 of the partial derivatives of a distribution of $\mathcal{D}'(\mathbb{R}^d)$ in 1945 [68, p. 61]. He extended it to distributions of $\mathcal{D}'(\mathbb{R}^d; E)$ in 1957 [72, p. 68].
Jean LERAY had previously used the same definition in 1934 to define the *quasi-derivatives* in $L^2(\mathbb{R}^3)$ of functions of $L^2(\mathbb{R}^3)$ [53, p. 205], and more precisely of the functions that possess such quasi-derivatives.
Sergei SOBOLEV, too, had previously defined in 1936 [86, p. 62] derivatives that coincide with the derivatives in the distribution sense.

(a) *For $i \in [\![1, d]\!]$, we define the* **derivative** $\partial_i f \in \mathcal{D}'(\Omega; E)$ *by: for every* $\varphi \in \mathcal{D}(\Omega)$,

$$\langle \partial_i f, \varphi \rangle \stackrel{\text{def}}{=} -\langle f, \partial_i \varphi \rangle.$$

(b) *For $\beta \in \mathbb{N}^d$, we define $\partial^\beta f \in \mathcal{D}'(\Omega; E)$ by:*

$$\partial^\beta f \stackrel{\text{def}}{=} \partial_1^{\beta_1} \cdots \partial_d^{\beta_d} f, \quad \partial^0 f \stackrel{\text{def}}{=} f.$$

(c) *We define the* **gradient** $\nabla f \in \mathcal{D}'(\Omega; E^d)$ *by: for every* $\varphi \in \mathcal{D}(\Omega)$,

$$\langle \nabla f, \varphi \rangle \stackrel{\text{def}}{=} (\langle \partial_1 f, \varphi \rangle, \ldots, \langle \partial_d f, \varphi \rangle). \blacksquare$$

Justification. **1.** *Justification of (a).* It is necessary to verify that, as we claimed, $\partial_i f$ is a distribution.

The right-hand side, $\langle f, \partial_i \varphi \rangle$ makes sense since $\partial_i \varphi \in \mathcal{D}(\Omega)$ from Theorem 2.6.

Due to the characterization of distributions from Theorem 3.3, for every semi-norm $\| \ \|_{E;\nu}$ of E, there exist $p \in \mathcal{C}^+(\Omega)$ and $c \in \mathbb{R}$ such that

$$\|\langle f, \varphi \rangle\|_{E;\nu} \le c \|\varphi\|_{\mathcal{D}(\Omega);p}.$$

The definition of $\partial_i f$ therefore gives

$$\|\langle \partial_i f, \varphi \rangle\|_{E;\nu} = \|\langle f, \partial_i \varphi \rangle\|_{E;\nu} \le c \|\partial_i \varphi\|_{\mathcal{D}(\Omega);p} \le c \|\varphi\|_{\mathcal{D}(\Omega);p+1}. \tag{5.2}$$

Indeed, by Definition 2.5 of the semi-norms of $\mathcal{D}(\Omega)$,

$$\|\partial_i \varphi\|_{\mathcal{D}(\Omega);p} = \sup_{x \in \Omega} \sup_{0 \le |\beta| \le p(x)} p(x) |\partial^\beta \partial_i \varphi(x)| \le$$

$$\le \sup_{x \in \Omega} \sup_{0 \le |\beta| \le p(x)+1} (p(x)+1) |\partial^\beta \varphi(x)| = \|\varphi\|_{\mathcal{D}(\Omega);p+1}.$$

Since the mapping $\partial_i f$ is linear (because f is linear), inequality (5.2) implies, again due to the characterization from Theorem 3.3, that we indeed have

$$\partial_i f \in \mathcal{D}'(\Omega; E).$$

2. *Justification of (c).* It is necessary here to verify that ∇f is a distribution, with values in E^d. The definition of ∇f, Definition (5.1) of the semi-norms of E^d and inequality (5.2) give

$$\|\langle \nabla f, \varphi \rangle\|_{E^d;\nu} = \left(\sum_{i=1}^d \|\langle \partial_i f, \varphi \rangle\|_{E;\nu}^2 \right)^{1/2} \le c d^{1/2} \|\varphi\|_{\mathcal{D}(\Omega);p+1}.$$

Therefore, again due to the characterization from Theorem 3.3,

$$\nabla f \in \mathcal{D}'(\Omega; E^d). \ \square$$

The partial derivatives $\partial_i f$ are often denoted by $\partial f/\partial x_i$. In dimension $d = 1$, we denote by $\mathrm{d}f/\mathrm{d}x$ the unique partial derivative and we call it *the derivative*.

Underlying idea of Definition 5.4 (a) of $\partial_i f$. Since the derivation of distributions must generalize that of functions, as was explained in the comment *Compatibility with the operations on functions*, p. 53, it is necessary that, if $f \in C^1(\Omega; E)$, we have, for all $\varphi \in \mathcal{D}(\Omega)$,

$$\langle \partial_i f, \varphi \rangle = \int \partial_i f \, \varphi.$$

Since φ is zero in the neighborhood of the boundary of Ω, by integrating the right-hand side by parts, it becomes $-\int f \, \partial_i \varphi$, that is here $-\langle f, \partial_i \varphi \rangle$, which gives

$$\langle \partial_i f, \varphi \rangle = -\langle f, \partial_i \varphi \rangle.$$

In defining $\partial_i f$ by this equality for all $f \in \mathcal{D}'(\Omega; E)$, we indeed generalize the derivative of functions, as we will verify in Theorem 5.5.

Observe that this equality is necessarily satisfied by all $f \in \mathcal{D}'(\Omega; E)$, since $C^1(\Omega; E)$ is dense in $\mathcal{D}'(\Omega; E)$, see Theorem 8.18. □

Space of gradients. We can, with Definition 5.1, separate the gradient into its components

$$\nabla f = (\partial_1 f, \ldots, \partial_d f),$$

i.e. represent it as an element of $(\mathcal{D}'(\Omega; E))^d$, since this is isomorphic to $\mathcal{D}'(\Omega; E^d)$ (Theorem 5.3). We prefer to use the latter space, in order to remain in a distribution space. □

For Definition 5.4 to be compatible with the identification of continuous functions with distributions made in Paragraph 3.3, it is necessary to check that for a function it gives back the classical partial derivatives. This is the purpose of the following result[2].

Theorem 5.5.– *If $f \in C^1(\Omega; E)$, its partial derivatives $\partial_i f$ in the distribution sense given by Definition 5.4 coincide with those in the function sense given by Definition 1.16.*

That is to say that, with identification (3.6), p. 52, of f with a distribution, Definition 5.4 again gives $\partial_i f \in C(\Omega; E)$ and, for every $\varphi \in \mathcal{D}(\Omega)$,

$$\langle \partial_i f, \varphi \rangle = \int_\Omega (\partial_i f) \varphi = -\int_\Omega f \, \partial_i \varphi. \ \blacksquare$$

2. **History of Theorem 5.5.** Laurent SCHWARTZ proved in 1945 [68, p. 61] that the partial derivatives of a continuously differentiable real function coincide with the partial derivatives in the distribution sense.

Proof. **1.** *Coincidences of the notion of derivation.* Let $f \in \mathcal{C}^1(\Omega; E)$. To avoid confusion, denote here $\overline{f}$ the distribution associated with this function by Theorem 3.5, and $\overline{\partial_i f}$ the distribution associated with the function $\partial_i f$, which is the derivative of f in the function sense. We have to check that

$$\overline{\partial_i f} = \partial_i \overline{f}. \tag{5.3}$$

Thus, let $\varphi \in \mathcal{D}(\Omega)$. Definition 5.4 (a) of the derivative in the distribution sense gives, with the definition of $\overline{f}$ from Theorem 3.5,

$$\langle \partial_i \overline{f}, \varphi \rangle = -\langle \overline{f}, \partial_i \varphi \rangle = -\int_\Omega f\, \partial_i \varphi. \tag{5.4}$$

On the other hand, $f\varphi \in \mathcal{C}^1(\Omega; E)$ (Theorem A.60) and its support is compact (Theorem 2.2 (a), since it is included in that of φ which is a compact subset of Ω), so, due to Orstogradsky's formula which we will prove in Theorem 5.6,

$$\int_\Omega \partial_i (f\varphi) = 0_E. \tag{5.5}$$

Leibniz's formula $\partial_i (f\varphi) = (\partial_i f)\varphi + f\, \partial_i \varphi$ (Theorem A.60, again) then gives

$$\int_\Omega (\partial_i f)\varphi = \int_\Omega \partial_i (f\varphi) - \int_\Omega f\, \partial_i \varphi = -\int_\Omega f\, \partial_i \varphi.$$

That is, with the definition of $\overline{\partial_i f}$ from Theorem 3.5,

$$\langle \overline{\partial_i f}, \varphi \rangle = \int_\Omega (\partial_i f)\varphi = -\int_\Omega f\, \partial_i \varphi. \tag{5.6}$$

Hence, with (5.4),

$$\langle \overline{\partial_i f}, \varphi \rangle = \langle \partial_i \overline{f}, \varphi \rangle.$$

This proves (5.3), i.e. that the derivative $\overline{\partial_i f}$ calculated in the function sense coincides with that, $\partial_i \overline{f}$, calculated in the distribution sense.

2. *Equalities.* The stated equalities are a simple rewriting of (5.6) with identification (3.6), p. 52, of the distribution $\overline{\partial_i f}$ with the function $\partial_i f$. □

It remains to prove equality (5.5). It is a particular case of the following version of **Ostrogradsky's formula**[3], with values in a Neumann space.

3. **History of Ostrogradsky's formula.** Mikhail Vasilyevitch OSTROGRADSKY proved in 1831 [58] that $\int_\Omega \partial_1 f_1 + \partial_2 f_2 + \partial_3 f_3 \, dx = \int_{\Omega} n_1 f_1 + n_2 f_2 + n_3 f_3 \, ds$ which gives the formula from Theorem 5.6 by taking $f_i = f$ and $f_j = 0$ if $j \neq i$.

Theorem 5.6.– *Let f be a function of $\mathcal{C}^1(\Omega; E)$ with compact support, where Ω is an open subset of $\mathbb{R}^d$ and E is a Neumann space, and $i \in [\![1, d]\!]$. Then,*

$$\int_\Omega \partial_i f = 0_E. \blacksquare$$

Proof. As f has compact support, its extension $\widetilde{f}$ by 0_E belongs to $\mathcal{C}^1(\mathbb{R}^d; E)$. Let

$$\Lambda = \{x \in \mathbb{R}^d : a_j < x_j < b_j, \ \forall j\}$$

be a parallelepiped containing the support of f. By Definition 1.16 of the partial derivative of a function, $\partial_i f = \mathrm{d}f/\mathrm{d}x_i$, thus the expression for the integral of a derivative from Theorem A.79 gives

$$\int_{a_i}^{b_i} \partial_i \widetilde{f}(x_1, \ldots, x_d) \, \mathrm{d}x_i = \int_{a_i}^{b_i} \frac{\mathrm{d}\widetilde{f}(x_1, \ldots, x_d)}{\mathrm{d}x_i} \, \mathrm{d}x_i =$$

$$= \widetilde{f}(x_1, \ldots, x_{i-1}, b_i, x_{i+1}, \ldots, x_d) - \widetilde{f}(x_1, \ldots, x_{i-1}, a_i, x_{i+1}, \ldots, x_d) = 0_E.$$

Integrating next with respect to the other x_j, we obtain, since integrals can be regrouped and commute (Theorem A.89),

$$\int_\Lambda \partial_i \widetilde{f} = \int_{a_d}^{b_d} \cdots \int_{a_1}^{b_1} \partial_i \widetilde{f} = \int_{a_d}^{b_d} \cdots \int_{a_{i+1}}^{b_{i+1}} \int_{a_{i-1}}^{b_{i-1}} \cdots \int_{a_1}^{b_1} \int_{a_i}^{b_i} \partial_i \widetilde{f} = 0_E.$$

Since the sets on which the function is zero do not contribute to the integral (Theorem A.77),

$$\int_\Omega \partial_i f = \int_{\mathbb{R}^d} \partial_i \widetilde{f} = \int_\Lambda \partial_i \widetilde{f} = 0_E. \ \square$$

Some fundamental consequences. It follows from Definition 5.4 (b) of successive derivatives that[4]:

$$\textbf{\textit{Every distribution is infinitely derivable.}} \tag{5.7}$$

In particular, with identification (3.6), p. 52, of continuous functions with distributions:

Every continuous function is infinitely derivable in the distribution sense. □

Converse of Theorem 5.5: distributions with continuous derivatives. If the partial derivatives of a distribution are continuous, then it is a continuously differentiable function (Theorem 12.4). □

4. **History of derivability of distributions.** Laurent SCHWARTZ mentioned in 1950 [69, Chap. II, § 1, p. 35] that every distribution is infinitely differentiable.

Distribution with zero derivatives. If the partial derivatives of a distribution are zero, then it is a locally constant function (Theorem 12.3). This property is easier to state than to prove, see the comment *Elementary, my dear Watson?*, p. 251. □

Now let us observe that derivatives commute [5].

Theorem 5.7.– *Let $f \in \mathcal{D}'(\Omega; E)$, where Ω is an open subset of $\mathbb{R}^d$ and E is a Neumann space, and i and j belong to $[\![1, d]\!]$. Then,*

$$\partial_i \partial_j f = \partial_j \partial_i f. \blacksquare$$

Proof. Reiterating Definition 5.4 (a) of a partial derivative, we obtain, for every $\varphi \in \mathcal{D}(\Omega)$,

$$\langle \partial_i \partial_j f, \varphi \rangle = -\langle \partial_j f, \partial_i \varphi \rangle = \langle f, \partial_i \partial_j \varphi \rangle.$$

The derivatives of φ in the function sense commute according to Schwartz's theorem (Theorem 1.17), thus

$$\langle \partial_i \partial_j f, \varphi \rangle = \langle f, \partial_i \partial_j \varphi \rangle = \langle f, \partial_j \partial_i \varphi \rangle = \langle \partial_j \partial_i f, \varphi \rangle. \ \square$$

Now let us provide the expressions for the higher-order derivatives.

Theorem 5.8.– *Let $f \in \mathcal{D}'(\Omega; E)$ and $\varphi \in \mathcal{D}(\Omega)$, where Ω is an open subset of $\mathbb{R}^d$ and E is a Neumann space, and $\beta \in \mathbb{N}^d$. Then,*

$$\langle \partial^\beta f, \varphi \rangle = (-1)^{|\beta|} \langle f, \partial^\beta \varphi \rangle. \blacksquare$$

Proof. Definition 5.4 (b) of $\partial^\beta f$ gives, by reiterating Definition 5.4 (a) of ∂_i,

$$\langle \partial^\beta f, \varphi \rangle = \langle \partial_1^{\beta_1} \partial_2^{\beta_2} \cdots \partial_d^{\beta_d} f, \varphi \rangle = (-1)^{\beta_1} \langle \partial_2^{\beta_2} \cdots \partial_d^{\beta_d} f, \partial_1^{\beta_1} \varphi \rangle =$$
$$= \cdots = (-1)^{\beta_1 + \cdots + \beta_d} \langle f, \partial_d^{\beta_d} \cdots \partial_2^{\beta_2} \partial_1^{\beta_1} \varphi \rangle.$$

We conclude by observing that, since the derivatives in the function sense commute due to Schwartz's theorem (Theorem 1.17),

$$\partial_d^{\beta_d} \cdots \partial_2^{\beta_2} \partial_1^{\beta_1} \varphi = \partial_1^{\beta_1} \partial_2^{\beta_2} \cdots \partial_d^{\beta_d} \varphi = \partial^\beta \varphi. \ \square$$

5. **History of the commutation of derivatives.** Laurent SCHWARTZ proved in 1950 [69, Chap. II, § 1, p. 36] that we can invert the order of the partial derivatives of a distribution. He also provided the expression in Theorem 5.8 [69, Formula (II, 1; 7), p. 35].

Let us show that derivation is a continuous mapping[6].

Theorem 5.9.– *Let Ω be an open subset of $\mathbb{R}^d$, E a Neumann space, $i \in [\![1, d]\!]$ and $\beta \in \mathbb{N}^d$.*

Then, ∂_i and ∂^β are continuous linear mappings from $\mathcal{D}'(\Omega; E)$ into itself, and ∇ is a continuous linear mapping from $\mathcal{D}'(\Omega; E)$ into $\mathcal{D}'(\Omega; E^d)$.

These mappings are sequentially continuous. ▮

Proof. Let $\{\| \|_{E;\nu} : \nu \in \mathcal{N}_E\}$ be the family of semi-norms of E.

1. *First-order derivatives.* By Definition 5.4 (a) of $\partial_i f$, the mapping $f \mapsto \partial_i f$ has values in $\mathcal{D}'(\Omega; E)$ and is obviously linear. By Definition 3.1 of the semi-norms of $\mathcal{D}'(\Omega; E)$, for every $\varphi \in \mathcal{D}(\Omega)$ and $\nu \in \mathcal{N}_E$,

$$\|\partial_i f\|_{\mathcal{D}'(\Omega;E);\varphi,\nu} = \|\langle \partial_i f, \varphi\rangle\|_{E;\nu} = \|\langle f, \partial_i \varphi\rangle\|_{E;\nu} = \|f\|_{\mathcal{D}'(\Omega;E);\partial_i\varphi,\nu}.$$

This implies the continuity of the mapping $f \mapsto \partial_i f$, due to the characterization of continuous linear mapping from Theorem 1.12 (a).

2. *Higher-order derivatives.* By using the expression for $\partial^\beta f$ from Theorem 5.8, we similarly obtain, for every $\varphi \in \mathcal{D}(\Omega)$ and $\nu \in \mathcal{N}_E$,

$$\|\partial^\beta f\|_{\mathcal{D}'(\Omega;E);\varphi,\nu} = \|\langle \partial^\beta f, \varphi\rangle\|_{E;\nu} = \|(-1)^{|\beta|}\langle f, \partial^\beta \varphi\rangle\|_{E;\nu} = \|f\|_{\mathcal{D}'(\Omega;E);\partial^\beta\varphi,\nu},$$

which implies the continuity of the mapping $f \mapsto \partial^\beta f$.

3. *Gradient.* From Definitions 5.4 (c) of the gradient and (5.1), p. 81, of the semi-norms of E^d, we have, for every $\varphi \in \mathcal{D}(\Omega)$ and $\nu \in \mathcal{N}_E$,

$$\|\nabla f\|_{\mathcal{D}'(\Omega;E^d);\varphi,\nu} = \|\langle \nabla f, \varphi\rangle\|_{E^d;\nu} = \Big(\sum_{1\le i\le d} \|\langle \partial_i f, \varphi\rangle\|^2_{E;\nu}\Big)^{1/2} \le$$

$$\le \sum_{1\le i\le d} \|\langle \partial_i f, \varphi\rangle\|_{E;\nu} = \sum_{1\le i\le d} \|\langle f, \partial_i \varphi\rangle\|_{E;\nu} = \sum_{1\le i\le d} \|f\|_{\mathcal{D}'(\Omega;E);\partial_i\varphi,\nu}.$$

This implies the continuity of the mapping $f \mapsto \nabla f$, again due to the characterization from Theorem 1.12 (a).

4. *Sequential continuity.* These mappings are sequentially continuous, as is any continuous mapping (Theorem 1.10). □

6. **History of the continuity of derivation.** Laurent SCHWARTZ proved in 1945 [68, Theorem 2, p. 71] that the derivation of distributions is continuous from $\mathcal{D}'(\mathbb{R}^d)$ into itself. He proved in 1957 [72, p. 68] that it is continuous linear from $\mathcal{D}'(\mathbb{R}^d; E)$-unif into itself.

5.3. Image under a linear mapping

Let us define the image of a distribution under a sequentially continuous linear mapping.

Definition 5.10.– *Let $f \in \mathcal{D}'(\Omega; E)$, where Ω is an open subset of $\mathbb{R}^d$ and E is a Neumann space, and L a continuous, or sequentially continuous, linear mapping from E into a Neumann space F.*

We define the ***image*** *$Lf \in \mathcal{D}'(\Omega; F)$ by: for every $\varphi \in \mathcal{D}(\Omega)$,*

$$\langle Lf, \varphi \rangle \stackrel{\text{def}}{=} L(\langle f, \varphi \rangle). \blacksquare$$

Justification. In order for Lf to be a distribution, it suffices, due to the characterization of distributions from Theorem 3.4 (b), that it is sequentially continuous from $\mathcal{D}(\Omega)$ into F, as it is linear. Therefore let

$$\varphi_n \to \varphi \ \text{ in } \mathcal{D}(\Omega).$$

Due to the characterization from Theorem 3.4 (b),

$$\langle f, \varphi_n \rangle \to \langle f, \varphi \rangle \ \text{ in } E.$$

Since L is sequentially continuous (if it is continuous, this is given by Theorem 1.10), $L(\langle f, \varphi_n \rangle) \to L(\langle f, \varphi \rangle)$ in F, i.e.

$$\langle Lf, \varphi_n \rangle \to \langle Lf, \varphi \rangle \ \text{ in } F.$$

Therefore, we indeed have

$$Lf \in \mathcal{D}'(\Omega; F). \ \square$$

It is necessary, like for any operation on distributions, to check its compatibility with the identification of continuous functions with distributions. Let us thus show that for a function we recover the classical image.

Theorem 5.11.– *If $f \in \mathcal{C}(\Omega; E)$, its image Lf in the distribution sense given by Definition 5.10 coincides with that in the function sense.*

That is to say that, with identification (3.6), p. 52, of f with a distribution, Definition 5.10 also gives $Lf \in \mathcal{C}(\Omega; F)$ and, for every $x \in \Omega$,

$$(Lf)(x) = L(f(x)). \blacksquare$$

Proof. The point is to check that the distribution $\bar{f}$ associated with the function f by Theorem 3.5 satisfies

$$L\bar{f} = \overline{Lf}.$$

This is indeed the case since Definition 5.10 of $L\bar{f}$, the expression for $\langle \bar{f}, \varphi \rangle$ from Theorem 3.5, the commutation of the Cauchy integral with continuous linear mappings (Theorem A.75), and again Theorem 3.5 successively give, for every $\varphi \in \mathcal{D}(\Omega)$,

$$\langle L\bar{f}, \varphi \rangle = L(\langle \bar{f}, \varphi \rangle) = L\Big(\int_\Omega f\varphi \Big) = \int_\Omega L(f\varphi) = \int_\Omega (Lf)\varphi = \langle \overline{Lf}, \varphi \rangle. \ \square$$

Let us show that linear mappings commute with the derivation of distributions.

Theorem 5.12.– *Let $f \in \mathcal{D}'(\Omega; E)$, where Ω is an open subset of $\mathbb{R}^d$ and E is a Neumann space, L a sequentially continuous linear mapping from E into a Neumann space, and $\beta \in \mathbb{N}^d$. Then,*

$$\partial^\beta (Lf) = L(\partial^\beta f). \ \blacksquare$$

Proof. For $i \in [\![1, d]\!]$, Definitions 5.4 (a) of the derivatives and 5.10 of the image under L give, for every $\varphi \in \mathcal{D}(\Omega)$,

$$\langle \partial_i (Lf), \varphi \rangle = -\langle Lf, \partial_i \varphi \rangle = -L(\langle f, \partial_i \varphi \rangle) = L(\langle \partial_i f, \varphi \rangle) = \langle L(\partial_i f), \varphi \rangle.$$

Which proves that

$$\partial_i Lf = L\partial_i f.$$

By reiterating this, Definition 5.4 (b) of $\partial^\beta f$ then gives $\partial^\beta Lf = L\partial^\beta f$. $\square$

Observe that the image of a distribution under a composition of mappings is the image of its image.

Theorem 5.13.– *Let $f \in \mathcal{D}'(\Omega; E)$, where Ω is an open subset of $\mathbb{R}^d$ and E is a Neumann space, L a sequentially continuous linear mapping from E into a Neumann space F, and M a sequentially continuous linear mapping from F into a Neumann space. Then,*

$$(M \circ L)f = M(Lf). \ \blacksquare$$

Proof. The composite mapping $M \circ L$ is sequentially continuous (Theorem A.30) and Definition 5.10 of the image by a linear mapping gives, for every $\varphi \in \mathcal{D}(\Omega)$,

$$\langle (M \circ L)f, \varphi \rangle = M(L(\langle f, \varphi \rangle)) = M(\langle Lf, \varphi \rangle) = \langle M(Lf), \varphi \rangle. \ \square$$

We now provide continuity and injectivity properties of the mapping $f \mapsto Lf$.

Theorem 5.14.– *Let L be a sequentially continuous linear mapping from a Neumann space E into another F, and Ω an open subset of $\mathbb{R}^d$. Then:*

(a) *The mapping $f \mapsto Lf$ is linear and sequentially continuous from $\mathcal{D}'(\Omega; E)$ into $\mathcal{D}'(\Omega; F)$.*

(b) *If L is continuous, the mapping $f \mapsto Lf$ is continuous from $\mathcal{D}'(\Omega; E)$ into $\mathcal{D}'(\Omega; F)$.*

(c) *If L is injective, the mapping $f \mapsto Lf$ is injective from $\mathcal{D}'(\Omega; E)$ into $\mathcal{D}'(\Omega; F)$.* ▮

Proof. **1.** *Property (a): sequential continuity.* By Definition 5.10 of Lf, the mapping $f \mapsto Lf$ has values in $\mathcal{D}'(\Omega; F)$ and is obviously linear.

Let

$$f_n \to f \ \text{ in } \mathcal{D}'(\Omega; E).$$

For every $\varphi \in \mathcal{D}(\Omega)$, the characterization of convergent sequences of distributions from Theorem 4.3 (c) gives $\langle f_n - f, \varphi \rangle \to 0_E$ in E. From which

$$\langle Lf_n - Lf, \varphi \rangle = L(\langle f_n - f, \varphi \rangle) \to 0_F \ \text{ in } F.$$

And so, again from Theorem 4.3 (c),

$$Lf_n \to Lf \ \text{ in } \mathcal{D}'(\Omega; F).$$

This proves that the mapping $f \mapsto Lf$ is sequentially continuous.

2. *Property (b): continuity.* Suppose that L is continuous from E into F and denote respectively by $\{\| \ \|_{E;\nu} : \nu \in \mathcal{N}_E\}$ and $\{\| \ \|_{F;\mu} : \mu \in \mathcal{N}_F\}$ the families of semi-norms of E and F.

Due to the characterization of continuous linear mappings from Theorem 1.12 (a), for every $\mu \in \mathcal{N}_F$, there exist $c \in \mathbb{R}$ and a finite subset N of $\mathcal{N}_E$ such that, for every $u \in E$,

$$\|Lu\|_{F;\mu} \leq c \sup_{\nu \in N} \|u\|_{E;\nu}.$$

Definition 3.1 of the semi-norms of $\mathcal{D}'(\Omega; E)$ then gives, for every $\varphi \in \mathcal{D}(\Omega)$,

$$\|Lf\|_{\mathcal{D}'(\Omega;F);\varphi,\mu} = \|\langle Lf, \varphi\rangle\|_{F;\mu} = \|L(\langle f, \varphi\rangle)\|_{F;\mu} = \\ \leq c \sup_{\nu \in N} \|\langle f, \varphi\rangle\|_{E;\nu} = c \sup_{\nu \in N} \|f\|_{\mathcal{D}'(\Omega;E);\varphi,\nu}.$$

Which implies, again due to the characterization from Theorem 1.12 (a), that the mapping $f \mapsto Lf$ is continuous.

3. *Property (c): injectivity.* Suppose that L is injective, and let $f \in \mathcal{D}'(\Omega; E)$ be such that

$$Lf = 0.$$

For every $\varphi \in \mathcal{D}(\Omega)$, we have

$$L(\langle f, \varphi\rangle) = \langle Lf, \varphi\rangle = 0_F.$$

From which, $\langle f, \varphi\rangle = 0_E$ since L is injective. Therefore,

$$f = 0.$$

Which proves that the mapping $f \mapsto Lf$ is injective. □

Image by a non-linear mapping. It is generally not possible to define the image $T(f)$ of a distribution f under a non-linear mapping T, in such a way that the mapping $f \mapsto T(f)$ is sequentially continuous and that $T(f)(x) = T(f(x))$ if f is continuous. This holds even if T is very regular. For example:

$$\left\{\begin{array}{l}\textit{The mapping } f \mapsto f^2 \textit{ has no sequentially continuous} \\ \textit{extension from } \mathcal{D}'(\mathbb{R}^d) \textit{ into itself.}\end{array}\right. \tag{5.8}$$

Proof. The regularizing function ρ_n defined by (8.1), p. 169, namely $\rho_n(x) = cn^d\rho(nx)$, satisfies (Theorem 8.3), when $n \to \infty$,

$$\rho_n \to \delta_0 \text{ in } \mathcal{D}'(\mathbb{R}^d).$$

However, ρ_n^2 does not have a limit in $\mathcal{D}'(\mathbb{R}^d)$. Indeed, if $\varphi \in \mathcal{D}(\Omega)$ equals 1 in the ball $B(0,1)$,

$$\int_{\mathbb{R}^d} \rho_n^2 \varphi = c^2 n^{2d} \int_{\mathbb{R}^d} \rho^2(nx)\,dx = \frac{c^2 n^{2d}}{n^d} \int_{\mathbb{R}^d} \rho^2(y)\,dy = c' n^d \to \infty. \ \square$$

5.4. Product with a regular function

This section is dedicated to the product of a distribution with an infinitely differentiable function. We start with its definition [7].

7. **History of the definition of the product.** Laurent SCHWARTZ defined in 1945 [68, Formula (17), p. 66] the product of a distribution of $\mathcal{D}'(\mathbb{R}^d)$ with an infinitely differentiable function.

Definition 5.15.– *Let*

$$f \in \mathcal{D}'(\Omega; E) \ \textit{and} \ \alpha \in \mathcal{C}^\infty(\Omega),$$

where Ω is an open subset of $\mathbb{R}^d$ and E is a Neumann space.

*We define the **product** $\alpha f \in \mathcal{D}'(\Omega; E)$ by: for every $\varphi \in \mathcal{D}(\Omega)$,*

$$\langle \alpha f, \varphi \rangle \stackrel{\text{def}}{=} \langle f, \alpha\varphi \rangle. \ ■$$

Justification. **1.** *Meaning of $\langle f, \alpha\varphi \rangle$.* This term is well-defined since $\alpha\varphi \in \mathcal{D}(\Omega)$, as:

$$\alpha \in \mathcal{C}^\infty(\Omega), \ \varphi \in \mathcal{D}(\Omega) \quad \Rightarrow \quad \alpha\varphi \in \mathcal{D}(\Omega). \tag{5.9}$$

Indeed, the product $\alpha\varphi$ is infinitely differentiable due to Theorem A.60, and its support is compact due to Theorem 2.2 (a) since it is included in that of φ which is a compact subset of Ω.

2. *Obtaining a distribution.* So that

$$\alpha f \in \mathcal{D}'(\Omega; E),$$

it suffices, due to the characterization of distributions from Theorem 3.4 (d), since $\langle f, \alpha\varphi \rangle$ linearly depends on φ, that, for every compact subset K of Ω,

$$\langle f, \alpha\varphi \rangle \text{ depends continuously on } \varphi \text{ in } \mathcal{C}^\infty_K(\Omega).$$

Thus, let $\| \ \|_{E;\nu}$ be a semi-norm of E. Due to the characterization of distributions from Theorem 3.3, there exist $p \in \mathcal{C}^+(\Omega)$ and $c \in \mathbb{R}$ such that, for every $\phi \in \mathcal{D}(\Omega)$,

$$\|\langle f, \phi \rangle\|_{E;\nu} \leq c\|\phi\|_{\mathcal{D}(\Omega);p}.$$

And, as we will verify in Lemma 5.16, there exist $m \in \mathbb{N}$ and $c' \in \mathbb{R}$ (depending on p and K) such that, for every $\varphi \in \mathcal{C}^\infty_K(\Omega)$,

$$\|\alpha\varphi\|_{\mathcal{D}(\Omega);p} \leq c' \|\alpha\|_{\mathcal{C}^\infty(\Omega);m,K} \|\varphi\|_{\mathcal{C}^\infty_{\mathrm{b}}(\Omega);m}. \tag{5.10}$$

Hence, denoting $c'' = cc' \|\alpha\|_{\mathcal{C}^\infty(\Omega);m,K}$,

$$\|\langle f, \alpha\varphi \rangle\|_{E;\nu} \leq c'' \|\varphi\|_{\mathcal{C}^\infty_{\mathrm{b}}(\Omega);m}.$$

The characterization of continuous linear mappings from Theorem 1.12 (a) then shows that the mapping $\varphi \mapsto \langle f, \alpha\varphi \rangle$ is indeed continuous from $\mathcal{C}^\infty_K(\Omega)$ into E, since $\mathcal{C}^\infty_K(\Omega)$ is, by Definition 2.3 (a), endowed with the semi-norms of $\mathcal{C}^\infty_{\mathrm{b}}(\Omega)$. □

To fully justify Definition 5.15, it remains to verify inequality (5.10), that is the following property.

Lemma 5.16.– *Let Ω be an open subset of $\mathbb{R}^d$, K a compact subset of Ω, and $p \in \mathcal{C}^+(\Omega)$.*

Then, there exist $m \in \mathbb{N}$ and $c \in \mathbb{R}$ such that, for every $\alpha \in \mathcal{C}^\infty(\Omega)$ and $\varphi \in \mathcal{C}^\infty(\Omega)$ such that $\operatorname{supp}(\alpha\varphi) \subset K$,

$$\|\alpha\varphi\|_{\mathcal{D}(\Omega);p} \leq c\|\alpha\|_{\mathcal{C}^\infty(\Omega);m,K}\|\varphi\|_{\mathcal{C}^\infty(\Omega);m,K} \leq c\|\alpha\|_{\mathcal{C}^\infty(\Omega);m,K}\|\varphi\|_{\mathcal{C}_b^\infty(\Omega);m}. \blacksquare$$

Proof. **1.** *First inequality.* Let $m \in \mathbb{N}$ be an upper bound of $\sup_{x\in K} p(x)$ (it exists since continuous functions are bounded on compact sets due to Theorem A.32). Since $\alpha\varphi$ is zero outside of K, Definition 2.5 of the semi-norms of $\mathcal{D}(\Omega)$ gives

$$\|\alpha\varphi\|_{\mathcal{D}(\Omega);p} = \sup_{x\in\Omega}\sup_{|\beta|\leq p(x)} p(x)|\partial^\beta(\alpha\varphi)(x)| \leq \sup_{x\in K}\sup_{|\beta|\leq m} m|\partial^\beta(\alpha\varphi)(x)|$$

Using the bound on the norm of the derivatives of the image by a bilinear mapping from Theorem A.63 (with $T(\alpha,\varphi) = \alpha\varphi$ and so $I = 2$), we get

$$|\partial^\beta(\alpha\varphi)(x)| \leq 2^{|\beta|}\sup_{|\sigma|\leq|\beta|}|\partial^\sigma\alpha(x)|\sup_{|\sigma|\leq|\beta|}|\partial^\sigma\varphi(x)|.$$

Hence, with Definition 1.19 (a) of the semi-norms of $\mathcal{C}^\infty(\Omega)$,

$$\|\alpha\varphi\|_{\mathcal{D}(\Omega);p} \leq m2^m \sup_{|\sigma|\leq m}\sup_{x\in K}|\partial^\sigma\alpha(x)| \sup_{|\sigma|\leq m}\sup_{x\in K}|\partial^\sigma\varphi(x)| =$$

$$= m2^m\|\alpha\|_{\mathcal{C}^\infty(\Omega);m,K}\|\varphi\|_{\mathcal{C}^\infty(\Omega);m,K}.$$

2. *Second inequality.* It results from the fact that, by Definition 1.19 of the semi-norms of $\mathcal{C}^\infty(\Omega)$ and $\mathcal{C}_b^\infty(\Omega)$,

$$\|\varphi\|_{\mathcal{C}^\infty(\Omega);m,K} \leq \|\varphi\|_{\mathcal{C}_b^\infty(\Omega);m}. \square$$

Utility of the first inequality of Lemma 5.16. We have only used the second inequality to prove Theorem 5.15. The first inequality will be used to prove Theorems 5.18 and 5.24. □

As always, let us check the compatibility with the identification of continuous functions with distributions, namely that for a function we recover the classical image.

Theorem 5.17.– *If $f \in \mathcal{C}(\Omega; E)$, its product with $\alpha \in \mathcal{C}^\infty(\Omega)$ in the distribution sense given by Definition 5.15 coincides with that in the sense of functions.*

That is to say that, with identification (3.6), p. 52, of f with a distribution, Definition 5.15 again gives $\alpha f \in \mathcal{C}(\Omega; E)$ and, for every $x \in \Omega$,

$$(\alpha f)(x) = \alpha(x) f(x). \blacksquare$$

Proof. The point is to check that the distribution $\bar{f}$ associated with the function f by Theorem 3.5 satisfies

$$\alpha \bar{f} = \overline{\alpha f}.$$

This is the case, since Definition 5.15 of $\alpha \bar{f}$ and the expression for $\langle \bar{f}, \alpha\varphi \rangle$ from Theorem 3.5 give, for every $\varphi \in \mathcal{D}(\Omega)$,

$$\langle \alpha \bar{f}, \varphi \rangle = \langle \bar{f}, \alpha\varphi \rangle = \int_\Omega f(\alpha\varphi) = \int_\Omega (\alpha f)\varphi = \langle \overline{\alpha f}, \varphi \rangle. \ \square$$

Now let us prove that the product is sequentially continuous [8].

Theorem 5.18.– *Let Ω be an open subset of $\mathbb{R}^d$ and E a Neumann space. Then:*

(a) *For every $\alpha \in \mathcal{C}^\infty(\Omega)$, the mapping $f \mapsto \alpha f$ is continuous linear from $\mathcal{D}'(\Omega; E)$ into itself.*

(b) *For every $f \in \mathcal{D}'(\Omega; E)$, the mapping $\alpha \mapsto \alpha f$ is continuous linear from $\mathcal{C}^\infty(\Omega)$ into $\mathcal{D}'(\Omega; E)$.*

(c) *The mapping $(\alpha, f) \mapsto \alpha f$ is sequentially continuous bilinear from the product $\mathcal{C}^\infty(\Omega) \times \mathcal{D}'(\Omega; E)$ into $\mathcal{D}'(\Omega; E)$.* $\blacksquare$

Proof. Let $\{\| \ \|_{E;\nu} : \nu \in \mathcal{N}_E\}$ be the family of semi-norms of E.

1. Property (a). Given $\varphi \in \mathcal{D}(\Omega)$ and $\nu \in \mathcal{N}_E$, Definitions 3.1 of the semi-norms of $\mathcal{D}'(\Omega; E)$ and 5.15 of the product by a regular function give

$$\|\alpha f\|_{\mathcal{D}'(\Omega;E);\varphi,\nu} = \|\langle \alpha f, \varphi \rangle\|_{E;\nu} = \|\langle f, \alpha\varphi \rangle\|_{E;\nu} = \|f\|_{\mathcal{D}'(\Omega;E);\alpha\varphi,\nu}.$$

---

8. **History of the continuity of the product.** Laurent SCHWARTZ proved in 1950 [69, Chap. V, § 2, Theorem III, p. 119] that the mapping $(\alpha, f) \mapsto f$ is hypocontinuous from $\mathcal{C}^\infty(\mathbb{R}^d) \times \mathcal{D}'(\mathbb{R}^d)$-unif into $\mathcal{D}'(\mathbb{R}^d)$-unif, which implies that it is sequentially continuous.

Which implies the continuity of the mapping $f \mapsto \alpha f$, due to the characterization of continuous linear mappings from Theorem 1.12 (a).

2. Property (b). Given $\varphi \in \mathcal{D}(\Omega)$ and $\nu \in \mathcal{N}_E$, we have, as above,

$$\|\alpha f\|_{\mathcal{D}'(\Omega;E);\varphi,\nu} = \|\langle f, \alpha\varphi\rangle\|_{E;\nu}.$$

For a given f, due to the characterization of distributions from Theorem 3.3, there exist $p \in \mathcal{C}^+(\Omega)$ and $c \in \mathbb{R}$ such that, for every $\phi \in \mathcal{D}(\Omega)$,

$$\|\langle f, \phi\rangle\|_{E;\nu} \le c\|\phi\|_{\mathcal{D}(\Omega);p}.$$

Due to the first inequality of Lemma 5.16 with $K = \operatorname{supp}\varphi$, there exist $m \in \mathbb{N}$ and $c' \in \mathbb{R}$ such that

$$\|\alpha\varphi\|_{\mathcal{D}(\Omega);p} \le c'\|\varphi\|_{\mathcal{C}^\infty(\Omega);m,K}\|\alpha\|_{\mathcal{C}^\infty(\Omega);m,K}.$$

Denoting $c'' = \|\varphi\|_{\mathcal{C}^\infty(\Omega);m,K}$, we therefore have, for every $\alpha \in \mathcal{C}^\infty(\Omega)$,

$$\|\alpha f\|_{\mathcal{D}'(\Omega;E);\varphi,\nu} \le cc'c''\|\alpha\|_{\mathcal{C}^\infty(\Omega);m,K}.$$

Which implies the continuity of the mapping $\alpha \mapsto \alpha f$, again due to the characterization from Theorem 1.12 (a).

3. Property (c). Let

$$\alpha_n \to \alpha \text{ in } \mathcal{C}^\infty(\Omega), \quad f_n \to f \text{ in } \mathcal{D}'(\Omega; E)$$

and $\varphi \in \mathcal{D}(\Omega)$. The second inequality of Lemma 5.16 with $K = \operatorname{supp}\varphi$ implies

$$\alpha_n\varphi \to \alpha\varphi \text{ in } \mathcal{D}(\Omega).$$

Hence, due to the sequential continuity of the mapping $\langle\ ,\ \rangle$ (Theorem 4.4 (c)),

$$\langle f_n, \alpha_n\varphi\rangle \to \langle f, \alpha\varphi\rangle \text{ in } E.$$

By Definition 5.15 of the product by a regular function, this is written

$$\langle \alpha_n f_n, \varphi\rangle \to \langle \alpha f, \varphi\rangle \text{ in } E.$$

Whence, due to the characterization of convergent sequences of distributions from Theorem 4.3 (c),

$$\alpha_n f_n \to \alpha f \text{ in } \mathcal{D}'(\Omega; E).$$

Which proves the sequential continuity of the mapping $(\alpha, f) \mapsto \alpha f$. $\square$

We now give the Leibniz formula for the derivatives of the product[9] of a distribution with a regular function and we calculate the **Laplacian** $\Delta = \partial_1^2 + \cdots + \partial_d^2$ of the product.

Theorem 5.19.– *Let $f \in \mathcal{D}'(\Omega; E)$ and $\alpha \in \mathcal{C}^\infty(\Omega)$, where Ω is an open subset of $\mathbb{R}^d$ and E is a Neumann space, and $i \in [\![1, d]\!]$. Then,*

$$\partial_i(\alpha f) = \alpha\,\partial_i f + \partial_i\alpha\, f$$

and

$$\Delta(\alpha f) = \alpha\,\Delta f + 2\nabla\alpha \cdot \nabla f + \Delta\alpha\, f. \ ▮$$

Proof. **1.** *Derivative of the product.* Definitions 5.4 (a) of the partial derivatives and 5.15 of the product with a regular function give, for every $\varphi \in \mathcal{D}(\Omega)$,

$$\langle \partial_i(\alpha f), \varphi \rangle = -\langle \alpha f, \partial_i \varphi \rangle = -\langle f, \alpha\,\partial_i \varphi \rangle.$$

Due to the Leibniz formula for functions (Theorem A.60), $\partial_i(\alpha\varphi) = \alpha\,\partial_i\varphi + \partial_i\alpha\,\varphi$, so the last term of the above equality is worth

$$-\langle f, \partial_i(\alpha\varphi) - \partial_i\alpha\,\varphi \rangle = \langle \partial_i f, \alpha\varphi \rangle + \langle f, \partial_i\alpha\,\varphi \rangle = \langle \alpha\,\partial_i f, \varphi \rangle + \langle \partial_i\alpha\, f, \varphi \rangle.$$

Which proves that

$$\partial_i(\alpha f) = \alpha\,\partial_i f + \partial_i\alpha\, f.$$

2. *Laplacian of the product.* By reiterating the above equality, we obtain

$$\partial_i^2(\alpha f) = \alpha\,\partial_i^2 f + 2\partial_i\alpha\,\partial_i f + \partial_i^2\alpha\, f.$$

By summing over i from 1 to d, we obtain the stated expression for $\Delta(\alpha f)$. □

Let us give the expression for the image of the product under a linear mapping.

Theorem 5.20.– *Let $f \in \mathcal{D}'(\Omega; E)$ and $\alpha \in \mathcal{C}^\infty(\Omega)$, where Ω is an open subset of $\mathbb{R}^d$ and E is a Neumann space, and let L be a sequentially continuous linear mapping from E into a Neumann space. Then,*

$$L(\alpha f) = \alpha L f. \ ▮$$

9. **History of the derivative of the product.** Gottfried von LEIBNIZ gave in 1675 the expression for a the derivative of a product of functions, in the form $\mathrm{d}\,xv = x\,\mathrm{d}v + v\,\mathrm{d}x$ [51, p. 467].
Laurent SCHWARTZ extended it in 1945 to the product of a distribution with a regular function [68, p. 66].

Proof. Definitions 5.10 of the image by L and 5.15 of the product with a regular function give, for every $\varphi \in \mathcal{D}(\Omega)$,

$$\langle L(\alpha f), \varphi\rangle = L(\langle \alpha f, \varphi\rangle) = L(\langle f, \alpha\varphi\rangle) = \langle Lf, \alpha\varphi\rangle = \langle \alpha Lf, \varphi\rangle. \; \Box$$

Product of two distributions. It is **not possible** to define the product of two arbitrary distributions in a way that it is sequentially continuous and gives back the usual product of continuous functions. Else, the product of a distribution with itself would be sequentially continuous, which is not true as we have established in Property (5.8), p. 94.

It is **also not possible** to define the product of two distributions in a way that is associative and gives back the product with a regular function. Else, as was shown by Laurent SCHWARTZ [70], in $\mathcal{D}'(\mathbb{R})$, the distribution called main value of $1/x$, denoted by Mv, and the function defined by $\mathrm{Id}(x) = x$ would satisfy

$$(\delta_0 \mathrm{Id}) \times \mathrm{Mv} = \delta_0 \times (\mathrm{Id}\, \mathrm{Mv}).$$

Since $\delta_0 \mathrm{Id} = 0$ and $\mathrm{Id}\, \mathrm{Mv} = 1$, we would get

$$0 = \delta_0. \; \Box$$

Product of certain distributions with certain functions. In Volume 5, we will define the product of a distribution in $W^{-m,r}(\Omega)$ with a function in $W^{m,s}(\Omega)$, where $1/r + 1/s \leq 1$. We can define the product of many other particular classes of distributions with other classes of regular functions, such as the product of a distribution of order m in $\mathcal{D}'(\Omega)$ with a function in $\mathcal{C}^m(\Omega)$. $\Box$

Generalization of the Leibniz formula? Although it is formally equal to the Leibniz formula for continuously differentiable functions (Theorem A.60), the equality $\partial_i(\alpha f) = \alpha\partial_i f + \partial_i\alpha\, f$ from Theorem 5.19 does not generalize it since, if f is more general, α is less so. This compensates for that. $\Box$

5.5. Change of variables

The image of a distribution under a change of variable T involves its determinant **Jacobian determinant**, or **Jacobian**, $\det[\nabla T]$. Definition A.71 of the determinant $\det[v] = \det[v^1, \ldots, v^d]$ of d vectors v^i in $\mathbb{R}^d$ and its properties are recalled in the Appendix. We use the brackets [ ] for $d \times d$ **matrices** and often for the families of d vector in $\mathbb{R}^d$, to highlight that these families can be represented by a square table (this convention is not very usual). So,

$$[\nabla T] = [\partial_1 T, \ldots, \partial_d T] = [\partial_j T_i]_{1 \leq i \leq d,\, 1 \leq j \leq d}.$$

We denote by $\mathcal{C}^\infty(\Lambda; \Omega)$ the set of functions from Λ into Ω which are infinitely differentiable (from Λ into $\mathbb{R}^d$). Note that this is not a vector space.

Let us define the image of a distribution under an infinitely differentiable bijection [10].

10. **History of the change the variables in a distribution.** Laurent SCHWARTZ expressed in the 1966 edition of his *Théorie des distributions* [69, p. 31] the principle of the definition of the change of variable by transposition, which is used here. He detailed it [69, Chap. IX, § 1, p. 320 and § 5, p. 373–396] in the framework of the theory of real-valued currents on a manifold, very general... and no less abstruse.

Definition 5.21.– *Let $f \in \mathcal{D}'(\Omega; E)$, where Ω is an open subset of $\mathbb{R}^d$ and E is a Neumann space, and let T be a bijection from an open subset Λ of $\mathbb{R}^d$ onto Ω such that*

$$T \in \mathcal{C}^\infty(\Lambda; \Omega), \quad T^{-1} \in \mathcal{C}^\infty(\Omega; \Lambda).$$

*We define the **image** $f \circ T \in \mathcal{D}'(\Lambda; E)$ by: for every $\phi \in \mathcal{D}(\Lambda)$,*

$$\langle f \circ T, \phi \rangle_\Lambda \overset{\text{def}}{=} \langle f, |\det[\nabla T^{-1}]|\, \phi \circ T^{-1} \rangle_\Omega. \blacksquare$$

It is necessary to verify that $f \circ T$ is indeed a distribution, i.e. that the right-hand side in its definition depends continuously on ϕ. Let us first check that it is well-defined, which follows from the following property.

Theorem 5.22.– *Let T be a bijection from an open subset Λ of $\mathbb{R}^d$ onto an open subset Ω of $\mathbb{R}^d$ such that $T \in \mathcal{C}^\infty(\Lambda; \Omega)$ and $T^{-1} \in \mathcal{C}^\infty(\Omega; \Lambda)$. Then,*

$$|\det[\nabla T^{-1}]| \in \mathcal{C}^\infty(\Omega)$$

and, for every $\phi \in \mathcal{D}(\Lambda)$,

$$|\det[\nabla T^{-1}]|\, \phi \circ T^{-1} \in \mathcal{D}(\Omega). \blacksquare$$

Proof. **1.** *Regularity of* $|\det[\nabla T^{-1}]|$. The image $\det[\nabla T^{-1}]$ of ∇T^{-1} under the multilinear mapping det is infinitely differentiable on Ω due to Theorem A.62. Furthermore, in each connected component (Definition 10.20) of Ω, the sign ϵ of $\det[\nabla T^{-1}]$ remains constant, which we will verify in Lemma 5.23, and therefore

$$|\det[\nabla T^{-1}]| = \epsilon \det[\nabla T^{-1}].$$

Thus, $|\det[\nabla T^{-1}]|$ is infinitely differentiable on each connected component of Ω, and hence on all of Ω. Which proves that

$$|\det[\nabla T^{-1}]| \in \mathcal{C}^\infty(\Omega). \tag{5.11}$$

2. *Regularity of* $|\det[\nabla T^{-1}]|\, \phi \circ T^{-1}$. Let $\phi \in \mathcal{D}(\Lambda)$. The composite mapping $\phi \circ T^{-1}$ is infinitely differentiable according to Theorem A.65. Its product with $|\det[\nabla T^{-1}]|$ is then also according to Theorem A.60, i.e.

$$|\det[\nabla T^{-1}]|\, \phi \circ T^{-1} \in \mathcal{C}^\infty(\Omega).$$

Let us examine its support. By Definition 2.1 of the support of a function, ϕ is zero outside of its support K, which is compact. Thus, $\phi \circ T^{-1}$ is zero outside of $T^{-1}(K)$, which is compact, as is every image of a compact set under a continuous mapping (Theorem A.31). The function $|\det[\nabla T^{-1}]|\, \phi \circ T^{-1}$ is also zero outside of this compact subset of Ω, so its support is compact due to Theorem 2.2 (b). Whence,

$$|\det[\nabla T^{-1}]|\, \phi \circ T^{-1} \in \mathcal{D}(\Omega). \square$$

Let us verify the sign property of the Jacobian that we have used to establish (5.11).

Lemma 5.23.– *Let T be a bijection from an open subset Λ of $\mathbb{R}^d$ onto a connected open subset Ω of $\mathbb{R}^d$ such that $T \in \mathcal{C}^1(\Lambda; \Omega)$ and $T^{-1} \in \mathcal{C}^1(\Omega; \Lambda)$. Then, for every $y \in \Lambda$,*

$$\det[\nabla T^{-1}(T(y))] \det[\nabla T(y)] = 1,$$

and $\det[\nabla T^{-1}]$ *does not change sign on* Ω. ▮

Proof. **1.** *Calculation of the product of determinants.* Let us differentiate the equality $(T_k^{-1} \circ T)(y) = y_k$. The expression for a change of variable in a derivative from Theorem A.64 (a) gives

$$\partial_i (T_k^{-1} \circ T)(y) = \sum_{j=1}^{d} \partial_j T_k^{-1}(T(y))\, \partial_i T_j(y) = \mathrm{I}_k^i,$$

where $\mathrm{I}_k^i = 0$ if $i \neq k$ and $\mathrm{I}_i^i = 1$. That is, by definition of the matrix product (p. 362),

$$[\nabla T^{-1}(T(y))]\, [\nabla T(y)] = [\mathrm{I}] = [\mathbf{e}_1, \ldots, \mathbf{e}_d].$$

Since the determinant of a matrix product is the product of the determinants (Theorem A.73) and the determinant of the identity matrix is 1 (Theorem A.72 (a)),

$$\det[\nabla T^{-1}(T(y))] \det[\nabla T(y)] = 1.$$

2. *Sign of* $\det[\nabla T^{-1}]$. So,

$$\det[\nabla T^{-1}(T(y))] \neq 0. \tag{5.12}$$

Consider the set

$$U \stackrel{\text{def}}{=} \det[\nabla T^{-1}](\Omega).$$

It is connected in $\mathbb{R}$, since every image of a connected set under a continuous mapping is (Theorem A.31). If it contained both a point $a < 0$ and a point $b > 0$, it would be covered (since it does not contain 0 from (5.12)) by the disjoint open sets $(-\infty, 0)$ and $(0, +\infty)$ whose intersections with it would be non empty, which would contradict its connectedness (Definition 10.20). Thus,

either $\det[\nabla T^{-1}]$ is > 0 on all of Ω, or it is < 0 on all of Ω. □

The justification of Definition 5.21 will also use the following property of composed functions, which was proven in Volume 2 [82, Theorem 3.13].

Theorem 5.24.– *Let $f \in \mathcal{C}^m(\Omega)$ and $T \in \mathcal{C}^m(\Lambda;\Omega)$, where Ω is an open subset of $\mathbb{R}^d$, Λ is an open subset of $\mathbb{R}^\ell$ and $m \in \mathbb{N}$.*

Then, $f \circ T \in \mathcal{C}^m(\Lambda)$ and, for every compact subset Q of Λ and for $K = T(Q)$,

$$\|f \circ T\|_{\mathcal{C}^m(\Lambda);Q} \leq 2^{m(m-1)/2} \|f\|_{\mathcal{C}^m(\Omega);K} \sup\Big\{1, d \sup_{1\leq i\leq d} \|T_i\|_{\mathcal{C}^m(\Lambda);Q}\Big\}^m. \blacksquare$$

We are now able to verify that the mapping $f \circ T$ is indeed a distribution.

Justification of Definition 5.21. According to the characterization of distributions from Theorem 3.4 (d), it suffices to check that, for every compact subset Q of Λ,

$$\begin{cases} \phi \mapsto \langle f, |\det[\nabla T^{-1}]|\, \phi \circ T^{-1}\rangle_\Omega \textit{ is a continuous} \\ \textit{linear mapping from } \mathcal{C}_Q^\infty(\Lambda) \textit{ into } E. \end{cases}$$

Observe that this quantity is well-defined, since $|\det[\nabla T^{-1}]|\, \phi \circ T^{-1} \in \mathcal{D}(\Omega)$ due to Theorem 5.22, and that it depends linearly on ϕ, since f is linear. It remains to verify that it depends continuously on ϕ.

Let $\{\|\ \|_{E;\nu} : \nu \in \mathcal{N}_E\}$ be the family of semi-norms of E and

$$K \stackrel{\text{def}}{=} T(Q).$$

It is a compact set, as is any image of a compact set under a continuous mapping (Theorem A.31), included in Ω. Due to the characterization of distributions from Theorem 3.3, for every $\nu \in \mathcal{N}_E$, there exist $p \in \mathcal{C}^+(\Omega)$ and $b \in \mathbb{R}$ such that, for every $\varphi \in \mathcal{D}(\Omega)$,

$$\|\langle f, \varphi\rangle_\Omega\|_{E;\nu} \leq b\|\varphi\|_{\mathcal{D}(\Omega);p}.$$

Moreover, due to the first inequality in Lemma 5.16, there exist $c \in \mathbb{R}$ and $m \in \mathbb{N}$ such that, for all functions $\alpha \in \mathcal{C}^\infty(\Omega)$ and $\varphi \in \mathcal{C}_K^\infty(\Omega)$,

$$\|\alpha\varphi\|_{\mathcal{D}(\Omega);p} \leq c\|\alpha\|_{\mathcal{C}^\infty(\Omega);m,K}\|\varphi\|_{\mathcal{C}^\infty(\Omega);m,K}.$$

Let us finally use the estimate of the $\mathcal{C}^m$-norm of a composed function recalled in Theorem 5.24, for T^{-1}. It gives, since $\|\ \|_{\mathcal{C}^\infty(\Omega);m,K} = \|\ \|_{\mathcal{C}^m(\Omega);K}$ according to Definitions 1.18 (a) and 1.19 (a) of these semi-norms, for every $\phi \in \mathcal{C}_Q^\infty(\Lambda)$,

$$\|\phi \circ T^{-1}\|_{\mathcal{C}^\infty(\Omega);m,K} \leq e\|\phi\|_{\mathcal{C}^\infty(\Lambda);m,Q},$$

where $e = 2^{m(m-1)/2} \sup\{1, d \sup_{1\leq i\leq d} \|(T^{-1})_i\|_{\mathcal{C}^m(\Omega);K}\}^m$.

These three inequalities, with $\alpha = |\det[\nabla T^{-1}]|$ which belongs to $\mathcal{C}^\infty(\Omega)$ due to Theorem 5.22, give

$$\|\langle f, |\det[\nabla T^{-1}]|\, \phi \circ T^{-1}\rangle_\Omega\|_{E;\nu} \leq b' \|\phi\|_{\mathcal{C}^\infty(\Lambda);m,Q} \leq b' \|\phi\|_{\mathcal{C}_b^\infty(\Lambda);m},$$

where $b' = bce \|\det[\nabla T^{-1}]\|_{\mathcal{C}^\infty(\Lambda);m,K}$.

Since $\mathcal{C}_Q^\infty(\Lambda)$ is by Definition 2.3 (a) endowed with the semi-norms of $\mathcal{C}_b^\infty(\Lambda)$, the characterization of continuous linear mappings from Theorem 1.12 (a) then shows that the mapping $\phi \mapsto \langle f, |\det[\nabla T^{-1}]|\, \phi \circ T^{-1}\rangle_\Omega$ is continuous from $\mathcal{C}_Q^\infty(\Lambda)$ into E. Thus it is indeed a distribution of $\mathcal{D}'(\Lambda; E)$, due to Theorem 3.4 (d), i.e.

$$f \circ T \in \mathcal{D}'(\Lambda; E). \ \square$$

Now that we have checked that $f \circ T$ is a distribution, we ensure that, when f is a function, we recover the classical definition.

Theorem 5.25.– *If $f \in \mathcal{C}(\Omega; E)$, its image $f \circ T$ in the distribution sense given by Definition 5.21 coincides with that in the function sense.*

That is to say that, with identification (3.6), p. 52, of f with a distribution, Definition 5.21 again gives $f \circ T \in \mathcal{C}(\Lambda; E)$ and, for every $y \in \Lambda$,

$$(f \circ T)(y) = f(T(y)). \ \blacksquare$$

Proof. The point is to check that the distribution $\overline{f}$ associated with the function f by Theorem 3.5 satisfies

$$\overline{f} \circ T = \overline{f \circ T}. \tag{5.13}$$

Definition 5.21 of $\overline{f} \circ T$ and the expression for $\langle \overline{f}, \varphi\rangle$ from Theorem 3.5 give, for every $\phi \in \mathcal{D}(\Lambda)$,

$$\langle \overline{f} \circ T, \phi\rangle_\Lambda = \langle \overline{f}, |\det[\nabla T^{-1}]|\, \phi \circ T^{-1}\rangle_\Omega = \int_\Omega f\, |\det[\nabla T^{-1}]|\, \phi \circ T^{-1}.$$

Due to Theorem A.85 (c) on the change of variable in a Cauchy integral,

$$\int_\Omega f\, |\det[\nabla T^{-1}]|\, \phi \circ T^{-1} = \int_\Lambda (f \circ T)\, \phi. \tag{5.14}$$

These equalities give, again with the expression from Theorem 3.5, here for $\overline{f \circ T}$,

$$\langle \overline{f} \circ T, \phi\rangle_\Lambda = \int_\Lambda (f \circ T)\, \phi = \langle \overline{f \circ T}, \phi\rangle_\Lambda.$$

This proves (5.13), except that Theorem A.85 (c) used to establish (5.14) assumes that T and T^{-1} are uniformly continuous, which is not the case here.

We come back to this case by replacing Λ by the interior λ of the support of ϕ, which does not change the values of the integrals (Theorem A.77) since ϕ and $\phi \circ T^{-1}$ are zero respectively outside of λ and $T(\lambda)$. Then, Theorem A.85 applies since T is uniformly continuous on the compact set $\overline{\lambda}$, and so on λ, from Heine's theorem (Theorem A.32) and similarly T^{-1} is uniformly continuous on $\overline{T(\lambda)}$ and thus on $T(\lambda)$. This implies (5.14), and therefore (5.13). □

Let us prove that the change of variable is an isomorphism.

Theorem 5.26.– *Let T be a bijection from an open subset Λ of $\mathbb{R}^d$ onto another, Ω, such that $T \in \mathcal{C}^\infty(\Lambda; \Omega)$ and $T^{-1} \in \mathcal{C}^\infty(\Omega; \Lambda)$, and let E be a Neumann space.*

Then, the mapping $f \mapsto f \circ T$ is an isomorphism from $\mathcal{D}'(\Omega; E)$ onto $\mathcal{D}'(\Lambda; E)$. ▮

Proof. ***1.** Bijectivity.* Definition 5.21 of the change of variables successively for T^{-1} and T gives, for every $f \in \mathcal{D}'(\Omega; E)$ and $\varphi \in \mathcal{D}(\Omega)$,

$$\langle (f \circ T) \circ T^{-1}, \varphi \rangle_\Omega = \langle f \circ T, |\det[\nabla T]|\, \varphi \circ T \rangle_\Lambda =$$
$$= \langle f, |\det[\nabla T^{-1}]|\, (|\det[\nabla T]| \circ T^{-1})\, \varphi \rangle_\Omega = \langle f, \varphi \rangle_\Omega,$$

since the product of the Jacobian determinants equals 1 due to Lemma 5.23. So,

$$(f \circ T) \circ T^{-1} = f.$$

Likewise, for every distribution $g \in \mathcal{D}'(\Lambda; E)$,

$$(g \circ T^{-1}) \circ T = g,$$

which proves that the mapping $f \mapsto f \circ T$ is bijective and that its inverse mapping is $g \mapsto g \circ T^{-1}$.

***2.** Bicontinuity.* It remains to verify that the mappings $f \mapsto f \circ T$ and $g \mapsto g \circ T^{-1}$ are continuous.

Let $\{\| \; \|_{E;\nu} : \nu \in \mathcal{N}_E\}$ be the family of semi-norms of E. Due to Definitions 3.1 of the semi-norms of $\mathcal{D}'(\Lambda; E)$ and 5.21 of $f \circ T$, we have, for every $\phi \in \mathcal{D}(\Lambda)$ and $\nu \in \mathcal{N}_E$,

$$\|f \circ T\|_{\mathcal{D}'(\Lambda;E);\phi,\nu} = \|\langle f \circ T, \phi \rangle_\Lambda\|_{E;\nu} =$$
$$= \|\langle f, |\det[\nabla T^{-1}]|\, \phi \circ T^{-1} \rangle_\Omega\|_{E;\nu} = \|f\|_{\mathcal{D}'(\Omega;E);\varphi,\nu},$$

where $\varphi = |\det[\nabla T^{-1}]|\, \phi \circ T^{-1}$, which belongs to $\mathcal{D}(\Omega)$ due to Theorem 5.22. This implies the continuity of the mapping $f \mapsto f \circ T$ due to the characterization of continuous linear mappings from Theorem 1.12 (a).

Similarly, the mapping $g \mapsto g \circ T^{-1}$ is continuous from $\mathcal{D}'(\Lambda; E)$ into $\mathcal{D}'(\Omega; E)$. □

Let us calculate the derivative of the image of a distribution under a change of variables.

Theorem 5.27.– *Let $f \in \mathcal{D}'(\Omega; E)$, where Ω is an open subset of $\mathbb{R}^d$ and E is a Neumann space, and let T be a bijection from an open subset Λ of $\mathbb{R}^d$ onto Ω such that $T \in \mathcal{C}^\infty(\Lambda; \Omega)$ and $T^{-1} \in \mathcal{C}^\infty(\Omega; \Lambda)$. Then,*

$$\partial_i(f \circ T) = \sum_{j=1}^{d} \partial_i T_j\, (\partial_j f) \circ T. \tag{5.15}$$

■

Proof. **1.** *Values for a test function ϕ.* Definitions 5.4 of the derivative and 5.21 of the change of variables give, for every $\phi \in \mathcal{D}(\Lambda)$,

$$\langle \partial_i(f \circ T), \phi \rangle_\Lambda = -\langle f \circ T, \partial_i \phi \rangle_\Lambda = -\langle f, J\, \partial_i \phi \circ T^{-1} \rangle_\Omega, \tag{5.16}$$

where

$$J \stackrel{\text{def}}{=} |\det[\nabla T^{-1}]|.$$

On the other hand, Definition 5.15 of the product with a regular function, in this case $\partial_i T_j$, and again Definitions 5.21 and 5.4 give

$$\langle \partial_i T_j\, (\partial_j f) \circ T, \phi \rangle_\Lambda = \langle (\partial_j f) \circ T, \partial_i T_j\, \phi \rangle_\Lambda =$$
$$= \langle \partial_j f, J\, (\partial_i T_j\, \phi) \circ T^{-1} \rangle_\Omega = -\langle f, \partial_j (J\, (\partial_i T_j\, \phi) \circ T^{-1}) \rangle_\Omega. \tag{5.17}$$

Therefore,

$$\langle \partial_i(f \circ T) - \sum_{j=1}^{d} \partial_i T_j\, (\partial_j f) \circ T, \phi \rangle_\Lambda = -\langle f, X \rangle_\Omega, \tag{5.18}$$

where

$$X = J\, \partial_i \phi \circ T^{-1} - \sum_{j=1}^{d} \partial_j (J\, (\partial_i T_j\, \phi) \circ T^{-1}).$$

2. *Conclusion.* If now $f \in \mathcal{C}^1(\Omega)$, equality (5.15) is satisfied (Theorem A.64 (a)), thus (5.18) gives, with the expression for the identification of f with a distribution from Theorem 3.9,

$$\int_\Omega f X = 0_E.$$

According to the Du Bois-Reymond lemma (Theorem 3.7), it follows that $X = 0$. Equality (5.18) then gives (5.15). □

5.6. Some particular changes of variables

We start with **translation**. The translated function $\tau_x f$ of a function f defined on Ω is defined on $\Omega + x$ by

$$(\tau_x f)(y) = f(y - x).$$

Let us now define the translation of a distribution.

Theorem 5.28.– *Let $f \in \mathcal{D}'(\Omega; E)$, where Ω is an open subset of $\mathbb{R}^d$ and E is a Neumann space, and $x \in \mathbb{R}^d$.*

We define $\tau_x f \in \mathcal{D}'(\Omega + x; E)$ by: for every $\phi \in \mathcal{D}(\Omega + x)$,

$$\langle \tau_x f, \phi \rangle_{\Omega + x} \stackrel{\text{def}}{=} \langle f, \tau_{-x} \phi \rangle_\Omega.$$

The mapping $f \mapsto \tau_x f$ is an isomorphism from $\mathcal{D}'(\Omega; E)$ onto $\mathcal{D}'(\Omega + x; E)$. ▮

Proof. The definition of $\tau_x f$ is a particular case of Definition 5.21 of the image under a change of variable. More precisely $\tau_x f = f \circ T$ where $T(y) = y - x$, since $T^{-1}(y) = y + x$, $[\nabla T^{-1}] = [\mathbf{e}_1, \ldots, \mathbf{e}_d]$ and (Theorem A.72 (a))

$$\det[\nabla T^{-1}] = \det[\mathbf{e}_1, \ldots, \mathbf{e}_d] = 1.$$

Thus we indeed have

$$\tau_x f \in \mathcal{D}'(\Omega + x; E)$$

and $f \mapsto \tau_x f$ is an isomorphism by Theorem 5.26. □

We consider now the case of a **linear change of variable**.

Theorem 5.29.– *Let $f \in \mathcal{D}'(\Omega; E)$, where Ω is an open subset of $\mathbb{R}^d$ and E is a Neumann space, and let T be a linear bijection from $\mathbb{R}^d$ onto itself.*

Then, $f \circ T \in \mathcal{D}'(T^{-1}(\Omega); E)$ and, for every $\phi \in \mathcal{D}(T^{-1}(\Omega))$,

$$\langle f \circ T, \phi \rangle_{T^{-1}(\Omega)} = \kappa^{-1} \langle f, \phi \circ T^{-1} \rangle_\Omega, \tag{5.19}$$

where

$$\kappa = |\det[T(\mathbf{e}_1), \ldots, T(\mathbf{e}_d)]|$$

and $\kappa^{-1} = 1/\kappa = |\det[T^{-1}(\mathbf{e}_1), \ldots, T^{-1}(\mathbf{e}_d)]|$.

The mapping $f \mapsto f \circ T$ *is an isomorphism from* $\mathcal{D}'(\Omega; E)$ *onto* $\mathcal{D}'(T^{-1}(\Omega); E)$.

Proof. **1.** *Differentiability of* T. Observe that each linear mapping T from $\mathbb{R}^d$ into itself is infinitely differentiable and, for every $x \in \mathbb{R}^d$,

$$[\nabla T](x) = [T(\mathbf{e}_1), \ldots, T(\mathbf{e}_d)].$$

Indeed, it satisfies Definition 1.15 of differentiability since, for every $z \in \mathbb{R}^d$,

$$T(x+z) - T(x) - \sum_{i=1}^{d} z_i T(\mathbf{e}_i) = 0,$$

because $T(z) = T(\sum_{i=1}^d z_i \mathbf{e}_i) = \sum_{i=1}^d z_i T(\mathbf{e}_i)$. Since ∇T is hence independent of x, it has successive gradients $\nabla^m T$ which are all zero.

2. *Equalities.* By hypothesis, T is invertible. Then, T^{-1} is also linear and Definition 5.21 of the image under a change of variable gives

$$f \circ T \in \mathcal{D}'(T^{-1}(\Omega); E)$$

and equality (5.19) with

$$\kappa^{-1} = |\det[\nabla T^{-1}]| = |\det[T^{-1}(\mathbf{e}_1), \ldots, T^{-1}(\mathbf{e}_d)]|.$$

The product of the Jacobian determinants of T and T^{-1} equals 1 due to Lemma 5.23, and more precisely

$$\det[\nabla T^{-1}] \det[\nabla T] \circ T^{-1} = 1 \text{ on } \mathbb{R}^d.$$

So,

$$\kappa = 1/\kappa^{-1} = |\det[\nabla T]| \circ T^{-1} = |\det[T(\mathbf{e}_1), \ldots, T(\mathbf{e}_d)]|.$$

3. *Isomorphism.* The mapping $f \mapsto f \circ T$ is an isomorphism due to Theorem 5.26. □

We finish this section with the **permutation of variables** for a distribution given on a product $\Omega_1 \times \Omega_2$.

Theorem 5.30.– *Let $f \in \mathcal{D}'(\Omega_1 \times \Omega_2; E)$, where Ω_1 is an open subset of $\mathbb{R}^{d_1}$, Ω_2 is an open subset of $\mathbb{R}^{d_2}$ and E is a Neumann space.*

We define $\check{f} \in \mathcal{D}'(\Omega_2 \times \Omega_1; E)$ by: for every $\varphi \in \mathcal{D}(\Omega_2 \times \Omega_1)$,

$$\langle \check{f}, \varphi \rangle_{\Omega_2 \times \Omega_1} \stackrel{\text{def}}{=} \langle f, \hat{\varphi} \rangle_{\Omega_1 \times \Omega_2},$$

where $\hat{\varphi}(x_1, x_2) = \varphi(x_2, x_1)$.

The mapping $f \mapsto \check{f}$ is an isomorphism from $\mathcal{D}'(\Omega_1 \times \Omega_2; E)$ onto $\mathcal{D}'(\Omega_2 \times \Omega_1; E)$. ▮

Proof. This is a particular case of Theorem 5.29 on the image under a linear change of variables.

More precisely, $\check{f} = f \circ T$ where $T(x_2, x_1) = (x_1, x_2)$ and $\Omega = \Omega_1 \times \Omega_2$. Indeed, $(T(\mathbf{e}_1), \ldots, T(\mathbf{e}_d))$, where $d = d_1 + d_2$, is the family $(\mathbf{e}_1, \ldots, \mathbf{e}_d)$ arranged in a different order, so (Theorem A.72 (b) and (a))

$$\kappa = |\det[T(\mathbf{e}_1), \ldots, T(\mathbf{e}_d)]| = |\det[\mathbf{e}_1, \ldots, \mathbf{e}_d]| = 1. \ \square$$

Notation. We do not denote the permutation of the variables of f and that of φ in the same way, since their variables are not in the same order: $\check{f}$ is defined on $\Omega_2 \times \Omega_1$, like φ, while $\hat{\varphi}$ is defined on $\Omega_1 \times \Omega_2$, like f. □

Separation of variables. The separation of variables is much more delicate than their permutation. This is the subject of the *kernel theorem* (Theorem 15.10). □

5.7. Positive distributions

Let us define positive real distributions [11].

Definition 5.31.– *Let $f \in \mathcal{D}'(\Omega)$, where Ω is an open subset of $\mathbb{R}^d$.*

*We say that f is **positive** and we denote it by $f \geq 0$ if, for every $\varphi \in \mathcal{D}(\Omega)$ such that $\varphi \geq 0$,*

$$\langle f, \varphi \rangle \geq 0$$

(i.e. if this is true for every $\varphi \in \mathcal{D}(\Omega)$ such that $\varphi(x) \geq 0$ for every $x \in \Omega$).

We denote $f \geq g$ or $g \leq f$ if $f - g \geq 0$. ▮

11. **History of positive distributions.** Laurent SCHWARTZ gave Definition 5.31 of a positive distribution in 1945 [68, p. 66].

Let us check that this definition is compatible with that of a positive function.

Theorem 5.32.– *If $f \in C(\Omega)$, positivity in the distribution sense given by Definition 5.31 coincides with that in the function sense.*

That is to say that, with identification (3.6), p. 52, of f with a distribution, Definition 5.31 is equivalent to:

$$f(x) \geq 0, \ \forall x \in \Omega. \ \blacksquare$$

Proof. We denote here $\bar{f}$ the distribution that is associated with the function f by Theorem 3.5 and $\succeq$ the positivity in the distribution sense. It is a question of checking that $\bar{f} \succeq 0$ is equivalent to $f \geq 0$.

1. *First implication.* Suppose

$$\bar{f} \succeq 0.$$

If f was not positive, there would exist $x \in \Omega$ such that $f(x) = a < 0$ and therefore, thanks to its continuity, there would exist $r > 0$ such that the open ball $\overset{\circ}{B} = \{y \in \mathbb{R}^d : |y - x| < r\}$ is included in Ω and

$$f \leq \frac{a}{2} \ \text{ on } \overset{\circ}{B}.$$

Let $\varphi \in \mathcal{D}(\Omega)$ have its support in $\overset{\circ}{B}$ and satisfy

$$\varphi \geq 0, \quad \varphi \neq 0$$

(for example $\varphi(y) = \rho((y - x)/2r)$ where ρ is given by (2.4), p. 23).

Then, definition of $\bar{f}$ from Theorem 3.5 would give, with the properties of the real integral from Theorems A.77 and A.76 (a) and (c),

$$\langle \bar{f}, \varphi \rangle = \int_\Omega f\varphi = \int_{\overset{\circ}{B}} f\varphi \leq \frac{a}{2} \int_{\overset{\circ}{B}} \varphi < 0,$$

which would contradict $\bar{f} \succeq 0$. Thus,

$$f \geq 0.$$

2. *Converse implication.* Suppose now

$$f \geq 0.$$

Then, for each test function $\varphi \geq 0$, we have $f\varphi \geq 0$, from which, again according to Definition 3.5 and Theorem A.76 (a),

$$\langle \bar{f}, \varphi \rangle = \int_\Omega f\varphi \geq 0.$$

And thus

$$\bar{f} \succeq 0. \ \square$$

We call the **positive cone** the set of positive distributions, that we denote

$$\mathcal{D}'^{+}(\Omega) \stackrel{\text{def}}{=} \{f \in \mathcal{D}'(\Omega) : f \geq 0\}.$$

Let us show that it is closed in $\mathcal{D}'(\Omega)$.

Theorem 5.33.– *The set $\mathcal{D}'^{+}(\Omega)$ is closed in $\mathcal{D}'(\Omega)$, for every open subset Ω of $\mathbb{R}^d$.* ▌

Proof. By Definition A.7 (b) of a closed set, it is a question of proving that the complement of $\mathcal{D}'^{+}(\Omega)$ is open (Definition A.7 (a)). So let

$$f \in \mathcal{D}'(\Omega) \setminus \mathcal{D}'^{+}(\Omega).$$

Then there exists $\varphi \in \mathcal{D}(\Omega)$ such that $\varphi \geq 0$ and $\langle f, \varphi \rangle = a < 0$. Therefore, if $g \in \mathcal{D}'(\Omega)$ satisfies $|\langle g - f, \varphi \rangle| \leq a/2$, then $\langle g, \varphi \rangle < 0$, i.e.

$$\|g - f\|_{\mathcal{D}'(\Omega);\varphi} \leq a/2 \quad \Rightarrow \quad g \in \mathcal{D}'(\Omega) \setminus \mathcal{D}'^{+}(\Omega).$$

This proves that $\mathcal{D}'(\Omega) \setminus \mathcal{D}'^{+}(\Omega)$ is open. □

Now let us prove that every positive distribution is a measure [12].

Theorem 5.34.– *Every positive distribution on an open subset of $\mathbb{R}^d$ is a measure.* ▌

Complement. Denoting $\mathcal{M}^{+}(\Omega)$ the set of positive measures, we have

$$\mathcal{D}'^{+}(\Omega) = \mathcal{M}^{+}(\Omega).$$

Indeed, every measure is a distribution from identification (3.15), p. 61, and, if it is positive in the sense of measures, it is too in the sense of distributions; and the converse is given by Theorem 5.34. □

12. **History of Theorem 5.34.** Laurent SCHWARTZ proved in 1945 [68, Theorem, p. 66] that every positive distribution is a measure.

Proof of Theorem 5.34. **1.** *A sufficient condition.* Let $f \in \mathcal{D}'^+(\Omega)$. It suffices to check the condition given in Theorem 3.19 for a distribution to be a measure, that is to construct a function p of $\mathcal{C}^+(\Omega)$ such that, for every $\varphi \in \mathcal{D}(\Omega)$,

$$|\langle f, \varphi \rangle| \leq \sup_{x \in \Omega} p(x)|\varphi(x)|. \tag{5.20}$$

2. *Local upper bounded.* Let $(\kappa_n)_{n\in\mathbb{N}}$ be the cover of Ω by crown-shaped sets (see Definition 2.18) and $(\alpha_n)_{n\in\mathbb{N}}$ be a subordinate partition of unity (it exists due to Theorem 2.21).

Let $\varphi \in \mathcal{D}(\Omega)$. By Definition 2.20 of a partition of unity, α_n is zero outside of κ_n and is positive, so

$$\alpha_n \varphi \leq c_n \alpha_n, \text{ where } c_n = \sup_{x\in\kappa_n} |\varphi(x)|.$$

Therefore, $c_n\alpha_n - \alpha_n\varphi \geq 0$, so the positivity of f gives

$$\langle f, c_n\alpha_n - \alpha_n\varphi \rangle \geq 0,$$

i.e.

$$\langle f, \alpha_n\varphi \rangle \leq c_n \langle f, \alpha_n \rangle.$$

Likewise, $-\alpha_n\varphi \leq c_n\alpha_n$ from which

$$-\langle f, \alpha_n\varphi \rangle \leq c_n \langle f, \alpha_n \rangle.$$

Thus,

$$|\langle f, \alpha_n\varphi \rangle| \leq \langle f, \alpha_n \rangle \sup_{x\in\kappa_n} |\varphi(x)|. \tag{5.21}$$

3. *Global upper bound.* Again according to Definition 2.20 of a partition of unity, $\sum_{n\in\mathbb{N}} \alpha_n = 1$, hence

$$\varphi = \sum_{n\in N_\varphi} \alpha_n \varphi,$$

where N_φ is the set of the n for which $\alpha_n\varphi$ is not zero. This set N_φ is finite, since only a finite number of the α_n are not zero in the support of φ (again due to the definition of a partition of unity, with $K = \operatorname{supp}\varphi$). Hence, with (5.21),

$$|\langle f, \varphi \rangle| = \Big| \sum_{n\in N_\varphi} \langle f, \alpha_n\varphi \rangle \Big| \leq \sum_{n\in N_\varphi} \langle f, \alpha_n \rangle \sup_{x\in\kappa_n} |\varphi(x)|.$$

By bounding this sum from above by a weighted supremum thanks to Lemma 2.24, it follows that

$$|\langle f, \varphi \rangle| \leq \sup_{n\in\mathbb{N}} 2^{n+1} \langle f, \alpha_n \rangle \sup_{x\in\kappa_n} |\varphi(x)|. \tag{5.22}$$

4. Construction of p. We define $p \in \mathcal{C}^+(\Omega)$ by: for every $x \in \Omega$,

$$p(x) \stackrel{\text{def}}{=} \sum_{n \in \mathbb{N}} 2^{n+1} \langle f, \alpha_n \rangle \, a_n(x), \tag{5.23}$$

where, denoting $\alpha_{-1} = 0$,

$$a_n \stackrel{\text{def}}{=} \alpha_{n-1} + \alpha_n + \alpha_{n+1}.$$

The sum in (5.23) makes sense because only a finite number of its terms are not zero. Indeed, due to Theorem 2.19 (b), the crown-shaped set κ_n only intersects κ_{n-1} and κ_{n+1}, so at most α_{n-1}, α_n and α_{n+1} are not zero on κ_n, and so at most a_{n-2}, a_{n-1}, a_n, a_{n+1} and a_{n+2} are not zero there. Hence, if $x \in \kappa_n$,

$$p(x) = \sum_{i=n-2}^{n+2} 2^{i+1} \langle f, \alpha_i \rangle \, a_i(x). \tag{5.24}$$

Each of these five functions a_i is continuous and positive, since the α_i are, hence p is continuous on every κ_n, and thus on their union Ω.

5. Inequality. Again by Definition 2.20 of a partition of unity, we have, still for $x \in \kappa_n$,

$$a_n(x) = \alpha_{n-1}(x) + \alpha_n(x) + \alpha_{n+1}(x) = \sum_{i \in \mathbb{N}} \alpha_i(x) = 1.$$

Moreover, the $\langle f, \alpha_i \rangle$ are positive, from the positivity hypothesis on f. So, (5.24) gives

$$p(x) \geq 2^{n+1} \langle f, \alpha_n \rangle.$$

Inequality (5.22) then gives

$$|\langle f, \varphi \rangle| \leq \sup_{n \in \mathbb{N}} \sup_{x \in \kappa_n} p(x) |\varphi(x)| = \sup_{x \in \Omega} p(x) |\varphi(x)|.$$

This proves (5.20) which, as we have seen, implies $f \in \mathcal{M}(\Omega)$. □

5.8. Distributions with values in a product space

Here we focus on the distributions with values in a product $E_1 \times E_2 \times \cdots \times E_I$ of Neumann spaces. Recall that, if the E_i are Neumann spaces, their product is a Neumann space, as we have established in Volume 1 [81, Theorem 6.7], so $\mathcal{D}'(\Omega; E_1 \times \cdots \times E_I)$ is well-defined.

The properties are analogous to those of fields considered in § 5.1, namely to those of distributions with values in E^d, but the manipulation is a little bit more laborious, since the semi-norms here are indexed by several ν_i, rather than only just one ν, and, for their components, by several pairs (φ_i, ν_i) rather than just one pair (φ, ν).

Let us define the **components** f_i of a distribution f with values in a product.

Definition 5.35.– *Let*

$$f \in \mathcal{D}'(\Omega; E_1 \times \cdots \times E_I),$$

where Ω is an open subset of $\mathbb{R}^d$ and $E_1, \ldots, E_I$ are Neumann spaces. We define

$$(f_1, \ldots, f_I) \in \mathcal{D}'(\Omega; E_1) \times \cdots \times \mathcal{D}'(\Omega; E_I)$$

by: for every $i \in [\![1, I]\!]$ and $\varphi \in \mathcal{D}(\Omega)$,

$$\langle f_i, \varphi \rangle \stackrel{\text{def}}{=} (\langle f, \varphi \rangle)_i. \ ■$$

Justification. The components f_i are indeed distributions of the $\mathcal{D}'(\Omega; E_i)$, by Definition 5.10 of the image of the distribution f under a continuous linear mapping, here $(e_1, \ldots, e_I) \mapsto e_i$, from $E_1 \times \cdots \times E_I$ into E_i. □

Let us complete Definition 5.35 by specifying the families of semi-norms of the spaces which play a role there. Recall that, denoting by $\{\| \ \|_{E_i;\nu_i} : \nu_i \in \mathcal{N}_{E_i}\}$ the family of semi-norms of E_i, the **product** $E_1 \times \cdots \times E_I$ is endowed with the semi-norms, indexed by $\nu_1 \in \mathcal{N}_{E_1}, \ldots, \nu_I \in \mathcal{N}_{E_I}$,

$$\|(u_1, \ldots, u_I)\|_{E_1 \times \cdots \times E_I;\nu_1,\ldots,\nu_I} \stackrel{\text{def}}{=} (\|u_1\|^2_{E_1;\nu_1} + \cdots + \|u_I\|^2_{E_I;\nu_I})^{1/2}. \tag{5.25}$$

Definition 5.36.– *Let Ω be an open subset of $\mathbb{R}^d$, $E_1, \ldots, E_I$ Neumann spaces, and $\{\| \ \|_{E_i;\nu_i} : \nu_i \in \mathcal{N}_{E_i}\}$ the family of semi-norms of E_i.*

(a) *The space $\mathcal{D}'(\Omega; E_1 \times \cdots \times E_I)$ is endowed with the semi-norms, indexed by $\varphi \in \mathcal{D}(\Omega)$ and $\nu_1 \in \mathcal{N}_{E_1}, \ldots, \nu_I \in \mathcal{N}_{E_I}$,*

$$\|f\|_{\mathcal{D}'(\Omega;E_1\times\cdots\times E_I);\varphi,\nu_1,\ldots,\nu_I} \stackrel{\text{def}}{=} \|\langle f, \varphi \rangle\|_{E_1\times\cdots\times E_I;\nu_1,\ldots,\nu_I} \stackrel{\text{def}}{=}$$
$$\stackrel{\text{def}}{=} \left(\|(\langle f, \varphi \rangle)_1\|^2_{E_1;\nu_1} + \cdots + \|(\langle f, \varphi \rangle)_I\|^2_{E_I;\nu_I}\right)^{1/2}.$$

(b) *The space $\mathcal{D}'(\Omega; E_1) \times \cdots \times \mathcal{D}'(\Omega; E_I)$ is endowed with the semi-norms, indexed by $\varphi_1 \in \mathcal{D}(\Omega)$, $\nu_1 \in \mathcal{N}_{E_1}, \ldots, \varphi_I \in \mathcal{D}(\Omega)$, $\nu_I \in \mathcal{N}_{E_I}$,*

$$\|(f_1, \ldots, f_I)\|_{\mathcal{D}'(\Omega;E_1)\times\cdots\times\mathcal{D}'(\Omega;E_I);(\varphi_1,\nu_1),\ldots,(\varphi_I,\nu_I)} \stackrel{\text{def}}{=}$$
$$\stackrel{\text{def}}{=} \left(\|f_1\|^2_{\mathcal{D}'(\Omega;E_1);\varphi_1,\nu_1} + \cdots + \|f_I\|^2_{\mathcal{D}'(\Omega;E_I);\varphi_I,\nu_I}\right)^{1/2} \stackrel{\text{def}}{=}$$
$$\stackrel{\text{def}}{=} \left(\|\langle f_1, \varphi_1 \rangle\|^2_{E_1;\nu_1} + \cdots + \|\langle f_I, \varphi_I \rangle\|^2_{E_I;\nu_I}\right)^{1/2}. \ ■$$

Justification. These expressions follow from Definition 3.1 of the semi-norms of $\mathcal{D}'(\Omega; E)$, due to the definition of the semi-norms of a product (equality (5.25)). □

Separation into components in the function sense. For a function with values in a product space, the separation into components of Definition 5.35 coincides with the separation in the function sense since the image under the continuous linear mapping $(e_1, \ldots, e_I) \mapsto e_i$ coincides with the image in the function sense due to Theorem 5.11. □

Separation into components of a field. For a field $q \in \mathcal{D}'\Omega; E^d)$, the separation into components of Definition 5.35 coincides with that of Definition 5.1 since, as we established in Volume 1 [81, Theorem 6.11], the family of semi-norms of E^d given by (5.25) is equivalent (but not identical) to the family defined by (5.1), p. 81, i.e.

$$E^d \mathrel{\overline{\overline{\rightleftarrows}}} \underbrace{E \times \cdots \times E}_{d \text{ times}}. \ \square$$

Utility of distributions with values in a product. Distributions with values in a product of semi-normed spaces is, for example, useful when we consider a velocity and pressure solution of the Navier-Stokes equations and we separate time and space variables; then, for suitable data, see Volume 7, (u, p) lives in $\mathcal{D}'((0, T); (H^1(\Omega))^d \times W^{-1,\infty}(\Omega))$.

Such distributions are equally useful for the study of anisotropic problems. □

Let us prove that separation into components is an isomorphism. Although this result is natural, it is not unhelpful to detail the required utilization of semi-norms.

Theorem 5.37.– *Let Ω be an open subset of $\mathbb{R}^d$ and $E_1, \ldots, E_I$ Neumann spaces.*

The separation into components is an isomorphism from $\mathcal{D}'(\Omega; E_1 \times \cdots \times E_I)$ onto $\mathcal{D}'(\Omega; E_1) \times \cdots \times \mathcal{D}'(\Omega; E_I)$. ▮

Proof. Let $\{\| \, \|_{E_i;\nu_i} : \nu_i \in \mathcal{N}_{E_i}\}$ be the family of semi-norms of E_i.

1. *Injectivity of $f \mapsto (f_1, \ldots, f_I)$.* Let $f \in \mathcal{D}'(\Omega; E_1 \times \cdots \times E_I)$ be such that

$$(f_1, \ldots, f_I) = 0.$$

For every $i \in [\![1, I]\!]$ and $\varphi \in \mathcal{D}(\Omega)$, we have $\langle f_i, \varphi \rangle = 0_{E_i}$, that is $(\langle f, \varphi \rangle)_i = 0_{E_i}$ by Definition 5.35 of f_i. Thus, $\langle f, \varphi \rangle = 0$ in $E_1 \times \cdots \times E_I$, hence

$$f = 0.$$

This proves that the mapping $f \mapsto (f_1, \ldots, f_I)$ is injective.

2. *Surjectivity.* Let

$$(f_1, \ldots, f_I) \in \mathcal{D}'(\Omega; E_1) \times \cdots \times \mathcal{D}'(\Omega; E_I).$$

We define a mapping f from $\mathcal{D}(\Omega)$ into $E_1 \times \cdots \times E_I$ by: for every $\varphi \in \mathcal{D}(\Omega)$,

$$\langle f, \varphi \rangle \stackrel{\text{def}}{=} (\langle f_1, \varphi \rangle, \ldots, \langle f_I, \varphi \rangle).$$

Due to the characterization of distributions from Theorem 3.3, for every $i \in [\![1, I]\!]$ and $\nu_i \in \mathcal{N}_{E_i}$, there exist $p_{\nu_i} \in \mathcal{C}^+(\Omega)$ and $c_{\nu_i} \in \mathbb{R}$ such that, for every $\varphi \in \mathcal{D}(\Omega)$,

$$\|\langle f_i, \varphi \rangle\|_{E_i;\nu_i} \leq c_{\nu_i} \|\varphi\|_{\mathcal{D}(\Omega);p_{\nu_i}}.$$

Hence,

$$\begin{aligned}\|\langle f, \varphi \rangle\|_{E_1 \times \cdots \times E_I;\nu_1,\ldots,\nu_I} &= (\|\langle f_1, \varphi \rangle\|^2_{E_1;\nu_1} + \cdots + \|\langle f_I, \varphi \rangle\|^2_{E_I;\nu_I})^{1/2} \leq \\ &\leq (c_{\nu_1} + \cdots + c_{\nu_I})(\|\varphi\|_{\mathcal{D}(\Omega);p_{\nu_1}} + \cdots + \|\varphi\|_{\mathcal{D}(\Omega);p_{\nu_I}}).\end{aligned}$$

So, due to the characterization of continuous linear mappings from Theorem 1.12 (a), the mapping f is continuous from $\mathcal{D}(\Omega)$ into $E_1 \times \cdots \times E_I$, i.e.

$$f \in \mathcal{D}'(\Omega; E_1 \times \cdots \times E_I).$$

This proves that the mapping $f \mapsto (f_1, \ldots, f_I)$ is surjective. It is therefore bijective due to step 1.

3. *Continuity.* Given $f \in \mathcal{D}'(\Omega; E_1 \times \cdots \times E_I)$, Definitions 5.36 (b) of the semi-norms of $\mathcal{D}'(\Omega; E_1) \times \cdots \times \mathcal{D}'(\Omega; E_I)$ and 5.35 of the components successively give, for all pairs $\varphi_1 \in \mathcal{D}(\Omega)$, $\nu_1 \in \mathcal{N}_{E_1}, \ldots$ and $\varphi_I \in \mathcal{D}(\Omega)$, $\nu_I \in \mathcal{N}_{E_I}$,

$$\begin{aligned}&\|(f_1, \ldots, f_I)\|_{\mathcal{D}'(\Omega;E_1)\times\cdots\times\mathcal{D}'(\Omega;E_I);(\varphi_1,\nu_1),\ldots,(\varphi_I,\nu_I)} = \\ &\qquad = \Big(\sum_{i=1}^{I} \|\langle f_i, \varphi_i \rangle\|^2_{E_i;\nu_i}\Big)^{1/2} = \Big(\sum_{i=1}^{I} \|(\langle f, \varphi_i \rangle)_i\|^2_{E_i;\nu_i}\Big)^{1/2}.\end{aligned}$$

By Definition 5.36 (a) of the semi-norms of $\mathcal{D}'(\Omega; E_1 \times \cdots \times E_I)$, this is bounded by

$$\leq \sum_{j=1}^{I} \Big(\sum_{i=1}^{I} \|(\langle f, \varphi_j \rangle)_i\|^2_{E_i;\nu_i}\Big)^{1/2} = \sum_{j=1}^{I} \|f\|_{\mathcal{D}'(\Omega;E_1\times\cdots\times E_I);\varphi_j,\nu_1,\ldots,\nu_I}.$$

The mapping $f \mapsto (f_1, \ldots, f_I)$, being linear, is therefore continuous, again due to the characterization from Theorem 1.12 (a).

4. *Isomorphism.* Conversely, for every $\varphi \in \mathcal{D}(\Omega)$, $\nu_1 \in \mathcal{N}_{E_1}, \ldots, \nu_I \in \mathcal{N}_{E_I}$,

$$\|f\|_{\mathcal{D}'(\Omega;E_1\times\cdots\times E_I);\varphi,\nu_1,\ldots,\nu_I} = \|(f_1, \ldots, f_I)\|_{\mathcal{D}'(\Omega;E_1)\times\cdots\times\mathcal{D}'(\Omega;E_I);(\varphi,\nu_1),\ldots,(\varphi,\nu_I)}.$$

Hence the continuity of the inverse mapping $(f_1, \ldots, f_I) \mapsto f$, still according to Theorem 1.12 (a).

The mapping $f \mapsto (f_1, \ldots, f_d)$ is therefore an isomorphism. □

Chapter 6

Restriction, Gluing and Support

We show here the *local character* of distributions: if two distributions are equal on various open subsets ω_i of their domains of definition, they are equal on the union of the ω_i. It is the *gluing theorem for equalities* (Theorem 6.10). *Equal on* ω_i meaning that their restrictions to ω_i are equal, we first study restriction (§ 6.1).

We also give the *gluing property* for distributions: if distributions f_i, respectively defined on open sets ω_i, satisfy $f_i = f_j$ on $\omega_i \cap \omega_j$ for all i and j, there exists a unique distribution, defined on the union of the ω_i, which is equal to f_i on ω_i for each i. This is the *gluing theorem for distributions* (Theorem 6.16). The main tool for this is the *localization–extension*, which we study beforehand (§ 6.4).

We finally look at the support of a distribution (§ 6.6) and at the space $\mathcal{D}'_K(\Omega; E)$ of distributions which have their support in a closed subset K of Ω (§ 6.8). We show that a distribution whose support is closed can be extended by 0 (Theorem 6.29).

In this chapter, as in the previous one, results are "natural" and proofs are simple.

6.1. Restriction

Let us define the **restriction of a distribution**.

Definition 6.1.– *Let $f \in \mathcal{D}'(\Omega; E)$, where Ω is an open subset of $\mathbb{R}^d$ and E is a Neumann space, and ω an open subset of Ω.*

We define the **restriction** $f|_\omega \in \mathcal{D}'(\omega; E)$ *by: for every $\varphi \in \mathcal{D}(\omega)$,*

$$\langle f|_\omega, \varphi \rangle_\omega \overset{\text{def}}{=} \langle f, \widetilde{\varphi} \rangle_\Omega,$$

where

$$\widetilde{\varphi} \in \mathcal{D}(\Omega)$$

is the extension of φ by 0. ∎

Justification. **1.** *Test function* $\widetilde{\varphi}$. The function $\widetilde{\varphi}$ is infinitely differentiable on ω (since it is equal to φ there) and on the open set $\Omega \setminus \operatorname{supp} \varphi$ (since it is zero there), and hence also on their union Ω.

Since its support is compact, φ is zero outside of a compact subset of ω due to Theorem 2.2 (a). The function $\widetilde{\varphi}$ is zero outside of this compact set, which is included in Ω, so its support is compact, again due to Theorem 2.2 (a). Thus,

$$\widetilde{\varphi} \in \mathcal{D}(\Omega). \tag{6.1}$$

2. *Distribution* $f|_\omega$. The linear mapping $f|_\omega$ is indeed a distribution on ω since, due to the characterization of distributions from Theorem 3.3, for every semi-norm $\| \ \|_{E;\nu}$ of E, there exist $p \in \mathcal{C}^+(\Omega)$ and $c \in \mathbb{R}$ such that

$$\|\langle f|_\omega, \varphi \rangle_\omega\|_{E;\nu} = \|\langle f, \widetilde{\varphi} \rangle_\Omega\|_{E;\nu} \leq c\|\widetilde{\varphi}\|_{\mathcal{D}(\Omega);p} = c\|\varphi\|_{\mathcal{D}(\omega);p|_\omega}.$$

The last equality results from the fact that, by Definition 2.5 of the semi-norms of $\mathcal{D}(\Omega)$,

$$\|\widetilde{\varphi}\|_{\mathcal{D}(\Omega);p} = \sup_{x \in \Omega} \sup_{|\beta| \leq p(x)} p(x)|\partial^\beta \widetilde{\varphi}(x)| =$$
$$= \sup_{x \in \omega} \sup_{|\beta| \leq p(x)} p(x)|\partial^\beta \varphi(x)| = \|\varphi\|_{\mathcal{D}(\omega);p|_\omega}. \ \square$$

Lighter notation. We often omit the symbol $|_\omega$ when there is no ambiguity, for example when denoting $f \in \mathcal{C}(\omega; E)$ to express the continuity of the restriction to ω of a distribution f on Ω. $\square$

Let us check that we recover the restriction of a function.

Theorem 6.2.– *If $f \in \mathcal{C}(\Omega; E)$, its restriction in the distribution sense given by Definition 6.1 coincides with its restriction in the function sense.*

That is to say that, with identification (3.6), p. 52, of f with a distribution, Definition 6.1 again gives, $f|_\omega \in \mathcal{C}(\omega; E)$ and, for every $x \in \omega$,

$$(f|_\omega)(x) = f(x). \ \blacksquare$$

Proof. It is a question of checking that the distribution $\overline{f}$ associated with the function f by Theorem 3.5 satisfies

$$\overline{f}|_\omega = \overline{f|_\omega}.$$

It is indeed the case, since Definition 6.1 of $\overline{f}|_\omega$, the expression for $\langle \overline{f}, \widetilde{\varphi} \rangle$ from Theorem 3.5, the fact that the sets where the function is zero do not contibute to the integral (Theorem A.77), and again Theorem 3.5 give, for every $\varphi \in \mathcal{D}(\omega)$,

$$\langle \overline{f}|_\omega, \varphi \rangle_\omega = \langle \overline{f}, \widetilde{\varphi} \rangle_\Omega = \int_\Omega f \widetilde{\varphi} = \int_\omega f|_\omega \varphi = \langle \overline{f|_\omega}, \varphi \rangle_\omega. \ \square$$

Let us show that the restriction of distributions is continuous.

Theorem 6.3.– *Let Ω and ω be two open subsets of $\mathbb{R}^d$ such that $\omega \subset \Omega$, and E a Neumann space.*

Then, the mapping $f \mapsto f|_\omega$ is continuous, and hence sequentially continuous, linear from $\mathcal{D}'(\Omega; E)$ into $\mathcal{D}'(\omega; E)$. ▮

Proof. Definitions 3.1 of the semi-norms of $\mathcal{D}'(\Omega; E)$ and 6.1 of restriction give, for every semi-norm $\| \ \|_{E;\nu}$ of E,

$$\|f|_\omega\|_{\mathcal{D}'(\omega;E),\varphi,\nu} = \|\langle f|_\omega, \varphi \rangle_\omega\|_{E;\nu} = \|\langle f, \widetilde{\varphi} \rangle_\Omega\|_{E;\nu} = \|f\|_{\mathcal{D}'(\Omega;E),\widetilde{\varphi},\nu}.$$

This implies the continuity of the mapping $f \mapsto f|_\omega$, due to the characterization of continuous linear mappings from Theorem 1.12 (a).

This mapping is hence sequentially continuous, like every continuous mapping (Theorem 1.10). $\square$

Let us verify that restriction commutes with the previously defined operations. We begin with differentiation.

Theorem 6.4.– *Let $f \in \mathcal{D}'(\Omega; E)$, where Ω is an open subset of $\mathbb{R}^d$ and E is a Neumann space, ω an open subset of Ω, and $\beta \in \mathbb{N}^d$. Then,*

$$\partial^\beta (f|_\omega) = (\partial^\beta f)|_\omega. \ ▮$$

Proof. The expression for the derivative ∂^β from Theorem 5.8 and Definition 6.1 of the restriction give, for every $\varphi \in \mathcal{D}(\omega)$,

$$\langle \partial^\beta (f|_\omega), \varphi \rangle_\omega = (-1)^{|\beta|} \langle f|_\omega, \partial^\beta \varphi \rangle_\omega = (-1)^{|\beta|} \langle f, \widetilde{\partial^\beta \varphi} \rangle_\Omega.$$

However $\widetilde{\partial^\beta \varphi}$ is equal to $\partial^\beta \widetilde{\varphi}$ on ω (since both are equal to $\partial^\beta \varphi$ there) and on the open set $\Omega \setminus \operatorname{supp} \varphi$ (since both are zero there from Theorem 2.6), and so on their union Ω. Hence it follows, again by Theorem 5.8 and Definition 6.1,

$$\langle \partial^\beta (f|_\omega), \varphi \rangle_\omega = (-1)^{|\beta|} \langle f, \partial^\beta \widetilde{\varphi} \rangle_\Omega = \langle \partial^\beta f, \widetilde{\varphi} \rangle_\Omega = \langle (\partial^\beta f)|_\omega, \varphi \rangle_\omega. \ \square$$

Now we come to commutation with linear mappings.

Theorem 6.5.– *Let $f \in \mathcal{D}'(\Omega; E)$, where Ω is an open subset of $\mathbb{R}^d$ and E is a Neumann space, ω an open subset of Ω, and L a sequentially continuous linear mapping from E into a Neumann space. Then,*

$$L(f|_\omega) = (Lf)|_\omega. \blacksquare$$

Proof. Definitions 5.10 of the image under L and 6.1 of the restriction give, for every $\varphi \in \mathcal{D}(\omega)$,

$$\langle L(f|_\omega), \varphi\rangle_\omega = L(\langle f|_\omega, \varphi\rangle_\omega) = L(\langle f, \widetilde{\varphi}\rangle_\Omega) = \langle Lf, \widetilde{\varphi}\rangle_\Omega = \langle (Lf)|_\omega, \varphi\rangle_\omega. \square$$

We continue with the commutation with the change of variables.

Theorem 6.6.– *Let $f \in \mathcal{D}'(\Omega; E)$, where Ω is an open subset of $\mathbb{R}^d$ and E is a Neumann space, ω an open subset of Ω, and T a bijection from an open subset Λ of $\mathbb{R}^d$ onto Ω such that $T \in \mathcal{C}^\infty(\Lambda; \Omega)$ and $T^{-1} \in \mathcal{C}^\infty(\Omega; \Lambda)$. Then,*

$$f|_\omega \circ T = (f \circ T)|_{T^{-1}(\omega)}. \blacksquare$$

Proof. Denoting $\lambda = T^{-1}(\omega)$, Definitions 5.21 of the image under T and 6.1 of restriction give, for every $\phi \in \mathcal{D}(\lambda)$,

$$\langle f|_\omega \circ T, \phi\rangle_\lambda = \langle f|_\omega, |\det[\nabla T^{-1}]|\, \phi \circ T^{-1}\rangle_\omega = \langle f, (|\det[\nabla T^{-1}]|\, \phi \circ T^{-1})^{\sim}\rangle_\Omega,$$

which, again with Definitions 6.1 and 5.21, is equal to

$$= \langle f, |\det[\nabla T^{-1}]|\, \widetilde{\phi} \circ T^{-1}\rangle_\Omega = \langle f \circ T, \widetilde{\phi}\rangle_\Lambda = \langle (f \circ T)|_\lambda, \phi\rangle_\lambda. \square$$

We finish with the commutation with the product by a regular function.

Theorem 6.7.– *Let $f \in \mathcal{D}'(\Omega; E)$ and $\alpha \in \mathcal{C}^\infty(\Omega)$, where Ω is an open subset of $\mathbb{R}^d$ and E is a Neumann space, and let ω be an open subset of Ω. Then,*

$$(\alpha f)|_\omega = \alpha|_\omega f|_\omega. \blacksquare$$

Proof. Definitions 6.1 of restriction and 5.15 of the product by α give, for every $\varphi \in \mathcal{D}(\omega)$,

$$\langle (\alpha f)|_\omega, \varphi\rangle_\omega = \langle \alpha f, \widetilde{\varphi}\rangle_\Omega = \langle f, \alpha\widetilde{\varphi}\rangle_\Omega =$$

$$= \langle f, \widetilde{\alpha|_\omega \varphi}\rangle_\Omega = \langle f|_\omega, \alpha|_\omega \varphi\rangle_\omega = \langle \alpha|_\omega f|_\omega, \varphi\rangle_\omega. \square$$

6.2. Additivity with respect to the domain

Let us give an additivity property with respect to disjoint sets.

Theorem 6.8.– *Let Ω be an open subset and $(\omega_i)_{i\in\mathcal{I}}$ a family of open subsets of $\mathbb{R}^d$ such that*

$$\Omega = \bigcup_{i\in\mathcal{I}} \omega_i, \quad \omega_i \cap \omega_j = \emptyset \text{ if } i \neq j.$$

Let moreover $\varphi \in \mathcal{D}(\Omega)$. Then, $\varphi|_{\omega_i}$ belongs to $\mathcal{D}(\omega_i)$ for every $i \in \mathcal{I}$ and is zero except for a finite number of i.

In addition, let $f \in \mathcal{D}'(\Omega; E)$, where E is a Neumann space. Then,

$$\langle f, \varphi\rangle_\Omega = \sum_{i\in\mathcal{I}} \langle f|_{\omega_i}, \varphi|_{\omega_i}\rangle_{\omega_i}. \blacksquare$$

Proof. **1.** *Restriction of φ.* Let us check that the support of $\varphi|_{\omega_i}$ is compact. Since the support of φ is compact, the strong inclusion theorem (Theorem A.22) provides $r > 0$ such that

$$\operatorname{supp} \varphi + B(0, r) \subset \Omega.$$

Therefore,

$$\{x \in \omega_i : \varphi(x) \neq 0\} + B(0, r) \subset \omega_i,$$

since, if not, there would exist $x \in \omega_i$ and $y \in B(x, r) \cap \omega_j$ with $j \neq i$, and therefore the ball $B(x, r)$ would be covered by the disjoint open sets ω_i and $\bigcup_{j\neq i} \omega_j$ and would have non-empty intersections with them, which would contradict its connectedness (which follows from its convexity due to Theorem A.15).

Thus, $\overline{\{x \in \omega_i : \varphi(x) \neq 0\}}$ is a compact subset of ω_i, which implies, due to Theorem 2.2 (b), that the support of $\varphi|_{\omega_i}$ is compact. Since this function is infinitely differentiable,

$$\varphi|_{\omega_i} \in \mathcal{D}(\omega_i).$$

2. *Vanishing restrictions.* The ω_i form an open cover of the support of φ, which is compact, therefore there exists a finite subcover. If ω_i does not belong to this subcover, φ is zero there and hence $\varphi|_{\omega_i} = 0$.

3. *Additivity.* The extension $\widetilde{\varphi|_{\omega_i}}$ of $\varphi|_{\omega_i}$ by 0 on Ω belongs to $\mathcal{D}(\Omega)$ (we have seen this in Definition 6.1) and

$$\varphi = \sum_{i\in\mathcal{I}} \widetilde{\varphi|_{\omega_i}},$$

where only a finite number of terms in the sum are not zero. Hence, according to Definition 6.1 of restriction,

$$\langle f,\varphi\rangle_\Omega=\sum_{i\in\mathcal{I}}\langle f,\widetilde{\varphi|_{\omega_i}}\rangle_\Omega=\sum_{i\in\mathcal{I}}\langle f|_{\omega_i},\varphi|_{\omega_i}\rangle_{\omega_i}.\ \square$$

6.3. Local character

Let us define the local equality for two distributions.

Definition 6.9.– *Let $f\in\mathcal{D}'(\Omega;E)$ and $g\in\mathcal{D}'(\Lambda;E)$, where Ω and Λ are two open subsets of $\mathbb{R}^d$ and E is a Neumann space, and ω an open subset of $\Omega\cap\Lambda$.*

*We say that f and g are **equal on** ω, and we denote "$f=g$ on ω", if*

$$f|_\omega=g|_\omega.$$

That is to say, by Definition 6.1 of the restriction, if the extensions by 0 on Ω and Λ of every test function $\varphi\in\mathcal{D}(\omega)$ satisfy

$$\langle f,\widetilde{\varphi}^\Omega\rangle_\Omega=\langle g,\widetilde{\varphi}^\Lambda\rangle_\Lambda.\qquad(6.2)$$

■

We often denote "$f=0_E$ on ω" to express that f is equal to the zero distribution there, instead of, for example, "$f=0_{\mathcal{D}'(\Lambda;E)}$ on ω", which is too cumbersome.

We now show that, if two distributions coincide on some open sets, then they coincide on their union. This is the **gluing theorem for equalities** [1].

Theorem 6.10.– *Let $f\in\mathcal{D}'(\Omega;E)$ and $g\in\mathcal{D}'(\Lambda;E)$, where Ω and Λ are two open subsets of $\mathbb{R}^d$ and E is a Neumann space, and let $(\omega_i)_{i\in\mathcal{I}}$ be a family of open subsets of $\Omega\cap\Lambda$. If*

$$f=g\ \text{ on } \omega_i,\ \ \forall i\in\mathcal{I},$$

then

$$f=g\ \text{ on } \bigcup_{i\in\mathcal{I}}\omega_i.\ ■$$

1. **History of the gluing of equalities.** Laurent SCHWARTZ proved in 1950 [69, Chap. I, § 3, p. 27, last line] that, if two real-valued distributions coincide on some open sets, they coincide on their union.

Local character of distributions. A distribution is hence determined by its restrictions to arbitrarily small open sets ω_i covering Ω. This "local" character is analogous to the "pointwise" character for functions. $\square$

Gluing theorem for equalities versus gluing theorem for distributions. In the *gluing theorem for equalities* (Theorem 6.10), it is a question of showing that **distributions f and g given on the union of the ω_i** are equal there if they are on each of the ω_i.

In the *gluing theorem for distributions* (Theorem 6.16) it is a question of showing that **there exists a distribution f on the union of the ω_i** which, on each ω_i, is equal to a distribution f_i given on ω_i (this requires that the f_i are pairwise equal on the sets where they are both defined). $\square$

Proof of Theorem 6.10. **1.** *A particular case.* Let us consider the case where

$$\Omega = \Lambda = \bigcup_{i \in \mathcal{I}} \omega_i.$$

Let $(\alpha_i)_{i \in \mathcal{I}}$ be a subordinate partition of unity (see Theorem 2.21) and $\varphi \in \mathcal{D}(\Omega)$.

As we will verify in Lemma 6.11 below,

$$(\alpha_i \varphi)|_{\omega_i} \in \mathcal{D}(\omega_i), \tag{6.3}$$

therefore $\alpha_i \varphi$ is its extension by 0. The local equality hypothesis $f = g$ on ω_i then gives, from its characterization (6.2),

$$\langle f, \alpha_i \varphi \rangle_\Omega = \langle g, \alpha_i \varphi \rangle_\Omega.$$

By Definition 2.20 of a partition of unity, only a finite number of the α_i are not zero on the support of φ, since this one is compact. Denote N_φ the finite set of these indices i. On the other hand, again by Definition 2.20, $\sum_{i \in \mathcal{I}} \alpha_i = 1$. Therefore,

$$\varphi = \sum_{i \in N_\varphi} \alpha_i \varphi,$$

because this equality is satisfied both on the support of φ (since the other α_i are zero there) and outside of it (since φ is zero there). Thus,

$$\langle f, \varphi \rangle_\Omega = \sum_{i \in N_\varphi} \langle f, \alpha_i \varphi \rangle_\Omega = \sum_{i \in N_\varphi} \langle g, \alpha_i \varphi \rangle_\Omega = \langle g, \varphi \rangle_\Omega.$$

2. *General case.* Denote

$$\lambda \stackrel{\text{def}}{=} \bigcup_{i \in \mathcal{I}} \omega_i.$$

Definition 6.1 of restriction implies $f|_\lambda|_{\omega_i} = f|_{\omega_i}$. The local equality hypothesis on f and g implies therefore that on their restrictions to λ, namely

$$f|_\lambda|_{\omega_i} = f|_{\omega_i} = g|_{\omega_i} = g|_\lambda|_{\omega_i}.$$

From which, due to step 1,

$$f|_\lambda = g|_\lambda.$$

That is, by Definition 6.9 of local equality,

$$f = g \text{ on } \lambda. \ \square$$

It remains to establish Property (6.3), that is the following result.

Lemma 6.11.– *Let $\varphi \in \mathcal{D}(\Omega)$ and $\alpha \in \mathcal{C}^\infty(\Omega)$ be such that*

$$\operatorname{supp} \alpha \subset \omega \subset \Omega,$$

where Ω and ω are open subsets of $\mathbb{R}^d$. Then,

$$(\alpha\varphi)|_\omega \in \mathcal{D}(\omega). \ \blacksquare$$

Proof. **1.** *Support.* Denote

$$D \stackrel{\text{def}}{=} \overline{\{x \in \Omega : \alpha(x) \neq 0\}}.$$

By Definition 2.1 of the support of a function, $\operatorname{supp} \alpha = D \cap \Omega$ and

$$\operatorname{supp}(\alpha\varphi)|_\omega \subset \operatorname{supp} \alpha \cap \operatorname{supp} \varphi \subset D \cap \operatorname{supp} \varphi.$$

However,

$$D \cap \operatorname{supp} \varphi \text{ is compact}$$

due to the Borel–Lebesgue theorem (Theorem A.26 (b)), since it is closed (as is every intersection of closed sets) and bounded (as is $\operatorname{supp} \varphi$) in $\mathbb{R}^d$. And,

$$D \cap \operatorname{supp} \varphi \subset (\operatorname{supp} \alpha \cup \partial\Omega) \cap \operatorname{supp} \varphi \subset \operatorname{supp} \alpha \subset \omega,$$

(since $D \subset \operatorname{supp} \alpha \cup \partial\Omega$ and $\partial\Omega \cap \operatorname{supp} \varphi = \emptyset$).

Being thus included in a compact subset of ω, the support of $\alpha\varphi|_\omega$ is compact due to Theorem 2.2 (a).

2. *Differentiability.* The restriction $(\alpha\varphi)|_\omega$ is infinitely differentiable on ω, since $\alpha\varphi$ is so on Ω (Theorem A.60). It thus belongs to $\mathcal{D}(\omega)$ since its support is compact by step 1. $\square$

Behaviors of α and φ at the boundary of ω. Let us go back to the hypotheses of Lemma 6.11.
— The hypothesis $\omega \subset \Omega$ implies that a subset γ_1 of the boundary $\partial\omega$ of ω is included in Ω, and that the rest γ_2 of $\partial\omega$ is included in $\partial\Omega$ (γ_1 and γ_2 may be empty).
— The hypothesis $\operatorname{supp} \alpha \subset \omega$ implies that α is zero outside of ω, and in a neighborhood of γ_1 (i.e. in a ball around each of its points).
— The hypothesis $\varphi \in \mathcal{D}(\Omega)$ implies that φ is zero in a neighborhood of γ_2 and outside of a bounded set.

When these hypotheses are met, $\alpha\varphi$ is zero in a neighborhood of $\partial\omega$, outside of ω, and outside of a bounded set. $\square$

6.4. Localization-extension

We intend to show that, after multiplication by a suitable function, every distribution is extendable by 0. That is the object for the following definition.

Definition 6.12.– *Let $f \in \mathcal{D}'(\Omega; E)$ and*

$$\alpha \in \mathcal{C}^\infty(\widetilde{\Omega}) \text{ such that } \operatorname{supp} \alpha \subset \Omega \subset \widetilde{\Omega}, \tag{6.4}$$

where Ω and $\widetilde{\Omega}$ are two open subsets of $\mathbb{R}^d$ and E is a Neumann space.

We define the **localized extension** $\widetilde{\alpha f} \in \mathcal{D}'(\widetilde{\Omega}; E)$ *by: for every $\phi \in \mathcal{D}(\widetilde{\Omega})$,*

$$\langle \widetilde{\alpha f}, \phi \rangle_{\widetilde{\Omega}} \stackrel{\text{def}}{=} \langle f, (\alpha\phi)|_\Omega \rangle_\Omega. \tag{6.5}$$

■

Behavior of α at the boundary of Ω. Let us look at the consequences of hypothesis (6.4), thanks to the observations (up to notation) following the proof of Lemma 6.11.

— The hypothesis $\Omega \subset \widetilde{\Omega}$ implies that a subset γ_1 of the boundary $\partial\Omega$ of Ω is included in $\widetilde{\Omega}$, and that the rest γ_2 of $\partial\Omega$ is included in $\partial\widetilde{\Omega}$ (γ_1 and γ_2 may be empty).

— The hypothesis $\operatorname{supp} \alpha \subset \Omega$ implies that α is zero outside of Ω, and in a neighborhood of γ_1 (that is in a ball around each of its points).

Observe that, although only the restriction of α to Ω is involved in equality (6.5), **we cannot** replace hypothesis (6.4) by

$$\alpha \in \mathcal{C}^\infty(\Omega) \text{ and } \operatorname{supp} \alpha \subset \Omega \subset \widetilde{\Omega}.$$

Indeed, this would not impose any behavior on α in the neighbourhood of $\partial\Omega$ (since the support of a function defined on Ω is always included in this one, by Definition 2.1). □

Justification of Definition 6.12. The right-hand side of (6.5) make sense since, due to Lemma 6.11,

$$(\alpha\phi)|_\Omega \in \mathcal{D}(\Omega).$$

For

$$\widetilde{\alpha f} \in \mathcal{D}'(\widetilde{\Omega}; E),$$

it suffices, due to the characterization of distributions from Theorem 3.4 (d), since $\langle f, (\alpha\phi)|_\Omega \rangle_\Omega$ depends linearly on ϕ, that, for each compact subset Q of $\widetilde{\Omega}$,

$$\langle f, (\alpha\phi)|_\Omega \rangle_\Omega \text{ depends continuously on } \phi \text{ in } \mathcal{C}^\infty_Q(\widetilde{\Omega}). \tag{6.6}$$

So let $\| \ \|_{E;\nu}$ be a semi-norm of E. Due to the characterization of distributions from Theorem 3.3, there exist $p \in \mathcal{C}^+(\Omega)$ and $b \in \mathbb{R}$ such that, for every $\varphi \in \mathcal{D}(\Omega)$,

$$\|\langle f, \varphi \rangle_\Omega\|_{E;\nu} \leq b \|\varphi\|_{\mathcal{D}(\Omega);p}.$$

In addition let Q be a compact subset of $\widetilde{\Omega}$ and

$$K \stackrel{\text{def}}{=} \operatorname{supp}\alpha \cap Q.$$

For every $\phi \in \mathcal{C}_Q^\infty(\widetilde{\Omega})$, by Definition 2.1 of the support of a function,

$$\operatorname{supp}(\alpha\phi)|_\Omega \subset \operatorname{supp}\alpha \cap \operatorname{supp}\phi \subset K$$

and K is a compact set (since $\operatorname{supp}\alpha = \mathcal{F} \cap \widetilde{\Omega}$ where $\mathcal{F} = \overline{\{x \in \widetilde{\Omega} : \alpha(x) \neq 0\}}$ is closed, and so $K = \mathcal{F} \cap \widetilde{\Omega} \cap Q = \mathcal{F} \cap Q$) included in Ω.

Lemma 5.16 therefore provides the existence of $m \in \mathbb{N}$ and $c \in \mathbb{R}$, independent of ϕ, such that

$$\|(\alpha\phi)|_\Omega\|_{\mathcal{D}(\Omega);p} \le c\|\alpha\|_{\mathcal{C}^\infty(\Omega);m,K}\|\phi\|_{\mathcal{C}_\mathrm{b}^\infty(\Omega);m}.$$

Hence, denoting $a = \|\alpha\|_{\mathcal{C}^\infty(\Omega);m,K}$,

$$\|\langle f, (\alpha\phi)|_\Omega\rangle_\Omega\|_{E;\nu} \le bca\,\|\phi\|_{\mathcal{C}_\mathrm{b}^\infty(\widetilde{\Omega});m}.$$

The characterization of continuous linear mappings from Theorem 1.12 (a) then shows that the mapping $\phi \mapsto \langle f, (\alpha\phi)|_\Omega\rangle_\Omega$ is indeed continuous from $\mathcal{C}_Q^\infty(\widetilde{\Omega})$ into E, since $\mathcal{C}_Q^\infty(\widetilde{\Omega})$ is, by Definition 2.3 (a), endowed with the semi-norms of $\mathcal{C}_\mathrm{b}^\infty(\widetilde{\Omega})$.

This proves (6.6), and therefore $\widetilde{\alpha f} \in \mathcal{D}'(\widetilde{\Omega}; E)$. □

Terminology. We say that αf is a *localization* of f, since multiplying it by α produces a distribution αf whose support is compact (or at least closed).

We say that $\widetilde{\alpha f}$ is a *localized extension* since, due to Theorem 6.13, it is the extension of αf by 0_E. □

Utility of localization-extension. This operation will be used to glue a family $(f_i)_{i\in\mathcal{I}}$ of distributions (Theorem 6.16) into a distribution f defined by

$$f = \sum_{i\in\mathcal{I}} \widetilde{\alpha_i f_i},$$

where $(\alpha_i)_{i\in\mathcal{I}}$ is a partition of unity subordinate to the cover of Ω by the domains of definition of the f_i.

It also provides an expression for the global regularization of a distribution f (Theorem 8.14 (a)), namely $R_n f = (\widetilde{\alpha_n f} \diamond \rho_n)|_\Omega$, where α_n is a *localizing* function (Definition 8.11). □

Let us specify the value of the localized extension on Ω and outside of the support of α.

Theorem 6.13.– *The localized extension $\widetilde{\alpha f}$ given by Definition 6.12 satisfies*

$$\widetilde{\alpha f} = \begin{cases} \alpha f & \text{on } \Omega, \\ 0_E & \text{on } \widetilde{\Omega} \setminus \operatorname{supp}\alpha. \end{cases} \blacksquare$$

Proof. **1.** *Value on* Ω. Definitions 6.1 of the restriction, 6.12 of the localized extension and 5.15 of the product by α give, for every $\varphi \in \mathcal{D}(\Omega)$,

$$\langle (\widetilde{\alpha f})|_\Omega, \varphi\rangle_\Omega = \langle \widetilde{\alpha f}, \widetilde{\varphi}\rangle_{\widetilde{\Omega}} = \langle f, \alpha\varphi\rangle_\Omega = \langle \alpha f, \varphi\rangle_\Omega.$$

This proves that $(\widetilde{\alpha f})|_\Omega = \alpha f$, i.e.

$$\widetilde{\alpha f} = \alpha f \text{ on } \Omega.$$

2. *Value outside of the support of* α. By Definition 6.9 of the local equality, it is necessary that the set

$$\mathcal{O} \stackrel{\text{def}}{=} \widetilde{\Omega} \setminus \operatorname{supp} \alpha$$

is open. Which is the case since, by Definition 2.1 of the support of a function, here α,

$$\mathcal{O} = \widetilde{\Omega} \setminus \left(\overline{\{x \in \widetilde{\Omega} : \alpha(x) \neq 0\}} \cap \widetilde{\Omega}\right) = \widetilde{\Omega} \cap \left(\mathbb{R}^d \setminus \overline{\{x \in \widetilde{\Omega} : \alpha(x) \neq 0\}}\right),$$

which is open, as is every finite intersection of open sets (Theorem A.11).

Definitions 6.1 and 6.12, again, give, for every $\varphi \in \mathcal{D}(\mathcal{O})$,

$$\langle (\widetilde{\alpha f})|_\mathcal{O}, \varphi\rangle_\mathcal{O} = \langle \widetilde{\alpha f}, \widetilde{\varphi}\rangle_{\widetilde{\Omega}} = \langle f, \alpha\widetilde{\varphi}|_\Omega\rangle_\Omega,$$

which is zero, because $\alpha\widetilde{\varphi}|_\Omega = 0$ since $\alpha = 0$ on $\mathcal{O}$ and $\widetilde{\varphi} = 0$ outside of $\mathcal{O}$. This proves that $(\widetilde{\alpha f})|_\mathcal{O} = 0_E$, i.e.

$$\widetilde{\alpha f} = 0_E \text{ on } \widetilde{\Omega} \setminus \operatorname{supp} \alpha. \;\square$$

Let us show that the localization-extension is a continuous operation.

Theorem 6.14.– *Under the hypotheses of Definition 6.12,* $f \mapsto \widetilde{\alpha f}$ *is a continuous, and thus sequentially continuous, linear mapping from* $\mathcal{D}'(\Omega; E)$ *into* $\mathcal{D}'(\widetilde{\Omega}; E)$. ▮

Proof. Let $\{\|\ \|_{E;\nu} : \nu \in \mathcal{N}_E\}$ be the family of semi-norms of E. Definitions 3.1 of the semi-norms of $\mathcal{D}'(\widetilde{\Omega}; E)$ and 6.12 of the localized extension give, for every $\phi \in \mathcal{D}(\widetilde{\Omega})$ and $\nu \in \mathcal{N}_E$,

$$\|\widetilde{\alpha f}\|_{\mathcal{D}'(\widetilde{\Omega};E);\phi,\nu} = \|\langle \widetilde{\alpha f}, \phi\rangle_{\widetilde{\Omega}}\|_{E;\nu} = \|\langle f, (\alpha\phi)|_\Omega\rangle_\Omega\|_{E;\nu} = \|f\|_{\mathcal{D}'(\Omega;E);(\alpha\phi)|_\Omega,\nu}.$$

The characterization of continuous linear mappings from Theorem 1.12 (a) then shows that the mapping $f \mapsto \widetilde{\alpha f}$ is continuous.

It is thus sequentially continuous, like every continuous mapping (Theorem 1.10). $\square$

Let us observe that, for a continuous function, we recover the extension by 0 of the product of functions.

Theorem 6.15.– *If $f \in \mathcal{C}(\Omega; E)$, its localized extension $\widetilde{\alpha f}$ given by Definition 6.12 belongs to $\mathcal{C}(\widetilde{\Omega}; E)$ and is the extension by 0_E of αf.* ▮

Proof. Let f be a continuous function. The extension by 0_E of the function αf is continuous on $\widetilde{\Omega}$, since it is at every point of Ω (Theorem A.60, since there it equals αf) and also of $\widetilde{\Omega} \setminus \operatorname{supp} \alpha$ (since this is an open set where it is zero).

It is a question of checking that the distribution $\overline{\alpha f}$ associated with this extended function satisfies

$$\widetilde{\alpha f} = \overline{\alpha f} \quad \text{on } \widetilde{\Omega}.$$

This equality holds on Ω, since these two distributions are equal there to the function αf due to Theorem 6.13 and since the restriction of distributions yields that of functions (Theorem 6.2). It also holds on $\widetilde{\Omega} \setminus \operatorname{supp} \alpha$, since these two distributions are zero there (for the same reasons).

This equality thus holds on the union $\widetilde{\Omega}$ of Ω and $\widetilde{\Omega} \setminus \operatorname{supp} \alpha$ due to the gluing theorem for equalities (Theorem 6.10). □

6.5. Gluing

Distributions have the gluing property [2] expressed in the following **gluing theorem**.

Theorem 6.16.– *Let $(\omega_i)_{i \in \mathcal{I}}$ be a family of open subsets of $\mathbb{R}^d$ and*

$$\Omega \stackrel{\text{def}}{=} \bigcup_{i \in \mathcal{I}} \omega_i.$$

For every $i \in \mathcal{I}$, let $f_i \in \mathcal{D}'(\omega_i; E)$, where E is a Neumann space, such that, for every $j \in \mathcal{I}$,

$$f_i = f_j \quad \text{on } \omega_i \cap \omega_j.$$

*Then, there exists a unique distribution $f \in \mathcal{D}'(\Omega; E)$, what we call the **gluing** of the f_i, such that, for every $i \in \mathcal{I}$,*

$$f = f_i \quad \text{on } \omega_i.$$ ▮

2. **History of the gluing theorem.** Laurent SCHWARTZ proved in 1950 [69, Chap. I, § 3, Theorem IV, p. 27] the gluing property for real distributions.

Proof. Let us proceed step-by-step.

1. *Uniqueness of f.* Two possible gluings are equal (to f_i, and therefore between themselves) on each ω_i, and hence on their union Ω, due to the gluing theorem for equalities (Theorem 6.10).

2. *Construction of f.* Let $(\alpha_i)_{i\in\mathcal{I}}$ be a partition of unity (given by Theorem 2.21) subordinate to the cover $(\omega_i)_{i\in\mathcal{I}}$ of Ω. For every $\varphi \in \mathcal{D}(\Omega)$, let

$$\langle f, \varphi\rangle_\Omega \overset{\text{def}}{=} \sum_{i\in\mathcal{I}} \langle f_i, (\alpha_i\varphi)|_{\omega_i}\rangle_{\omega_i}.$$

Observe that $\langle f_i, (\alpha_i\varphi)|_{\omega_i}\rangle_{\omega_i}$ is defined since $(\alpha_i\varphi)|_{\omega_i} \in \mathcal{D}(\omega_i)$ (Lemma 6.11).

This sum over i makes sense even if $\mathcal{I}$ is infinite, since these terms are zero except for a finite subset I_φ of $\mathcal{I}$ (as only a finite number of the α_i is not zero on the compact set $\operatorname{supp}\varphi$, by Definition 2.20 of a partition of unity). That is, more precisely,

$$\langle f, \varphi\rangle_\Omega = \sum_{i\in I_\varphi} \langle f_i, (\alpha_i\varphi)|_{\omega_i}\rangle_{\omega_i}. \tag{6.7}$$

3. *Belonging to $\mathcal{D}'(\Omega; E)$.* Let K be a compact subset of Ω. There exists (again by Definition 2.20) a finite set I_K such that $I_\varphi \subset I_K$ for every $\varphi \in C_K^\infty(\Omega)$. So, by Definition 6.12 of the localized extension $\widetilde{\alpha_i f_i} \in \mathcal{D}'(\Omega; E)$,

$$\langle f, \varphi\rangle_\Omega = \sum_{i\in I_K} \langle f_i, (\alpha_i\varphi)|_{\omega_i}\rangle_{\omega_i} = \sum_{i\in I_K} \langle \widetilde{\alpha_i f_i}, \varphi\rangle_\Omega.$$

Due to the characterization of distributions from Theorem 3.4 (d), $\widetilde{\alpha_i f_i}$ is continuous linear from $C_K^\infty(\Omega)$ into E. This holds for each i in I_K, which is finite.

The mapping $f \mapsto \langle f, \varphi\rangle_\Omega$ is thus continuous linear from $C_K^\infty(\Omega)$ into E. This holds for each compact subset K of Ω, therefore, again by Theorem 3.4 (d),

$$f \in \mathcal{D}'(\Omega; E).$$

4. *Equality $f = f_j$ on ω_j.* Let

$$\phi \in \mathcal{D}(\omega_j).$$

Its extension $\widetilde{\phi}$ to Ω by 0 belongs to $\mathcal{D}(\Omega)$.

The support of $\alpha_i\widetilde{\phi}$ is included in $\omega_i \cap \omega_j$, hence its restriction belongs to $\mathcal{D}(\omega_i \cap \omega_j)$. Definition 6.1 of the restriction, here of f_i to $\omega_i \cap \omega_j$, therefore gives

$$\langle f_i|_{\omega_i\cap\omega_j}, (\alpha_i\widetilde{\phi})|_{\omega_i\cap\omega_j}\rangle_{\omega_i\cap\omega_j} = \langle f_i, (\alpha_i\widetilde{\phi})|_{\omega_i}\rangle_{\omega_i}.$$

Likewise,

$$\langle f_j|_{\omega_i\cap\omega_j}, (\alpha_i\widetilde{\phi})|_{\omega_i\cap\omega_j}\rangle_{\omega_i\cap\omega_j} = \langle f_j, (\alpha_i\widetilde{\phi})|_{\omega_j}\rangle_{\omega_j}.$$

Since, by hypothesis, $f_i = f_j$ on $\omega_i \cap \omega_j$, the left-hand sides of these two equalities are equal. The equality of the right-hand sides gives, with (6.7),

$$\langle f, \widetilde{\phi}\rangle_\Omega = \sum_{i\in I_{\widetilde{\phi}}} \langle f_i, (\alpha_i\widetilde{\phi})|_{\omega_i}\rangle_{\omega_i} = \sum_{i\in I_{\widetilde{\phi}}} \langle f_j, (\alpha_i\widetilde{\phi})|_{\omega_j}\rangle_{\omega_j} = \left\langle f_j, \left(\sum_{i\in I_{\widetilde{\phi}}} \alpha_i\widetilde{\phi}\right)_{\omega_j}\right\rangle_{\omega_j}.$$

However, again by Definition 2.20 of a partition of unity,

$$\sum_{i\in I_{\widetilde{\phi}}} \alpha_i = \sum_{i\in \mathcal{I}} \alpha_i = 1 \text{ on the support of } \widetilde{\phi}.$$

Therefore, $\sum_{i\in I_{\widetilde{\phi}}} \alpha_i\widetilde{\phi}$ is equal to $\widetilde{\phi}$ on the whole Ω, and thus is equal to ϕ on ω_j. Hence, finally,

$$\langle f, \widetilde{\phi}\rangle_\Omega = \langle f_j, \phi\rangle_{\omega_j}.$$

That is, by Definition 6.9 of local equality,

$$f = f_j \text{ on } \omega_j. \ \square$$

6.6. Annihilation domain and support

We introduce the *annihilation domain* of a distribution, which is nothing other than the complement of its support, because many proofs of properties of the support are simplified by using it.

Definition 6.17.– *The* ***annihilation domain*** *of a distribution $f \in \mathcal{D}'(\Omega; E)$, where Ω is an open subset of $\mathbb{R}^d$ and E is a Neumann space, is the largest open subset of Ω, denoted by* nihil f, *such that*

$$f = 0_E \text{ on nihil } f.$$

In other words, nihil f *is the largest of the open sets $\mathcal{O}$ such that*

$$f|_\mathcal{O} = 0_E, \tag{6.8}$$

i.e. such that

$$\langle f, \varphi\rangle_\Omega = 0_E, \text{ for every } \varphi \in \mathcal{D}(\Omega) \text{ such that supp}\,\varphi \subset \mathcal{O}. \tag{6.9}$$

■

Abuse of notation. Equality (6.8) means that "$f|_{\mathcal{O}}$ is the zero distribution on $\mathcal{O}$ with values in E", which we could denote "$f|_{\mathcal{O}} = 0_{\mathcal{D}'(\mathcal{O};E)}$", but these two formulations lose in clarity what they bring in exactitude, while there is no ambiguity here. □

Justification of Definition 6.17. There does indeed exist a largest open set nihil f, eventually empty, on which $f = 0_E$, namely the union of the open sets $\mathcal{O}$ on which $f = 0_E$. Indeed, this set is open, as is every union of open sets (Theorem A.11), and f is zero there due to the gluing theorem for equalities (Theorem 6.10).

It is characterized by Property (6.9), by Definition 6.9 of local equality. □

Caution with terminology. It would be misleading to use the term *cancellation* here in the place of *annihilation*.

Indeed, when f is a continuous function, its *cancellation domain* can be larger than its *annihilation domain* (or, more precisely that the annihilation domain of the distribution to which it is identified). For example, for the function I defined on $\mathbb{R}$ by $I(x) = x$,

$$\{x \in \mathbb{R} : I(x) = 0\} = \{0\} \quad \text{while} \quad \operatorname{nihil} I = \emptyset.$$

In terms of points, we can speak of a distribution, or of a continuous function, that *annihilates* at a point x if it is zero in a neighborhood of it, namely if it (is defined and) is zero in a ball centered at x:

$$x \in \operatorname{nihil} f \;\Leftrightarrow\; [\text{ there exists } r > 0 \text{ such that } f = 0_E \text{ on } \mathring{B}(x,r) \text{ }]. \tag{6.10}$$

□

Let us now define the support of a distribution [3].

Definition 6.18.– *The* **support** *of a distribution $f \in \mathcal{D}'(\Omega; E)$, where Ω is an open subset of $\mathbb{R}^d$ and E is a Neumann space, is the set*

$$\operatorname{supp} f \stackrel{\text{def}}{=} \Omega \setminus \operatorname{nihil} f. \;\blacksquare$$

Cancellation of $\langle f, \varphi\rangle$. If

$$\operatorname{supp} f \cap \operatorname{supp} \varphi = \emptyset, \tag{6.11}$$

then $\operatorname{supp} \varphi \subset \Omega \setminus \operatorname{supp} f = \operatorname{nihil} f$, thus characterization (6.9) of the annihilation domain gives

$$\langle f, \varphi\rangle = 0_E.$$

Condition (6.11) supposes, see (6.10), that φ is zero on a ball around each point of the support of f, and therefore that all its derivatives $\partial^\beta \varphi$ are zero on such a ball. We will see in Theorem 16.18 that, when E is a Banach space, it suffices that φ and the $\partial^\beta \varphi$ are zero on the support of f. □

Let us check that we recover the support of a function.

3. **History of the support of a distribution.** Laurent SCHWARTZ defined the support of a distribution in 1945 [68, p. 60], calling it then the *kernel*. He baptized it the *support* in 1950 [69].

Theorem 6.19.– *If $f \in C(\Omega; E)$, its support in the distribution sense given by Definition 6.18 coincides with its support in the function sense given by Definition 2.1.*

That is to say that, with identification (3.6), p. 52, of f with a distribution, Definition 6.18 also gives

$$\operatorname{supp} f = \overline{\{x \in \Omega : f(x) \neq 0_E\}} \cap \Omega. \blacksquare$$

Proof. It is a question of checking that that the support of the distribution $\overline{f}$ associated with the function f by Theorem 3.5 coincides with that of f.

Definitions 6.18 of the support and 6.17 of the annihilation domain give

$$\operatorname{supp} \overline{f} = \Omega \setminus \mathcal{O} \text{ where } \mathcal{O} \text{ is the largest open subset of } \mathbb{R}^d \text{ on which } \overline{f} = 0_E.$$

On an open set $\mathcal{O}$, the equality $\overline{f} = 0_E$ (Definition 6.9) coincides with the equality $f = 0_E$, since the restriction $\overline{f}|_{\mathcal{O}}$ in the distribution sense coincides with that of $f|_{\mathcal{O}}$ in the function sense due to Theorem 6.2. The equality of the supports of $\overline{f}$ and f therefore follows from the fact that, as we verify in the next Theorem 6.20,

$$\operatorname{supp} f = \Omega \setminus \mathcal{O} \text{ where } \mathcal{O} \text{ is the largest open subset of } \mathbb{R}^d \text{ on which } f = 0_E. \quad (6.12)$$

The stated equality follows, by Definition 2.1 of the support of a function. □

It remains to prove Property (6.12) of the support of a function.

Theorem 6.20.– *Let f be a function from an open subset Ω of $\mathbb{R}^d$ into a separated semi-normed space E. Then,*

$$\operatorname{supp} f = \Omega \setminus \mathcal{O} \text{ where } \mathcal{O} \text{ is the largest open subset of } \mathbb{R}^d \text{ on which } f = 0_E. \blacksquare$$

Proof. Let

$$\mathcal{O} \stackrel{\text{def}}{=} \Omega \setminus \operatorname{supp} f = \Omega \cap \left(\mathbb{R}^d \setminus \overline{\{x \in \Omega : f(x) \neq 0_E\}}\right).$$

It is an open set, as is every finite intersection of open sets (Theorem A.11), on which the function f is zero.

If $\mathcal{U}$ is another open subset of Ω on which f is zero, then $\{x \in \Omega : f(x) \neq 0_E\}$ is included in $\mathbb{R}^d \setminus \mathcal{U}$, which is closed. So, by Definition 2.1 of the support of a function,

$$\operatorname{supp} f \subset \overline{\{x \in \Omega : f(x) \neq 0_E\}} \subset \overline{\mathbb{R}^d \setminus \mathcal{U}} = \mathbb{R}^d \setminus \mathcal{U}.$$

From which,

$$\mathcal{U} \subset \Omega \cap (\mathbb{R}^d \setminus \operatorname{supp} f) = \mathcal{O}.$$

Therefore, $\mathcal{O}$ is indeed the largest of these sets $\mathcal{U}$. □

6.7. Properties of the annihilation domain and support

We propose to examine the effect of the various operations on the annihilation domain and the support. Let us begin with the effect of partial differentiation.

Theorem 6.21.– *Let $f \in \mathcal{D}'(\Omega; E)$, where Ω is an open subset of $\mathbb{R}^d$ and E is a Neumann space, and $\beta \in \mathbb{N}^d$. Then,*

$$\operatorname{nihil} \partial^\beta f \supset \operatorname{nihil} f,$$

$$\operatorname{supp} \partial^\beta f \subset \operatorname{supp} f. \blacksquare$$

Proof. **1.** *Annihilation domains.* Denote

$$\mathcal{O} \stackrel{\text{def}}{=} \operatorname{nihil} f.$$

By Definition 6.17, it is the largest open set $\mathcal{O}$ such that $f|_{\mathcal{O}} = 0_E$. Since restriction commutes with differentiation from Theorem 6.4,

$$(\partial^\beta f)|_{\mathcal{O}} = \partial^\beta (f|_{\mathcal{O}}) = 0_E.$$

Hence, again by definition of the annihilation domain (here, of $\partial^\beta f$), $\mathcal{O} \subset \operatorname{nihil} \partial^\beta f$, i.e.

$$\operatorname{nihil} \partial^\beta f \supset \operatorname{nihil} f. \tag{6.13}$$

2. *Supports.* Since $\operatorname{supp} f = \Omega \setminus \operatorname{nihil} f$ by Definition 6.18, (6.13) gives

$$\operatorname{supp} \partial^\beta f \subset \operatorname{supp} f. \quad \square$$

Example of strict inclusion. The support of $\partial^\beta f$ can be strictly included in that of f. For example, if $f = 1$ on Ω, then $\partial^\beta f = 0$ as soon as $\beta \neq 0$, thus $\operatorname{supp} \partial^\beta f = \emptyset$, while $\operatorname{supp} f = \Omega$. $\square$

Now let us continue with the effect of a linear mapping.

Theorem 6.22.– *Let $f \in \mathcal{D}'(\Omega; E)$, where Ω is an open subset of $\mathbb{R}^d$ and E is a Neumann space, and L a sequentially continuous linear mapping from E into a Neumann space. Then,*

$$\operatorname{nihil} Lf \supset \operatorname{nihil} f,$$

$$\operatorname{supp} Lf \subset \operatorname{supp} f.$$

If L is injective,

$$\operatorname{nihil} Lf = \operatorname{nihil} f,$$

$$\operatorname{supp} Lf = \operatorname{supp} f. \blacksquare$$

Proof. Denote by F the space of values of L.

1. *General case.* Denote again

$$\mathcal{O} \stackrel{\text{def}}{=} \operatorname{nihil} f.$$

By Definition 6.17 of the annihilation domain, it is the largest open set $\mathcal{O}$ on which $f|_{\mathcal{O}} = 0_E$. Since restriction commutes with linear mappings from Theorem 6.5,

$$(Lf)|_{\mathcal{O}} = L(f|_{\mathcal{O}}) = 0_F.$$

Hence, again by definition of the annihilation domain (here of Lf), $\mathcal{O} \subset \operatorname{nihil} Lf$, i.e.

$$\operatorname{nihil} Lf \supset \operatorname{nihil} f. \tag{6.14}$$

The inclusion of supports follows since, by Definition 6.18, $\operatorname{supp} f = \Omega \setminus \operatorname{nihil} f$.

2. *Injective case.* Now let

$$\mathcal{U} \stackrel{\text{def}}{=} \operatorname{nihil} Lf.$$

Then, $(Lf)|_{\mathcal{U}} = 0_F$. Again due to Theorem 6.5, we have, in $\mathcal{D}'(\mathcal{U}; F)$,

$$L(f|_{\mathcal{U}}) = (Lf)|_{\mathcal{U}} = 0_F.$$

If L is injective from E into F, the mapping $f \mapsto Lf$ is injective from $\mathcal{D}'(\mathcal{U}; E)$ into $\mathcal{D}'(\mathcal{U}; F)$ due to Theorem 5.14 (c), therefore

$$f|_{\mathcal{U}} = 0_E.$$

Hence, still by definition of the annihilation domain (here of f), $\mathcal{U} \subset \operatorname{nihil} f$, i.e.

$$\operatorname{nihil} Lf \subset \operatorname{nihil} f.$$

It is the inverse inclusion of (6.14), which is therefore an equality. The equality of supports follows. □

Example of strict inclusion. If L is not injective, the support of Lf may be strictly included in that of f. For example, if $f = 1$ on Ω and $L = 0$, then $\operatorname{supp} Lf = \emptyset$, while $\operatorname{supp} f = \Omega$. □

Let us continue with the effect of a change of variables.

Theorem 6.23.– *Let $f \in \mathcal{D}'(\Omega; E)$, where Ω is an open subset of $\mathbb{R}^d$ and E is a Neumann space, and let T be a bijection from an open subset Λ of $\mathbb{R}^d$ onto Ω such that $T \in \mathcal{C}^\infty(\Lambda; \Omega)$ and $T^{-1} \in \mathcal{C}^\infty(\Omega; \Lambda)$. Then,*

$$\operatorname{nihil}(f \circ T) = T^{-1}(\operatorname{nihil} f),$$

$$\operatorname{supp}(f \circ T) = T^{-1}(\operatorname{supp} f). \blacksquare$$

Proof. **1.** *Inclusion of the annihilation domains.* Denote, here again,

$$\mathcal{O} \stackrel{\text{def}}{=} \text{nihil}\, f.$$

By Definition 6.17, it is the largest open set on which $f|_{\mathcal{O}} = 0_E$. Since restriction commutes with a change of variables due to Theorem 6.6,

$$(f \circ T)|_{T^{-1}(\mathcal{O})} = f|_{\mathcal{O}} = 0_E.$$

The set $T^{-1}(\mathcal{O})$ is open, as is every preimage of an open set under a continuous mapping (Theorem A.28), hence the definition of the annihilation domain (here, of $f \circ T$) gives $T^{-1}(\mathcal{O}) \subset \text{nihil}(f \circ T)$, i.e.

$$\text{nihil}(f \circ T) \supset T^{-1}(\text{nihil}\, f). \tag{6.15}$$

2. *Equality of the annihilation domains.* Inclusion (6.15) for $f \circ T$ and T^{-1} gives

$$\text{nihil}\, f = \text{nihil}((f \circ T) \circ T^{-1}) \subset T(\text{nihil}(f \circ T)).$$

This gives the inverse inclusion of (6.15), which is therefore an equality.

3. *Equality of the supports.* According to Definition 6.18 of support, the equality on annihilation domains and the bijectivity of T give

$$\begin{aligned}\text{supp}(f \circ T) = \Lambda \setminus \text{nihil}(f \circ T) = T^{-1}(\Omega) \setminus T^{-1}(\text{nihil}\, f) = \\ = T^{-1}(\Omega \setminus \text{nihil}\, f) = T^{-1}(\text{supp}\, f). \quad \square\end{aligned}$$

Now we arrive at the effect of the product with a regular function.

Theorem 6.24.– *Let $f \in \mathcal{D}'(\Omega; E)$ and $\alpha \in \mathcal{C}^\infty(\Omega)$, where Ω is an open subset of $\mathbb{R}^d$ and E is a Neumann space. Then,*

$$\text{nihil}(\alpha f) \supset \text{nihil}\, \alpha \cup \text{nihil}\, f,$$

$$\text{supp}(\alpha f) \subset \text{supp}\, \alpha \cap \text{supp}\, f,$$

where nihil α *is the annihilation domain of the distribution associated with α, i.e. the largest open set on which α is zero (it is characterized by (6.10), p. 131).* ▮

Proof. **1.** *Annihilation domain of α.* Denote by $\overline{\alpha}$ the distribution associated with the function α by Theorem 3.5. By Definition 6.17 of the annihilation domain, nihil $\overline{\alpha}$ is the largest open set on which $\overline{\alpha}$ is zero. Therefore, since the restriction of distributions gives that of continuous functions (Theorem 6.2),

$$\text{nihil}\,\overline{\alpha} \text{ is the largest open set on which } \alpha \text{ is zero.}$$

2. *Annihilation domain of αf.* Denote

$$\mathcal{O} \stackrel{\text{def}}{=} \text{nihil}\, f, \quad \mathcal{U} \stackrel{\text{def}}{=} \text{nihil}\,\overline{\alpha},$$

Since the restriction commutes with the product due to Theorem 6.7,

$$(\alpha f)|_{\mathcal{O}} = \alpha|_{\mathcal{O}} f|_{\mathcal{O}} = \alpha|_{\mathcal{O}} 0_E = 0_E \text{ on } \mathcal{O}.$$

Likewise,

$$(\alpha f)|_{\mathcal{U}} = \alpha|_{\mathcal{U}} f|_{\mathcal{U}} = 0 f|_{\mathcal{U}} = 0_E \text{ on } \mathcal{U}.$$

Therefore, due to the gluing theorem for equalities (Theorem 6.10),

$$\alpha f = 0_E \text{ on } \mathcal{U} \cup \mathcal{O}.$$

Hence, by definition of the annihilation domain (here of αf), $\mathcal{U} \cup \mathcal{O} \subset \text{nihil}(\alpha f)$, i.e.

$$\text{nihil}(\alpha f) \supset \text{nihil}\,\alpha \cup \text{nihil}\, f. \tag{6.16}$$

3. *Supports.* With Definition 6.18 of support, (6.16) gives

$$\begin{aligned} \text{supp}(\alpha f) = \Omega \setminus \text{nihil}(\alpha f) \subset \Omega \setminus (\text{nihil}\,\alpha \cup \text{nihil}\, f) = \\ = (\Omega \setminus \text{nihil}\,\alpha) \cap (\Omega \setminus \text{nihil}\, f) = \text{supp}\,\alpha \cap \text{supp}\, f. \ \square \end{aligned}$$

Example of strict inclusion. The inclusions from Theorem 6.24 can be strict, for example for the Dirac measure δ_0 on $\mathbb{R}$ and for $\alpha(x) = x$, we have $\text{supp}\,\delta_0 = \{0\}$ and $\text{supp}\,\alpha = \mathbb{R}$, but $\alpha\delta_0 = 0$ and therefore $\text{supp}(\alpha\delta_0) = \emptyset$, while $\text{supp}\,\alpha \cap \text{supp}\,\delta_0 = \{0\}$.

We can prove that

$$\text{nihil}\,\alpha \cup \text{nihil}\, f \subset \text{nihil}(\alpha f) \subset \{x \in \Omega : \alpha(x) = 0\} \cup \text{nihil}\, f,$$

but the right inclusion can also be strict. For example, this is the case if $f(x) = 1$ and $\alpha(x) = x$ on $\mathbb{R}$, since then $\text{nihil}(\alpha f) = \emptyset$ while $\{x \in \Omega : \alpha(x) = 0\} = \{0\}$. □

We finish with the effect of restriction.

Theorem 6.25.– *Let $f \in \mathcal{D}'(\Omega; E)$, where Ω is an open subset of $\mathbb{R}^d$ and E is a Neumann space, and ω an open subset of Ω. Then,*

$$\text{nihil}(f|_\omega) = \omega \cap \text{nihil}\, f,$$

$$\text{supp}(f|_\omega) = \omega \cap \text{supp}\, f. \ \blacksquare$$

Proof. **1.** *Annihilation domain.* By Definition 6.17 of the annihilation domain, f is zero on the open set $\operatorname{nihil} f$, therefore $f|_\omega$ is zero on the open set $\omega \cap \operatorname{nihil} f$, hence

$$\omega \cap \operatorname{nihil} f \subset \operatorname{nihil}(f|_\omega). \tag{6.17}$$

Conversely, $f|_\omega$ is zero on the open set $\operatorname{nihil}(f|_\omega)$, thus f is zero there, and so $\operatorname{nihil}(f|_\omega) \subset \operatorname{nihil} f$. This gives the converse inclusion of (6.17) (as $\operatorname{nihil}(f|_\omega) \subset \omega$).

2. *Support.* Definition 6.18 of support gives, with the equality on the annihilation domains,

$$\operatorname{supp}(f|_\omega) = \omega \setminus \operatorname{nihil}(f|_\omega) = \omega \setminus (\omega \cap \operatorname{nihil} f) = \omega \setminus \operatorname{nihil} f = \operatorname{supp} f. \ \square$$

6.8. The space $\mathcal{D}'_K(\Omega; E)$

Let us define the distribution spaces with support in a fixed closed set.

Definition 6.26.– *Let Ω be an open subset of $\mathbb{R}^d$, K a closed subset of Ω and E a Neumann space. We denote*

$$\mathcal{D}'_K(\Omega; E) \stackrel{\text{def}}{=} \{f \in \mathcal{D}'(\Omega; E) : \operatorname{supp} f \subset K\}$$

endowed with the semi-norms of $\mathcal{D}'(\Omega; E)$. ▮

Let us show that $\mathcal{D}'_K(\Omega; E)$ is closed in $\mathcal{D}'(\Omega; E)$.

Theorem 6.27.– *Let Ω be an open subset of $\mathbb{R}^d$, K be a closed subset of Ω and E be a Neumann space.*

Then, $\mathcal{D}'_K(\Omega; E)$ is a Neumann space, and it is a topologically closed, and thus sequentially closed, subspace of $\mathcal{D}'(\Omega; E)$. ▮

Proof. **1.** A characterization of $\mathcal{D}'_K(\Omega; E)$. The property

$$f \in \mathcal{D}'_K(\Omega; E)$$

is equivalent, according to Definition 6.18 of support, to $\Omega \setminus \operatorname{nihil} f \subset K$, and hence to $\Omega \setminus K \subset \operatorname{nihil} f$, i.e., due to Definition 6.17 of the annihilation domain, to

$$\langle f, \varphi \rangle_\Omega = 0_E, \ \forall \varphi \in \mathcal{D}(\Omega) \text{ such that } \operatorname{supp} \varphi \subset \Omega \setminus K. \tag{6.18}$$

2. *Subspace.* The set $\mathcal{D}'_K(\Omega; E)$ is a vector subspace of $\mathcal{D}'(\Omega; E)$ since, if f and g satisfy characterization (6.18), it is the same then of $f + g$, and of tf for every $t \in \mathbb{R}$.

Since $\mathcal{D}'_K(\Omega; E)$ is, by definition, endowed with the semi-norms of $\mathcal{D}'(\Omega; E)$, it is therefore a topological subspace (Definition 1.7 (d)) of it.

3. *Closure.* Let us check that $\mathcal{D}'_K(\Omega; E)$ is closed, i.e. that its complement $\mathcal{D}'(\Omega; E) \setminus \mathcal{D}'_K(\Omega; E)$ is open. Therefore, let

$$f \notin \mathcal{D}'_K(\Omega; E).$$

From characterization (6.18), there exists $\varphi \in \mathcal{D}(\Omega)$ such that $\operatorname{supp} \varphi \subset \Omega \setminus K$ for which

$$\langle f, \varphi \rangle_\Omega \neq 0_E.$$

Since E is separated (Definition 1.1) as is every Neumann space (Definition 1.4), there exists a semi-norm $\| \ \|_{E;\nu}$ of E such that

$$\| \langle f, \varphi \rangle_\Omega \|_{E;\nu} = a > 0.$$

That is, by Definition 3.1 of the semi-norms of $\mathcal{D}'(\Omega; E)$,

$$\| f \|_{\mathcal{D}'(\Omega;E);\varphi,\nu} = a.$$

For every $g \in \mathcal{D}'(\Omega; E)$ such that

$$\| g - f \|_{\mathcal{D}'(\Omega;E);\varphi,\nu} \leq a/2,$$

we have $\| g \|_{\mathcal{D}'(\Omega;E);\varphi,\nu} = \| \langle g, \varphi \rangle_\Omega \|_{E;\nu} \geq a/2$, hence $\langle g, \varphi \rangle_\Omega \neq 0_E$ and thus

$$g \notin \mathcal{D}'_K(\Omega; E).$$

So, $\mathcal{D}'(\Omega; E) \setminus \mathcal{D}'_K(\Omega; E)$ satisfies Definition A.7 (a) of an open set, and thus

$$\mathcal{D}'_K(\Omega; E) \text{ is closed.}$$

It is therefore also sequentially closed, as is every closed set (Theorem A.10).

4. *Sequential completion.* Since $\mathcal{D}'(\Omega; E)$ is a Neumann space from Theorem 4.5, $\mathcal{D}'_K(\Omega; E)$ is too, as is every closed topological subspace of a Neumann space due to Theorem A.24. □

Let us show that, if the support of a distribution is included in a closed or compact subset of Ω, it is itself closed or compact.

Theorem 6.28.– *Let $f \in \mathcal{D}'(\Omega; E)$, where Ω is an open subset of $\mathbb{R}^d$ and E is a Neumann space, and K such that*

$$\operatorname{supp} f \subset K \subset \Omega.$$

(a) *If K is closed, then the support of f is closed.*

(b) *If K is compact, then the support of f is compact.* ▮

Proof. **1.** *Property (a).* Since $\operatorname{supp} f \subset K \subset \Omega$, we have

$$\mathbb{R}^d \setminus \operatorname{supp} f = (\mathbb{R}^d \setminus K) \cup (\Omega \setminus \operatorname{supp} f).$$

By Definition 6.18 of support, $\Omega \setminus \operatorname{supp} f = \operatorname{nihil} f$, which is open by Definition 6.17.

If K is closed, then $\mathbb{R}^d \setminus K$ is open, thus $\mathbb{R}^d \setminus \operatorname{supp} f$ is too, as is every union of open sets (Theorem A.11); and therefore the support of f is closed.

2. *Property (b).* If K is compact, it is closed and bounded due to the Borel–Lebesgue theorem (Theorem A.26 (b)). The support of f is therefore closed due to (a), and it is bounded, and therefore it is compact, again due to the Borel–Lebesgue theorem. □

Let us show finally that every distribution which has a closed support, i.e. that is zero in a neighborhood of the boundary, is extendable by 0.

Theorem 6.29.– *Let $f \in \mathcal{D}'(\Omega; E)$, where Ω is an open subset of $\mathbb{R}^d$ and E is a Neumann space, and $\widetilde{\Omega}$ an open set containing Ω.*

If the support of f is closed, then there exists a unique distribution $\widetilde{f} \in \mathcal{D}'(\widetilde{\Omega}; E)$ such that

$$\widetilde{f} = \begin{cases} f & \text{on } \Omega, \\ 0_E & \text{on } \widetilde{\Omega} \setminus \operatorname{supp} f. \end{cases}$$ ▮

Proof. Let

$$\mathcal{U} \overset{\text{def}}{=} \widetilde{\Omega} \setminus \operatorname{supp} f.$$

It is open, since it is equal to $\widetilde{\Omega} \cap (\mathbb{R}^d \setminus \operatorname{supp} f)$, which is open as is every finite intersection of open sets (Theorem A.11), because $\mathbb{R}^d \setminus \operatorname{supp} f$ is open since the support of f is closed, by hypothesis.

Observe that, with Definition 6.18 of the support,

$$\Omega \cap \mathcal{U} = \Omega \cap (\widetilde{\Omega} \setminus \operatorname{supp} f) = \Omega \setminus \operatorname{supp} f = \operatorname{nihil} f.$$

By Definition 6.17 of the annihilation domain, $f = 0_E$ on $\operatorname{nihil} f$, i.e. on $\Omega \cap \mathcal{U}$. Thus, denoting by g the zero distribution of $\mathcal{D}'(\mathcal{U}; E)$,

$$f = 0_E = g \;\text{ on } \Omega \cap \mathcal{U}.$$

The gluing theorem for distributions (Theorem 6.16) then shows that f and g have a gluing $\widetilde{f}$ which has the stated properties. □

Notation. It is possible to use the same notation $\widetilde{\ }$ in Theorem 6.29 as in Definition 6.12 of the localized extension $\widetilde{\alpha f}$, since the latter is the extension of αf in the sense of Theorem 6.29. □

Behavior of f at the boundary of Ω. The closure hypothesis on its support implies that f is zero on a neighborhood of the boundary $\partial\Omega$ of Ω. Indeed, for every $x \in \partial\Omega$, there exists $r > 0$ such that the ball $B(x, r)$ does not intersect the support of f, in which case f is zero on the open set $\mathring{B}(x, r) \cap \Omega$. □

An extendibility condition. A distribution with values in a Banach space is extendable if (Theorem 16.16), on every bounded set, it is the finite order derivative of a uniformly continuous function. □

Chapter 7

Weighting

In this chapter, we construct the weighted distribution $f \diamond \mu$ of a distribution $f \in \mathcal{D}'(\Omega; E)$ by a weight which has compact support $\mu \in \mathcal{D}'_D(\mathbb{R}^d)$. Its role will be central in the construction of primitives (Chapter 13) and in the separation of variables (Chapter 15).

Weighting plays an analogous role for Ω to that played by the convolution for $\mathbb{R}^d$. But, since f is not necessarily extendable, $f \diamond \mu$ is only defined on the open set $\Omega_D = \{x : x + D \subset \Omega\}$. We call it *weighting* since, in the case of functions, $(f \diamond \mu)(x) = \int_{\mathring{D}} f(x+y)\mu(y)\, dy$, i.e. the integral of f in the neighborhood $x + \mathring{D}$ of x with the weight μ.

The general definition of $f \diamond \mu$, by $\langle f \diamond \mu, \varphi \rangle_{\Omega_D} = \langle f, \mu \diamond \check{\varphi} \rangle_\Omega$ (Definition 7.12), uses the case where the weight is regular to make sense of $\mu \diamond \check{\varphi}$, namely $(\mu \diamond \check{\varphi})(x) = \langle \mu, \tau_x \check{\varphi} \rangle$ where $\tau_x \check{\varphi}(y) = \varphi(x - y)$. Before the general case, we thus study weighting by a regular function in § 7.1 to 7.3.

Then we come to general weighting.

— We define it in § 7.5, and check in § 7.6 that it coincides with the weighting by a regular function and also with the weighting of functions.

— We show in § 7.7 that it is sequentially continuous and, in § 7.8 and 7.9, we study its derivatives, its support and its image under a linear mapping, and we show various properties of which $f \diamond \delta_0 = f$.

7.1. Weighting by a regular function

Let us define the weighting of a distribution by a regular function, which is a step in the general construction of weighting [1] that is given in Definition 7.12.

Definition 7.1.– *Let*

$$f \in \mathcal{D}'(\Omega; E) \text{ and } \mu \in \mathcal{C}^\infty_D(\mathbb{R}^d),$$

where Ω is an open subset of $\mathbb{R}^d$, E is a Neumann space and D is a compact subset of $\mathbb{R}^d$.

1. **History of weighting of distributions.** The history is given in Note 5, p. 153.

*The **weighted distribution** of f by the weight μ is the function $f \diamond \mu$ defined on the set*

$$\Omega_D \stackrel{\text{def}}{=} \{x \in \mathbb{R}^d : x + D \subset \Omega\}$$

by

$$(f \diamond \mu)(x) \stackrel{\text{def}}{=} \langle f, \tau_x \mu \rangle_\Omega,$$

*where the **translated function** $\tau_x \mu \in \mathcal{D}(\Omega)$ is defined, for every $y \in \Omega$, by*

$$(\tau_x \mu)(y) = \mu(y - x). \ \blacksquare$$

Justification. The function $\tau_x \mu$ indeed belongs to $\mathcal{D}(\Omega)$, since its support is included in the compact set $x + D$, and hence is compact due to Theorem 2.2 (a). □

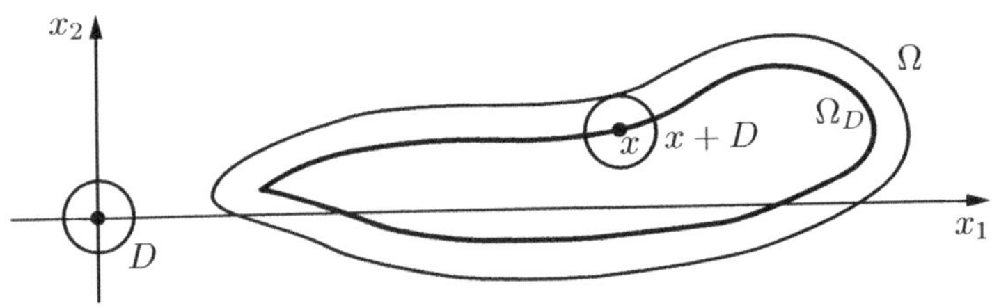

Figure 7.1. *Domain of definition Ω_D of the weighted distribution $f \diamond \mu$*

Terminology. *Weighting* is a **new term**, so the reader should recall its definition when using it. We say that $f \diamond \mu$ is the weighted distribution of f by μ since, as we will see in Theorem 7.4, when f is a continuous function, $(f \diamond \mu)(x)$ is the integral $\int_{\mathring{D}} f(x + y)\mu(y)\, dy$ of f in the neighborhood $x + \mathring{D}$ of x with the weight μ. □

A shortcut. We denote $\langle f, \tau_x \mu \rangle_\Omega$ instead of $\langle f, \tau_x \mu|_\Omega \rangle_\Omega$ which would be more rigorous since $\tau_x \mu$ is defined on all of $\mathbb{R}^d$, as is $\tau_x \mu$. But this would be less clear and there is no risk of ambiguity here. □

Domain Ω_D. The domain of definition Ω_D of $f \diamond \mu$ is represented in Figure 7.1 above. It is "relatively" smaller than Ω, in the sense that it can be included in it after a suitable translation. If $0 \in D$, the translation is not needed since then $\Omega_D \subset \Omega$.

If Ω is bounded, Ω_D cannot strictly contain Ω, even after a translation; and it cannot be equal to it except if D is empty or reduces to a point, in which case $\mu = 0$. In contrast, if Ω is not bounded and $0 \notin D$, then Ω_D can contain Ω: for example, if $\Omega = (0, \infty)$ and $D = [1, 2]$, then $\Omega_D = (-1, \infty)$.

In order for Ω_D to not be empty, it is necessary that, up to a translation, D is smaller than Ω. □

Choice of the "domain" D of μ. In this volume, D is always a ball centered around 0, in which case Ω_D is Ω with a neighborhood of its boundary removed, as in Figure 7.1 above.

In contrast, in the following volumes, to define the weighting in the neighborhood of the boundary, we will use a conic section K, in which case Ω_K goes up to a section of the boundary if this is regular, as in Figure 7.2 hereafter. This is explained in the comments *Choice of D_n* and *Explicit primitive up to the boundary*, p. 172 and 256. □

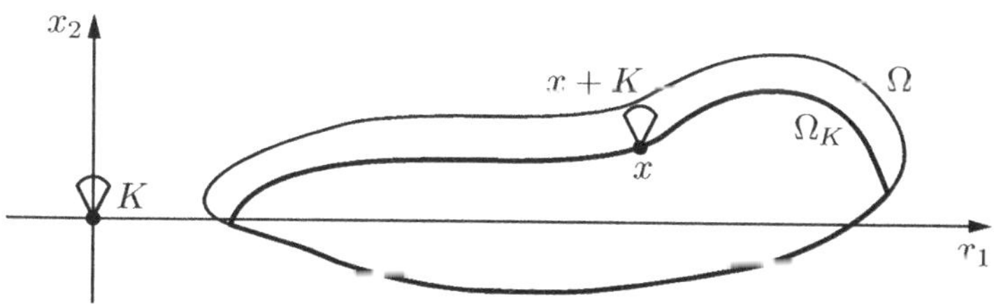

Figure 7.2. *Domain Ω_K going up to a portion of the boundary. The contour of Ω_K is in bold*

Motivation to study the weighting by a regular function. The weighting of a distribution f by a regular function μ serves for the general Definition 7.12 of the weighting of f by a not necessarily regular distribution μ. On the other hand, the weighting by a regularizing function ρ_n locally provides a regular approximation $f \diamond \rho_n$ of a distribution (Theorem 8.4). □

Let us give properties of the domain of definition Ω_D of the weighted distribution.

Theorem 7.2.– *Let Ω be an open subset of $\mathbb{R}^d$ and D and K compact subsets of $\mathbb{R}^d$. Then:*

$$\Omega_D \text{ is open}; \quad \Omega_D + D \subset \Omega; \quad (\Omega_D)_K = \Omega_{D+K}. \blacksquare$$

Proof. **1.** *Openness.* Let $x \in \Omega_D$. Then, $x + D$ is a compact subset of Ω, so the strong inclusion theorem (Theorem A.22) provides us with a closed ball $B(0, r)$ such that $x + D + B(0, r) \subset \Omega$. Then, $B(x, r)$ is included in Ω_D, which is therefore open.

2. *Inclusion.* If $x \in \Omega_D$ and $z \in D$, then $x + z \in \Omega$ by definition of Ω_D.

3. *Equality.* The property $x \in (\Omega_D)_K$ is equivalent to $x + K \subset \Omega_D$ and therefore to $x + D + K \subset \Omega$, i.e. to $x \in \Omega_{D+K}$. □

Remark. The inclusion $\Omega_D + D \subset \Omega$ may be strict. For example, this is the case if Ω is the union of a large ball with a plate of smaller thickness and D is a ball that can be placed inside the ball but not inside the plate, as such:

$$\Omega = \{x \in \mathbb{R}^d : |x| < 2 \text{ or } 0 < x_1 < 1/2\} \text{ and } D = \{x \in \mathbb{R}^d : |x| \leq 1\},$$

since then $\Omega_D \subset \{x \in \mathbb{R}^d : |x| \leq 2\}$ hence $\Omega_D + D \subset \{x \in \mathbb{R}^d : |x| \leq 3\}$ and thus $\Omega \not\subset \Omega_D + D$. □

Let us recall that, in Volume 2, we defined the weighting of a continuous function as follows.

Definition 7.3.– *Let*

$$f \in \mathcal{C}(\Omega; E) \ \ \textit{and} \ \ \mu \in \mathcal{C}_D(\mathbb{R}^d),$$

where Ω is an open subset of $\mathbb{R}^d$, E a Neumann space and D is a compact subset of $\mathbb{R}^d$.

The ***weighted function*** *of f by μ is the function $f \diamond \mu$ defined on Ω_D by*

$$(f \diamond \mu)(x) \stackrel{\text{def}}{=} \int_{\mathring{D}} f(x+y)\mu(y)\,\mathrm{d}y. \ \blacksquare$$

Let us check that this weighting coincides with that of distributions when their validity conditions are combined, allowing us to use the same notation for both.

Theorem 7.4.– *For $f \in \mathcal{C}(\Omega; E)$ and $\mu \in \mathcal{C}_D^\infty(\mathbb{R}^d)$, the weighted distribution $f \diamond \mu$ given by Definition 7.1 coincides with the weighted function given by Definition 7.3.*

That is to say that, with identification (3.6), p. 52, of f with a distribution, Definition 7.1 also gives, for every $x \in \Omega_D$,

$$(f \diamond \mu)(x) = \int_{\mathring{D}} f(x+y)\mu(y)\,\mathrm{d}y. \ \blacksquare$$

Proof. Denoting by $\bar{f}$ the distribution associated with the function f by Theorem 3.5, Definition 7.1 may be written, by performing a translation by x in the integral (see Theorem A.86) and observing that only $\mathring{D}$ contributes to the integral (Theorem A.77) since μ is zero outside of it (Theorem A.45),

$$(\bar{f} \diamond \mu)(x) = \int_{\Omega} f(z)\mu(z-x)\,\mathrm{d}z =$$

$$= \int_{\Omega - x} f(x+y)\mu(y)\,\mathrm{d}y = \int_{\mathring{D}} f(x+y)\mu(y)\,\mathrm{d}y. \ \square$$

7.2. Regularizing character of the weighting by a regular function

Let us show that the weighting by a regular function is **regularizing**[2].

2. **History of the regularizing characters of weighting. Convolution.** Laurent SCHWARTZ showed in 1950 [69, Chap. VI, § 3, Theorem XI, p. 166] that, if $f \in \mathcal{D}'(\mathbb{R}^d)$, then $f \star \mu \in \mathcal{C}^\infty(\mathbb{R}^d)$, which is equivalent to $f \diamond \mu \in \mathcal{C}^\infty(\mathbb{R}^d_D)$, since $f \star \mu = f \diamond \check{\mu}$ from (7.15), p. 155, and $\mathbb{R}^d_D = \mathbb{R}^d$.

Weighting. The properties of Theorems 7.5 and 7.8 were given in 1993 [SIMON, 78, Theorem 8, p. 12] for a Banach space E.

Theorem 7.5.– *For every*

$$f \in \mathcal{D}'(\Omega; E) \ \ \textit{and} \ \ \mu \in \mathcal{C}^{\infty}_{D}(\mathbb{R}^d),$$

where Ω is an open subset of $\mathbb{R}^d$, E is a Neumann space and D is a compact subset of $\mathbb{R}^d$,

$$f \diamond \mu \in \mathcal{C}^{\infty}(\Omega_D; E). \ \blacksquare$$

The proof is based on the following property of differentiability of a regular function with respect to translations, which we assume for the moment.

Theorem 7.6.– *Let $\mu \in \mathcal{C}^{\infty}_{D}(\mathbb{R}^d)$, where D is a compact subset of $\mathbb{R}^d$, and Ω an open subset of $\mathbb{R}^d$. Then, the function $x \mapsto \tau_x \mu|_{\Omega}$ is infinitely differentiable from Ω_D into $\mathcal{D}(\Omega)$, and*

$$\frac{\partial}{\partial x_i}(\tau_x \mu|_{\Omega}) = -(\tau_x \partial_i \mu)|_{\Omega}. \tag{7.1}$$

$\blacksquare$

Proof of Theorem 7.5. The function $x \mapsto \tau_x \mu|_{\Omega}$ is infinitely differentiable from Ω_D into $\mathcal{D}(\Omega)$, due to Theorem 7.6, and the mapping f is continuous linear from $\mathcal{D}(\Omega)$ into E, by Definition 3.1 of a distribution. The composition $x \mapsto \langle f, \tau_x \mu \rangle_{\Omega}$ is then infinitely differentiable due to Theorem A.59. That is, by Definition 7.1 of the weighting by a regular function,

$$f \diamond \mu \in \mathcal{C}^{\infty}(\Omega_D; E). \ \square$$

It remains to prove Theorem 7.6. For this, we will use the following result of the differentiability with values in $\mathcal{C}^{\infty}_{\mathrm{b}}(\mathbb{R}^d)$ established in Volume 2 [82, Theorem 3.18].

Theorem 7.7.– *Given $\mu \in \mathcal{C}^{\infty}_{\mathrm{b}}(\mathbb{R}^d)$, the function $x \mapsto \tau_x \mu$ is infinitely differentiable from $\mathbb{R}^d$ into $\mathcal{C}^{\infty}_{\mathrm{b}}(\mathbb{R}^d)$ and*

$$\frac{\partial \tau_x \mu}{\partial x_i} = -\tau_x \partial_i \mu. \ \blacksquare$$

We are now in the position to give the aforementioned proof.

Proof of Theorem 7.6. **1.** *Preliminaries.* Let

$$\mu \in C_D^\infty(\Omega)$$

and B be a closed ball included in Ω_D, i.e. $B + D \subset \Omega$, and let $x \in B$. Then, $B + D$ is a compact set, as is any sum of compact subsets of $\mathbb{R}^d$ (Theorem A.27), and

$$\operatorname{supp} \tau_x \mu = x + \operatorname{supp} \mu \subset B + D.$$

For every $\beta \in \mathbb{N}^d$, we have $\operatorname{supp} \partial^\beta \tau_x \mu \subset \operatorname{supp} \tau_x \mu \subset B + D$ and thus

$$\partial^\beta \tau_x \mu \in C_{B+D}^\infty(\mathbb{R}^d).$$

We denote

$$F(x) = \tau_x \mu|_\Omega.$$

2. *Differentiation with values in* $C_b^\infty(\Omega)$. According to Theorem 7.7, the function $x \mapsto \tau_x \mu$ is infinitely differentiable from $\mathbb{R}^d$, and hence from $\mathring{B}$, into $C_b^\infty(\mathbb{R}^d)$; since the restriction is continuous linear from $C_b^\infty(\mathbb{R}^d)$ into $C_b^\infty(\mathring{B})$, the composite function satisfies (Theorem A.59):

$$F \text{ is infinitely differentiable from } \mathring{B} \text{ into } C_b^\infty(\Omega)$$

and its derivatives are given by (7.1).

3. *Differentiability with values in* $C_{B+D}^\infty(\Omega)$. Let $\beta \in \mathbb{N}^d$ and $x \in \mathring{B}$. The differentiability of $\partial^\beta F$ at the point x with values in $C_b^\infty(\Omega)$ is equivalent, by Definition 1.15, to the existence, for every $m \in \mathbb{N}$ and $\epsilon > 0$, of $\eta > 0$ such that, if $z \in \mathbb{R}^d$, $|z| \leq \eta$ and $x + z \in \mathring{B}$, then

$$\|\partial^\beta F(x+z) - \partial^\beta F(x) - z \cdot \nabla \partial^\beta F(x)\|_{C_b^\infty(\Omega);m} \leq \epsilon |z|. \qquad (7.2)$$

Due to step 1,

$$\partial^\beta F(x), \partial^\beta F(x+z) \text{ and } z \cdot \nabla \partial^\beta F(x) \text{ belong to } C_{B+D}^\infty(\Omega).$$

(For the third term, this follows from $z \cdot \nabla \partial^\beta F(x) = \sum_{i=1}^d z_i \partial^{\beta + \mathrm{e}_i} F(x)$.)

Inequality (7.2) therefore provides the differentiability of $\partial^\beta F$ at the point x with values in $C_{B+D}^\infty(\Omega)$, since this is, by Definition 2.3 (a), endowed with the semi-norms of $C_b^\infty(\Omega)$. Thus,

$$F \text{ is infinitely differentiable from } \mathring{B} \text{ into } C_{B+D}^\infty(\Omega).$$

4. *Differentiability with values in* $\mathcal{D}(\Omega)$. Since $C_{B+D}^\infty(\Omega) \subsetneq \mathcal{D}(\Omega)$ (Theorem 2.10), differentiability with values in $C_{B+D}^\infty(\Omega)$ implies that in $\mathcal{D}(\Omega)$ (Theorem A.58), hence

$$F \text{ is infinitely differentiable from } \mathring{B} \text{ into } \mathcal{D}(\Omega).$$

This holds for every ball B included in Ω_D, therefore

$$F \text{ is infinitely differentiable from } \Omega_D \text{ into } \mathcal{D}(\Omega). \quad \square$$

Let us supplement Theorem 7.5 in proving that, when Ω is bounded and the support of the weight is strictly included in D, we get uniform regularity on Ω_D.

Theorem 7.8.– *Let $f \in \mathcal{D}'(\Omega; E)$ and $\mu \in \mathcal{C}_D^\infty(\mathbb{R}^d)$, where Ω is an open subset of $\mathbb{R}^d$, E is a Neumann space and D is a compact subset of $\mathbb{R}^d$, be such that*

$$\Omega \textit{ is bounded and } \operatorname{supp} \mu \subset \mathring{D}.$$

Then,

$$f \diamond \mu \in \mathbf{C}_\mathrm{b}^\infty(\Omega_D; E). \quad \blacksquare$$

Proof. Denote by K the support of μ. It is a compact subset of the open set $\mathring{D}$, so, due to the strong inclusion theorem (Theorem A.22), there exists a ball B such that

$$K + B \subset \mathring{D} \subset D.$$

If $x \in \Omega_D$, that is if $x + D \subset \Omega$, then $x + B + K \subset \Omega$, that is $x + B \in \Omega_K$. So,

$$\Omega_D + B \subset \Omega_K.$$

Since the closure $\overline{\Omega_D}$ is included in $\Omega_D + B$, it is therefore included in Ω_K. It is bounded, like Ω_D, since Ω is bounded by assumption, and hence compact from the Borel–Lebesgue theorem (Theorem A.26 (b)). So,

$$\overline{\Omega_D} \text{ is a compact subset of } \Omega_K.$$

We can replace D by K in Theorem 7.5, whence

$$f \diamond \mu \in \mathcal{C}^\infty(\Omega_K; E).$$

Therefore, from Heine's theorem (Theorem A.32), $f \diamond \mu$ and all its derivatives are uniformly continuous and bounded on the compact set $\overline{\Omega_D}$, and *a fortiori* on Ω_D, i.e.

$$f \diamond \mu \in \mathbf{C}_\mathrm{b}^\infty(\Omega_D; E). \quad \square$$

7.3. Derivatives and support of distributions weighted by a regular weight

Let us give properties of a distribution weighted by a regular function, which are steps in their generalization to an arbitrary weight, which will be done in § 7.8. We begin by calculating the derivatives [3].

Theorem 7.9.– *Let $f \in \mathcal{D}'(\Omega; E)$ and $\mu \in C_D^\infty(\mathbb{R}^d)$, where Ω is an open subset of $\mathbb{R}^d$, E is a Neumann space and D is a compact subset of $\mathbb{R}^d$. Then, for every $i \in [\![1, d]\!]$,*

$$\partial_i (f \diamond \mu) = \partial_i f \diamond \mu = -f \diamond \partial_i \mu$$

and, for every $\beta \in \mathbb{N}^d$,

$$\partial^\beta (f \diamond \mu) = \partial^\beta f \diamond \mu = (-1)^{|\beta|} f \diamond \partial^\beta \mu. \blacksquare$$

Sense of the partial derivatives. These are derivatives in the function sense, since $f \diamond \mu$ is an infinitely differentiable function due to Theorem 7.5. Recall that they then coincide with the derivatives in the distribution sense due to Theorem 5.5. □

Proof of Theorem 7.9. **1.** *First-order derivative: first expression.* The derivative of the function $x \mapsto \tau_x \mu|_\Omega$ from Ω_D into $\mathcal{D}(\Omega)$ being given by Theorem 7.6, the derivative of its image under f, which is linear and continuous from $\mathcal{D}(\Omega)$ into E by Definition 3.1 of a distribution, equals (Theorem A.59)

$$\frac{\partial}{\partial x_i} \langle f, \tau_x \mu|_\Omega \rangle = \left\langle f, \frac{\partial}{\partial x_i} \tau_x \mu|_\Omega \right\rangle = -\langle f, \tau_x \partial_i \mu|_\Omega \rangle.$$

That is, by Definition 7.1 of the weighting by a regular function,

$$\partial_i (f \diamond \mu)(x) = -(f \diamond \partial_i \mu)(x).$$

2. *First-order derivative: second expression.* As $\tau_x \partial_i \mu|_\Omega = \partial_i (\tau_x \mu|_\Omega)$, we similarly have, with Definition 5.4 of the derivative $\partial_i f$,

$$\partial_i (f \diamond \mu)(x) = -\langle f, \tau_x \partial_i \mu|_\Omega \rangle = -\langle f, \partial_i (\tau_x \mu|_\Omega) \rangle = \langle \partial_i f, \tau_x \mu|_\Omega \rangle = (\partial_i f \diamond \mu)(x).$$

3. *Derivatives of arbitrary order.* The expressions for $\partial^\beta (f \diamond \mu)$ follow from those of $\partial_i (f \diamond \mu)$, since $\partial^\beta = \partial_1^{\beta_1} \cdots \partial_d^{\beta_d}$ and $|\beta| = \beta_1 + \cdots \beta_d$. □

---

3. **History of the properties of derivatives and support.** Laurent SCHWARTZ established properties of the convolution analogous to Theorems 7.10 and 7.9, see Note 8, p. 161.

Let us come to the support of a distribution weighted by a regular weight.

Theorem 7.10.– *Let* $f \in \mathcal{D}'(\Omega; E)$ *and* $\mu \in \mathcal{C}^\infty_D(\mathbb{R}^d)$*, where* Ω *is an open subset of* $\mathbb{R}^d$*,* E *is a Neumann space and* D *is a compact subset of* $\mathbb{R}^d$*. Then,*

$$\operatorname{supp}(f \diamond \mu) \subset \operatorname{supp} f - \operatorname{supp} \mu,$$

where $\operatorname{supp} f - \operatorname{supp} \mu = \{y - z : y \in \operatorname{supp} f,\ z \in \operatorname{supp} \mu\}$. ∎

Proof. **1.** *A sufficient condition.* According to the characterization of the support of a function from Theorem 6.20,

$$\operatorname{supp}(f \diamond \mu) = \Omega_D \setminus \mathcal{O},$$

where $\mathcal{O}$ is the largest open set on which $f \diamond \mu$ is zero. The stated inclusion is thus equivalent to

$$\Omega_D \setminus (\operatorname{supp} f - \operatorname{supp} \mu) \subset \mathcal{O}.$$

Therefore, denoting by S_f the support of f and S_μ that of μ, it suffices to show that

$$\Omega_D \setminus (S_f - S_\mu) \text{ is an open set on which } f \diamond \mu \text{ is zero.} \qquad (7.3)$$

2. *Cancellation of* $f \diamond \mu$ *on* $\Omega_D \setminus (S_f - S_\mu)$. Let

$$x \in \Omega_D \setminus (S_f - S_\mu).$$

Then,

$$x + S_\mu \subset \Omega \setminus S_f, \qquad (7.4)$$

because $x + S_\mu \subset x + D \subset \Omega$ (since $x \in \Omega_D$) and $(x + S_\mu) \cap S_f = \emptyset$ (since $x \notin S_f - S_\mu$). Since $S_f = \Omega \setminus \operatorname{nihil} f$ (by Definition 6.18 of support), (7.4) gives

$$\operatorname{supp} \tau_x\mu \subset \operatorname{nihil} f.$$

By Definition 6.17 of the annihilation domain, it follows that

$$\langle f, \tau_x\mu \rangle_\Omega = 0_E,$$

i.e., by Definition 7.1 of $f \diamond \mu$,

$$(f \diamond \mu)(x) = 0_E. \qquad (7.5)$$

3. *Openness of* $\Omega_D \setminus (S_f - S_\mu)$. Again, let

$$x \in \Omega_D \setminus (S_f - S_\mu).$$

In inclusion (7.4), the set $x + S_\mu$ is compact (since S_μ is), and $\Omega \setminus S_f$ is open (since it is the set nihil f). Therefore, from the strong inclusion theorem (Theorem A.22), there exists a ball $B(0, \epsilon)$, where $\epsilon > 0$, such that

$$x + S_\mu + B(0, \epsilon) \subset \Omega \setminus S_f.$$

Which is to say that

$$B(x, \epsilon) \subset \Omega \setminus (S_f - S_\mu).$$

We can choose ϵ so that $B(x, \epsilon)$ is in addition included in Ω_D (since this is open due to Theorem 7.2), and hence in $\Omega_D \setminus (S_f - S_\mu)$. Which proves that

$$\Omega_D \setminus (S_f - S_\mu) \text{ is open.}$$

Along with (7.5), this finishes the proof of the sufficient condition (7.3). □

7.4. Continuity of the weighting by a regular function

Let us give continuity properties of the weighting by a regular function [4].

Theorem 7.11.– *Let Ω be an open subset of $\mathbb{R}^d$, E be a Neumann space and D be a compact subset of $\mathbb{R}^d$. Then:*

(a) *For every $\mu \in \mathcal{C}_D^\infty(\mathbb{R}^d)$, the mapping $f \mapsto f \diamond \mu$ is sequentially continuous from $\mathcal{D}'(\Omega; E)$ into $\mathcal{C}^\infty(\Omega_D; E)$.*

(b) *For every $f \in \mathcal{D}'(\Omega; E)$, the mapping $\mu \mapsto f \diamond \mu$ is continuous, and therefore sequentially continuous, from $\mathcal{C}_D^\infty(\mathbb{R}^d)$ into $\mathcal{C}^\infty(\Omega_D; E)$.* ▌

(c) *The mapping $\diamond$ is sequentially continuous from the product $\mathcal{D}'(\Omega; E) \times \mathcal{C}_D^\infty(\mathbb{R}^d)$ into $\mathcal{C}^\infty(\Omega_D; E)$.* ▌

Proof. Let $\{\| \ \|_{E;\nu} : \nu \in \mathcal{N}_E\}$ be the family of semi-norms of E.

1. *Property (a): sequential continuity of $f \mapsto f \diamond \mu$.* Let

$$f_n \to f \ \text{ in } \mathcal{D}'(\Omega; E).$$

4. **History of the continuity properties of weighting.** Laurent SCHWARTZ proved in 1950 [69, Chap. VI, § 3, Theorem XII, p. 167] that the convolution product is hypocontinuous, i.e. continuous on the bounded sets, from $\mathcal{D}'(\mathbb{R}^d)$-unif $\times$ $\mathcal{D}(\mathbb{R}^d)$ into $\mathcal{C}^\infty(\mathbb{R}^d)$. Whence we can deduce the properties from Theorem 7.11 for $\Omega = \mathbb{R}^d$ and $E = \mathbb{R}$.

It is a question of proving that

$$f_n \diamond \mu \to f \diamond \mu \ \text{ in } \mathcal{C}^\infty(\Omega_D; E). \tag{7.6}$$

Which is to say, by Definitions 1.19 (a) of the semi-norms of $\mathcal{C}^\infty(\Omega_D; E)$ and 1.3 (a) of convergence, that, for every compact subset K of Ω_D and every $m \in \mathbb{N}$ and $\nu \in \mathcal{N}_E$,

$$\|(f_n - f) \diamond \mu\|_{\mathcal{C}^\infty(\Omega_D;E);m,K,\nu} = \sup_{|\beta| \leq m} \sup_{x \in K} \|\partial^\beta((f_n - f) \diamond \mu)(x)\|_{E;\nu} \to 0. \tag{7.7}$$

Let us prove this by contradiction, by supposing that there exist K, m and ν such that (7.7) is not satisfied. Then, there would exist $\beta \in \mathbb{N}^d$, $a > 0$, a subsequence of $(f_n)_{n \in \mathbb{N}}$ and points $x_n \in K$ such that, for every $n \in \mathbb{N}$,

$$\|\partial^\beta((f_n - f) \diamond \mu)(x_n)\|_{E;\nu} \geq a.$$

That is, since $\partial^\beta((f_n - f) \diamond \mu) = \partial^\beta(f_n - f) \diamond \mu$ (Theorem 7.9), by Definition 7.1 of the weighting by a regular function,

$$\|\langle \partial^\beta(f_n - f), \tau_{x_n}\mu \rangle\|_{E;\nu} \geq a. \tag{7.8}$$

Due to the Bolzano–Weierstrass theorem (Theorem A.26 (c)), we could extract a subsequence such that $x_n \to x$ in $\mathbb{R}^d$. For this subsequence, the continuity of the mapping $x \mapsto \tau_x\mu$ (which follows, due to Theorem A.54, from its differentiability established in Theorem 7.6), would imply

$$\tau_{x_n}\mu \to \tau_x\mu \ \text{ in } \mathcal{D}(\Omega).$$

On the other hand, from the sequential continuity of the derivative (Theorem 5.9)

$$\partial^\beta f_n \to \partial^\beta f \ \text{ in } \mathcal{D}'(\Omega; E).$$

The sequential continuity of the mapping $\langle \ , \ \rangle$ (Theorem 4.4 (c)) would then give

$$\|\langle \partial^\beta(f_n - f), \tau_{x_n}\mu \rangle\|_{E;\nu} \to 0.$$

This would contradict inequality (7.8). Therefore, (7.7), and hence (7.6), are true. This proves that the mapping

$$f \mapsto f \diamond \mu \text{ is sequentially continuous from } \mathcal{D}'(\Omega; E) \text{ into } \mathcal{C}^\infty(\Omega_D; E).$$

2. *Property (b): continuity of* $\mu \mapsto f \diamond \mu$. Let K be a compact subset of Ω_D, $m \in \mathbb{N}$ and $\nu \in \mathcal{N}_E$. Definition 1.19 (a) of the semi-norms of $\mathcal{C}^\infty$, the second expression for $\partial^\beta(f \diamond \mu)$ from Theorem 7.9, and Definition 7.1 of the weighting by a regular function successively give

$$\begin{aligned}\|f \diamond \mu\|_{\mathcal{C}^\infty(\Omega_D;E);m,K,\nu} &= \sup_{|\beta| \leq m} \sup_{x \in K} \|\partial^\beta(f \diamond \mu)(x)\|_{E;\nu} = \\ &= \sup_{|\beta| \leq m} \sup_{x \in K} \|(f \diamond \partial^\beta\mu)(x)\|_{E;\nu} = \sup_{|\beta| \leq m} \sup_{x \in K} \|\langle f, \tau_x \partial^\beta\mu \rangle\|_{E;\nu}.\end{aligned} \tag{7.9}$$

Since the set $K + D$ is compact, as is every sum of compact subsets of $\mathbb{R}^d$ (Theorem A.27), and included in Ω, the characterization of distributions from Theorem 3.3 and the inequality from Theorem 2.11 show that there exist $p \in \mathcal{C}^+(\Omega)$ and $c \in \mathbb{R}$ such that, for every $\varphi \in \mathcal{C}^\infty_{K+D}(\Omega)$,

$$\|\langle f, \varphi \rangle\|_{E;\nu} \leq c\|\varphi\|_{\mathcal{D}(\Omega);p} \leq cm'\|\varphi\|_{\mathcal{C}^{m'}_{\mathrm{b}}(\Omega)}, \tag{7.10}$$

where m' is an integer that bounds p from above on $K + D$.

For $x \in K$, the support of $\tau_x \mu$ is $x + \operatorname{supp} \mu$ which is included in $K + D$. Thus $\tau_x \partial^\beta \mu \in \mathcal{C}^\infty_{K+D}(\Omega)$ and (7.9) give, with (7.10),

$$\|f \diamond \mu\|_{\mathcal{C}^\infty(\Omega_D;E);m,K,\nu} \leq cm' \sup_{|\beta| \leq m} \sup_{x \in K} \|\tau_x \partial^\beta \mu\|_{\mathcal{C}^{m'}_{\mathrm{b}}(\Omega)} \leq cm'\|\mu\|_{\mathcal{C}^{m+m'}_{\mathrm{b}}(\mathbb{R}^d)}.$$

Since $\mathcal{C}^\infty_D(\mathbb{R}^d)$ is (by Definition 2.3 (a)) endowed with the semi-norms of $\mathcal{C}^\infty_{\mathrm{b}}(\mathbb{R}^d)$ and since $\|\mu\|_{\mathcal{C}^{m+m'}_{\mathrm{b}}(\mathbb{R}^d)} = \|\mu\|_{\mathcal{C}^\infty_{\mathrm{b}}(\mathbb{R}^d);m+m'}$, this inequality shows, according to the characterization of continuous linear mappings from Theorem 1.12 (a), that

the mapping $\mu \mapsto f \diamond \mu$ is continuous from $\mathcal{C}^\infty_D(\mathbb{R}^d)$ into $\mathcal{C}^\infty(\Omega_D; E)$.

It is therefore also sequentially continuous, as is every continuous mapping (Theorem 1.10).

3. *Property (c): sequential continuity of* $\diamond$. Let

$$f_n \to f \text{ in } \mathcal{D}'(\Omega; E) \text{ and } \mu_n \to \mu \text{ in } \mathcal{C}^\infty_D(\Omega; E).$$

Decompose

$$f_n \diamond \mu_n - f \diamond \mu = (f_n - f) \diamond \mu + f_n \diamond (\mu_n - \mu).$$

From (7.6), $(f_n - f) \diamond \mu \to 0$ in $\mathcal{C}^\infty(\Omega_D; E)$, hence it remains to show that

$$f_n \diamond (\mu_n - \mu) \to 0 \text{ in } \mathcal{C}^\infty(\Omega_D; E). \tag{7.11}$$

The sequence of the f_n is bounded, as is every convergent sequence (Theorem A.5). Theorem 4.2 (a) then shows that it is equicontinuous in $\mathcal{C}^\infty_K(\Omega)$; more precisely, given a compact subset K of Ω_D and $\nu \in \mathcal{N}_E$, it provides (since $K + D$ is a compact subset of Ω, as observed in (b)) the existence of $m' \in \mathbb{N}$ and $c \in \mathbb{R}$ such that, for every $n \in \mathbb{N}$ and $\varphi \in \mathcal{C}^\infty_{K+D}(\Omega)$,

$$\|\langle f_n, \varphi \rangle\|_{E;\nu} \leq c\|\varphi\|_{\mathcal{C}^{m'}_{\mathrm{b}}(\Omega)}.$$

With (7.9), it follows (since $\tau_x \mu \in \mathcal{C}^\infty_{K+D}(\Omega)$ for every $x \in K$ as we saw in (b)) that

$$\|f_n \diamond (\mu_n - \mu)\|_{\mathcal{C}^\infty(\Omega_D;E);m,K,\nu} \leq c\|\mu_n - \mu\|_{\mathcal{C}^{m+m'}_{\mathrm{b}}(\mathbb{R}^d)}.$$

Since the right-hand side converges to 0 by hypothesis, this proves (7.11) and so that

$\diamond$ is sequentially continuous from $\mathcal{D}'(\Omega; E) \times \mathcal{C}^\infty_D(\mathbb{R}^d)$ into $\mathcal{C}^\infty(\Omega_D; E)$. $\square$

7.5. Weighting by a distribution

Let us now give the general definition of weighting[5], i.e. the weighting of an arbitrary distribution by another (the weight) which has a compact support.

Definition 7.12.– *Let*

$$f \in \mathcal{D}'(\Omega; E) \;\; \textit{and} \;\; \mu \in \mathcal{D}'_D(\mathbb{R}^d),$$

where Ω is an open subset of $\mathbb{R}^d$, E is a Neumann space and D is a compact subset of $\mathbb{R}^d$.

The ***weighted distribution*** *of f by the weight μ is the distribution*

$$f \diamond \mu \in \mathcal{D}'(\Omega_D; E),$$

where

$$\Omega_D \overset{\text{def}}{=} \{x \in \mathbb{R}^d : x + D \subset \Omega\},$$

defined by: for every $\varphi \in \mathcal{D}(\Omega_D)$,

$$\langle f \diamond \mu, \varphi \rangle_{\Omega_D} \overset{\text{def}}{=} \langle f, \mu \diamond \check{\varphi} \rangle_{\Omega},$$

where $\check{\varphi} \in \mathcal{D}(\mathbb{R}^d)$ is defined by $\check{\varphi}(x) = \varphi(-x)$ if $-x \in \Omega_D$ and $\check{\varphi}(x) = 0$ otherwise, and where

$$\mu \diamond \check{\varphi} \in \mathcal{D}(\Omega)$$

is given by Definition 7.1 of the weighting by a regular function, i.e.

$$(\mu \diamond \check{\varphi})(x) = \langle \mu, \tau_x \check{\varphi} \rangle_{\mathbb{R}^d}. \;\blacksquare$$

---

5. **History of the weighting of distributions.** Definition 7.12 of the weighting was given for a Banach space E in 1993 [SIMON, 79, Definition 3, p. 205]. It generalizes the definition of the convolution to an open subset Ω of $\mathbb{R}^d$, up to a symmetry on the weight μ.

Convolution. In 1945, Laurent SCHWARTZ extended the convolution product of functions to two distributions of $\mathcal{D}'(\mathbb{R}^d)$ of which one has compact support [68, p. 68], see also [69, Chap. VI, Theorem I, p. 155]. He extended it in 1957 [72, § 3, p. 72–73] to the case where one of the two distributions has values in a *locally convex separated topological vector space*, which is equivalent to *separated semi-normed space*.

Methods. The method used here to define weighting, which consists of reducing via transposition to the weighting by a regular function, had been used for the convolution on $\mathcal{D}'(\mathbb{R}^d)$ by François TREVES in 1967 [90, Definition 27.3, p. 293].
That of Laurent SCHWARTZ consists in reducing to a tensor product, in showing that there exists a unique distribution $f \star \mu$ such that $\langle f \star \mu, \varphi \rangle_{\mathbb{R}^d} = \langle f(x) \otimes \mu(y), \varphi(x+y) \rangle_{\mathbb{R}^{2d}}$, for every $\varphi \in \mathcal{D}(\mathbb{R}^d)$ [69, Chap. VI, § 1, Theorem I, p. 155].

A shortcut. In the expressions $\langle f, \mu \diamond \check{\varphi} \rangle_\Omega$ and $\mu \diamond \check{\varphi} \in \mathcal{D}(\Omega)$, it is in reality the restriction $(\mu \diamond \check{\varphi})|_\Omega$ of the function $\mu \diamond \check{\varphi}$, since this is defined on all of $\mathbb{R}^d$. But our notation covers the essential without being ambiguous. □

Justification of Definition 7.12. **1.** *Notation.* To avoid confusion with Definition 7.1 of the weighting by a regular function, we denote this by $\diamond\!\!\!\cdot$, i.e.

$$\langle f \diamond \mu, \varphi \rangle_{\Omega_D} \stackrel{\text{def}}{=} \langle f, \mu \,\diamond\!\!\!\cdot\, \check{\varphi} \rangle_\Omega,$$

where $(\mu \,\diamond\!\!\!\cdot\, \check{\varphi})(x) = \langle \mu, \tau_x \check{\varphi} \rangle_{\mathbb{R}^d}$.

2. *Domain.* It is possible to define a distribution on Ω_D, since this set is open due to Theorem 7.2.

3. *Test function.* We indeed have

$$\mu \,\diamond\!\!\!\cdot\, \check{\varphi} \in \mathcal{D}(\Omega).$$

because:

— since the weighting by a regular function is regularizing (Theorem 7.5), $\mu \,\diamond\!\!\!\cdot\, \check{\varphi}$ belongs to $\mathcal{C}^\infty(\mathbb{R}^d_{-K})$, where K is the support of φ, that is $\mathcal{C}^\infty(\mathbb{R}^d)$ as $\mathbb{R}^d_{-K} = \mathbb{R}^d$;

— due to the inclusion of the support from Theorem 7.10 and again to Theorem 7.2,

$$\operatorname{supp}(\mu \,\diamond\!\!\!\cdot\, \check{\varphi}) \subset \operatorname{supp} \mu - \operatorname{supp} \check{\varphi} \subset D + \Omega_D \subset \Omega. \tag{7.12}$$

4. *Distribution.* Now, let us prove that

$$f \diamond \mu \in \mathcal{D}'(\Omega_D; E). \tag{7.13}$$

It suffices, due to the characterization of distributions from Theorem 3.4 (b), to check that is it a sequentially continuous mapping, i.e. that the mapping

$$\varphi \mapsto \langle f, \mu \,\diamond\!\!\!\cdot\, \check{\varphi} \rangle_\Omega \text{ is sequentially continuous from } \mathcal{D}(\Omega_D) \text{ into } E. \tag{7.14}$$

Thus, let

$$\varphi_n \to \varphi \text{ in } \mathcal{D}(\Omega_D).$$

Due to the characterization of convergent sequences in $\mathcal{D}(\Omega_D)$ from Theorem 2.13 (a), there exists a compact subset K of Ω_D such that this convergence occurs in $\mathcal{C}^\infty_K(\mathbb{R}^d)$, hence

$$\check{\varphi}_n \to \check{\varphi} \text{ in } \mathcal{C}^\infty_{-K}(\mathbb{R}^d).$$

Since weighting is (Theorem 7.11 (b)) sequentially continuous with respect to the weight from $\mathcal{C}^\infty_{-K}(\mathbb{R}^d)$ into $\mathcal{C}^\infty(\mathbb{R}^d_{-K})$, and since $\mathbb{R}^d_{-K}$ is equal to $\mathbb{R}^d$,

$$\mu \,\diamond\!\!\!\cdot\, \check{\varphi}_n \to \mu \,\diamond\!\!\!\cdot\, \check{\varphi} \text{ in } \mathcal{C}^\infty(\mathbb{R}^d).$$

Moreover, from (7.12), $\mathrm{supp}(\mu \diamond \check{\varphi}_n) \subset \Omega$, hence, again due to the characterization of convergent sequences of test functions (here in $\mathcal{D}(\Omega)$) from Theorem 2.13 (a),

$$\mu \diamond \check{\varphi}_n \to \mu \diamond \check{\varphi} \text{ in } \mathcal{D}(\Omega).$$

Hence,

$$\langle f, \mu \diamond \check{\varphi}_n \rangle_\Omega \to \langle f, \mu \diamond \check{\varphi} \rangle_\Omega.$$

This proves (7.14), and consequently (7.13). □

Interest in weighting. Weighting allows us to extend to distributions defined on an open set Ω the fundamental properties that convolution gives for distributions defined on all of $\mathbb{R}^d$. Its role is essential because a distribution on Ω is not generally extendable to all of $\mathbb{R}^d$.

We will use it extensively in order to construct primitives of distributions in Chapter 13, to prove the *kernel theorem* on separation of variables in Chapter 15, to represent distributions with values in a Banach space as derivatives of continuous functions in Chapter 16, and to study Lebesgue and Sobolev spaces in the following volumes. □

Weighting versus convolution. When $\Omega = \mathbb{R}^d$, the **convolution** [6] $\star$ is, up to a symmetry, the weighting:

$$f \star \mu = f \diamond \check{\mu}, \tag{7.15}$$

where $\check{\mu}(x) = \mu(-x)$. It produces the pleasant formulae

$$\partial_i (f \star \mu) = \partial_i f \star \mu = f \star \partial_i \mu, \quad f \star \mu = \mu \star f, \tag{7.16}$$

whereas weighting produces (Theorems 7.17 and 8.25) the ugly formulae

$$\partial_i (f \diamond \mu) = \partial_i f \diamond \mu = -f \diamond \partial_i \mu, \quad f \diamond \mu = (\mu \diamond f)\check{} = \check{\mu} \diamond \check{f}. \tag{7.17}$$

We could have avoided these asymmetries by replacing μ by $\check{\mu}$ in Definition 7.12 of weighting, which would have generalized the convolution by an operation $\tilde{\star}$ from $\mathcal{D}'(\Omega, E) \times \mathcal{D}'_D(\mathbb{R}^d)$ into $\mathcal{D}'(\Omega_{-D}; E)$.

Why did we not do it? Because this would have been very inconvenient when D is not symmetric, which is essential for some applications, such as immersion and extension properties for Sobolev spaces, which we obtain with the help of integrals of the form $(f \diamond \mu)(x) = \int_D f(x+y)\mu(y)\,\mathrm{d}y$ where D is a **cone interior to** Ω. With $\tilde{\star}$ in the place of $\diamond$, the weight would need to have its support within the opposite cone, $-D$, that is **exterior**! Or it would be necessary to use $f \,\tilde{\star}\, \check{\mu}$, which cause the asymmetries to reappear.

Moreover, as soon as Ω differs from $\mathbb{R}^d$, the roles of f and of μ no longer can be the same, since f is not necessarily extendable outside of Ω, which is our essential interest in weighting, while μ must be extendable by 0 outside of D. In addition f is vector-valued, whereas μ is not. In fact, in the applications we will make, the roles of f and of μ will be quite distinct and the commutativity formulae of (7.17) will be of little use. □

---

6. **History of the convolution.** Vito VOLTERRA introduced the convolution of real functions on $\mathbb{R}$ in 1913 [91]. He called it the *composition of the first kind*. The convolution of functions on $\mathbb{R}^d$ was used in 1934, by Jean LERAY [53, Formula (1.18), p. 206].

7.6. Comparison of the definitions of weighting

Now let us first check that the weighting by a distribution recovers that by a regular function, so that we can use the same notation. We next check that it recovers the weighting of functions.

Theorem 7.13.– *If $\mu \in \mathcal{C}_D^\infty(\mathbb{R}^d)$, the weighted distribution $f \diamond \mu$ by the distribution μ given by Definition 7.12 coincides with that by the regular function μ given by Definition 7.1.*

That is to say that, with identification (3.6), p. 52, of μ with a distribution, Definition 7.12 also gives $f \diamond \mu \in \mathcal{C}^\infty(\Omega_D; E)$ and, for every $x \in \Omega_D$,

$$(f \diamond \mu)(x) = \langle f, \tau_x \mu \rangle_\Omega. \blacksquare$$

Proof. **1.** *Notation.* We denote by $\diamond\!\!\!\cdot$ the weighting by a regular function given by Definition 7.1 and by $\diamond$ the weighting by a distribution given by Definition 7.12, i.e.

$$\langle f \diamond \mu, \varphi \rangle_{\Omega_D} \stackrel{\text{def}}{=} \langle f, \mu \,\diamond\!\!\!\cdot\, \check{\varphi} \rangle_\Omega. \tag{7.18}$$

2. *Equivalence of the definitions.* It is a question of showing that

$$f \,\diamond\!\!\!\cdot\, \mu = f \diamond \mu. \tag{7.19}$$

Let $\varphi \in \mathcal{D}(\Omega_D)$. Since $f \,\diamond\!\!\!\cdot\, \mu$ is a continuous function (due to the regularizing property of the weighting by a regular function from Theorem 7.5), its identification with a distribution (Theorem 3.9) and Definition 7.1 successively give

$$\langle f \,\diamond\!\!\!\cdot\, \mu, \varphi \rangle_{\Omega_D} = \int_{\Omega_D} (f \,\diamond\!\!\!\cdot\, \mu)(x) \varphi(x) \,\mathrm{d}x = \int_{\Omega_D} \langle f, \tau_x \mu \rangle_\Omega \varphi(x) \,\mathrm{d}x. \tag{7.20}$$

Moreover, the linearity of f gives

$$\langle f, \tau_x \mu \rangle_\Omega \varphi(x) = \langle f, \tau_x \mu \varphi(x) \rangle_\Omega. \tag{7.21}$$

The Cauchy integral commuting with continuous linear mappings,

$$\int_{\Omega_D} \langle f, \tau_x \mu \varphi(x) \rangle_\Omega \,\mathrm{d}x = \left\langle f, \int_{\Omega_D} \tau_x \mu \varphi(x) \,\mathrm{d}x \right\rangle_\Omega. \tag{7.22}$$

More precisely, this equality is given by Theorem A.75 for the mapping $L = \langle f, \boldsymbol{\cdot} \rangle_\Omega$, which is continuous from $\mathcal{D}(\Omega)$ into E, applied to the function ϕ defined by $\phi(x) = \tau_x \mu \varphi(x)$, which belongs to $\mathcal{B}(\Omega_D; \mathcal{D}(\Omega))$, i.e. that is uniformly continuous and bounded from Ω_D into $\mathcal{D}(\Omega)$ (this property is given by

Theorem A.33, corollary of Heine's theorem, since ϕ is continuous due to Theorem 7.6 and has a compact support included in Ω_D, because it is that of φ).

Denoting by K the support of φ, only its interior $\overset{\circ}{K}$ contributes to the integral due to Theorem A.77 (since φ is zero outside of $\overset{\circ}{K}$ by Theorem A.45), i.e., for every $y \in \Omega$,

$$\Big(\int_{\Omega_D} \tau_x\mu\,\varphi(x)\,\mathrm{d}x\Big)(y) = \int_{\Omega_D} \mu(y-x)\,\varphi(x)\,\mathrm{d}x = \int_{\overset{\circ}{K}} \mu(y-x)\,\varphi(x)\,\mathrm{d}x.$$

By performing a symmetry in the integral (Theorem A.87) and then using the expression from Theorem 7.4 for the weighting of a continuous function by a regular weight (since $\check{\varphi} \in \mathcal{C}^\infty_{-K}(\mathbb{R}^d)$), it therefore comes to

$$\Big(\int_{\Omega_D} \tau_x\mu\,\varphi(x)\,\mathrm{d}x\Big)(y) = \int_{-\overset{\circ}{K}} \mu(y+z)\,\check{\varphi}(z)\,\mathrm{d}z = (\mu \diamond \check{\varphi})(y). \tag{7.23}$$

Equalities (7.20), (7.21), (7.22) and (7.23) give, along with Definition (7.18) of $\diamond$,

$$\langle f \diamond \mu, \varphi\rangle_{\Omega_D} = \langle f, \mu \diamond \check{\varphi}\rangle_{\Omega} = \langle f \diamond \mu, \varphi\rangle_{\Omega_D}.$$

This proves (7.19).

3. *Continuity and expression for $f \diamond \mu$.* So, $f \diamond \mu$ belongs to $\mathcal{C}^\infty(\Omega_D; E)$, since $f \diamond \mu$ belongs to it due to the regularizing property of the weighting by a regular function (Theorem 7.5). And, from Definition 7.1 of $\diamond$, it is given by

$$(f \diamond \mu)(x) = \langle f, \tau_x\mu\rangle_{\Omega}. \ \square$$

Let us now check that the weighting of distributions equally recovers that of continuous functions, which again permits us to use the same notation for both.

Theorem 7.14.– *If $f \in \mathcal{C}(\Omega; E)$ and $\mu \in \mathcal{C}_D(\mathbb{R}^d)$, the weighted distribution $f \diamond \mu$ given by Definition 7.12 coincides with the weighted function given by Definition 7.3.*

That is to say that, with identification (3.6), p. 52, of f and μ with distributions, Definition 7.12 again gives $f \diamond \mu \in \mathcal{C}(\Omega_D; E)$ and, for every $x \in \Omega_D$,

$$(f \diamond \mu)(x) = \int_{\overset{\circ}{D}} f(x+y)\mu(y)\,\mathrm{d}y. \ \blacksquare$$

Necessity for a proof. It does not suffice to observe that the weighting $\diamond$ by a distributions reproduces the weighting $\diamond$ by a regular function (Theorem 7.13) and that this reproduces the weighting $\square$ of functions (Theorem 7.4). Indeed, these theorems respectively give

$$f \diamond \mu = f \diamond \mu \quad \text{if } f \in \mathcal{D}'(\Omega; E) \text{ and } \mu \in \mathcal{C}^\infty_D(\mathbb{R}^d);$$
$$f \diamond \mu = f \,\square\, \mu \quad \text{if } f \in \mathcal{C}(\Omega; E) \text{ and } \mu \in \mathcal{C}^\infty_K(\mathbb{R}^d).$$

This gives $f \diamond \mu = f \,\square\, \mu$ only for $f \in \mathcal{C}(\Omega; E)$ and $\mu \in \mathcal{C}^\infty_K(\mathbb{R}^d)$, hence it remains to establish this equality for $f \in \mathcal{C}(\Omega; E)$ and $\mu \in \mathcal{C}_K(\mathbb{R}^d)$. $\square$

Proof of Theorem 7.14. **1.** *Notation.* We continue to denote by $\diamond\!\!\!\cdot$ the weighting by a regular function given by Definition 7.1 and by $\diamond$ the weighting by a distribution given by Definition 7.12, and we denote by $\square$ the weighting of functions given by Definition 7.3.

2. *Continuity of $f \square \mu$.* Let us show that

$$f \square \mu \in \mathcal{C}(\Omega_D; E). \tag{7.24}$$

Let $\{\| \|_{E;\nu} : \nu \in \mathcal{N}_E\}$ be the family of semi-norms of E. For every $x \in \Omega_D$, $x' \in \Omega_D$ and $\nu \in \mathcal{N}_E$, Definition 7.3 of $\square$ and the bound of the semi-norms of the integral from Theorem 1.23 (b) give

$$\|(f \square \mu)(x') - (f \square \mu)(x)\|_{E;\nu} = \left\| \int_{\mathring{D}} (f(x' + y) - f(x + y))\, \mu(y)\, \mathrm{d}y \right\|_{E;\nu} \leq$$

$$\leq |\mathring{D}| \sup_{y \in D} \|f(x' + y) - f(x + y)\|_{E;\nu} \sup_{y \in D} |\mu(y)|. \tag{7.25}$$

For a given x, let $r > 0$ be such that $B = \{x' \in \mathbb{R}^d : |x' - x| \leq r\}$ is included in Ω_D. Then, $B + D$ is a compact subset of Ω, so f is uniformly continuous on it due to Heine's theorem (Theorem A.32), and hence the last term in (7.25) converges to 0 when $x' \to x$. This proves (7.24).

3. *Equivalence of definitions.* It is a question of showing that

$$f \diamond \mu = f \square \mu.$$

That is to say that, for every $\varphi \in \mathcal{D}(\Omega_D)$,

$$\langle f \diamond \mu, \varphi \rangle_{\Omega_D} = \langle f \square \mu, \varphi \rangle_{\Omega_D}. \tag{7.26}$$

For that, we will express these two terms as double integrals whose equality will be given by a permutation of variables. We detail these calculations because they involve integration with values in a Neumann space that is new and so must be handled with caution.

On one hand, Definition 7.12 of $f \diamond \mu$ and $\check{\varphi}$ gives, with the expression for the identification of f with a distribution from Theorem 3.9 and the expression for $(\mu \diamond\!\!\!\cdot\, \check{\varphi})(x)$ from Theorem 7.4,

$$\langle f \diamond \mu, \varphi \rangle_{\Omega_D} = \langle f, \mu \diamond\!\!\!\cdot\, \check{\varphi} \rangle_{\Omega} = \int_{\Omega} f(x) \Big(\int_{\mathbb{R}^d} \mu(x + y)\, \widetilde{\varphi}(-y)\, \mathrm{d}y \Big)\, \mathrm{d}x. \tag{7.27}$$

The integral, here that with respect to y, is linear (Theorem A.74), invariant under the symmetry $y \mapsto -y$ (Theorem A.87) and can be restricted to a domain outside of which the integrand is zero (Theorem A.77), here Ω_D (since $\widetilde{\varphi}$ is zero outside of Ω_D), hence the last term of (7.27) equals

$$\int_\Omega \Big(\int_{\mathbb{R}^d} f(x)\,\mu(x-y)\,\widetilde{\varphi}(y)\,\mathrm{d}y \Big)\,\mathrm{d}x = \int_\Omega \Big(\int_{\Omega_D} f(x)\,\mu(x-y)\,\varphi(y)\,\mathrm{d}y \Big)\,\mathrm{d}x. \quad (7.28)$$

On the other hand, since $f \,\square\, \mu$ is continuous due to (7.24), the expression for its identification with a distribution from Theorem 3.9 and its Definition 7.3 give

$$\langle f \,\square\, \mu, \varphi \rangle_{\Omega_D} = \int_{\Omega_D} \Big(\int_{\mathring{D}} f(y+x)\,\mu(x)\,\mathrm{d}x \Big) \varphi(y)\,\mathrm{d}y. \quad (7.29)$$

The integral, now that with respect to x, is linear, invariant under the translation $x \mapsto x - y$ (Theorem A.86), and its domain of integration can be extended to Ω (again by Theorem A.77) since this contains $\mathring{D} + y$ and since $\mu(x-y) = 0$ if $x \notin \mathring{D} + y$, hence the right-hand side of (7.29) equals

$$\int_{\Omega_D} \Big(\int_{\mathring{D}+y} f(x)\,\mu(x-y)\,\varphi(y)\,\mathrm{d}x \Big)\,\mathrm{d}y = \int_{\Omega_D} \Big(\int_{\Omega} f(x)\,\mu(x-y)\,\varphi(y)\,\mathrm{d}x \Big)\,\mathrm{d}y.$$

It only remains to observe that the right-hand side is equal to that of (7.28) due to Theorem A.89 on the permutation of variables in an integral for the function defined on $\Omega \times \Omega_D$ by

$$F(x,y) \stackrel{\text{def}}{=} f(x)\mu(x-y)\varphi(y).$$

The said Theorem A.89 requires that F is uniformly continuous, which is the case here due to a corollary (Theorem A.33) of Heine's theorem, since its support is included in the compact set $(D+K) \times K$, where K is the support of φ, which is included in $\Omega \times \Omega_D$ (because $D + K \subset \Omega$, since $K \subset \Omega_D$).

The right-hand sides of (7.27) and (7.29) are thus equal, which proves (7.26), i.e.

$$f \diamond \mu = f \,\square\, \mu.$$

4. *Continuity of $f \diamond \mu$ and integral expression.* So, $f \diamond \mu$ is continuous from Ω_D into E, since $f \,\square\, \mu$ is from (7.24). And, from Definition 7.3 of $\square$, it is given by

$$(f \diamond \mu)(x) = \int_{\mathring{D}} f(x+y)\,\mu(y)\,\mathrm{d}y. \ \square$$

7.7. Continuity of the weighting by a distribution

Let us give continuity properties of the weighting by a distribution [7].

7. **History of the continuity properties in Theorem 7.15. Convolution.** Laurent SCHWARTZ proved in 1950 [69, Chap. VI, § 3, Theorem V, p. 157] that the convolution product of distributions is continuous from

Theorem 7.15.– *Let Ω be an open subset of $\mathbb{R}^d$, D a compact subset of $\mathbb{R}^d$ and E a Neumann space. Then:*

(a) *The mapping $\diamond$ is sequentially continuous bilinear from $\mathcal{D}'(\Omega; E) \times \mathcal{D}'_D(\mathbb{R}^d)$ into $\mathcal{D}'(\Omega_D; E)$.*

(b) *The mapping $f \mapsto f \diamond \mu$ is continuous, and hence sequentially continuous, from $\mathcal{D}'(\Omega; E)$ into $\mathcal{D}'(\Omega_D; E)$.* ▮

Proof. Here again we denote by $\diamond$ the weighting by a regular function given by Definition 7.1 and by $\diamond$ the weighting by a distribution given by Definition 7.12, i.e.

$$\langle f \diamond \mu, \varphi \rangle_{\Omega_D} \stackrel{\text{def}}{=} \langle f, \mu \diamond \check{\varphi} \rangle_{\Omega},$$

where $(\mu \diamond \check{\varphi})(x) = \langle \mu, \tau_x \check{\varphi} \rangle_{\mathbb{R}^d}$.

1. *Property (a).* **1.a.** *Bilinearity of* $\diamond$. This follows from the fact that:

— The mapping $f \mapsto f \diamond \mu$ is linear, by Definition (1.1), p. 8, of the addition and scalar multiplication of mappings.

— The mapping $\mu \mapsto f \diamond \mu$ is linear, since $\mu \mapsto \mu \diamond \check{\varphi}$ is linear by Definition 7.1 of the weighting by a regular function, and since f is linear by Definition 3.1 of a distribution.

1.b. *Sequential continuity of* $\diamond$. Let

$$f_n \to f \text{ in } \mathcal{D}'(\Omega; E) \quad \text{and} \quad \mu_n \to \mu \text{ in } \mathcal{D}'_D(\mathbb{R}^d).$$

Let $\varphi \in \mathcal{D}(\Omega_D)$ and K be the support of φ. By the sequential continuity of the weighting by a regular function (Theorem 7.11 (a)),

$$\mu_n \diamond \check{\varphi} \to \mu \diamond \check{\varphi} \text{ in } \mathcal{C}^\infty(\mathbb{R}^d{}_{-K}).$$

Moreover, $\mathbb{R}^d{}_{-K} = \mathbb{R}^d$. Furthermore (Theorem 7.10), $\text{supp}(\mu_n \diamond \check{\varphi}) \subset D + K$, which is a compact subset of Ω. Therefore, due to the characterization of convergence in $\mathcal{D}(\Omega)$ from Theorem 2.13 (a),

$$\mu_n \diamond \check{\varphi} \to \mu \diamond \check{\varphi} \text{ in } \mathcal{D}(\Omega).$$

$\mathcal{D}'(\mathbb{R}^d)$-unif $\times\, \mathcal{E}'_K(\mathbb{R}^d)$ into $\mathcal{D}'(\mathbb{R}^d)$-unif, i.e. for the uniform convergence on the bounded subsets of $\mathcal{D}(\Omega)$, where $\mathcal{E}'(\mathbb{R}^d)$ is the dual of $\mathcal{C}^\infty(\mathbb{R}^d)$ and $\mathcal{E}'_K(\mathbb{R}^d)$ is its subspace of elements with support in K. From which we can deduce the properties of Theorem 7.15 for $\Omega = \mathbb{R}^d$ and $E = \mathbb{R}$.

Method. Our method of proof is distinct from that of SCHWARTZ, since it defines the convolution via the tensor product, see Note 5, p. 153.

Since $\langle\ ,\ \rangle$ is sequentially continuous (Theorem 4.4 (c)),

$$\langle f_n, \mu_n \diamond \check{\varphi} \rangle_\Omega \to \langle f, \mu \diamond \check{\varphi} \rangle_\Omega.$$

That is, $\langle f_n \diamond \mu_n, \varphi \rangle_{\Omega_D} \to \langle f \diamond \mu, \varphi \rangle_{\Omega_D}$. Which proves that

$$f_n \diamond \mu_n \to f \diamond \mu \text{ in } \mathcal{D}'(\Omega_D; E),$$

and thus the sequential continuity of $\diamond$.

2. *Property (b): continuity of* $f \mapsto f \diamond \mu$. Let $\{\|\ \|_{E;\nu} : \nu \in \mathcal{N}_E\}$ be the family of semi-norms of E. Definition 3.1 of the semi-norms of $\mathcal{D}'(\Omega; E)$ gives, for every $\varphi \in \mathcal{D}(\Omega_D)$ and $\nu \in \mathcal{N}_E$,

$$\|f \diamond \mu\|_{\mathcal{D}'(\Omega_D;E);\varphi,\nu} = \|\langle f \diamond \mu, \varphi \rangle_{\Omega_D}\|_{E;\nu} =$$
$$= \|\langle f, \mu \diamond \check{\varphi} \rangle_\Omega\|_{E;\nu} = \|f\|_{\mathcal{D}'(\Omega;E);\mu \diamond \check{\varphi},\nu}.$$

Due to the characterization of continuous linear mappings from Theorem 1.12 (a), this equality shows that the mapping $f \mapsto f \diamond \mu$ is continuous from $\mathcal{D}'(\Omega; E)$ into $\mathcal{D}'(\Omega_D; E)$. □

Continuity of weighting in spaces of differentiable functions. Let us recall, for the record, some properties established in Volume 2 [82, Theorem 7.5].

Theorem 7.16.– *Given Ω an open subset of $\mathbb{R}^d$, E a Neumann space, $m \in \mathbb{N}$, $\ell \in \mathbb{N}$ and D a compact subset of $\mathbb{R}^d$, the weighting $\diamond$ is continuous, and hence sequentially continuous:*

from $\mathcal{C}^m(\Omega; E) \times \mathcal{C}^\ell_D(\mathbb{R}^d)$ into $\mathcal{C}^{m+\ell}(\Omega_D; E)$;

from $\mathcal{C}^m_b(\Omega; E) \times \mathcal{C}^\ell_D(\mathbb{R}^d)$ into $\mathcal{C}^{m+\ell}_b(\Omega_D; E)$;

from $\mathcal{C}^m_b(\Omega; E) \times \mathcal{C}^\ell_D(\mathbb{R}^d)$ into $\mathcal{C}^{m+\ell}_b(\Omega_D; E)$. □

7.8. Derivatives and support of a weighted distribution

Let us calculate the derivatives of a weighted distribution [8], by reducing to the results for a regular weight given in Theorem 7.9.

8. **History of the properties of weighted distributions of Theorems 7.17 to 7.21.** Laurent SCHWARTZ established in 1950 properties of the convolution of distributions in $\mathcal{D}'(\mathbb{R}^d)$ that are equivalent, for such distributions, to the properties of Theorems 7.17, 7.18 and 7.19, see respectively [69, Chap. VI, § 3: (VI,3,10), p. 160; Theorem II, p. 156; (VI,3,4), p. 159].
The properties of weighting in Theorems 7.17 to 7.21 were stated in 1993 [SIMON, 78, Theorem 7, p. 12] for E a Banach space.

Methods. Our methods of proof differ from those of Laurent SCHWARTZ, since he defines the convolution via the tensor product, see the Note 5, p. 153.

Theorem 7.17.– *Let $f \in \mathcal{D}'(\Omega; E)$ and $\mu \in \mathcal{D}'_D(\mathbb{R}^d)$, where Ω is an open subset of $\mathbb{R}^d$, E is a Neumann space and D is a compact subset of $\mathbb{R}^d$.*

Then, for every $i \in [\![1, d]\!]$,

$$\partial_i(f \diamond \mu) = \partial_i f \diamond \mu = -f \diamond \partial_i \mu,$$

and, for every $\beta \in \mathbb{N}^d$,

$$\partial^\beta(f \diamond \mu) = \partial^\beta f \diamond \mu = (-1)^{|\beta|} f \diamond \partial^\beta \mu. \blacksquare$$

Proof. Denote by $\mathbin{\diamond\!\!\!\!\cdot}$ the weighting by a regular function given by Definition 7.1 and by $\diamond$ the weighting by a distribution given by Definition 7.12, i.e.

$$\langle f \diamond \mu, \varphi \rangle_{\Omega_D} \stackrel{\text{def}}{=} \langle f, \mu \mathbin{\diamond\!\!\!\!\cdot} \check{\varphi} \rangle_\Omega. \tag{7.30}$$

1. *First-order derivative, first equality.* Definition 5.4 of the derivative of a distribution gives, with (7.30),

$$\langle \partial_i(f \diamond \mu), \varphi \rangle_{\Omega_D} = -\langle f \diamond \mu, \partial_i \varphi \rangle_{\Omega_D} = -\langle f, \mu \mathbin{\diamond\!\!\!\!\cdot} (\partial_i \varphi)\check{} \rangle_\Omega. \tag{7.31}$$

Since $-(\partial_i \varphi)\check{} = \partial_i \check{\varphi}$, the second expression for the derivative of a distribution weighted by a regular function from Theorem 7.9 gives

$$\mu \mathbin{\diamond\!\!\!\!\cdot} (\partial_i \varphi)\check{} = -\mu \mathbin{\diamond\!\!\!\!\cdot} \partial_i \check{\varphi} = \partial_i(\mu \mathbin{\diamond\!\!\!\!\cdot} \check{\varphi}).$$

Therefore (7.31) gives, again with Definition 5.4 and (7.30),

$$\langle \partial_i(f \diamond \mu), \varphi \rangle_{\Omega_D} = -\langle f, \partial_i(\mu \mathbin{\diamond\!\!\!\!\cdot} \check{\varphi}) \rangle_\Omega = \langle \partial_i f, \mu \mathbin{\diamond\!\!\!\!\cdot} \check{\varphi} \rangle_\Omega = \langle \partial_i f \diamond \mu, \varphi \rangle_{\Omega_D}.$$

This proves that

$$\partial_i(f \diamond \mu) = \partial_i f \diamond \mu.$$

2. *First-order derivative, second equality.* Theorem 7.9 also gives

$$\mu \mathbin{\diamond\!\!\!\!\cdot} (\partial_i \varphi)\check{} = -\mu \mathbin{\diamond\!\!\!\!\cdot} \partial_i \check{\varphi} = \partial_i \mu \mathbin{\diamond\!\!\!\!\cdot} \check{\varphi}.$$

Therefore (7.31) also gives, with (7.30),

$$\langle \partial_i(f \diamond \mu), \varphi \rangle_{\Omega_D} = -\langle f, \partial_i \mu \mathbin{\diamond\!\!\!\!\cdot} \check{\varphi} \rangle_\Omega = -\langle f \diamond \partial_i \mu, \varphi \rangle_{\Omega_D}.$$

This proves that

$$\partial_i(f \diamond \mu) = -f \diamond \partial_i \mu.$$

3. *Derivatives of order β.* The stated equalities follow from those of order one, by iteration. $\square$

Let us come to the support of a weighted distribution, by reducing to the result established in Theorem 7.10 for a regular weight.

Theorem 7.18.– *Let* $f \in \mathcal{D}'(\Omega; E)$ *and* $\mu \in \mathcal{D}'_D(\mathbb{R}^d)$*, where* Ω *is an open subset of* $\mathbb{R}^d$*,* E *is a Neumann space and* D *is a compact subset of* $\mathbb{R}^d$*. Then,*

$$\operatorname{supp}(f \diamond \mu) \subset \operatorname{supp} f - \operatorname{supp} \mu,$$

where $\operatorname{supp} f - \operatorname{supp} \mu = \{y - z : y \in \operatorname{supp} f,\ z \in \operatorname{supp} \mu\}$. ▮

Example of a strict inclusion. The support of $f \diamond \mu$ can be strictly included in $\operatorname{supp} f - \operatorname{supp} \mu$, and even in $(\operatorname{supp} f - \operatorname{supp} \mu) \cap \Omega_D$.

For example, if $f = 1$ on Ω and μ is a continuous function whose integral is zero, then $f \diamond \mu = 0$, so $\operatorname{supp}(f \diamond \mu) = \emptyset$, while $\operatorname{supp} f - \operatorname{supp} \mu = \Omega - D$. □

Proof of Theorem 7.18. We continue to denote by $\diamond\!\!\!\cdot$ the weighting by a regular function given by Definition 7.1, and by $\diamond$ the weighting by a distribution given by Definition 7.12, i.e.

$$\langle f \diamond \mu, \varphi \rangle_{\Omega_D} \stackrel{\text{def}}{=} \langle f, \mu \diamond\!\!\!\cdot\, \check{\varphi} \rangle_{\Omega}.$$

1. *Sufficient condition.* By Definition 6.18 of the support of a distribution,

$$\operatorname{supp}(f \diamond \mu) = \Omega_D \setminus \operatorname{nihil}(f \diamond \mu).$$

Denoting S_f the support of f and S_μ that of μ, it is then a question of showing that

$$\Omega_D \setminus \operatorname{nihil}(f \diamond \mu) \subset S_f - S_\mu.$$

This inclusion is equivalent to

$$\Omega_D \setminus (S_f - S_\mu) \subset \operatorname{nihil}(f \diamond \mu). \tag{7.32}$$

By Definition 6.17 of the annihilation domain, $\operatorname{nihil}(f \diamond \mu)$ is the largest open set $\mathcal{O}$ such that, for every $\varphi \in \mathcal{D}(\Omega_D)$ such that $\operatorname{supp} \varphi \subset \mathcal{O}$,

$$\langle f \diamond \mu, \varphi \rangle_{\Omega_D} = 0_E. \tag{7.33}$$

To obtain (7.32), it therefore suffices to show that the set $\Omega_D \setminus (S_f - S_\mu)$ is open and that, if

$$\operatorname{supp} \varphi \subset \Omega_D \setminus (S_f - S_\mu), \tag{7.34}$$

then (7.33) is satisfied, i.e.

$$\langle f, \mu \diamond\!\!\!\cdot\, \check{\varphi} \rangle_{\Omega} = 0_E. \tag{7.35}$$

2. Verification of (7.35). Denoting S_φ the support of φ, condition (7.34) is equivalent to

$$S_\varphi \subset \Omega_D \quad \text{and} \quad S_\varphi \cap (S_f - S_\mu) = \emptyset.$$

These conditions respectively imply

$$S_\varphi + S_\mu \subset S_\varphi + D \subset \Omega \quad \text{and} \quad (S_\varphi + S_\mu) \cap S_f = \emptyset.$$

Which are equivalent to

$$S_\varphi + S_\mu \subset \Omega \setminus S_f = \text{nihil}\, f.$$

With the inclusion of the support of a distribution weighted by a regular function from Theorem 7.10, it follows that

$$\text{supp}(\mu \diamond \check{\varphi}) \subset S_\mu - S_{\check{\varphi}} = S_\mu + S_\varphi \subset \text{nihil}\, f.$$

Which implies (7.35), due to characterization (6.9) (in Definition 6.17) of the annihilation domain of f.

3. Openness of $\Omega_D \setminus (S_f - S_\mu)$. Let

$$x \in \Omega_D \setminus (S_f - S_\mu),$$

i.e. $x \in \Omega_D$ and $x \notin S_f - S_\mu$. These conditions respectively imply

$$x + S_\mu \subset x + D \subset \Omega \quad \text{and} \quad (x + S_\mu) \cap S_f = \emptyset.$$

Which are equivalent to

$$x + S_\mu \subset \Omega \setminus S_f.$$

Here, $x + S_\mu$ is compact (since the support of μ is compact) and $\Omega \setminus S_f$ is open (it is the annihilation domain of f, which is open by Definition 6.17). The strong inclusion theorem (Theorem A.22) therefore shows that there exists a ball $B(0, r)$ such that

$$x + S_\mu + B(0, r) \subset \Omega \setminus S_f.$$

That is to say,

$$B(x, r) \subset \Omega \setminus (S_f - S_\mu).$$

Since Ω_D is open (Theorem 7.2), we can choose r small enough so that this ball is also included in Ω_D, and hence in $\Omega_D \setminus (S_f - S_\mu)$, which proves that this set is open. $\square$

7.9. Miscellanous properties of weighting

First, let us observe that the Dirac mass is the neutral element of weighting.

Theorem 7.19.– *Let $f \in \mathcal{D}'(\Omega; E)$, where Ω is an open subset of $\mathbb{R}^d$ and E is a Neumann space. Then,*

$$f \diamond \delta_0 = f. \blacksquare$$

Proof. The Dirac mass δ_0 introduced in Definition 3.14 has the set $\{0\}$ for its support. The weighted distribution $f \diamond \delta_0$ is therefore, by Definition 7.12, the distribution on

$$\Omega_{\{0\}} = \{x \in \mathbb{R}^d : x + 0 \in \Omega\} = \Omega,$$

given, for every $\varphi \in \mathcal{D}(\Omega)$, by

$$\langle f \diamond \delta_0, \varphi \rangle_\Omega = \langle f, \delta_0 \diamond \check{\varphi} \rangle_\Omega.$$

In the right-hand side, by Definition 7.1 of the weighting by a regular function,

$$(\delta_0 \diamond \check{\varphi})(x) = \langle \delta_0, \tau_x \check{\varphi} \rangle_{\mathbb{R}^d} = \tau_x \check{\varphi}(0) = \check{\varphi}(-x) = \varphi(x).$$

Hence,

$$\langle f \diamond \delta_0, \varphi \rangle_\Omega = \langle f, \varphi \rangle_\Omega. \square$$

Now we consider the image under a linear mapping.

Theorem 7.20.– *Let $f \in \mathcal{D}'(\Omega; E)$ and $\mu \in \mathcal{D}'_D(\mathbb{R}^d)$, where Ω is an open subset of $\mathbb{R}^d$, E is a Neumann space and D is a compact subset of $\mathbb{R}^d$, and let L be a continuous, or sequentially continuous, linear mapping from E into a Neumann space. Then,*

$$L(f \diamond \mu) = Lf \diamond \mu. \blacksquare$$

Proof. By Definitions 5.10 of the image under L and 7.12 of the weighting, we have, for every $\varphi \in \mathcal{D}(\Omega)$,

$$\langle L(f \diamond \mu), \varphi \rangle_{\Omega_D} = L(\langle f \diamond \mu, \varphi \rangle_{\Omega_D}) = L(\langle f, \mu \diamond \check{\varphi} \rangle_\Omega) =$$

$$= \langle Lf, \mu \diamond \check{\varphi} \rangle_\Omega = \langle Lf \diamond \mu, \varphi \rangle_{\Omega_D}. \square$$

Let us show that the weighted distribution of two positive distributions is positive.

Theorem 7.21.– *Let $f \in \mathcal{D}'(\Omega)$ and $\mu \in \mathcal{D}'_D(\mathbb{R}^d)$, where Ω is an open subset of $\mathbb{R}^d$ and D is a compact subset of $\mathbb{R}^d$, such that*

$$f \geq 0 \;\; \textit{and} \;\; \mu \geq 0.$$

Then,

$$f \diamond \mu \geq 0. \;\blacksquare$$

Proof. By Definitions 7.12 of the weighting and 5.31 of a positive distribution, it is a question of showing that, for every positive test function $\varphi \in \mathcal{D}(\Omega_D)$,

$$\langle f \diamond \mu, \varphi \rangle_{\Omega_D} = \langle f, \mu \diamond \check{\varphi} \rangle_\Omega \geq 0, \tag{7.36}$$

where, for every $x \in \Omega$,

$$(\mu \diamond \check{\varphi})(x) = \langle \mu, \tau_x \check{\varphi} \rangle_{\mathbb{R}^d}.$$

The test function $\tau_x \check{\varphi}$ is positive, therefore the positivity of μ implies that of the right-hand side, and hence

$$\mu \diamond \check{\varphi} \geq 0.$$

The positivity of f then implies (7.36). □

Let us give a condition for a distribution to be zero.

Theorem 7.22.– *Let $f \in \mathcal{D}'(\Omega; E)$, where Ω is an open subset of $\mathbb{R}^d$ and E is a Neumann space, such that, for every test function $\varphi \in \mathcal{D}(\Omega)$,*

$$f \diamond \varphi = 0_E \;\; \textit{in} \; \Omega_{\operatorname{supp} \varphi}.$$

Then,

$$f = 0. \;\blacksquare$$

Proof. By Definition 7.1 of the domain of a weighted distribution,

$$\Omega_{\operatorname{supp} \varphi} = \{x \in \mathbb{R}^d : x + \operatorname{supp} \varphi \subset \Omega\},$$

hence this set contains 0. Hence, by hypothesis,

$$(f \diamond \varphi)(0) = 0_E.$$

Due to the expression for the weighting by a regular function from Theorem 7.13,

$$(f \diamond \varphi)(0) = \langle f, \tau_0 \varphi \rangle_\Omega = \langle f, \varphi \rangle_\Omega.$$

Therefore, $\langle f, \varphi \rangle_\Omega = 0_E$ for every $\varphi \in \mathcal{D}(\Omega)$, thus f is identically zero. □

Another condition for cancellation. We will see in Theorem 8.9 that it is sufficient that the condition of Theorem 7.22 be satisfied by a regularizing sequence, namely that $f \diamond \rho_n = 0_E$ for every $n \in \mathbb{N}^*$ $\square$

Other properties. Commutativity and associativity properties are given in Theorems 8.25 and 8.26, and regularizing properties are given in Theorems 8.4 and 8.5. $\square$

Let us give an expression for the weighting by a continuous function.

Theorem 7.23.– *Let $f \in \mathcal{C}(\Omega; E)$ and $\mu \in \mathcal{C}_D(\mathbb{R}^d)$, where Ω is an open subset of $\mathbb{R}^d$ and E is a Neumann space. Then, $f \diamond \mu \in \mathcal{C}(\Omega_D; E)$ and, for every $x \in \Omega_D$,*

$$(f \diamond \mu)(x) = \int_\Omega f(y)\mu(y - x)\, dy. \blacksquare$$

Proof. According to Theorem 7.14, the weighting of distributions reproduces that of continuous functions and, more precisely, $f \diamond \mu \in \mathcal{C}(\Omega_D; E)$ and

$$(f \diamond \mu)(x) = \int_{\mathring{D}} f(x + z)\mu(z)\, dz.$$

Since the integral is invariant under a translation (Theorem A.86), here the change of variable $z = y - x$, and the domain where the function cancels does not contribute (Theorem A.77), it follows

$$(f \diamond \mu)(x) = \int_{\mathring{D}+x} f(y)\mu(y - x)\, dy = \int_\Omega f(y)\mu(y - x)\, dy,$$

because $\mathring{D} + x \subset \Omega$ and $\mu(y - x) = 0$ if $y \notin \mathring{D} + x$ (since μ is zero outside of $\mathring{D}$ due to Theorem A.45). $\square$

Here we give an integral equality which extends to the case where f, μ and φ are continuous functions of the equality $\langle f \diamond \mu, \varphi\rangle_{\Omega_D} = \langle f, \varphi \diamond \check{\mu}\rangle_\Omega$ from Definition 7.12.

Theorem 7.24.– *Let $f \in \mathcal{C}(\Omega; E)$, $\mu \in \mathcal{C}_D(\mathbb{R}^d)$ and $\varphi \in \mathcal{K}(\Omega_D)$, where Ω is an open subset of $\mathbb{R}^d$, E is a Neumann space and D is a compact subset of $\mathbb{R}^d$. Then, denoting $\check{\mu}(x) = \mu(-x)$,*

$$\int_{\Omega_D} (f \diamond \mu)\varphi = \int_\Omega f\,(\varphi \diamond \check{\mu}). \blacksquare$$

Proof. Let K be the support of φ and F the function defined on $\Omega \times \Omega_D$ by

$$F(x,y) \stackrel{\text{def}}{=} f(y)\mu(y-x)\varphi(x).$$

This function is continuous, and its support is included in $(K+D) \times K$ which is a compact subset of $\Omega \times \Omega_D$. It is therefore uniformly continuous from a corollary (Theorem A.33) of Heine's theorem. Theorem A.89 on the separation and permutation of variables in an integral then gives

$$\int_{\Omega_D \times \Omega} F = \int_{\Omega_D} \Big(\int_\Omega f(y)\mu(y-x)\,\mathrm{d}y \Big) \varphi(x)\,\mathrm{d}x = \\ = \int_\Omega f(y) \Big(\int_{\Omega_D} \mu(y-x)\varphi(x)\,\mathrm{d}x \Big)\,\mathrm{d}y.$$

That is, with the expression for the weighting of continuous functions from Theorem 7.23 for $(f \diamond \mu)(x)$ and $(\varphi \diamond \check{\mu})(y)$ respectively,

$$\int_{\Omega_D} (f \diamond \mu)(x)\,\varphi(x)\,\mathrm{d}x = \int_\Omega f(y)\,(\varphi \diamond \check{\mu})(y)\,\mathrm{d}y. \ \square$$

We finish with an estimate for weighted continuous functions, which will serve in the proof of the theorem of control of test function (Theorem 15.5).

Lemma 7.25.– *Let $\phi \in \mathcal{C}_\mathrm{b}^m(\Omega)$ and $\mu \in \mathcal{C}_D(\mathbb{R}^d)$, where Ω is an open subset of $\mathbb{R}^d$, D is a compact subset of $\mathbb{R}^d$ and $m \in \mathbb{N}$. Then,*

$$\|\phi \diamond \mu\|_{\mathcal{C}_\mathrm{b}^m(\Omega_D)} \le \|\phi\|_{\mathcal{C}_\mathrm{b}^m(\Omega)} \int_{\mathbb{R}^d} |\mu|. \ \blacksquare$$

Proof. The equality $\partial^\beta(\phi \diamond \mu) = \partial^\beta \phi \diamond \mu$ from Theorem 7.17 and the expression for the weighting of continuous functions from Theorem 7.23 give

$$\partial^\beta(\phi \diamond \mu)(x) = (\partial^\beta \phi \diamond \mu)(x) = \int_\Omega \partial^\beta \phi(y)\mu(y-x)\,\mathrm{d}y.$$

Hence, with the bound of the semi-norms (here, its absolute value) of the integral (Theorem 1.23 (a)), the growth of the real integral (Theorem A.76 (a) and (b)) and its linearity,

$$|\partial^\beta(\phi \diamond \mu)(x)| \le \int_\Omega |\partial^\beta \phi(y)\mu(y-x)|\,\mathrm{d}y \le \sup_{y \in \Omega} |\partial^\beta \phi(y)| \int_{\mathbb{R}^d} |\mu(y)|\,\mathrm{d}y.$$

By taking the upper bound with respect to x and β, it follows

$$\sup_{0 \le |\beta| \le m} \sup_{x \in \Omega_D} |\partial^\beta(\phi \diamond \mu)(x)| \le \sup_{0 \le |\beta| \le m} \sup_{y \in \Omega} |\partial^\beta(x)| \int_{\mathbb{R}^d} |\mu(y)|\,\mathrm{d}y,$$

which is the stated inequality, due to Definition 1.18 (b) of the semi-norms (here, the norm) of $\mathcal{C}_\mathrm{b}^m(\Omega)$. $\square$

Chapter 8

Regularization and Applications

We construct here regularizations, namely sequences of regular approximations, of a distribution f. We construct two types: local, or global.

First of all, we construct local approximations $f \diamond \rho_n$, where ρ_n is a regularizing function, namely regular with support in the ball $B(0, 1/n)$ and whose integral is 1. So, $f \diamond \rho_n$ is defined on the open set $\Omega_{1/n}$ obtained by removing from Ω a neighborhood of its boundary of "width" $1/n$. It is regular and locally converges to f.

In § 8.3, we construct global approximations $R_n f = \widetilde{\alpha_n f}$, where α_n is a localizing function, namely regular with compact support and equal to 1 except in a neighborhood of the boundary, and where $\widetilde{}$ is the extension by 0. So, $R_n f$ is defined on the whole of Ω. It is regular and globally converges to f, at the price of a loss of approximation in the neighborhood of the boundary.

Finally, we establish, by using the regularization:

— associativity and commutativity properties of weighting, in § 8.6;

— convergence properties of $\langle f_n - f, \varphi\rangle$ uniformly in φ when $f_n \to f$, in § 8.7.

8.1. Local regularization

Before the regular local approximations, let us define *regularizing sequences* which we will use to construct them.

Definition 8.1.– *A* ***regularizing sequence*** *is a sequence* $(\rho_n)_{n\geq 1}$ *of functions such that:*

$$\rho_n \in \mathcal{D}(\mathbb{R}^d), \quad \rho_n(x) = 0 \textit{ if } |x| \geq \frac{1}{n}, \quad \rho_n \geq 0, \quad \int_{\mathbb{R}^d} \rho_n = 1. \ \blacksquare$$

Justification. There exists a regularizing sequence, given for example by

$$\rho_n(x) = cn^d \rho(nx), \tag{8.1}$$

where ρ is the function in $\mathcal{D}(\mathbb{R}^d)$ defined by (2.4), p. 23, and $c = 1/\int_{\mathbb{R}^d}\rho$. Indeed, from Theorem A.88,

$$\int_{\mathbb{R}^d} \rho(nx)\,\mathrm{d}x = n^{-d}\int_{\mathbb{R}^d} \rho(y)\,\mathrm{d}y = \frac{1}{cn^d}. \ \square$$

Let us now introduce the local regularization [1] of a distribution.

Definition 8.2.– *Let $f \in \mathcal{D}'(\Omega; E)$, where Ω is an open subset of $\mathbb{R}^d$ and E is a Neumann space, and let $(\rho_n)_{n\geq 1}$ be a regularizing sequence (Definition 8.1).*

We call ***regular local approximations*** *of f the functions*

$$f \diamond \rho_n \in \mathcal{C}^\infty(\Omega_{D_n}; E)$$

for the $n \in \mathbb{N}^*$, where $D_n = \operatorname{supp}\rho_n$ and $\Omega_{D_n} = \{x \in \mathbb{R}^d : x + D_n \subset \Omega\}$. ▮

Justification. Since regularizing sequences exist as we saw in the justification of Definition 8.1, local approximations exist. They belong to $\mathcal{C}^\infty(\Omega_{D_n}; E)$ due to the regularizing property of the weighting by a regular function (Theorem 7.5). □

Terminology. The functions $f \diamond \rho_n$ are called **local approximations** since they are only defined, when Ω is bounded, on a subset of Ω (indeed, Ω_{D_n} cannot then be equal to Ω nor contain it, see the comment *Domain Ω_D*, p. 142) and since they converge to f (Theorem 8.4) on it.

Global approximations, namely on all of Ω, are constructed in Definition 8.13. Although they converge to f on all of Ω, they can be very different in the neighborhood of the boundary, because they are zero there. Contrary to the local regularization, the global one does not commute with the derivatives. □

Regularization of functions. When $f \in \mathcal{C}(\Omega; E)$, Definition 8.2 reproduces the regularization of a function considered in Volume 2 [82, Definition 7.7], since it is equally given by $f \diamond \rho_n$ and since the weighting of a distribution by a regular function reproduces that of a function due to Theorem 7.4. □

The convergence of the approximations is based on the following property [2].

1. **History of the regularization of distributions.** Laurent SCHWARTZ indicated in 1950 [69, Chap. VI, § 4, p. 166] how to use convolution to construct regular approximations $f \star \rho_n \in \mathcal{C}^\infty(\mathbb{R}^d)$ of $f \in \mathcal{D}'(\mathbb{R}^d)$.

Precursor. Jean LERAY defined in 1934 the regular approximations $f \star \rho_n$ of a function $f \in L^2(\mathbb{R}^d)$ [53, Formula (1.18), p. 206], which coincides with $f \diamond \check{\rho}_n$ where $\check{\rho}_n(x) = \rho_n(-x)$. He showed that this approximation is infinitely differentiable and that the regularization commutes with the derivatives [53, Lemma 4, p. 208], namely $\partial_i(f \star \rho_n) = \partial_i f \star \rho_n$ which corresponds to our Theorem 8.8.

2. **History of the convergence of regularizing sequences.** Laurent SCHWARTZ observed that $\rho_n \to \delta_0$ in $\mathcal{D}'(\mathbb{R}^d)$ in 1950 [69, Chap. VI, § 4, p. 166].

Theorem 8.3.– *Let $(\rho_n)_{n\geq 1}$ be a regularizing sequence (Definition 8.1). Then, when $n \to \infty$,*

$$\rho_n \to \delta_0 \quad \text{in } \mathcal{D}'(\mathbb{R}^d). \blacksquare$$

Proof. Let $\varphi \in \mathcal{D}(\mathbb{R}^d)$. The identification of ρ_n with a distribution is expressed (Theorem 3.9) by

$$\langle \rho_n, \varphi \rangle = \int_{\mathbb{R}^d} \rho_n(x)\varphi(x)\, \mathrm{d}x.$$

On the other hand, since $\int_{\mathbb{R}^d} \rho_n = 1$,

$$\langle \delta_0, \varphi \rangle = \varphi(0) = \int_{\mathbb{R}^d} \rho_n(x)\varphi(0)\, \mathrm{d}x.$$

The bound of the semi-norms, i.e. here of the absolute value, of the integral from Theorem 1.23 (a) and the growth of the real integral (Theorem A.76 (a)) therefore give

$$|\langle \rho_n - \delta_0, \varphi \rangle| = \left| \int_{\mathbb{R}^d} \rho_n(x)(\varphi(x) - \varphi(0))\, \mathrm{d}x \right| \leq \sup_{|x|\leq 1/n} |\varphi(x) - \varphi(0)|.$$

When $n \to \infty$, this converges to 0 since φ is continuous. Hence, according to the characterization of convergent sequences of distributions from Theorem 4.3 (c),

$$\rho_n \to \delta_0. \ \square$$

Let us show that the approximations $f \diamond \rho_n$ converge to f locally, namely outside of a neighborhood of the boundary of Ω of arbitrary width $1/m$, which we will use often.

Theorem 8.4.– *Let $f \in \mathcal{D}'(\Omega; E)$, where Ω is an open subset of $\mathbb{R}^d$ and E is a Neumann space, $(\rho_n)_{n\geq 1}$ a regularizing sequence (Definition 8.1), $m \in \mathbb{N}^*$ and*

$$\Omega_{1/m} \stackrel{\text{def}}{=} \{x \in \Omega : B(x, 1/m) \subset \Omega\}.$$

where $B(x, 1/m) \stackrel{\text{def}}{=} \{x \in \mathbb{R}^d : |x| \leq 1/m\}$.

Then, when $n \to \infty$ and $n \geq m$,

$$f \diamond \rho_n \to f \quad \text{in } \mathcal{D}'(\Omega_{1/m}; E). \blacksquare$$

Proof. This is a particular case of Theorem 8.5 below, for $\omega = \Omega_{1/m}$. Indeed, for every $n \geq m$,

$$\Omega_{1/m} \subset \Omega_{D_n} = \{x \in \mathbb{R}^d : x + D_n \subset \Omega\},$$

since, by Definition 8.2, $D_n = \operatorname{supp} \rho_n \subset B(0, 1/n) \subset B(0, 1/m)$. □

Let us show that, more generally, local approximations converge on subsets ω of Ω that may extend up to (a part of) its boundary.

Theorem 8.5.– *Let $f \in \mathcal{D}'(\Omega; E)$, where Ω is an open subset of $\mathbb{R}^d$ and E is a Neumann space, $(\rho_n)_{n\geq 1}$ a regularizing sequence (Definition 8.1) and $D_n = \operatorname{supp} \rho_n$. Moreover, let ω be an open subset of Ω and $m \in \mathbb{N}^*$ be such that, for every $n \geq m$,*

$$\omega \subset \Omega_{D_n}.$$

Then, when $n \to \infty$ and $n \geq m$,

$$f \diamond \rho_n \to f \ \text{ in } \mathcal{D}'(\omega; E). \ ▮$$

Choice of D_n. In this volume, we always choose $D_n = B(0, 1/n)$, in which case Ω_{D_n} is equal to $\Omega_{1/n}$, i.e. to Ω with a neighborhood of its boundary of width $1/n$ removed, and the local convergence from Theorem 8.4 is sufficient for our purposes. But it does not allow to regularize *up to the boundary*.

In the following volumes, to regularize up to a part of the boundary, we will choose a regularizing sequence with support in a conic section $K_n = C \cap B(0, 1/n)$, where C is a cone of vertex 0, and $\omega = \Omega_{K_n}$. Since Ω_{K_n} can extend up to a part of the boundary as in Figure 7.2, p. 143, Theorem 8.5 then allows to regularize up to it.

In particular, in Volume 7, we will establish the uniform continuity in *Lions' lemma of continuity*, up to the bound 0 of the interval of definition $[0, T]$ by using an *asymmetric regularization* by ρ_n whose support is $D_n = [0, 1/n]$ instead of the ball $[-1/n, 1/n]$. This will provide an approximation on $(0, T - 1/n)$, which uniformly converges on $(0, t)$ for every $t < T$. And we will establish the uniform continuity up to T by using ρ_n whose support is $D_n = [-1/n, 0]$. □

Proof of Theorem 8.5. Let

$$K \stackrel{\text{def}}{=} \overline{\textstyle\bigcup_{n\geq m} D_n}.$$

It is a compact subset of $\mathbb{R}^d$ due to the Borel–Lebesgue theorem (Theorem A.26 (b)), since it is closed (by construction) and bounded (since it is included in $B(0, 1)$, by Definition 8.1 of ρ_n). It satisfies

$$\omega + K \subset \Omega. \tag{8.2}$$

Indeed, if $x \in \omega$, there exists $\epsilon > 0$ such that $B(x, \epsilon) \subset \omega$ and, according to the characterization of the closure (Theorem A.9), if $y \in K$, there exist $n \geq m$ and

$z \in D_n$ such that $|y - z| \leq \epsilon$, and hence $x + y \in x + z + B(0, \epsilon)$, which is included in $\omega + D_n$ and thus in Ω (from the hypothesis $\omega \subset \Omega_{D_n}$).

If $n \geq m$, then $\operatorname{supp} \rho_n \subset K$ and

$$\rho_n \in \mathcal{D}'_K(\mathbb{R}^d).$$

Since $\mathcal{D}'_K(\mathbb{R}^d)$ is, by Definition 6.26, endowed with the semi-norms of $\mathcal{D}'(\mathbb{R}^d)$, the convergence $\rho_n \to \delta_0$ in $\mathcal{D}'(\mathbb{R}^d)$ established in Theorem 8.3 gives

$$\rho_n \to \delta_0 \text{ in } \mathcal{D}'_K(\mathbb{R}^d).$$

The sequential continuity of weighting (Theorem 7.15 (a)) then implies

$$f \diamond \rho_n \to f \diamond \delta_0 \text{ in } \mathcal{D}'(\Omega_K; E).$$

Now, $f \diamond \delta_0 = f$, since the Dirac mass is the neutral element of weighting according to Theorem 7.19. And, the convergence takes place in $\mathcal{D}'(\omega; E)$, since $\omega \subset \Omega_K$ from (8.2) and since the restriction is continuous (Theorem 6.3). That is to say,

$$f \diamond \rho_n \to f \text{ in } \mathcal{D}'(\omega; E). \ \square$$

We observe that, for distributions on $\mathbb{R}^d$, local approximations converge globally.

Theorem 8.6.– *Let $f \in \mathcal{D}'(\mathbb{R}^d; E)$, where E is a Neumann space, $(\rho_n)_{n \geq 1}$ be a regularizing sequence, $m \in \mathbb{N}$ and K a closed subset of $\mathbb{R}^d$.*

(a) *Then, $f \diamond \rho_n \in \mathcal{C}^\infty(\mathbb{R}^d; E)$ and, when $n \to \infty$,*

$$f \diamond \rho_n \to f \text{ in } \mathcal{D}'(\mathbb{R}^d; E).$$

(b) *If $f \in \mathcal{D}'_K(\mathbb{R}^d; E)$, then $f \diamond \rho_n \in \mathcal{C}^\infty_{K+B(0,1/n)}(\mathbb{R}^d; E)$ and, when $n \to \infty$ and $n \geq m$,*

$$f \diamond \rho_n \to f \text{ in } \mathcal{D}'_{K+B(0,1/m)}(\mathbb{R}^d; E). \ \blacksquare$$

Proof. **1.** Property (a). Regularity is given by Theorem 7.5, since $\operatorname{supp} \rho_n \subset D_n$ and

$$\mathbb{R}^d_{D_n} = \{x \in \mathbb{R}^d : x + D_n \subset \mathbb{R}^d\} = \mathbb{R}^d.$$

Convergence is therefore given by Theorem 8.4 for $\Omega = \mathbb{R}^d$.

2. Property (b). Due to the inclusion of the support of a weighted distribution from Theorem 7.10, if $n \geq m$,

$$\operatorname{supp}(f \diamond \rho_n) \subset \operatorname{supp} f - \operatorname{supp} \rho_n \subset K + B(0, 1/n) \subset K + B(0, 1/m).$$

So, $f \diamond \rho_n \in \mathcal{D}'_{K+B(0,1/m)}(\mathbb{R}^d; E)$ and the convergence takes place in this space since it is, by Definition 6.26, endowed with the semi-norms of $\mathcal{D}'(\Omega; E)$. $\square$

Recall that, if f is continuous, its local approximations converge locally in the space of continuous functions.

Theorem 8.7.– *Let Ω be an open subset of $\mathbb{R}^d$, E a Neumann space, $m \in \mathbb{N}$, $k \in \mathbb{N}$, $\Omega_{1/k} \stackrel{\text{def}}{=} \{x \in \Omega : B(x, 1/k) \subset \Omega\}$ and $(\rho_n)_{n\geq 1}$ a regularizing sequence (Definition 8.1). Then, when $n \to \infty$ and $n \geq k$:*

(a) *If $f \in \mathcal{C}^m(\Omega; E)$, then $f \diamond \rho_n \to f$ in $\mathcal{C}^m(\Omega_{1/k}; E)$.*

(b) *If $f \in \mathbf{C}_\mathrm{b}^m(\Omega; E)$, then $f \diamond \rho_n \to f$ in $\mathbf{C}_\mathrm{b}^m(\Omega_{1/k}; E)$.* ▮

Proof. These properties are proven in Volume 2 [82, Theorem 7.9] for the weighting of continuous functions. They are thus equally satisfied for the weighting of distributions, since it reproduces the weighting of continuous functions according to Theorem 7.14. □

8.2. Properties of local approximations

Let us observe that the local regularization commutes with the derivatives.

Theorem 8.8.– *Let $f \in \mathcal{D}'(\Omega; E)$, where Ω is an open subset of $\mathbb{R}^d$ and E is a Neumann space. Its local approximations given by Definition 8.2 satisfy, for every $n \in \mathbb{N}^*$, $i \in [\![1, d]\!]$ and $\beta \in \mathbb{N}^d$,*

$$\partial_i(f \diamond \rho_n) = (\partial_i f) \diamond \rho_n,$$
$$\partial^\beta(f \diamond \rho_n) = (\partial^\beta f) \diamond \rho_n. \quad ▮$$

Proof. This is just that derivatives commute with weighting (Theorem 7.9 or 7.17). □

Let us show that a distribution is zero as soon as its local approximations are.

Theorem 8.9.– *Let $f \in \mathcal{D}'(\Omega; E)$, where Ω is an open subset of $\mathbb{R}^d$ and E is a Neumann space, and $(\rho_n)_{n\geq 1}$ a regularizing sequence. If, for every $n \in \mathbb{N}^*$,*

$$f \diamond \rho_n = 0_E \text{ on } \Omega_{1/n},$$

then

$$f = 0_E \text{ on } \Omega. \quad ▮$$

Proof. According to the convergence of local approximations (Theorem 8.4), for every $m \in \mathbb{N}^*$, when $n \geq m$ and $n \to \infty$,

$$f \diamond \rho_n \to f \ \text{ in } \mathcal{D}'(\Omega_{1/m}; E).$$

If the $f \diamond \rho_n$ are zero, then

$$f = 0_E \ \text{ on } \Omega_{1/m}.$$

From the gluing theorem for equalities (Theorem 6.10), f is therefore zero on the union of the $\Omega_{1/m}$, namely Ω. □

Let us give an integral expression for the local approximations of a continuous function.

Theorem 8.10.– *Let $f \in \mathcal{C}(\Omega; E)$, where Ω is an open subset of $\mathbb{R}^d$ and E is a Neumann space. Its local approximations given by Definition 8.2 satisfy, for every $n \in \mathbb{N}^*$ and $x \in \Omega_D$,*

$$(f \diamond \rho_n)(x) = \int_{\mathring{D}} f(x+y)\rho_n(y)\, \mathrm{d}y. \ ▮$$

Proof. This is the expression for the weighting of a continuous function by a regular function given in Theorem 7.4. □

8.3. Global regularization

Before defining global regularization, we define *localizing sequences* that we will use to *localize* the support of the distribution that we wish to regularize.

Definition 8.11.– *Let Ω be an open subset of $\mathbb{R}^d$. A **localizing sequence** is a sequence $(\alpha_n)_{n\geq 1}$ of functions such that, for every n:*

$$\alpha_n \in \mathcal{D}(\Omega), \quad 0 \leq \alpha_n \leq 1, \quad \alpha_n = \begin{cases} 1 & \text{on } \Omega^n_{3/n}, \\ 0 & \text{outside of } \Omega^n_{2/n}, \end{cases}$$

where

$$\Omega^n_r \overset{\text{def}}{=} \{x \in \Omega : |x| < n,\ B(x,r) \subset \Omega\}$$

and $B(x,r) \overset{\text{def}}{=} \{y \in \mathbb{R}^d : |y - x| \leq r\}$. ▮

Justification of Definition 8.11. Urysohn's theorem (Theorem 8.12, below) gives the existence of the functions α_n since $\Omega^n_{3/n} \Subset \Omega^n_{2/n} \Subset \Omega$. □

Recall the definition of the **compact inclusion** of a subset ω of $\mathbb{R}^d$ into another, Ω:

$$\omega \Subset \Omega \quad \overset{\text{def}}{\Leftrightarrow} \quad \omega \text{ is bounded and } \overline{\omega} \subset \overset{\circ}{\Omega}.$$

(This inclusion is called *compact* since $\overline{\omega}$ is compact here, by the Borel–Lebesgue theorem, since it is closed and bounded in $\mathbb{R}^d$).

Also recall **Urysohn's theorem**, proved in Volume 2 [82, Theorem 3.21].

Theorem 8.12.– *Given $\omega \Subset \Omega \subset \mathbb{R}^d$, there exists a function α such that:*

$$\alpha \in \mathcal{D}(\Omega), \quad 0 \leq \alpha \leq 1, \quad \alpha = 1 \text{ on } \omega. \ ▮$$

Let us now come to the global regularization, namely on all of Ω.

Definition 8.13.– *Let $f \in \mathcal{D}'(\Omega; E)$, where Ω is an open subset of $\mathbb{R}^d$ and E is a Neumann space, $(\alpha_n)_{n \geq 1}$ a localizing sequence (Definition 8.11) and $(\rho_n)_{n \geq 1}$ a regularizing sequence (Definition 8.1).*

We call (global) ***regular approximations*** *of f the functions $R_n f$ defined, for every $x \in \Omega$, by*

$$(R_n f)(x) \overset{\text{def}}{=} \langle f, \alpha_n \tau_x \rho_n \rangle_\Omega,$$

where $(\tau_x \rho_n)(y) = \rho_n(y - x)$. ▮

Justification. The existence of the ρ_n and of the α_n are established respectively in the justifications of Definitions 8.1 and 8.11.

The term $\langle f, \alpha_n \tau_x \rho_n \rangle_\Omega$ makes sense, since $\alpha_n \in \mathcal{D}(\Omega)$ and $\tau_x \rho_n \in C^\infty(\Omega)$, and thus their product $\alpha_n \tau_x \rho_n$ belongs to $\mathcal{D}(\Omega)$ due to Property (5.9), p. 95. □

Terminology. The functions $R_n f$ are called **regular approximations** since they are regular (Theorem 8.15) and they converge to f (Theorem 8.16). Eventually, we qualify them as **global** since they are defined on the entire domain Ω of definition of f, contrary to *local* approximations $f \diamond \rho_n$ which are only, in general, defined on subsets Ω_{D_n} of Ω, see comment *Domain* Ω_D, p. 142.

Global approximations $R_n f$ can be very different from f in the neighborhood of the boundary, since they are zero there. Hence, contrary to the local regularization, the global one does not commute with derivatives. □

Regularization of functions. When $f \in \mathcal{C}(\Omega; E)$, Definition 8.13 reproduces the regularization of a function considered in Volume 2 [82, Definition 7.13], since the definition of $R_n f = (\widetilde{\alpha_n f} \diamond \rho_n)|_\Omega$ which is given there is the one obtained here in Theorem 8.14 (a), and because localization-extension and weighting by a regular function of a distribution reproduce that of a continuous function due to Theorems 6.15 and 7.4. □

Let us relate global regularization, localization-extension and local regularization.

Theorem 8.14.– *Let $f \in \mathcal{D}'(\Omega; E)$, where Ω is an open subset of $\mathbb{R}^d$ and E is a Neumann space, and $R_n f$ a regular approximation given by Definition 8.13. Then:*

(a)
$$R_n f = (\widetilde{\alpha_n f} \diamond \rho_n)|_\Omega,$$

where $\widetilde{\alpha_n f} \in \mathcal{D}'(\mathbb{R}^d; E)$ is the localized extension given by Definition 6.12 (and where $\alpha_n \in \mathcal{D}(\mathbb{R}^d)$ is extended here by 0 outside of Ω).

(b)
$$R_n f = \begin{cases} f \diamond \rho_n & \text{on } \Omega^n_{4/n}, \\ 0_E & \text{outside of } \Omega^n_{1/n}. \end{cases}$$

where $\Omega^n_r \stackrel{\text{def}}{=} \{x \in \Omega : |x| < n,\ B(x, r) \subset \Omega\}$. ▮

Proof. **1.** *Equality (a).* Definition 8.13 of $R_n f(x)$ is written, with Definition 6.12 of the localized extension $\widetilde{\alpha_n f}$ and Definition 7.1 of the weighting by a regular function,

$$(R_n f)(x) = \langle f, \alpha_n \tau_x \rho_n \rangle_\Omega = \langle \widetilde{\alpha_n f}, \tau_x \rho_n \rangle_{\mathbb{R}^d} = (\widetilde{\alpha_n f} \diamond \rho_n)(x).$$

Therefore, $R_n f$ is the restriction to Ω of $\widetilde{\alpha_n f} \diamond \rho_n$, which is defined on all of $\mathbb{R}^d$.

2. *First equality of (b): values on $\Omega^n_{4/n}$.* If $x \in \Omega^n_{4/n}$,

$$(R_n f)(x) = \langle f, \alpha_n \tau_x \rho_n \rangle_\Omega = \langle f, \tau_x \rho_n \rangle_\Omega = (f \diamond \rho_n)(x).$$

Indeed, for such x, $\alpha_n \tau_x \rho_n = \tau_x \rho_n$ on all of Ω because:
— it is true on $\Omega^n_{3/n}$ since α_n equals 1 there (by Definition 8.11 of a localizing sequence);
— it is true outside of $\Omega^n_{3/n}$ since $\tau_x \rho_n$ is zero there (because $\rho_n = 0$ outside of $B(0, 1/n)$ by Definition 8.1 of a regularizing sequence, whence $\tau_x \rho_n = 0$ outside of the ball $B(x, 1/n)$, which is included in $\Omega^n_{3/n}$).

3. *Second equality of (b): value outside of $\Omega^n_{1/n}$.* If $x \notin \Omega^n_{1/n}$,

$$(R_n f)(x) = \langle f, \alpha_n \tau_x \rho_n \rangle_\Omega = 0_E.$$

Indeed, for such x, the function $\alpha_n \tau_x \rho_n$ is zero on all of Ω because:
— it is zero outside of $\Omega^n_{2/n}$, since α_n is zero there;
— it is zero on $\Omega^n_{2/n}$, since $\tau_x \rho_n$ is zero there (because it is zero outside of $B(x, 1/n)$, which does not intersect $\Omega^n_{2/n}$). □

Let us show that global approximations are regular and have compact support. We denote

$$\mathcal{K}^\infty(\Omega; E) \stackrel{\text{def}}{=} \left\{ \begin{array}{l} \textit{infinitely differentiable functions} \\ \textit{with compact support from } \Omega \textit{ into } E \end{array} \right\}.$$

Theorem 8.15.– *Let $f \in \mathcal{D}'(\Omega; E)$, where Ω is an open subset of $\mathbb{R}^d$ and E is a Neumann space. Every regular approximation $R_n f$ given by Definition 8.13 satisfies*

$$R_n f \in \mathcal{K}^\infty(\Omega; E),$$

and its support is a compact set included in the compact subset $\overline{\Omega^n_{1/n}}$ of Ω. ▮

Proof. Every global approximation $R_n f$ is infinitely differentiable since, according to Theorem 8.14 (a), it is the restriction to Ω of the weighted distribution $\widetilde{\alpha_n f} \diamond \rho_n$, which is infinitely differentiable on $\mathbb{R}^d$ due to Theorem 8.6 (a).

The support of the function $R_n f$ is included in $\overline{\Omega^n_{1/n}}$, since it is zero outside of $\Omega^n_{1/n}$ from Theorem 8.14 (b). And it is compact due to Theorem 2.2 (a), since it is thus included in a compact subset of Ω. □

8.4. Convergence of global approximations

Let us show that the global approximations $R_n f$ converge to f.

Theorem 8.16.– *Let $f \in \mathcal{D}'(\Omega; E)$, where Ω is an open subset of $\mathbb{R}^d$ and E is a Neumann space. Every sequence of regular approximations given by Definition 8.13 satisfies, when $n \to \infty$,*

$$R_n f \to f \ \textit{in } \mathcal{D}'(\Omega; E).$$ ▮

Proof. **1.** *An equality.* Given $\varphi \in \mathcal{D}(\Omega)$, let m be large enough so that

$$\operatorname{supp} \varphi \subset \Omega^m_{1/m}$$

(recall that $\Omega^m_{1/m} \overset{\text{def}}{=} \{x \in \Omega : |x| < m,\ B(x, 1/m) \subset \Omega\}$). Let moreover

$$n \geq 4m.$$

Identifying the function $R_n f$ with a distribution by Theorem 3.9 gives, since the domain where the function is zero does not contribute to the Cauchy integral (due to Theorem A.77) and since φ is zero outside of $\Omega^m_{1/m}$,

$$\langle R_n f, \varphi\rangle_\Omega = \int_\Omega R_n f\, \varphi = \int_{\Omega^m_{1/m}} R_n f\, \varphi.$$

Due to Theorem 8.14 (b), $R_n f$ equals $f \diamond \rho_n$ on $\Omega^n_{4/n}$, which contains $\Omega^m_{1/m}$, hence, again with the identification from Theorem 3.9,

$$\langle R_n f, \varphi\rangle_\Omega = \int_{\Omega^m_{1/m}} (f \diamond \rho_n)\, \varphi = \langle f \diamond \rho_n, \varphi\rangle_{\Omega^m_{1/m}}. \tag{8.3}$$

2. *Convergence*. Due to the convergence of local approximations from Theorem 8.4,

$$f \diamond \rho_n \to f \ \text{ in } \mathcal{D}'(\Omega_{1/m}; E) \text{ and hence in } \mathcal{D}'(\Omega^m_{1/m}; E).$$

Therefore, (8.3) implies (with Definition 6.1 of the restriction of f)

$$\langle R_n f, \varphi\rangle_\Omega = \langle f \diamond \rho_n, \varphi\rangle_{\Omega^m_{1/m}} \to \langle f, \varphi\rangle_{\Omega^m_{1/m}} = \langle f, \varphi\rangle_\Omega.$$

Which proves that $R_n f \to f$ in $\mathcal{D}'(\Omega; E)$. □

Let us show that the regular approximations $R_n f_n$ of approximations f_n of f are themselves approximations of f.

Theorem 8.17.– *Let $(f_n)_{n\in\mathbb{N}^*}$ be a sequence in $\mathcal{D}'(\Omega; E)$ and $f \in \mathcal{D}'(\Omega; E)$, where Ω is an open subset of $\mathbb{R}^d$ and E is a Neumann space, such that, when $n \to \infty$,*

$$f_n \to f \text{ in } \mathcal{D}'(\Omega; E),$$

and let $(R_n)_{n\in\mathbb{N}^*}$ be a sequence of regularizations given by Definition 8.13. Then,

$$R_n f_n \to f \text{ in } \mathcal{D}'(\Omega; E). \ ∎$$

Proof. In the proof of Theorem 8.16, we obtained equality (8.3) for every distribution f on Ω. It is therefore true for f_n. More precisely, given $\varphi \in \mathcal{D}(\Omega)$, m large enough so that $\Omega^m_{1/m}$ contains the support of φ, and $n > 4m$,

$$\langle R_n f_n, \varphi\rangle_\Omega = \langle f_n \diamond \rho_n, \varphi\rangle_{\Omega^m_{1/m}}.$$

Since $f_n \to f$ (by hypothesis) and $\rho_n \to \delta_0$ in $\mathcal{D}'_{B(0,1/m)}(\mathbb{R}^d)$ (due to Theorem 8.3), the sequential continuity of weighting (Theorem 7.15 (a)) gives

$$f_n \diamond \rho_n \to f \diamond \delta_0 \text{ in } \mathcal{D}'(\Omega_{B(0,1/m)}; E).$$

This convergence takes place in $\mathcal{D}'(\Omega^m_{1/m}; E)$, since $\Omega_{B(0,1/m)} = \Omega_{1/m} \subset \Omega^m_{1/m}$ and since the restriction is sequentially continuous (Theorem 6.3). Moreover, $f \diamond \delta_0 = f$ because δ_0 is the neutral element of weighting (Theorem 7.19). Thus,

$$\langle R_n f_n, \varphi \rangle_\Omega = \langle f_n \diamond \rho_n, \varphi \rangle_{\Omega^m_{1/m}} \to \langle f, \varphi \rangle_{\Omega^m_{1/m}} = \langle f, \varphi \rangle_\Omega.$$

Which proves that $R_n f_n \to f$ in $\mathcal{D}'(\Omega; E)$. □

We now observe that the set of regular functions with compact support is dense in the space of distributions [3].

Theorem 8.18.– *Let Ω be an open subset of $\mathbb{R}^d$ and E be a Neumann space. Then, $\mathcal{K}^\infty(\Omega; E)$ is sequentially dense, and therefore dense, in $\mathcal{D}'(\Omega; E)$.* ▮

Proof. Given $f \in \mathcal{D}'(\Omega; E)$, the approximations $R_n f$ given by Definition 8.13 belong to $\mathcal{K}^\infty(\Omega; E)$ (Theorem 8.15) and converge to f in $\mathcal{D}'(\Omega; E)$ (Theorem 8.16). Therefore, the set of all the $R_n f$, and hence $\mathcal{K}^\infty(\Omega; E)$, is sequentially dense in $\mathcal{D}'(\Omega; E)$, which implies (Theorem A.13) that it is dense. □

8.5. Properties of global approximations

Let us calculate the derivatives of the global approximations of a distribution.

Theorem 8.19.– *Let $f \in \mathcal{D}'(\Omega; E)$, where Ω is an open subset of $\mathbb{R}^d$ and E is a Neumann space, $R_n f$ a regular approximation given by Definition 8.13 and $\beta \in \mathbb{N}^d$. Then, for every $x \in \Omega$,*

$$(\partial^\beta R_n f)(x) = (-1)^{|\beta|} \langle f, \alpha_n \tau_x(\partial^\beta \rho_n) \rangle_\Omega$$

and

$$\partial^\beta (R_n f) = R_n(\partial^\beta f) \text{ on } \Omega^n_{4/n},$$

where $\Omega^n_{4/n} \stackrel{\text{def}}{=} \{x \in \Omega : |x| < n,\ B(x, 4/n) \subset \Omega\}$. ▮

3. **History of Theorem 8.18.** Laurent SCHWARTZ stated this for $\mathcal{D}'(\mathbb{R}^d)$ in 1945 [68, p. 72].

Proof. **1.** *Value on all of* Ω. The expression for the regular approximation from Theorem 8.14 (a), the commutation of the derivative with the restriction of a function, and the equality $\partial^\beta(g \diamond \mu) = (-1)^{|\beta|} g \diamond \partial^\beta \mu$ from Theorem 7.17 successively give

$$\partial^\beta R_n f = \partial^\beta((\widetilde{\alpha_n f} \diamond \rho_n)|_\Omega) = (\partial^\beta(\widetilde{\alpha_n f} \diamond \rho_n))|_\Omega = (-1)^{|\beta|}(\widetilde{\alpha_n f} \diamond \partial^\beta \rho_n)|_\Omega.$$

With the expression for the weighting by a regular function from Theorem 7.13 and Definition 6.12 of the localized extension, it comes, for $x \in \Omega$,

$$(\partial^\beta R_n f)(x) = (-1)^{|\beta|}\langle \widetilde{\alpha_n f}, \tau_x(\partial^\beta \rho_n)\rangle_{\mathbb{R}^d} = (-1)^{|\beta|}\langle f, \alpha_n \tau_x(\partial^\beta \rho_n)\rangle_\Omega.$$

2. *Value on* $\Omega^n_{4/n}$. On $\Omega^n_{4/n}$, the regularization coincides with the local regularization according to Theorem 8.14 (b) hence, now with the equality $\partial^\beta(g \diamond \mu) = \partial^\beta g \diamond \mu$ which is also given by Theorem 7.17,

$$\partial^\beta(R_n f) = \partial^\beta(f \diamond \rho_n) = (\partial^\beta f) \diamond \rho_n = R_n(\partial^\beta f). \ \square$$

Let us observe that the global regularization commutes with linear mappings.

Theorem 8.20.– *Let* $f \in \mathcal{D}'(\Omega; E)$, *where* Ω *is an open subset of* $\mathbb{R}^d$ *and* E *is a Neumann space,* $R_n f$ *a regular approximation given by Definition 8.13 and* L *a sequentially continuous linear mapping from* E *into a Neumann space. Then,*

$$L(R_n f) = R_n(Lf). \ ▮$$

Proof. By Definitions 8.13 of the regularization and 5.10 of the image Lf of f, we have, for every $x \in \Omega$,

$$L(R_n f)(x) = L(R_n f(x)) = L(\langle f, \alpha_n \tau_x \rho_n\rangle) = \langle Lf, \alpha_n \tau_x \rho_n\rangle = R_n(Lf)(x). \ \square$$

Let us show that all the regular approximations of a distribution with compact support have their support in the same compact subset of Ω.

Theorem 8.21.– *Let* Ω *be an open subset of* $\mathbb{R}^d$, E *a Neumann space, and*

$$f \in \mathcal{D}'(\Omega; E) \text{ a distribution with compact support.}$$

Then, f *and all of its regular approximations* $R_n f$ *given by Definition 8.13 have their support in the same compact subset* K *of* Ω.

More precisely, there exists such a compact set K *which only depends on* Ω *and on the support of* f. *And hence, which neither depends on* $n \in \mathbb{N}^*$, *nor on the choice of the localizing and regularizing sequences* $(\alpha_n)_{n\in\mathbb{N}^*}$ *and* $(\rho_n)_{n\in\mathbb{N}^*}$, *nor on* E. ▮

Proof. As always, denote $\Omega^n_r \stackrel{\text{def}}{=} \{x \in \Omega : |x| \le n,\ B(x,r) \subset \Omega\}$.

The support of f is included in Ω by Definition 6.18. Since it is compact by hypothesis, it is included in one of the open sets $\Omega^n_{2/n}$, because they cover Ω and grow with n. Say

$$\operatorname{supp} f \subset \Omega^m_{2/m}. \tag{8.4}$$

Let

$$K \stackrel{\text{def}}{=} \overline{\Omega^m_{1/m}}.$$

It is a compact subset of Ω which only depends, via m, on Ω and $\operatorname{supp} f$. It therefore suffices to check that

$$\operatorname{supp} R_n f \subset \overline{\Omega^m_{1/m}}. \tag{8.5}$$

For $n \le m$, this follows from the fact that, due to Theorem 8.15,

$$\operatorname{supp} R_n f \subset \overline{\Omega^n_{1/n}} \subset \overline{\Omega^m_{1/m}}.$$

For $n \ge m$, we use the following expression from Theorem 8.14 (a):

$$R_n f = (\widetilde{\alpha_n f} \diamond \rho_n)|_\Omega.$$

Since $\operatorname{supp}(f \diamond \mu) \subset \operatorname{supp} f - \operatorname{supp} \mu$ (by the inclusion of the support of a weighted distribution from Theorem 7.18) and since $\operatorname{supp} \widetilde{\alpha_n f} = \operatorname{supp} \alpha_n f \subset \operatorname{supp} f$ (by the inclusion of the support of a product from Theorem 6.24),

$$\operatorname{supp} R_n f \subset \operatorname{supp} \widetilde{\alpha_n f} - \operatorname{supp} \rho_n \subset \operatorname{supp} f + B(0, 1/n).$$

With (8.4), we again obtain (8.5) since

$$\operatorname{supp} R_n f \subset \Omega^m_{2/m} + B(0, 1/m) \subset \overline{\Omega^m_{1/m}}. \ \square$$

Let us show that the regular approximations of a positive distribution are positive.

Theorem 8.22.– *Let $f \in \mathcal{D}'(\Omega)$, where Ω is an open subset of $\mathbb{R}^d$, such that*

$$f \ge 0.$$

Then, every regular approximation $R_n f$ given by Definition 8.13 satisfies

$$R_n f \ge 0. \ \blacksquare$$

Proof. Let $x \in \Omega$. Since α_n and ρ_n are positive, the function $\alpha_n \tau_x \rho_n$ is too. By Definition 5.31 of a positive distribution, $f \geq 0$ therefore implies

$$\langle f, \alpha_n \tau_x \rho_n \rangle \geq 0,$$

that is $R_n f(x) \geq 0$ by Definition 8.13 of a regular approximation. □

Let us show that the global regularizations are sequentially continuous operations.

Theorem 8.23.– *Let Ω be an open subset of $\mathbb{R}^d$ and E a Neumann space.*

Then, every regularization R_n given by Definition 8.13 is a sequentially continuous linear mapping from $\mathcal{D}'(\Omega; E)$ into $\mathbf{C}^\infty_\mathrm{b}(\Omega; E)$. ▮

Proof. **1.** *Properties for values in $\mathcal{C}^\infty(\Omega; E)$.* The mapping R_n coincides, according to Theorem 8.14 (a), with the mapping $f \mapsto (\widetilde{\alpha_n f} \diamond \rho_n)|_\Omega$ which is the composition of the following three sequentially continuous linear mappings:

— $f \mapsto \widetilde{\alpha_n f}$ from $\mathcal{D}'(\Omega; E)$ into $\mathcal{D}'(\mathbb{R}^d; E)$ (Theorem 6.14);

— $g \mapsto g \diamond \rho_n$ from $\mathcal{D}'(\mathbb{R}^d; E)$ into $\mathcal{C}^\infty(\mathbb{R}^d; E)$ (Theorem 7.11 (a));

— $h \mapsto h|_\Omega$ from $\mathcal{C}^\infty(\mathbb{R}^d; E)$ into $\mathcal{C}^\infty(\Omega; E)$ (by Definition 1.19 (a) of $\mathcal{C}^\infty(\Omega; E)$).

Hence, as for every composition of such mappings (Theorem A.30),

$$R_n \text{ is sequentially continuous linear from } \mathcal{D}'(\Omega; E) \text{ into } \mathcal{C}^\infty(\Omega; E).$$

2. *Properties for values in $\mathbf{C}^\infty_\mathrm{b}(\Omega; E)$.* The support of $R_n f$ is, due to Theorem 8.15, included in the compact set $K = \overline{\Omega^n_{1/n}}$, which is included in Ω. Then,

$$R_n f \in \mathcal{C}^\infty_K(\Omega; E).$$

Since the space $\mathcal{C}^\infty_K(\Omega; E)$ is included in $\mathbf{C}^\infty_\mathrm{b}(\Omega; E)$ and the topologies of $\mathcal{C}^\infty(\Omega; E)$ and $\mathbf{C}^\infty_\mathrm{b}(\Omega; E)$ coincide there (Theorem A.51), it follows then that R_n is sequentially continuous from $\mathcal{D}'(\Omega; E)$ into $\mathbf{C}^\infty_\mathrm{b}(\Omega; E)$. □

8.6. Commutativity and associativity of weighting

Since weighting is commutative and associative only up to a symmetry on the weight, we begin by defining a **symmetry** of it.

Theorem 8.24.– *Given $\mu \in \mathcal{D}'(\mathbb{R}^d)$, we define $\check{\mu} \in \mathcal{D}'(\mathbb{R}^d)$ by: for all $\varphi \in \mathcal{D}(\mathbb{R}^d)$,*

$$\langle \check{\mu}, \varphi \rangle_{\mathbb{R}^d} \stackrel{\text{def}}{=} \langle \mu, \check{\varphi} \rangle_{\mathbb{R}^d},$$

where $\check{\varphi}(x) \stackrel{\text{def}}{=} \varphi(-x)$.

The mapping $\mu \mapsto \check{\mu}$ is an isomorphism from $\mathcal{D}'(\mathbb{R}^d)$ onto itself, and from $\mathcal{D}'_D(\mathbb{R}^d)$ onto $\mathcal{D}'_{-D}(\mathbb{R}^d)$ for every given compact subset D of $\mathbb{R}^d$. ▮

Proof. **1.** *Obtaining a distribution.* Theorem 5.29 on linear change of variables in a distribution, for $T(x) = -x$, gives

$$\check{\mu} \in \mathcal{D}'(\mathbb{R}^d),$$

since here $\kappa = |\det[-\mathbf{e}_1, \ldots, -\mathbf{e}_d]| = |\det[\mathbf{e}_1, \ldots, \mathbf{e}_d]| = 1$.

2. *Isomorphism on $\mathcal{D}'(\mathbb{R}^d)$.* Theorem 5.29 also gives the isomorphism from $\mathcal{D}'(\mathbb{R}^d)$ onto itself.

3. *Isomorphism on $\mathcal{D}'_D(\mathbb{R}^d)$.* By the effect of a change of variables on the support (Theorem 6.23),

$$\operatorname{supp} \check{\mu} = T^{-1}(\operatorname{supp} \mu) = -\operatorname{supp} \mu.$$

Therefore, $\mu \mapsto \check{\mu}$ is a bijection from $\mathcal{D}'_D(\mathbb{R}^d)$ onto $\mathcal{D}'_{-D}(\mathbb{R}^d)$. It is an isomorphism since $\mathcal{D}'_D(\mathbb{R}^d)$ is, by Definition 6.26, endowed with the semi-norms of $\mathcal{D}'(\mathbb{R}^d)$. □

Let us show that weighting is commutative, up to a symmetry.

Theorem 8.25.– *Let $\mu \in \mathcal{D}'_D(\mathbb{R}^d)$ and $\lambda \in \mathcal{D}'_K(\mathbb{R}^d)$, where D and K are compact subsets of $\mathbb{R}^d$. Then,*

$$\mu \diamond \lambda = (\lambda \diamond \mu)\check{} = \check{\lambda} \diamond \check{\mu},$$

where the symmetrized distribution $\check{\lambda} \in \mathcal{D}'_{-K}(\mathbb{R}^d)$ is defined in Theorem 8.24. These equalities take place in $\mathcal{D}'_{D-K}(\mathbb{R}^d)$. ▮

Dissymettry. Here, the roles of λ and μ are not symmetric like they are for the convolution, see (7.16), p. 155. This is the price to to pay to have a "weighted" integral over the support D of the weight μ (and not on $-D$, like with the convolution). This is explained in the comment *Weighting versus convolution*, p. 155. A small price however, since f and μ have quite distinct roles to play in the use of $f \diamond \mu$. □

Proof. We return by regularization to the case of continuous functions, for which it is a simple calculation of integrals.

1. Case of continuous functions: first equality. Assume for the moment that

$$\mu \in \mathcal{C}_D(\mathbb{R}^d) \text{ and } \lambda \in \mathcal{C}_K(\mathbb{R}^d).$$

Let $x \in \mathbb{R}^d$. Here the expression for the weighting of continuous functions from Theorem 7.14 gives, since the domain where the function is zero does not contribute to the integral (Theorem A.77), and since the integral on $\mathbb{R}^d$ is invariant under a translation (Theorem A.86), here the change of variable $y = z - x$,

$$(\mu \diamond \lambda)(x) = \int_{\mathring{K}} \mu(x+y)\,\lambda(y)\,\mathrm{d}y = \int_{\mathbb{R}^d} \mu(x+y)\,\lambda(y)\,\mathrm{d}y =$$
$$= \int_{\mathbb{R}^d} \lambda(z-x)\,\mu(z)\,\mathrm{d}z = \int_{\mathring{D}} \lambda(z-x)\,\mu(z)\,\mathrm{d}z = (\lambda \diamond \mu)(-x) = (\lambda \diamond \mu)\check{}\,(x).$$

2. Case of continuous functions: second equality. With the invariance of the integral under the change of variable $y = -x - z$ (for x fixed), i.e. by a translation and then a symmetry (Theorems A.86 and A.87), we obtain similarly

$$(\mu \diamond \lambda)(x) = \int_{\mathbb{R}^d} \mu(x+y)\,\lambda(y)\,\mathrm{d}y = \int_{\mathbb{R}^d} \mu(-z)\,\lambda(-x-z)\,\mathrm{d}z =$$
$$= \int_{\mathbb{R}^d} \check{\lambda}(x+z)\,\check{\mu}(z)\,\mathrm{d}z = (\check{\lambda} \diamond \check{\mu})(x).$$

3. General case. We arrive at the case where

$$\mu \in \mathcal{D}'_D(\mathbb{R}^d) \text{ and } \lambda \in \mathcal{D}'_K(\mathbb{R}^d).$$

Their local approximations $\mu_n = \mu \diamond \rho_n$ and $\lambda_n = \lambda \diamond \rho_n$ are, by Theorem 8.6 (b), continuous functions on $\mathbb{R}^d$ with support respectively in $D + B$ and $K + B$, where B is the unit ball in $\mathbb{R}^d$. From steps 1 and 2, we therefore have, in $\mathcal{C}(\mathbb{R}^d)$,

$$\mu_n \diamond \lambda_n = (\lambda_n \diamond \mu_n)\check{} = \check{\lambda}_n \diamond \check{\mu}_n.$$

Since $\mu_n \to \mu$ and $\lambda_n \to \lambda$ in $\mathcal{D}'(\mathbb{R}^d)$ due to Theorem 8.6 (a), and since weighting is sequentially continuous (Theorem 7.15 (a)) and the symmetrization is too (due to Theorem 8.24), this results in the limit, in $\mathcal{D}'(\mathbb{R}^d)$,

$$\mu \diamond \lambda = (\lambda \diamond \mu)\check{} = \check{\lambda} \diamond \check{\mu}.$$

4. Equalities in $\mathcal{D}'_{D-K}(\mathbb{R}^d)$. Since the support of $\mu \diamond \lambda$ is (Theorem 7.18) included in $\operatorname{supp}\mu - \operatorname{supp}\lambda$, and therefore in $D - K$, the distribution $\mu \diamond \lambda$ belongs to $\mathcal{D}'_{D-K}(\mathbb{R}^d)$ and these equalities indeed take place in this space. □

Now we show that the weighting is associative, up to a symmetry.

Theorem 8.26.– *Let $f \in \mathcal{D}'(\Omega; E)$, $\mu \in \mathcal{D}'_D(\mathbb{R}^d)$ and $\lambda \in \mathcal{D}'_K(\mathbb{R}^d)$, where Ω is an open subset of $\mathbb{R}^d$, E is a Neumann space, and D and K are compact subsets of $\mathbb{R}^d$. Then,*

$$(f \diamond \mu) \diamond \lambda = f \diamond (\mu \diamond \check{\lambda}) = f \diamond (\lambda \diamond \check{\mu}) = (f \diamond \lambda) \diamond \mu,$$

where the symmetrized distribution $\check{\lambda} \in \mathcal{D}'_{-K}(\mathbb{R}^d)$ is defined in Theorem 8.24. These equalities take place in $\mathcal{D}'(\Omega_{D+K}; E)$. ▮

Proof. We return by regularization to the case of continuous functions, for which it is a simple calculation of integrals. We denote

$$\mathcal{K}(\Omega; E) \stackrel{\text{def}}{=} \{\text{continuous functions with compact support from } \Omega \text{ into } E\}.$$

1. *Case of continuous functions: first equality.* Assume for the moment that

$$f \in \mathcal{K}(\Omega; E), \ \ \mu \in \mathcal{C}_D(\mathbb{R}^d) \ \text{ and } \ \lambda \in \mathcal{C}_K(\mathbb{R}^d),$$

The weighting of continuous functions (Theorem 7.14 or 7.23) successively gives $f \diamond \mu \in \mathcal{C}(\Omega_D; E)$ then (since $(\Omega_D)_K = \Omega_{D+K}$ from Theorem 7.2)

$$(f \diamond \mu) \diamond \lambda \in \mathcal{C}(\Omega_{D+K}; E).$$

Let $x \in \Omega_{D+K}$. Twice using the expression for weighting from Theorem 7.23, we successively obtain

$$((f \diamond \mu) \diamond \lambda)(x) = \int_{\Omega_D} (f \diamond \mu)(z) \lambda(z - x) \, \mathrm{d}z = \\ = \int_{\Omega_D} \Big(\int_\Omega f(y) \mu(y - z) \, \mathrm{d}y \Big) \lambda(z - x) \, \mathrm{d}z.$$

By permuting the order of integration in the last term via Theorem A.89 (that is permissible since the function $(y, z) \mapsto f(y)\mu(y - z)\lambda(z - x)$ is uniformly continuous with compact support), it follows

$$((f \diamond \mu) \diamond \lambda)(x) = \int_\Omega f(y) \Big(\int_{\Omega_D} \mu(y - z) \lambda(z - x) \, \mathrm{d}z \Big) \, \mathrm{d}y. \qquad (8.6)$$

In the right-hand side, we can replace Ω_D by $\mathbb{R}^d$, since the domain where the function is zero does not contribute to the integral (Theorem A.77) and since $\lambda(z - x) = 0$ if $z \notin \Omega_D$ (since then $z - x \notin K$, otherwise we would have $z = x + (z - x) \in \Omega_{D+K} + K$ which, from Theorem 7.2, is equal to $(\Omega_D)_K + K$ and thus is included in Ω_D).

With in addition the change of variable $z = x - u$, for x fixed (since the integral is invariant under translation and symmetry by Theorems A.86 and A.87), then the expression for the weighting of continuous functions from Theorem 7.4, we obtain

$$\int_{\Omega_D} \mu(y-z)\,\lambda(z-x)\,\mathrm{d}z = \int_{\mathbb{R}^d} \mu(y-x+u)\,\check{\lambda}(u)\,\mathrm{d}u = (\mu \diamond \check{\lambda})(y-x).$$

Thus, (8.6) gives, for every $x \in \Omega_{D+K}$,

$$((f \diamond \mu) \diamond \lambda)(x) = \int_\Omega f(y)\,(\mu \diamond \check{\lambda})(y-x)\,\mathrm{d}y = (f \diamond (\mu \diamond \check{\lambda}))(x). \tag{8.7}$$

2. *Case of continuous functions: all the equalities.* From Theorem 8.25,

$$\mu \diamond \check{\lambda} = \check{\check{\lambda}} \diamond \check{\mu} = \lambda \diamond \check{\mu}.$$

Using (8.7), (8.8) and again (8.7), we obtain successively, in Ω_{D+K},

$$(f \diamond \mu) \diamond \lambda = f \diamond (\mu \diamond \check{\lambda}) = f \diamond (\lambda \diamond \check{\mu}) = (f \diamond \lambda) \diamond \mu. \tag{8.8}$$

3. *General case: regularization.* We come to the case where

$$f \in \mathcal{D}'(\Omega; E), \;\; \mu \in \mathcal{D}'_D(\mathbb{R}^d) \;\text{ and }\; \lambda \in \mathcal{D}'_K(\mathbb{R}^d).$$

The global approximation $f_n \overset{\text{def}}{=} R_n f$ (Definition 8.13) satisfies (Theorem 8.15)

$$f_n \in \mathcal{K}(\Omega; E).$$

The local approximations $\mu_n \overset{\text{def}}{=} \mu \diamond \rho_n$ and $\lambda_n \overset{\text{def}}{=} \lambda \diamond \rho_n$ satisfy (Theorem 8.6 (b))

$$\mu_n \in \mathcal{C}_{D+B(0,1/n)}(\mathbb{R}^d), \quad \lambda_n \in \mathcal{C}_{K+B(0,1/n)}(\mathbb{R}^d).$$

Equalities (8.8) for continuous functions then gives

$$(f_n \diamond \mu_n) \diamond \lambda_n = f_n \diamond (\mu_n \diamond \check{\lambda}_n) = f_n \diamond (\lambda_n \diamond \check{\mu}_n) = (f_n \diamond \lambda_n) \diamond \mu_n,$$

in $\mathcal{C}(\omega_n; E)$, where, according to Theorem 7.2,

$$\omega_n \overset{\text{def}}{=} \Omega_{D+B(0,1/n)+K+B(0,1/n)} = (\Omega_{D+K})_{B(0,2/n)}.$$

4. *Convergence.* Let $m \in \mathbb{N}^*$. When $n \to \infty$ and $n \geq m$, the convergence of global approximations (Theorem 8.16) gives

$$f_n \to f \;\text{ in } \mathcal{D}'(\Omega; E)$$

and the convergence of local approximations (Theorem 8.6 (b)) gives

$$\mu_n \to \mu \;\text{ in } \mathcal{D}'_{D+B(0,1/m)}(\mathbb{R}^d), \quad \lambda_n \to \lambda \;\text{ in } \mathcal{D}'_{K+B(0,1/m)}(\mathbb{R}^d).$$

Since the weighting is sequentially continuous (Theorem 7.15 (a)) as well as the symmetrization (Theorem 8.24), in the limit we obtain

$$(f \diamond \mu) \diamond \lambda = f \diamond (\mu \diamond \check{\lambda}) = (f \diamond \lambda) \diamond \mu \quad \text{in } \mathcal{D}'(\omega_m; E).$$

The open sets $(\omega_m)_{m \in \mathbb{N}^*}$ covering Ω_{D+K}, the gluing theorem for equalities (Theorem 6.10) gives these equalities on all of Ω_{D+K}, i.e. in $\mathcal{D}'(\Omega_{D+K}; E)$. □

8.7. Uniform convergence of sequences of distributions

Let us give a property of uniform convergence[4] with respect to φ of convergent sequences of distributions whose proof uses the regularization of functions.

Theorem 8.27.– *Let $(f_n)_{n\in\mathbb{N}}$ be a sequence in $\mathcal{D}'(\Omega; E)$ and $f \in \mathcal{D}'(\Omega; E)$, where Ω is an open subset of $\mathbb{R}^d$ and E is a Neumann space. Then,*

$$f_n \to f \ \textit{in } \mathcal{D}'(\Omega; E)$$

if and only if, for every semi-norm $\| \ \|_{E;\nu}$ of E and every compact subset K of Ω, there exist $m \in \mathbb{N}$ and a decreasing real sequence $(c_n)_{n\in\mathbb{N}}$ converging to 0 such that: for every $\varphi \in C_K^\infty(\Omega)$ and $n \in \mathbb{N}$,

$$\|\langle f_n - f, \varphi\rangle\|_{E;\nu} \leq c_n \|\varphi\|_{C_b^m(\Omega)}. \tag{8.9}$$

∎

Proof. Since inequality (8.9) implies the convergence of f_n to f according to the characterization of convergent sequences of distributions from Theorem 4.3 (b), it remains to prove the converse.

Suppose then that

$$f_n \to f \ \text{ in } \mathcal{D}'(\Omega; E) \tag{8.10}$$

and let $\| \ \|_{E;\nu}$ be a semi-norm of E and K a compact subset of Ω. We will establish (8.9) by contradiction, in five steps.

1. Choice of m. Due to the strong inclusion theorem (Theorem A.22), there exists $r > 0$ such that $K + B(0, 2r) \subset \Omega$, where B designates the closed ball. Let

$$\omega \stackrel{\text{def}}{=} K + \mathring{B}(0, r) \ \text{ and } \ Q \stackrel{\text{def}}{=} \overline{\omega}.$$

These are respectively an open and a compact subset of $\mathbb{R}^d$.

The sequence $(f_n - f)_{n\in\mathbb{N}}$ is bounded in $\mathcal{D}'(\Omega; E)$, as is every convergent sequence (Theorem A.5). It is thus equicontinuous on $C_Q^\infty(\Omega)$ due to Theorem 4.2 (a).

4. **History of the uniform convergence with respect to test functions.** Theorem 8.27 is new, including for real-valued distributions: it is stronger than uniform convergence on the bounded subsets of $\mathcal{D}(\Omega)$, see the comment *Comparison of Theorems 8.27 and 8.28*, p. 190.
In the particular case of $\mathcal{D}'(\mathbb{R}^d)$, we could deduce Theorem 8.27 from Theorems XIII and XXIII of Laurent SCHWARTZ [69, Chap. III, § 3, p. 74, and § 6, p. 87] which state respectively that the convergence $f_n \to f$ in $\mathcal{D}'(\mathbb{R}^d)$ implies that in $\mathcal{D}'(\mathbb{R}^d)$-unif, and that this convergence implies, for every compact subset K of $\mathbb{R}^d$, the existence of $\beta \in \mathbb{N}^d$ and of a sequence $(g_n)_{n\in\mathbb{N}}$ such that $g_n \to g$ in $C_b(\mathbb{R}^d)$ and $f_n = \partial^\beta g_n$.

More precisely, this theorem gives the existence of $m \geq 1$ and $b \in \mathbb{R}$ such that, for every $\varphi \in \mathcal{C}_Q^\infty(\Omega)$,

$$\|\langle f_n - f, \varphi\rangle\|_{E;\nu} \leq b\|\varphi\|_{\mathcal{C}_b^{m-1}(\Omega)}. \tag{8.11}$$

2. *Extraction of a subsequence.* Should Property (8.9) not be satisfied, there would exist a subsequence, that we again denote by $(f_n)_{n\in\mathbb{N}}$, for which there would exist $a > 0$ and a sequence $(\phi_n)_{n\in\mathbb{N}}$ in $\mathcal{C}_K^\infty(\Omega)$ such that, for every n,

$$\|\phi_n\|_{\mathcal{C}_b^m(\Omega)} \leq 1, \quad \|\langle f_n - f, \phi_n\rangle\|_{E;\nu} \geq a. \tag{8.12}$$

Since every bounded sequence in $\mathcal{C}_K^m(\Omega)$ has a subsequence which converges in $\mathcal{C}_K^{m-1}(\Omega)$ (Theorem 2.15), we could extract a subsequence (of which we still denote by n the current index) such that

$$\phi_n \to \phi \ \text{ in } \mathcal{C}_K^{m-1}(\Omega). \tag{8.13}$$

3. *Contradiction of (8.12).* Decompose, for $\varphi \in \mathcal{D}(\Omega)$,

$$\langle f_n - f, \phi_n\rangle = \langle f_n - f, \phi_n - \varphi\rangle + \langle f_n - f, \varphi\rangle.$$

Assume for a moment that there exists $\varphi \in \mathcal{C}_Q^\infty(\Omega)$ (we will obtain it in step 4 by regularizing ϕ) such that

$$\|\varphi - \phi\|_{\mathcal{C}_b^{m-1}(\Omega)} \leq \frac{a}{2b}. \tag{8.14}$$

With inequality (8.11), we would then obtain

$$\|\langle f_n - f, \phi_n\rangle\|_{E;\nu} \leq b\|\phi_n - \varphi\|_{\mathcal{C}_b^{m-1}(\Omega)} + \|\langle f_n - f, \varphi\rangle\|_{E;\nu}.$$

and therefore, by decomposing $\phi_n - \varphi = (\phi_n - \phi) + (\phi - \varphi)$,

$$\|\langle f_n - f, \phi_n\rangle\|_{E;\nu} \leq \frac{a}{2} + b\|\phi_n - \phi\|_{\mathcal{C}_b^{m-1}(\Omega)} + \|\langle f_n - f, \varphi\rangle\|_{E;\nu}.$$

The last two terms on the right-hand side would tend to 0 by the convergences in equations (8.13) and (8.10) hence, for n large enough, inequality (8.12) would not be satisfied. This is the desired contradiction.

This proves Property (8.9), subject to check (8.14).

4. *Verification of (8.14).* The extension $\widetilde{\phi}$ of ϕ by 0 belongs to $\mathbf{C}_b^{m-1}(\mathbb{R}^d)$ thus its regular approximation $\widetilde{\phi} \diamond \rho_k$ belongs to $\mathbf{C}_b^\infty(\mathbb{R}^d)$ and satisfies (Theorem 8.7 (b))

$$\widetilde{\phi} \diamond \rho_k \to \widetilde{\phi} \ \text{ in } \mathbf{C}_b^{m-1}(\mathbb{R}^d).$$

Since this space is, by Definition 1.20, endowed with the semi-norms of $\mathcal{C}_b^{m-1}(\mathbb{R}^d)$, the restriction $\varphi \stackrel{\text{def}}{=} (\widetilde{\phi} \diamond \rho_k)|_\Omega$ satisfies (8.14) for k large enough.

Moreover, φ belongs to $\mathcal{C}_Q^\infty(\Omega)$ as soon as $k \geq 1/r$, since then the inclusion of the support of a weighted distribution from Theorem 7.10 gives

$$\operatorname{supp}(\widetilde{\phi} \diamond \rho_k) \subset \operatorname{supp} \widetilde{\phi} - \operatorname{supp} \rho_k \subset K + B(0,r) = Q.$$

5. Conclusion. This proves Property (8.9), except for the decrease of the sequence $(c_n)_{n\in\mathbb{N}}$. This is obtained by replacing c_n by $c'_n = \sup_{m\geq n} c_m$. □

Uniform convergence on the bounded subsets of $\mathcal{D}(\Omega)$. Let us show that the convergence in $\mathcal{D}'(\Omega; E)$ is equivalent to the uniform convergence on the bounded subsets of $\mathcal{D}(\Omega)$, namely to the convergence for the topology on $\mathcal{D}'(\Omega; E)$-unif used by Laurent SCHWARTZ[5].

Theorem 8.28.– *Let $(f_n)_{n\in\mathbb{N}}$ be a sequence in $\mathcal{D}'(\Omega; E)$ and $f \in \mathcal{D}'(\Omega; E)$, where Ω is an open subset of $\mathbb{R}^d$ and E is a Neumann space. Then,*

$$f_n \to f \text{ in } \mathcal{D}'(\Omega; E)$$

if and only if, for every bounded subset $\mathcal{B}$ of $\mathcal{D}(\Omega)$ and every semi-norm $\| \ \|_{E;\nu}$ of E,

$$\sup_{\varphi\in\mathcal{B}} \|\langle f_n - f, \varphi\rangle\|_{E;\nu} \to 0. \ \square$$

Proof. *1. Direct part.* Let $\mathcal{B}$ be a bounded subset of $\mathcal{D}(\Omega)$. It is bounded in $\mathcal{C}_b^\infty(\Omega)$ (since $\mathcal{D}(\Omega \subsetneq \mathcal{C}_b^\infty(\Omega)$ from Theorem 2.10) and hence in $\mathcal{C}_b^m(\Omega)$, for every $m \in \mathbb{N}$. So,

$$\sup_{\varphi\in\mathcal{B}} \|\varphi\|_{\mathcal{C}_b^m(\Omega)} = b_m < \infty.$$

All the φ of $\mathcal{B}$ have their support in the same compact subset K of Ω, due to Theorem 2.9. If $f_n \to f$ in $\mathcal{D}'(\Omega; E)$, Theorem 8.27 then provides $m \in \mathbb{N}$ and a sequence $(c_n)_{n\in\mathbb{N}}$ converging to 0 such that

$$\|\langle f_n - f, \varphi\rangle\|_{E;\nu} \leq c_n \|\varphi\|_{\mathcal{C}_b^m(\Omega)}.$$

Hence,

$$\sup_{\varphi\in\mathcal{B}} \|\langle f_n - f, \varphi\rangle\|_{E;\nu} \leq c_n b_m \to 0.$$

2. Converse. For every $\varphi \in \mathcal{D}(\Omega)$, we get $\langle f_n - f, \varphi\rangle \to 0$ by considering $\mathcal{B} = \{\varphi\}$. □

Comparison of Theorems 8.27 and 8.28. Given $\nu \in \mathcal{N}_E$, K a compact subset of Ω and $\mathcal{B} \subset \mathcal{C}_K^\infty(\Omega)$, Theorem 8.28 states (and states nothing more) that

$$\sup_{\varphi\in\mathcal{B}} \|\langle f_n - f, \varphi\rangle\|_{E;\nu} \to 0$$

if, **for every** $m \in \mathbb{N}$, we have $\sup_{\varphi\in\mathcal{B}} \|\varphi\|_{\mathcal{C}_b^m(\Omega)} < \infty$.

Theorem 8.27 states that it suffices for this to be realized **for one value of** m (unknown, depending on ν and K). It then provides in addition the uniform convergence on some unbounded subsets of $\mathcal{D}(\Omega)$. □

Utility of Theorem 8.27. We will not use this result, but its proof uses in a simple context arguments that we will use to prove an analogous characterization for distributions of distributions (Theorem 14.6), which will be used to prove the fundamental properties of the separation of variables of distributions given in the *kernel theorem* (Theorem 15.10). □

5. **History of Theorem 8.28.** Laurent SCHWARTZ showed in 1950 that, for distributions in $\mathcal{D}'(\mathbb{R}^d)$, convergence in $\mathcal{D}'(\mathbb{R}^d)$ in equivalent to the uniform convergence on the bounded subsets of $\mathcal{D}(\mathbb{R}^d)$ [69, Chap. III, § 3, Theorem XIII, p. 74].

Chapter 9

Potentials and Singular Functions

This chapter is dedicated to *potentials*, the most useful of which for the following is the *localized potential* γ, a solution of $\Delta\gamma + \eta = \delta_0$ in $\mathbb{R}^d$ with support in a ball $B = B(0, r)$ as small as we like it, where the *correction term* η is a regular function also with support in B (Theorem 9.17). This will be used to decompose (in Theorem 12.1) any distribution as $f = \sum_i \partial_i f \diamond \partial_i \gamma + f \diamond \eta$, on the domain of definition of f with an arbitrarily small neighborhood of its boundary removed.

We construct the *localized potential* γ in steps, as follows.

— We associate a distribution with each continuous function which may diverge to infinity at the origin while growing less quickly than $|x|^{-\lambda}$, where $\lambda < d$ (Theorem 9.5).

— We construct the *Newtonian potential* ξ, a solution of $-\Delta\xi = 0$ on all of $\mathbb{R}^d$ (Theorem 9.10).

— We *localize* ξ as $\gamma = \theta\xi$, where θ has its support in B and is equal to 1 on a small ball (Theorem 9.17).

On the other hand, we decompose the Dirac mass into a sum of derivatives of $\mathcal{C}^m$ functions with support in B (Theorem 9.18). And we show that the weighting by a weight μ which grows less rapidly than $|x|^{-\lambda}$ with $\lambda < d$ preserves the continuity of functions (Theorem 9.22).

We begin by recalling some facts about the integral over a sphere of a continuous function with values in a Neumann space.

9.1. Surface integral over a sphere

The construction of distributions associated with singular functions is facilitated by using surface integrals of continuous functions. Since this has not been defined for values in a Neumann space, we did it in the most elementary way possible in Volume 2 in the case of a sphere, which suffices for our needs. Let us recall this construction.

Definition 9.1.– *Let $f \in \mathcal{C}(S_r; E)$, where $S_r \stackrel{\text{def}}{=} \{x \in \mathbb{R}^d : |x| = r\}$, $d \geq 2$, $r > 0$ and E is a Neumann space.*

*The **surface integral** of f over S_r is the element of E defined by*

$$\int_{S_r} f \,\mathrm{d}s \stackrel{\text{def}}{=} \lim_{\epsilon \to 0} \frac{1}{\epsilon} \int_{C_{r,r+\epsilon}} f\Big(\frac{rx}{|x|}\Big)\,\mathrm{d}x,$$

where $C_{r,b} \stackrel{\text{def}}{=} \{x \in \mathbb{R}^d : r < |x| < b\}$. ▮

This definition is justified because, as we established in Volume 2 [82, Justification of Definition 10.1, p. 214–215]:
— the right-hand side Cauchy integral makes sense, since the integrand is uniformly continuous on the open crown $C_{r,r+\epsilon}$;
— the right-hand side has a limit when $\epsilon \to 0$.

The properties of the surface integral that are classical when E is a Banach space extend to the case of a Neumann space. In particular, we have the following **Stokes' formula** [1] established in Volume 2 [82, Theorem 10.8].

Theorem 9.2.– *Let* $f \in \mathcal{C}^1(\overline{C_{a,b}}; E)$, *where* $C_{a,b} \stackrel{\text{def}}{=} \{x \in \mathbb{R}^d : a < |x| < b\}$, $d \geq 2$, $0 < a < b < \infty$ *and* E *is a Neumann space. Then, for every* $i \in [\![1, d]\!]$,

$$\int_{C_{a,b}} \partial_i f = \int_{S_b} f n_i \,\mathrm{d}s + \int_{S_a} f n_i \,\mathrm{d}s,$$

where n *is the unit normal vector field on* S_a *and* S_b *directed outwards from* $C_{a,b}$, *i.e.*

$$n_i(x) = \sigma \frac{x_i}{|x|}, \quad \text{where } \sigma \text{ equals } 1 \text{ on } S_b \text{ and } -1 \text{ on } S_a. \;▮$$

The semi-norms of E take on the role played by the norm in the case of a Banach space. In particular, we have the following inequality, also established in volume 2 [82, Theorem 10.3].

Theorem 9.3.– *Let* $f \in \mathcal{C}(S_r; E)$, *where* $S_r \stackrel{\text{def}}{=} \{x \in \mathbb{R}^d : |x| = r\}$, $d \geq 2$, $r > 0$ *and* E *is a Neumann space, and let* υ_d *be the measure of the open unit ball in* $\mathbb{R}^d$. *Then, for every semi-norm* $\| \; \|_{E;\nu}$ *of* E,

$$\Big\| \int_{S_r} f \,\mathrm{d}s \Big\|_{E;\nu} \leq d\upsilon_d r^{d-1} \sup_{x \in S_r} \|f(x)\|_{E;\nu}. \;▮$$

We will also use the following classical calculation [82, Theorem 10.3 (b)].

Theorem 9.4.– *If* $S_r \stackrel{\text{def}}{=} \{x \in \mathbb{R}^d : |x| = r\}$, *where* $d \geq 2$ *and* $r > 0$, *and* υ_d *is the measure of the open unit ball of* $\mathbb{R}^d$,

$$\int_{S_r} 1 \,\mathrm{d}s = d\upsilon_d r^{d-1}. \;▮$$

1. **History of Stokes' formula.** This is mentioned in Note 3, p. 234.

9.2. Distribution associated with a singular function

Let us associate a distribution on all of $\mathbb{R}^d$ with a real function defined everywhere except at 0 that can become infinite in a neighborhood of it, yet growing strictly less rapidly than $|x|^{-d}$.

Theorem 9.5.– *Let $f \in \mathcal{C}(\mathbb{R}^d \setminus \{0\})$ be such that there exist real numbers $r_0 > 0$, $b \geq 0$ and*

$$\lambda < d$$

such that: for every $x \in \mathbb{R}^d$ such that $0 < |x| \leq r_0$,

$$|f(x)| \leq \frac{b}{|x|^\lambda}. \tag{9.1}$$

Then, we define $\overline{f} \in \mathcal{D}'(\mathbb{R}^d)$ by: for every $\varphi \in \mathcal{D}(\mathbb{R}^d)$,

$$\langle \overline{f}, \varphi \rangle \stackrel{\text{def}}{=} \lim_{\substack{\epsilon \to 0 \\ \epsilon > 0}} \int_{\{x \in \mathbb{R}^d : |x| > \epsilon\}} f\varphi. \quad \blacksquare$$

Proof. **1.** *Well-definedness of the integral.* Let $\varphi \in \mathcal{D}(\mathbb{R}^d)$. For $\epsilon > 0$, we denote

$$I_\epsilon \stackrel{\text{def}}{=} \int_{\{x \in \mathbb{R}^d : |x| > \epsilon\}} f\varphi.$$

This integral makes sense, by Definition 1.22 of the integral, because the restriction of $f\varphi$ to $\{x \in \mathbb{R}^d : |x| > \epsilon\}$ is:

— uniformly continuous, since $f\varphi$ is continuous on $\mathbb{R}^d \setminus \{0\}$ and thus is uniformly continuous on the compact set $\operatorname{supp} \varphi \cap \{x \in \mathbb{R}^d : |x| \geq \epsilon\}$ due to Heine's theorem (Theorem A.32) and since it is zero for $|x| \geq R$;

— equal to zero outside of a bounded set.

2. *Existence of the limit.* Let R be such that

$$\varphi(x) = 0 \ \text{ if } |x| \geq R,$$

and let $C_{\epsilon,R} \stackrel{\text{def}}{=} \{x \in \mathbb{R}^d : \epsilon < |x| < R\}$. Since the domain where the function is zero does not contribute to the integral (Theorem A.77),

$$I_\epsilon = \int_{C_{\epsilon,R}} f\varphi. \tag{9.2}$$

Given ϵ and η such that $0 < \eta < \epsilon \leq r_0$, the additivity of the real integral with respect to crowns, its upper bound and its growth (Theorems A.80 (a), 1.23 (a) and A.76 (a)), hypothesis (9.1) and the calculation for the integral of powers of $|x|$ on a crown (Theorem A.81) successively give

$$|I_\epsilon - I_\eta| = \left| \int_{C_{\eta,\epsilon}} f\varphi \right| \leq b \sup_{x\in\mathbb{R}^d} |\varphi(x)| \int_{C_{\eta,\epsilon}} |x|^{-\lambda}\, dx \leq \epsilon^{d-\lambda} \frac{bdv_d}{d-\lambda} \sup_{x\in\mathbb{R}^d} |\varphi(x)|. \tag{9.3}$$

This tends to 0 when $\epsilon \to 0$, since $\lambda < d$.

So, if $(\epsilon_n)_{n\in\mathbb{N}}$ is a sequence converging to 0, then $(I_{\epsilon_n})_{n\in\mathbb{N}}$ is a real Cauchy sequence, thus (Theorem A.26 (a)) it has a limit, that we denote by $\langle \bar{f}, \varphi \rangle$.

3. Obtaining a distribution. Since the mapping $\bar{f}$ just defined is linear from $\mathcal{D}(\mathbb{R}^d)$ into $\mathbb{R}$, it remains to check that it is continuous. For this, using (9.2), we bound

$$|I_{r_0}| = \left| \int_{C_{r_0,R}} f\varphi \right| \leq \sup_{x\in\mathbb{R}^d} |\varphi(x)| \int_{C_{r_0,R}} |f|$$

and therefore, with inequality (9.3) (for the pair (r_0, ϵ) in the place of (ϵ, η)),

$$|I_\epsilon| = |I_{r_0} - (I_{r_0} - I_\epsilon)| \leq a \sup_{x\in\mathbb{R}^d} |\varphi(x)|, \tag{9.4}$$

where $a = r_0^{d-\lambda} bdv_d/(d-\lambda) + \int_{C_{r_0,R}} |f|$.

Inequality (9.4) can be written, with Definition 2.5 of the semi-norms of $\mathcal{D}(\mathbb{R}^d)$ and denoting by $p_{1/2}$ the constant function equal to $1/2$, as

$$|I_\epsilon| \leq 2a \|\varphi\|_{\mathcal{D}(\mathbb{R}^d);p_{1/2}}.$$

In passing to the limit when $\epsilon \to 0$, it becomes

$$|\langle \bar{f}, \varphi \rangle| \leq 2a \|\varphi\|_{\mathcal{D}(\mathbb{R}^d);p_{1/2}}.$$

Which, due to the characterization of distributions from Theorem 3.3, proves that

$$\bar{f} \in \mathcal{D}'(\mathbb{R}^d). \ \square$$

Integrability. Under the hypotheses of Theorem 9.5, the distribution $\bar{f}$ is *locally integrable*, i.e. integrable on bounded sets, a notion that we will introduce in Volume 4. □

Let us show that two distinct singular functions give rise to distinct distributions, which allows us to identify every singular function with a distribution.

Theorem 9.6.– *The mapping $f \mapsto \bar{f}$ given by Theorem 9.5 is linear and injective.* ▮

Proof. **1.** *An equality.* Let $\varphi \in \mathcal{D}(\mathbb{R}^d \setminus \{0\})$ and $\widetilde{\varphi}$ be it extension by 0. Then $\widetilde{\varphi} \in \mathcal{D}(\mathbb{R}^d)$ (by (6.1), p. 118) and the definition of $\bar{f}$ from Theorem 9.5 gives

$$\langle \bar{f}, \widetilde{\varphi} \rangle = \int_{\mathbb{R}^d \setminus \{0\}} f\varphi, \tag{9.5}$$

since $\int_{x \in \mathbb{R}^d : |x| > \epsilon} f\varphi = \int_{\mathbb{R}^d \setminus \{0\}} f\varphi$ if ϵ is small enough (so that $\varphi(x) = 0$ if $|x| \leq \epsilon$).

2. *Injectivity.* If

$$\bar{f} = 0,$$

the integral in (9.5) is zero for every $\varphi \in \mathcal{D}(\mathbb{R}^d \setminus \{0\})$, which implies, according to the Du Bois-Reymond lemma (Theorem 3.7), that

$$f = 0.$$

Since the mapping $f \mapsto \bar{f}$ is linear, it follows that it is injective. □

Let us observe that, for functions which are non-singular at 0, we recover identification (3.6), p. 52, of a continuous function with a distribution.

Theorem 9.7.– *If f has a continuous extension $\widetilde{f}$ on all of $\mathbb{R}^d$, the distribution $\bar{f}$ given by Theorem 9.5 is the continuous function $\widetilde{f}$.*

More precisely, the distribution $\bar{f}$ associated with the singular function f by Theorem 9.5 coincides with the distribution $\bar{\widetilde{f}}$ associated with the continuous function $\widetilde{f}$ by Theorem 3.5. ▮

Proof. Suppose therefore that f has an extension $\widetilde{f} \in \mathcal{C}(\mathbb{R}^d)$. Given $\varphi \in \mathcal{D}(\mathbb{R}^d)$, using the additivity property from Theorem A.80 (b), we decompose

$$\int_{\mathbb{R}^d} \widetilde{f}\varphi = \int_{\{x \in \mathbb{R}^d : |x| < \epsilon\}} \widetilde{f}\varphi + \int_{\{x \in \mathbb{R}^d : |x| > \epsilon\}} \widetilde{f}\varphi.$$

The bound of the semi-norms (here, the absolute value) of the integral from Theorem 1.23 (b) and the expression for the measure of a ball from Theorem A.84 give

$$\left| \int_{\{x \in \mathbb{R}^d : |x| < \epsilon\}} \widetilde{f}\varphi \right| \leq |\mathring{B}(0, \epsilon)| \sup_{x \in \mathbb{R}^d} |\widetilde{f}(x)\varphi(x)| = c\epsilon^d,$$

where $c = v \sup_{x \in \mathbb{R}^d} |\widetilde{f}(x)\varphi(x)|$. Since this term goes to 0 when $\epsilon \to 0$, it remains

$$\int_{\mathbb{R}^d} \widetilde{f}\varphi = \lim_{\epsilon \to 0} \int_{\{x \in \mathbb{R}^d : |x| > \epsilon\}} \widetilde{f}\varphi.$$

That is, by the definitions of $\bar{\widetilde{f}}$ and $\bar{f}$ from Theorems 3.5 and 9.5,

$$\langle \bar{\widetilde{f}}, \varphi \rangle = \langle \bar{f}, \varphi \rangle.$$

This proves that $\bar{\widetilde{f}} = \bar{f}$. □

9.3. Derivatives of a distribution associated with a singular function

Let us give conditions for the derivative of a distribution associated with a singular function to be the distribution associated with its derivative.

Theorem 9.8.– *Let $f \in C^1(\mathbb{R}^d \setminus \{0\})$ and $i \in [\![1,d]\!]$ be such that there exist real numbers $r_0 > 0$, $b \geq 0$ and*

$$\lambda < d$$

such that: for every $x \in \mathbb{R}^d$ such that $0 < |x| \leq r_0$,

$$|f(x)| \leq b|x|^{1-\lambda}, \quad |\partial_i f(x)| \leq b|x|^{-\lambda}. \tag{9.6}$$

Then, denoting by $\overline{f}$ the distribution associated with f by Theorem 9.5,

$$\partial_i \overline{f} = \overline{\partial_i f}. \ \blacksquare$$

Proof. Let $\varphi \in \mathcal{D}(\mathbb{R}^d)$. Definition 5.4 (a) of the derivative in the distribution sense and the definition of $\overline{f}$ from Theorem 9.5 give

$$\langle \partial_i \overline{f}, \varphi \rangle = -\langle \overline{f}, \partial_i \varphi \rangle = -\lim_{\substack{\epsilon \to 0 \\ \epsilon > 0}} \int_{\{x \in \mathbb{R}^d : |x| > \epsilon\}} f \, \partial_i \varphi.$$

Let R be large enough so that

$$\varphi(x) = 0 \ \text{ si } |x| \geq R,$$

and let $C_{\epsilon,R} = \{x \in \mathbb{R}^d : \epsilon < |x| < R\}$. Since the domain where the function is zero does not contribute to the integral (Theorem A.77), it follows that

$$\langle \partial_i \overline{f}, \varphi \rangle = -\lim_{\substack{\epsilon \to 0 \\ \epsilon > 0}} \int_{C_{\epsilon,R}} f \, \partial_i \varphi. \tag{9.7}$$

Due to Leinbiz's formula (Theorem A.60),

$$-f \, \partial_i \varphi = \partial_i f \, \varphi - \partial_i (f\varphi). \tag{9.8}$$

Due to Stokes' formula (Theorem 9.2),

$$\int_{C_{\epsilon,R}} \partial_i (f\varphi) \, \mathrm{d}x = \int_{S_\epsilon} n_i f \varphi \, \mathrm{d}s + \int_{S_R} n_i f \varphi \, \mathrm{d}s, \tag{9.9}$$

where $S_\epsilon = \{x \in \mathbb{R}^d : |x| = \epsilon\}$ and n is the unit normal field on S_ϵ and S_R directed outwards from $C_{\epsilon,R}$. When $\epsilon \to 0$, the right-hand side of (9.9) tends to 0 because:

— the integral over S_R is zero, since φ is zero there;

— the absolute value of the integral over S_ϵ is bounded from above (Theorem 9.3) by $d v_d \epsilon^{d-1} b \epsilon^{1-\lambda} \sup_{x \in \mathbb{R}^d} |\varphi(x)|$ which tends to 0 since $\lambda < d$.

Equalities (9.7), (9.8) and (9.9) therefore give

$$\langle \partial_i \bar{f}, \varphi \rangle = \lim_{\substack{\epsilon \to 0 \\ \epsilon > 0}} \int_{C_{\epsilon,R}} \partial_i f \varphi.$$

We can again replace $C_{\epsilon,R}$ with $\{x \in \mathbb{R}^d : |x| > \epsilon\}$ due to Theorem A.77, which gives, with Definition 9.5 now for $\overline{\partial_i f}$,

$$\langle \partial_i \bar{f}, \varphi \rangle = \langle \overline{\partial_i f}, \varphi \rangle. \ \square$$

Let us give an analogous result for the Laplacian, $\Delta = \partial_1^2 + \cdots + \partial_d^2$.

Theorem 9.9.– *Let $f \in C^2(\mathbb{R}^d \setminus \{0\})$ be such that there exist real numbers $r_0 > 0$, $b \geq 0$ and*

$$\lambda < d$$

such that: for every $i \in [\![1, d]\!]$ and every $x \in \mathbb{R}^d$ such that $0 < |x| \leq r_0$,

$$|f(x)| \leq b|x|^{2-\lambda}, \quad |\partial_i f(x)| \leq b|x|^{1-\lambda}, \quad |\partial_i^2 f(x)| \leq b|x|^{-\lambda}. \tag{9.10}$$

Then, denoting by $\bar{f}$ the distribution associated with f by Theorem 9.5,

$$\Delta \bar{f} = \overline{\Delta f}. \ \blacksquare$$

Proof. Theorem 9.8 gives

$$\partial_i \bar{f} = \overline{\partial_i f}, \quad \partial_i \overline{\partial_i f} = \overline{\partial_i \partial_i f}.$$

Therefore, $\partial_i^2 \bar{f} = \partial_i \overline{\partial_i f} = \overline{\partial_i \partial_i f}$, hence

$$\Delta \bar{f} = \sum_{1 \leq i \leq d} \partial_i^2 \bar{f} = \sum_{1 \leq i \leq d} \overline{\partial_i^2 f} = \overline{\sum_{1 \leq i \leq d} \partial_i^2 f} = \overline{\Delta f}. \ \square$$

9.4. Elementary Newtonian potential

Let us construct the **elementary Newtonian potential**[2] ξ. It is a radial solution to the Laplace equation $-\Delta \xi = \delta_0$ on $\mathbb{R}^d$ (the only one that vanishes at infinity if $d \geq 3$).

2. **History of the calculation of the potential.** Laurent SCHWARTZ gave in 1945 [68, Property (22), p. 69] the expression for the potential (Theorem 9.10) in dimension three. In 1950, he calculated the potential in an arbitrary dimension [69, Properties (II,3,10) and (II,3,14), p. 45 and 46].

Theorem 9.10.– *Let $\xi \in \mathcal{D}'(\mathbb{R}^d)$ be the distribution associated by Theorem 9.5 with the function in $\mathcal{C}^\infty(\mathbb{R}^d \setminus \{0\})$ defined, for $x \neq 0$, by*

$$\xi(x) = \begin{cases} \dfrac{1}{(d-2)dv_d} \dfrac{1}{|x|^{d-2}} & \text{if } d \neq 2, \\ \dfrac{1}{2v_2} \log \dfrac{1}{|x|} & \text{if } d = 2, \end{cases}$$

where $v_d = |\mathring{B}_{\mathbb{R}^d}|$ is the measure of the unit ball in $\mathbb{R}^d$. Then,

$$-\Delta\xi = \delta_0.$$

Moreover, for every $i \in [\![1, d]\!]$, the derivative $\partial_i\xi \in \mathcal{D}'(\mathbb{R}^d)$ is the distribution associated by Theorem 9.5 with the function in $\mathcal{C}^\infty(\mathbb{R}^d \setminus \{0\})$ defined, for $x \neq 0$, by

$$(\partial_i\xi)(x) = -\frac{1}{dv_d} \frac{x_i}{|x|^d}. \blacksquare$$

Recall that $\Delta \stackrel{\text{def}}{=} \partial_1^2 + \cdots + \partial_d^2$ is the Laplacian and δ_0 is the Dirac mass at the point 0. Note that $v_2 = \pi$.

Terminology. The term "elementary Newtonian potential" is a shortening of the following two classic terms.

— The **Newtonian potential** of a distribution f defined on all of $\mathbb{R}^d$, $d \geq 3$, is the distribution $u = |x|^{2-d} \star f$. It is a particular solution to the Laplace equation, $-\Delta u = c_d f$. On $\mathbb{R}^2$, the **logarithmic potential** of f is the distribution $u = \log|x|^{-1} \star f$.

— An **elementary solution** to the Laplace equation is any solution u to $-\Delta u = \delta_0$.

The elementary potential ξ is therefore the Newtonian potential associated with δ_0/c_d. It is the unique elementary solution to the Laplace equation that vanishes at infinity. There exists infinitely many solutions to this equation, obtained by adding to ξ an arbitrary harmonic function, but no other one tends to 0 when $|x| \to \infty$. □

Utility of the potential. The potential ξ, once *localized* in the form $\gamma = \theta\xi$ in Theorem 9.17, plays an essential role which is explained in the comment *Utility of the localized potential*, p. 208. □

The proof of Theorem 9.10 will use the following calculation of the partial derivatives of powers of the Euclidean norm, in the function sense.

Lemma 9.11.– *The function $x \mapsto |x|^s$, where $s \in \mathbb{R}$, is infinitely differentiable from $\mathbb{R}^d \setminus \{0\}$ into $\mathbb{R}$ and, for every $i \in [\![1, d]\!]$ and $x \in \mathbb{R}^d$, $x \neq 0$:*

(a) $$\partial_i(|x|^s) = sx_i|x|^{s-2}.$$

(b) $$\partial_i^2(|x|^s) = s|x|^{s-2} + s(s-2)x_i^2|x|^{s-4}.$$

(c) $$\Delta(|x|^s) = s(d+s-2)|x|^{s-2}. \blacksquare$$

Proof. **1.** *Equality (a).* The function $x \mapsto x \cdot x$ is infinitely differentiable and the function $y \mapsto y^{s/2}$ is too away from the point $y = 0$ (Theorem A.67), hence their composition $x \mapsto |x|^s$ is as well away from the point $x = 0$ (Theorem A.65).

The chain rule theorem (Theorem A.64 (a) for $T(x) = x \cdot x$ and $\ell = 1$) gives, for $x \neq 0$,

$$\partial_i(|x|^s) = \partial_i\big((x \cdot x)^{s/2}\big) = \frac{s}{2}(x \cdot x)^{s/2-1}\partial_i(x \cdot x) = s x_i |x|^{s-2},$$

since $\mathrm{d}y^t/\mathrm{d}y = t|y|^{t-1}$ (Theorem A.67) and $\partial_i(x \cdot x) = \sum_{j=1}^d \partial_i x_j^2 = \partial_i x_i^2 = 2x_i$.

2. *Equality (b).* Differentiating equality (a) via Leibniz's formula (Theorem A.60) and using equality (a) for $s - 2$, we obtain

$$\partial_i^2(|x|^s) = s(\partial_i x_i |x|^{s-2} + x_i \partial_i(|x|^{s-2})) = s|x|^{s-2} + s(s-2)x_i^2|x|^{s-4}.$$

3. *Equality (c).* It follows by summing equality (b) over m from 1 to d. □

The proof of Theorem 9.10 will also use the following calculation of the partial derivatives of the logarithm of the norm, in the function sense.

Lemma 9.12.– *The function $x \mapsto \log|x|$ is infinitely differentiable from $\mathbb{R}^d \setminus \{0\}$ into $\mathbb{R}$ and, for every $i \in [\![1, d]\!]$ and $x \in \mathbb{R}^d$, $x \neq 0$:*

(a) $$\partial_i(\log|x|) = x_i|x|^{-2}.$$

(b) $$\partial_i^2(\log|x|) = |x|^{-2} - 2x_i^2|x|^{-4}.$$

(c) $$\Delta(\log|x|) = (d-2)|x|^{-2}. \blacksquare$$

Proof. **1.** *Equality (a).* The function $x \mapsto |x|$ is infinitely differentiable away from the point $x = 0$ according to Lemma 9.11, and the function log is too on $(0, \infty)$ (Theorem A.69), hence the composition $x \mapsto \log|x|$ is as well away from the point $x = 0$ (Theorem A.65).

From these statements, $\mathrm{d}|x|/\mathrm{d}x = x/|x|$ and $\mathrm{d}\log y/\mathrm{d}y = 1/y$, thus the chain rule theorem (Theorem A.64 (a) for $T = \log$ and $\ell = 1$) gives

$$\partial_i(\log|x|) = \frac{\partial_i(|x|)}{|x|} = x_i|x|^{-2}.$$

2. *Equality (b).* Differentiating equality (a) via Leibniz's formula (Theorem A.60) and using equality (a) of Theorem 9.11 for $s = -2$, we obtain

$$\partial_i^2(\log|x|) = \partial_i x_i|x|^{-2} + x_i\partial_i(|x|^{-2}) = |x|^{-2} - 2x_i^2|x|^{-4}.$$

3. *Equality (c).* It follows by summing equalities (b) over m from 1 to d. □

We are now able to establish the properties of the elementary potential.

Proof of Theorem 9.10. **1.** *Existence of the distribution* ξ. Here, we denote by $\underline{\xi}$ the function given in $\mathbb{R}^d \setminus \{0\}$, to distinguish it from the distribution ξ.

Theorem 9.5 indeed allows to associate a distribution $\xi \in \mathcal{D}'(\mathbb{R}^d)$ with the singular function $\underline{\xi}$ since its hypothesis (9.1), i.e. $|\underline{\xi}(x)| \leq b|x|^{-\lambda}$ where $\lambda < d$, is satisfied:
— if $d \neq 2$, with $\lambda = d - 2$;
— if $d = 2$, with $\lambda = 3/2$ since (Theorem A.69), if $|x| \leq 1$,

$$\left|\log \frac{1}{|x|}\right| \leq \frac{2}{|x|^{1/2}}. \tag{9.11}$$

2. *Calculation of* $\partial_i \xi$. Lemma 9.11 (a) if $d \neq 2$, and Lemma 9.12 (a) completed by the equality $\log(1/|x|) = -\log|x|$ (Theorem A.69) if $d = 2$, give, for $x \neq 0$,

$$\partial_i \underline{\xi}(x) = -\frac{x_i |x|^{-d}}{d v_d}. \tag{9.12}$$

Theorem 9.8 then shows that $\partial_i \xi$ is indeed the distribution associated with the function $\partial_i \underline{\xi}$ by Theorem 9.5, since its hypothesis (9.6) is satisfied by $\lambda = d - 1$ if $d \neq 2$, and by $\lambda = 1/2$ due to (9.11) if $d = 2$.

3. *Calculation of* $-\Delta\xi$. **3.a.** *Value on* φ. Let $\varphi \in \mathcal{D}(\mathbb{R}^d)$. By Definition 5.4 (a) of the derivatives in the distribution sense,

$$\langle -\Delta\xi, \varphi \rangle = \sum_{i=1}^{d} \langle -\partial_i^2 \xi, \varphi \rangle = \sum_{i=1}^{d} \langle \partial_i \xi, \partial_i \varphi \rangle. \tag{9.13}$$

The definition from Theorem 9.5 of the distribution (here $\partial_i \xi$) associated with a singular function (here $\underline{\xi}$) may be expressed, since the domain where the function is zero does not contribute to the integral (Theorem A.77), as

$$\langle \partial_i \xi, \varphi \rangle = \lim_{\substack{\epsilon \to 0 \\ \epsilon > 0}} \int_{\{x \in \mathbb{R}^d : |x| > \epsilon\}} \partial_i \underline{\xi}\, \varphi, = \lim_{\substack{\epsilon \to 0 \\ \epsilon > 0}} \int_{C_{\epsilon,R}} \partial_i \underline{\xi}\, \varphi,$$

where R is such that $\varphi(x) = 0$ if $|x| \geq R$ and $C_{\epsilon,R} = \{x \in \mathbb{R}^d : \epsilon < |x| < R\}$. Hence, (9.13) gives

$$\langle -\Delta\xi, \varphi \rangle = \lim_{\substack{\epsilon \to 0 \\ \epsilon > 0}} \int_{C_{\epsilon,R}} \sum_{i=1}^{d} \partial_i \underline{\xi}\, \partial_i \varphi. \tag{9.14}$$

3.b. *Transformation of the integral.* Due to Leibniz's formula (Theorem A.60),

$$\sum_{i=1}^{d} \partial_i \underline{\xi}\, \partial_i \varphi = \sum_{i=1}^{d} \left(\partial_i(\partial_i \underline{\xi}\, \varphi) - \partial_i^2 \underline{\xi}\, \varphi\right) = \sum_{i=1}^{d} \partial_i(\partial_i \underline{\xi}\, \varphi) - \Delta \underline{\xi}\, \varphi. \tag{9.15}$$

Using the value of the function ξ (given in the statement) and Lemma 9.11 (c) with $s = 2 - d$ if $d \geq 3$ or Lemma 9.12 (c) if $d = 2$, we obtain

$$\Delta\underline{\xi} = 0. \tag{9.16}$$

On the other hand, due to Stokes' formula (Theorem 9.2),

$$\int_{C_{\epsilon,R}} \sum_{i=1}^{d} \partial_i(\underline{\xi}\varphi)\,\mathrm{d}x = \int_{S_\epsilon} \sum_{i=1}^{d} n_i \partial_i \underline{\xi}\,\varphi\,\mathrm{d}s + \int_{S_R} \sum_{i=1}^{d} n_i \partial_i \underline{\xi}\,\varphi\,\mathrm{d}s. \tag{9.17}$$

where $S_\epsilon = \{x \in \mathbb{R}^d : |x| = \epsilon\}$ and n is the unit normal vector field on S_ϵ and S_R directed outwards from $C_{\epsilon,R}$. In the right-hand side:

— the integral over S_R is zero, since φ is zero there;

— the integral over S_ϵ equals, by (9.12) and since here $n_i = -x_i/|x|$ and $|x| = \epsilon$,

$$\int_{S_\epsilon} \sum_{i=1}^{d} n_i \partial_i \underline{\xi}\,\varphi\,\mathrm{d}s = \int_{S_\epsilon} \sum_{i=1}^{d} \frac{x_i^2}{dv_d|x|^{d+1}}\varphi\,\mathrm{d}s = \frac{1}{dv_d\epsilon^{d-1}} \int_{S_\epsilon} \varphi\,\mathrm{d}s. \tag{9.18}$$

Finally, the integral in (9.14) equals, by (9.15), (9.16), (9.17) and (9.18),

$$\int_{C_{\epsilon,R}} \sum_{i=1}^{d} \partial_i \underline{\xi}\,\partial_i \varphi = \frac{1}{dv_d\epsilon^{d-1}} \int_{S_\epsilon} \varphi\,\mathrm{d}s. \tag{9.19}$$

3.c. *Calculation of the limit.* Since $\int_{S_\epsilon} 1\,\mathrm{d}s = dv_d\epsilon^{d-1}$ (Theorem 9.4), we have

$$\left|\left(\frac{1}{dv_d\epsilon^{d-1}} \int_{S_\epsilon} \varphi\,\mathrm{d}s\right) - \varphi(0)\right| = \left|\frac{1}{dv_d\epsilon^{d-1}} \int_{S_\epsilon} \varphi - \varphi(0)\,\mathrm{d}s\right| \leq \sup_{x \in S_\epsilon} |\varphi(x) - \varphi(0)|$$

This tends to 0 when $\epsilon \to 0$ since φ is continuous at 0. Thus, (9.14) and (9.19) give

$$\langle -\Delta\xi, \varphi\rangle = \lim_{\substack{\epsilon \to 0 \\ \epsilon > 0}} \frac{1}{dv_d\epsilon^{d-1}} \int_{S_\epsilon} \varphi\,\mathrm{d}s = \varphi(0) = \langle \delta_0, \varphi\rangle.$$

Which proves that

$$-\Delta\xi = \delta_0. \quad \square$$

9.5. Newtonian potential of order n

Let us define the **Newtonian potential of order** n, denoted by ξ_n. It is a radial solution[3] to the equation $(-\Delta)^n \xi_n = \delta_0$ on $\mathbb{R}^d$.

3. **History of the calculation of the potential.** Laurent SCHWARTZ calculated the potential of order n in 1950 [69, Properties (II,3,16) and (II,3,17), p. 47].

Theorem 9.13.– *Let $\xi_n \in \mathcal{D}'(\mathbb{R}^d)$, where $n \in \mathbb{N}^*$, be the distribution associated by Theorem 9.5 with the function in $\mathcal{C}^\infty(\mathbb{R}^d \setminus \{0\})$ defined, for $x \neq 0$, by*

$$\xi_n(x) = \begin{cases} \dfrac{1}{\kappa_{n,d}} |x|^{2n-d} & \text{if } d \text{ is odd or } d > 2n, \\ \dfrac{1}{\kappa_{n,d}} |x|^{2n-d} \log |x| & \text{if } d \text{ is even and } d \leq 2n, \end{cases}$$

where, denoting by υ_d the measure of the unit ball in $\mathbb{R}^d$,

$$\kappa_{n,d} = (-1)^n (2n-d)(2n-2-d)\dots(2-d) \times 2^{n-1}(n-1)!\, d\upsilon_d,$$

product in which the terms which eventually are zero (there are some when $d = 2$, when $n = 1$ and when d is even and $\leq 2n$) must be replaced by 1. Then,

$$(-\Delta)^n \xi_n = \delta_0. \;\blacksquare$$

Utility of the potential of order n**.** The potential ξ_n is used in the proof of the *kernel theorem*, via the decomposition of the Dirac mass into a sum of the derivatives of continuous functions in Theorem 9.18, as it is explained in the comment *Utility of the decomposition...*, p. 210. □

Choice of signs. It might seem to be wise to suppress the coefficient $(-1)^n$ in the definition of κ_n, since it would also disappear from the equation satisfied by ξ_n, which would then become $\Delta^n \xi_n = \delta_0$. We do not do this because the operator $-\Delta$ is positive, meaning that $\langle -\Delta u, u\rangle \geq 0$ when $u \in H_0^1(\Omega)$, which gives $(-\Delta)^n$ some pleasant properties to use when studying PDEs.

That is why, for $n = 1$, we defined the elementary potential ξ (which is equal to ξ_1) in Theorem 9.10 as a solution of the equation $-\Delta\xi = \delta_0$.

Note that, unfortunately, $\kappa_{n,d}$ is not always positive; nor is $(-1)^n \kappa_{n,d}$. □

Some values of κ_n**.**

— When $n = 1$, then $\kappa_{1,d} = (d-2)d\upsilon_d$ if $d \neq 2$ and $\kappa_{1,2} = -2\upsilon_2$, thus we indeed have $\xi_1 = \xi$ where ξ is the elementary potential obtained in Theorem 9.10.

— When $d = 2$, then $\kappa_{n,2} = (-1)^n(2n-2)(2n-4)\dots(2) \times 2^{n-1}(n-1)!\, 2\upsilon_2$.

— When $d = 2n$, then $\kappa_{n,d} = (-1)^n(-2)\dots(4-d)(2-d) \times 2^{n-1}(n-1)!\, d\upsilon_d$.

— When d is even and $4 \leq d < 2n$, then

$$\kappa_{n,d} = (-1)^n(2n-d)(2n-2-d)\dots(2)(-2)\dots(4-d)(2-d) \times 2^{n-1}(n-1)!\, d\upsilon_d. \;\square$$

The proof of Theorem 9.13 will use the following calculations of partial derivatives in the function sense.

Lemma 9.14.– *The function $x \mapsto |x|^s \log |x|$, where $s \in \mathbb{R}$, is infinitely differentiable from $\mathbb{R}^d \setminus \{0\}$ into $\mathbb{R}$ and, for every $i \in [\![1, d]\!]$ and $x \in \mathbb{R}^d$, $x \neq 0$:*

(a) $$\partial_i(|x|^s \log |x|) = x_i |x|^{s-2}(1 + s \log |x|).$$

(b) $\partial_i^2(|x|^s \log|x|) = |x|^{s-4}(|x|^2 + (2s-2)x_i^2) + s(|x|^2 + (s-2)x_i^2)|x|^{s-4}\log|x|.$

(c) $$\Delta(|x|^s \log|x|) = (d+2s-2)|x|^{s-2} + s(d+s-2)|x|^{s-2}\log|x|. \blacksquare$$

Proof. **1.** *Equality (a).* Since the functions $x \mapsto |x|^s$ and $x \mapsto \log|x|$ are infinitely differentiable away from the point 0 (Lemmas 9.11 and 9.12), their product is too (Theorem A.60) and, by these results, its derivative equals

$$\partial_i(|x|^s \log|x|) = \partial_i(|x|^s)\log|x| + |x|^s\partial_i(\log|x|) = x_i|x|^{s-2}(1+s\log|x|).$$

2. *Equality (b).* This equality is obtained by differentiating equality (a), with again Leibniz's formula from Theorem A.60 and Lemmas 9.11 and 9.12.

3. *Equality (c).* It follows by summing equality (b) over m from 1 to d. □

Proof of Theorem 9.13. Since the case $n=1$ is given by Theorem 9.10, let us suppose that $n \geq 2$ and proceed inductively, denoting by $\overline{f}$ the distribution associated with a singular function f by Theorem 9.5.

1. *A preliminary calculation.* Let $s \in \mathbb{R}$. For $x \neq 0$, Lemma 9.11 (c) gives $\Delta(|x|^s) = s(d+s-2)|x|^{s-2}$. If $s > 2-d$, it follows from the calculation of the Laplacian of a singular function from Theorem 9.9 (whose hypothesis (9.10) is satisfied with $\lambda = 2-s$ according to Lemma 9.11 (a) and (b)) that

$$\Delta(\overline{|x|^s}) = s(d+s-2)\overline{|x|^{s-2}}.$$

Reiterating, it follows, for every $m \in \mathbb{N}^*$ and $s > 2m-d$,

$$\Delta^m(\overline{|x|^s}) = a_{s,m,d}\overline{|x|^{s-2m}}, \tag{9.20}$$

where

$$a_{s,m,d} \stackrel{\text{def}}{=} s(s-2)\dots(s-2m+2) \times (d+s-2)(d+s-4)\dots(d+s-2m).$$

2. *The case d odd or $> 2n$.* Equality (9.20) for $m = n-1$ and $s = 2n-d$ gives

$$\Delta^{n-1}(\overline{|x|^{2n-d}}) = a_{2n-d,n-1,d}\overline{|x|^{2-d}}.$$

Since (Theorem 9.10)

$$\Delta(\overline{|x|^{2-d}}) = (2-d)dv_d\delta_0 \tag{9.21}$$

and $a_{2n-d,n-1,d} = (2n-d)(2n-2-d)\dots(4-d) \times (2n-2)(2n-4)\dots2$, it follows

$$\Delta^n(\overline{|x|^{2n-d}}) = (-1)^n\kappa_{n,d}\delta_0.$$

From which, since here $\xi_n = \overline{|x|^{2n-d}}/\kappa_{n,d}$,

$$(-\Delta)^n \xi_n = \delta_0.$$

3. Another preliminary calculation. For $x \neq 0$, we have $\Delta(\log|x|) = (d-2)|x|^{-2}$ (Lemma 9.12 (c)). Hence, again by the calculation of the Laplacian of a singular function from Theorem 9.9, for $d \geq 3$,

$$\Delta\big(\overline{\log|x|}\big) = (d-2)\overline{|x|^{-2}}. \tag{9.22}$$

Indeed, hypothesis (9.10) of Theorem 9.9 is satisfied with $\lambda = 5/2$ by Lemma 9.12 (a) and (b) and since (Theorem A.69), if $|x| \leq 1$,

$$\big|\log|x|\big| \leq 2|x|^{-1/2}. \tag{9.23}$$

4. Another preliminary calculation. Let $s > 2-d$. The expression for $\Delta(|x|^s \log|x|)$ given in Lemma 9.14 (c) implies, again by the calculation of the Laplacian of a singular function from Theorem 9.9, whose hypothesis (9.10) is satisfied with $\lambda = 2-s$ by (9.23) and Lemma 9.12 (a) and (b),

$$\Delta\big(\overline{|x|^s \log|x|}\big) = s(d+2s-2)\overline{|x|^{s-2}} + s(d+s-2)\overline{|x|^{s-2}\log|x|}.$$

Reiterating, it follows with (9.20), for every $m \in \mathbb{N}^*$ and $s > 2m-d$,

$$\Delta^m\big(\overline{|x|^s \log|x|}\big) = b_{s,m,d}\overline{|x|^{s-2m}} + a_{s,m,d}\overline{|x|^{s-2m}\log|x|},$$

where $b_{s,m,d}$ is a real number whose value does not matter and $a_{s,m,d}$ is defined in step 1.

In particular, for d even, $s = 2n-d$ and $m = n-d/2$,

$$\Delta^{n-d/2}\big(\overline{|x|^{2n-d}\log|x|}\big) = b_{2n-d,n-d/2,d} + a_{2n-d,n-d/2,d}\overline{\log|x|}, \tag{9.24}$$

and

$$a_{2n-d,n-d/2,d} = (2n-d)(2n-d-2)\ldots 2 \times (2n-2)(2n-4)\ldots d. \tag{9.25}$$

5. The case $d = 2 \leq 2n$. The properties of the logarithm and the elementary potential (Theorems A.69 and 9.10) give

$$\Delta(\log|x|) = \Delta\Big(-\log\Big(\frac{1}{|x|}\Big)\Big) = -2\upsilon_2\Delta\xi = 2\upsilon_2\delta_0.$$

With (9.24) and (9.25), it follows

$$\Delta^n\big(\overline{|x|^{2n-d}\log|x|}\big) = a_{2n-2,n-1,2}2\upsilon_2\delta_0 = (-1)^n\kappa_{n,2}\delta_0.$$

Since now $\xi_n = \overline{|x|^{2n-d} \log |x|}/\kappa_{n,d}$, we have, again,

$$(-\Delta)^n \xi_n = \delta_0.$$

6. The case d even, $4 \leq d \leq 2n$. By (9.24) and (9.22),

$$\Delta^{n-d/2+1}\left(\overline{|x|^{2n-d} \log |x|} \right) = a_{2n-d,n-d/2,d}(d-2)\overline{|x|^{-2}}.$$

On the other hand, equality (9.20) with $m = d/2 - 2$ and $s = -2$ gives

$$\Delta^{d/2-2}\left(\overline{|x|^{-2}} \right) = a_{-2,d/2-2,d}\overline{|x|^{2-d}}.$$

These two equalities and (9.21) give

$$\Delta^n \left(\overline{|x|^{2n-d} \log |x|} \right) = (-1)^n \kappa_{n,d} \delta_0,$$

because, by (9.25),

$$a_{2n-d,n-d/2,d}(d-2)a_{-2,d/2-2,d}(2-d)dv_d = (-1)^n \kappa_{n,d}.$$

Since here again $\xi_n = \overline{|x|^{2n-d} \log |x|}/\kappa_{n,d}$, we still have

$$(-\Delta)^n \xi_n = \delta_0. \ \square$$

Now let us give a regularity property of the Newtonian potential of order n.

Theorem 9.15.– *If $2n > d+1$, the Newtonian potential $\zeta_n \in \mathcal{D}'(\mathbb{R}^d)$ defined in Theorem 9.13 satisfies*

$$\xi_n \in \mathcal{C}^{2n-d-1}(\mathbb{R}^d). \ \blacksquare$$

Proof. This follows from Lemma 9.16, below. $\square$

Lemma 9.16.– *Let $s \in \mathbb{R}$, $s > 0$. Then:*

The function $x \mapsto |x|^s$ is m times continuous differentiable from $\mathbb{R}^d$ into $\mathbb{R}$, if $m < s$.

The functions from $\mathbb{R}^d$ into $\mathbb{R}$ defined by $x \mapsto x_i|x|^{s-1}$, $x \mapsto |x|^s \log |x|$ and $x \mapsto x_i|x|^{s-1} \log |x|$ for $x \neq 0$ and being 0 for $x = 0$ have the same property. $\blacksquare$

Proof. Since these functions are infinitely differentiable on $\mathbb{R}^d \setminus \{0\}$ according to Lemmas 9.11 and 9.14, it remains to establish the differentiability of order m at 0, for each one.

1. *Function* $x \mapsto |x|^s$. Observe first that, for every $\beta \in \mathbb{N}^d$ and $x \in \mathbb{R}^d$, $x \neq 0$,

$$\partial^\beta(|x|^s) = \sum_{\sigma \in \mathbb{N}^d : |\sigma| = |\beta|} c_{s,\sigma} x^\sigma |x|^{s-2|\beta|}, \tag{9.26}$$

where $x^\sigma = x_1^{\sigma_1} \dots x_d^{\sigma_d}$ and $c_{s,\sigma} \in \mathbb{R}$.

Indeed, $\partial_i(|x|^s) = s x_i |x|^{s-2}$ (Lemma 9.11 (a)). Therefore, (9.26) is true for $|\beta| = 1$. And, if it is true for every β such that $|\beta| = k$, it is also true for every β such that $|\beta| = k + 1$ due to Leibniz's formula (Theorem A.60).

Now denote

$$g(x) \stackrel{\text{def}}{=} |x|^s$$

and let us show, by induction on m, that, if $|\beta| \leq m < s$, then

$$g \textit{ is } m \textit{ times differentiable at } 0 \textit{ and } \partial^\beta g(0) = 0. \tag{9.27}$$

Suppose (9.27) is true for every β such that $|\beta| \leq m - 1$. Then, with (9.26), if $x \neq 0$,

$$|\partial^\beta g(x) - \partial^\beta g(0)| = |\partial^\beta(|x|^s)| \leq c|t|^{s-|\beta|}. \tag{9.28}$$

By Definition 1.16 of differentiability, this shows, since $s - |\beta| > 1$, that $\partial^\beta g$ is differentiable at 0 and $\nabla \partial^\beta g(0) = 0$, and thus $\partial_i \partial^\beta(0) = 0$ for every $i \in [\![1, d]\!]$. Hence, (9.27) is true for every β such that $|\beta| = m$.

Being true for $m = 0$, Property (9.27) is therefore true for all $m < s$.

2. *Function* $x \mapsto x_i |x|^{s-1}$. If $s > m$, by step 1, the function

$$x \mapsto \frac{1}{s+1}|x|^{s+1} \text{ is } m+1 \text{ times continuously differentiable on } \mathbb{R}^d.$$

According to Lemma 9.11 (a), the function $x \mapsto x_i |x|^{s-1}$ is its partial derivative on $\mathbb{R}^d \setminus \{0\}$. This last function, also being continuous at 0, is also the partial derivative at this point.

So, the function $x \mapsto x_i |x|^{s-1}$ is, on all of $\mathbb{R}^d$, the partial derivative of the function $x \mapsto |x|^{s+1}/(s+1)$.

It is therefore m time continuously differentiable.

3. Function $x \mapsto |x|^s \log|x|$. Observe here that, for every $\beta \in \mathbb{N}^d$ and $x \in \mathbb{R}^d$, $x \neq 0$,

$$\partial^\beta(|x|^s \log|x|) = \sum_{\sigma \in \mathbb{N}^d : |\sigma| = |\beta|} x^\sigma |x|^{s-2|\beta|}(c_{s,\sigma} + c'_{s,\sigma} \log|x|), \tag{9.29}$$

where $c_{s,\sigma} \in \mathbb{R}$ and $c'_{s,\sigma} \in \mathbb{R}$.

Indeed, $\partial_i(|x|^s \log|x|) = x_i|x|^{s-2}(1 + s\log|x|)$ according to Lemma 9.14 (a). Therefore, equality (9.29) is true for $|\beta| = 1$. And, if it is true for every β such that $|\beta| = k$, it is true for every β such that $|\beta| = k + 1$ due to Leibniz's formula (Theorem A.60) and the equality $\partial_i(|x|^s) = sx_i|x|^{s-2}$ (Lemma 9.11 (a)).

Now let $m < s$, $\epsilon = (s - m)/2$ and β be such that $|\beta| \leq m - 1$. For $0 < |x| \leq 1$, we have (Theorem A.69)

$$\big|\log|x|\big| \leq \frac{1}{\epsilon}|x|^{-\epsilon}, \tag{9.30}$$

therefore (9.29) gives

$$|\partial^\beta(|x|^s \log|x|)| \leq c|x|^{s-|\beta|}(1 + \log|x|) \leq c(1 + 1/\epsilon)|x|^{1+\epsilon}. \tag{9.31}$$

The proof of (9.27) also applies to the function

$$g(x) \stackrel{\text{def}}{=} \begin{cases} |x|^s \log|x| & \text{if } x \neq 0, \\ 0 & \text{if } x = 0, \end{cases}$$

after replacing in (9.28) the inequality by (9.31).

Therefore, this function is also m times continuously differentiable.

4. Function $x \mapsto x_i|x|^{s-1} \log|x|$. If $s > m$, by step 3, the function

$$x \mapsto \frac{1}{s+1}|x|^{s+1} \log|x| \text{ is } m+1 \text{ times continuously differentiable on } \mathbb{R}^d.$$

Due to Lemma 9.14 (a), the function $x \mapsto x_i|x|^{s-1} \log|x|$ is its partial derivative on $\mathbb{R}^d \setminus \{0\}$. This last function, once extended by 0, is also continuous at 0 by (9.30). It is therefore also the partial derivative at this point.

So, the function $x \mapsto x_i|x|^{s-1} \log|x|$ is, on all of $\mathbb{R}^d$, the partial derivative of the function $x \mapsto |x|^{s+1} \log|x|/(s + 1)$.

It is therefore m times continuously differentiable. □

9.6. Localized potential

Let us define the localized elementary potential γ, namely the potential multiplied by a radial function with compact support, and give some of its properties.

Theorem 9.17.– *Let $\theta \in \mathcal{C}_B^\infty(\mathbb{R}^d)$ be such that $\theta(x)$ only depends on $|x|$ and:*

$$0 \leq \theta \leq 1, \;\; \theta(x) = 1 \textit{ if } |x| \leq r_0, \;\; \theta(x) = 0 \textit{ if } |x| \geq r_1,$$

where $0 < r_0 < r_1 < r$ and $B = \{x \in \mathbb{R}^d : |x| \leq r\}$.

(a) *We say that the* **localized elementary potential** *is the distribution*

$$\gamma \overset{\text{def}}{=} \theta\xi,$$

where ξ is the elementary potential defined in Theorem 9.10. It satisfies

$$\gamma \in \mathcal{D}'_B(\mathbb{R}^d), \;\; \gamma \in \mathcal{C}^\infty(\mathbb{R}^d \backslash \{0\}), \;\; \operatorname{supp} \gamma \subset \overset{\circ}{B}, \;\; \gamma(x) \textit{ only depending on } |x|,$$

and there exists a real number b, depending on d and r, such that: for all $i \in [\![1, d]\!]$ and $x \neq 0$,

$$|\gamma(x)| \leq b|x|^{1-d}, \;\; |\partial_i\gamma(x)| \leq b|x|^{1-d},$$

and, if $d \neq 2$,

$$|\gamma(x)| \leq b|x|^{2-d}.$$

(b) *We say that the* **correction term** *is the function*

$$\eta \overset{\text{def}}{=} 2\nabla\xi \cdot \nabla\theta + \xi\,\Delta\theta.$$

It satisfies

$$\eta \in \mathcal{D}_B(\mathbb{R}^d), \;\; \operatorname{supp} \eta \subset \overset{\circ}{B}, \;\; \int_{\mathbb{R}^d} \eta = 1.$$

(c)

$$-\Delta\gamma + \eta = \delta_0. \;\; \blacksquare$$

Terminology. We say that the function γ is the "localized potential", because it is equal to the elementary potential ξ in a neighborhood of its singularity. It is a particular case of a **parametrix** associated with the Laplace operator [SCHWARTZ, 69, p. 144 and 218]. A parametrix relating to an operator L is a distribution γ such that $L\gamma = \delta_0 + \eta$ where $\eta \in \mathcal{D}(\mathbb{R}^d)$. The name "parametrix" was introduced by Jacques HADAMARD. □

Utility of the localized potential. The localized potential γ is the key element in the local representation formula $f = f \diamond \eta + \sum_{i=1}^d \partial_i f \diamond \partial_i \gamma$ of a distribution f (Theorem 12.1), a formula which forms the basis of our proofs of the orthogonality and Poincaré's theorems (Theorems 13.5 and 13.7) on the existence of a primitive, of the construction of a sequentially continuous primitive mapping (Theorem 12.9), and of the explicit local expression for primitives (Theorem 12.7). □

Proof of Theorem 9.17. **1.** *Properties (a).* **1.a.** By Definition 5.15 of the product of a distribution with an infinitely differentiable function,

$$\gamma = \theta\xi \in \mathcal{D}'(\mathbb{R}^d).$$

1.b. The restriction to the open set $\omega = \mathbb{R}^d \setminus \{0\}$ equals (Theorem 6.7) $\gamma|_\omega = \theta|_\omega \xi|_\omega$, which coincides with the product in the function sense (Theorem 5.17). Since the restrictions $\theta|_\omega$ and $\xi|_\omega$ (Theorem 9.10) are infinitely differentiable, the same holds (Theorem A.60) for their product, namely

$$\gamma \in \mathcal{C}^\infty(\mathbb{R}^d \setminus \{0\}).$$

1.c. Since the support of a product is included in the intersection of their supports (Theorem 6.24),

$$\operatorname{supp}\gamma \subset \operatorname{supp}\theta \subset \mathring{B}.$$

1.d. For $x \neq 0$, since $\theta(x)$ and $\xi(x)$ only depend on x via $|x|$, their product

$$\gamma(x) \text{ only depends on } x \text{ via } |x|.$$

1.e. The expression for ξ (Theorem 9.10) and the conditions on θ give, if $d \neq 2$,

$$|\gamma(x)| \leq |\xi(x)| = c_d |x|^{2-d}.$$

1.f. If $d = 2$, Theorems 9.10 and A.69 (on the properties of the logarithm) give

$$|\gamma(x)| \leq \begin{cases} |\xi(x)| = c_d \log(|x|^{-1}) \leq c_d |x|^{-1} & \text{if } |x| \leq 1, \\ |\xi(x)| = c_d \log(|x|^{-1}) \leq c_d \log r \leq (c_d r \log r)|x|^{-1} & \text{if } 1 \leq |x| \leq r, \\ 0 & \text{if } |x| \geq r. \end{cases}$$

1.g. By Theorem 9.10,

$$\partial_i \xi(x) = -\frac{1}{dv_d} x_i |x|^{-d}, \tag{9.32}$$

so Leibniz's formula (Theorem A.60) gives, using $0 \leq \theta \leq 1$ and the above upper bounds of $|\xi(x)|$, and denoting by m an upper bound of the $|\partial_i \theta|$,

$$|\partial_i \gamma(x)| = |(\theta \partial_i \xi + \xi \partial_i \theta)(x)| \leq (1/dv_d + mc_d(1 + r|\log r|))|x|^{1-d}.$$

2. *Identity (c).* The expression for the Laplacian of a product from Theorem 5.19 and the equality $\Delta\xi = -\delta_0$ (Theorem 9.10) give, since $\theta = 1$ on a ball around 0,

$$\Delta\gamma = \theta\Delta\xi + 2\nabla\theta \cdot \nabla\xi + \xi\Delta\theta = -\theta\delta_0 + \eta = -\delta_0 + \eta. \tag{9.33}$$

3. *Properties (b).* **3.a.** The distribution $\eta = 2\nabla\xi \cdot \nabla\theta + \xi\Delta\theta$ is, as ξ and η are, infinitely differentiable away from the point 0 and it is zero outside of the support of $\nabla\theta$, i.e. for $|x| < r_0$ or $|x| > r_1$. Therefore,

$$\eta \in \mathcal{D}_B(\mathbb{R}^d), \quad \operatorname{supp}\eta \subset \mathring{B}.$$

3.b. Let us check that $\int_{\mathbb{R}^d}\eta = 1$. Denote $C_{r_0,r_1} = \{x \in \mathbb{R}^d : r_0 < |x| < r_1\}$; only this crown contributes to the integral (Theorem A.77) since $\eta = 0$ outside of it; on it, equality (9.33) reduces to $\eta = \Delta\gamma$. Stokes' formula from Theorem 9.2 thus gives

$$\int_{\mathbb{R}^d} \eta\,\mathrm{d}x = \int_{C_{r_0,r_1}} \Delta\gamma\,\mathrm{d}x = \int_{S_{r_1}} n \cdot \nabla\gamma\,\mathrm{d}s + \int_{S_{r_0}} n \cdot \nabla\gamma\,\mathrm{d}s,$$

where $S_r = \{x \in \mathbb{R}^d : |x| = r\}$ and n is the unit normal vector field on S_{r_1} and S_{r_0} directed outwards from C_{r_0,r_1}.

However $\nabla\gamma$ is equal to $\theta\nabla\xi + \xi\nabla\theta$, which is zero on S_{r_1} and which equals $\nabla\xi$ on S_{r_0}. Since $n(x) = -x/|x|$ on S_{r_0}, expression (9.32) of the $\partial_i\xi$ and the equality $\int_{S_{r_0}} 1\,\mathrm{d}s = dv_d r_0^{d-1}$ from Theorem 9.4 give

$$\int_{\mathbb{R}^d} \eta\,\mathrm{d}x = \int_{S_{r_0}} \Big(-\frac{x}{|x|}\Big) \cdot \Big(-\frac{1}{dv_d}\frac{x}{|x|^d}\Big)\,\mathrm{d}s = \int_{S_{r_0}} \frac{1}{dv_d r_0^{d-1}}\,\mathrm{d}s = 1. \ \square$$

9.7. Dirac mass as derivatives of continuous functions

Let us decompose the Dirac mass into a sum of derivatives of order $\leq m + d + 1$ of functions $\mathcal{C}^m$ with bounded support[4].

Theorem 9.18.– *For any integer $m \geq 1$ and any real number $r > 0$, there exist functions $\zeta_\beta \in \mathcal{C}_B^m(\mathbb{R}^d)$, where $B = \{x \in \mathbb{R}^d : |x| \leq r\}$, such that*

$$\delta_0 = \sum_{\beta \in \mathbb{N}^d :\, 0 \leq |\beta| \leq m+d+1} \partial^\beta \zeta_\beta. \ \blacksquare$$

Proof. The decomposition in Theorem 9.19, below, with $k = m + d + 1$ possesses the required properties, since $\eta_k \in \mathcal{C}_B^\infty(\mathbb{R}^d)$ and since $\gamma_\beta \in \mathcal{C}_B^m(\mathbb{R}^d)$ according to Theorem 9.20. $\square$

Utility of decomposing Dirac mass into a sum of derivatives. The decomposition of Theorem 9.18 will play an essential role in the proof of the surjectivity of the separation of variables of a distribution, given in the *kernel theorem* (Theorem 15.10), via the *tensorial control theorem* (Theorem 15.5). $\square$

More precisely, we have the following decomposition into a sum of derivatives of order $m + d + 1$, up to a regular remainder η_k.

4. **History of the Dirac mass decomposition.** Laurent SCHWARTZ established the decomposition of Theorem 9.18 in 1957 [72, Lemma, p. 86].

Theorem 9.19.– *For any integer $k \geq 1$ and any real number $r > 0$, there exists a decomposition*

$$\delta_0 = \eta_k + \sum_{\beta \in \mathbb{N}^d : |\beta| = k} \partial^\beta \gamma_\beta$$

such that

$$\eta_k \in C_B^\infty(\mathbb{R}^d), \quad \eta_k(x) = 0 \text{ if } |x| \leq r/2 \text{ or } |x| \geq r,$$

$$\gamma_\beta \in \mathcal{D}'_B(\mathbb{R}^d), \quad \gamma_\beta = \theta \chi_\beta,$$

where $B = \{x \in \mathbb{R}^d : |x| \leq r\}$, θ is an (arbitrary) function such that

$$\theta \in C_B^\infty(\mathbb{R}^d), \quad \theta(x) = 1 \text{ if } |x| \leq r/2, \quad \theta(x) = 0 \text{ if } |x| \geq r,$$

and χ_β is the distribution associated (by Theorem 9.5) with the function defined for $x \neq 0$ by:

— *for k even,*

$$\chi_\beta(x) = \begin{cases} c_\beta |x|^{k-d} & \text{if } d \text{ is odd or if } d > k, \\ c_\beta |x|^{k-d} \log |x| & \text{if } d \text{ is even and } \leq k; \end{cases}$$

— *for k odd,*

$$\chi_\beta(x) = \begin{cases} \sum_{1 \leq i \leq d} c_{\beta,i} x_i |x|^{k-1-d} & \text{if } d \text{ is odd or if } d > k, \\ \sum_{1 \leq i \leq d} c_{\beta,i} x_i |x|^{k-1-d} (1 + (k+1-d) \log |x|) & \text{if } d \text{ is even and } \leq k, \end{cases}$$

where $c_\beta \in \mathbb{R}$ and $c_{\beta,i} \in \mathbb{R}$. ▮

Moreover, we have the following continuity property.

Theorem 9.20.– *If $k \geq d+1$, then, for any decomposition given by Theorem 9.19,*

$$\gamma_\beta \in C_B^{k-1-d}(\mathbb{R}^d). \text{ ▮}$$

Some special cases of the Dirac mass decomposition. Theorem 9.19 gives the following decompositions.

— If $d = 1$, for every $k \geq 1$, we have $\delta_0 = \eta_k + \partial^k(\theta \chi_k)$, where $\chi_1(x) = c_1 x/|x|$, $\chi_2(x) = c_2|x|$, $\chi_3(x) = c_3 x|x|\ldots$ The function χ_1 is not continuous at $x = 0$; χ_2, χ_3 and the following ones are continuous.

In dimension $d = 2$, the term varies with k.

— If $k = 1$, then $\delta_0 = \eta_1 + \partial_1(\theta \chi_{1,0}) + \partial_2(\theta \chi_{0,1})$, where $\chi_\beta(x) = (c_{\beta,1} x_1 + c_{\beta,2} x_2)/|x|^2$.

— If $k = 2$, then $\delta_0 = \eta_2 + \partial_1^2(\theta \chi_{2,0}) + \partial_1 \partial_2(\theta \chi_{1,1}) + \partial_2^2(\theta \chi_{0,2})$, where $\chi_\beta(x) = c_\beta \log |x|$.

— If $k = 3$, then $\delta_0 = \eta_3 + \partial_1^3(\theta \chi_{3,0}) + \partial_1^2 \partial_2(\theta \chi_{2,1}) + \partial_1 \partial_2^2(\theta \chi_{1,2}) + \partial_2^3(\theta \chi_{0,3})$, where now $\chi_\beta(x) = (c_{\beta,1} x_1 + c_{\beta,2} x_2)(1 + 2 \log |x|)$.

The χ_β are not continuous at $x = 0$ if $k = 1$ or 2; they are continuous if $k \geq 3$. □

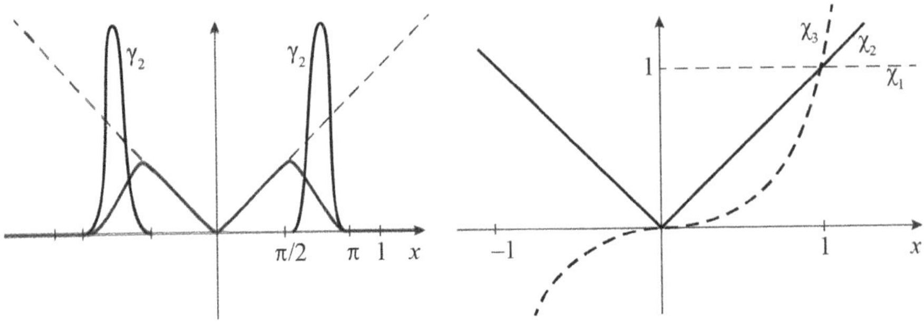

Figure 9.1. *Graph of the functions in the decomposition of δ_0 when $d = 1$. On the left, the graph of $\gamma_2 = \theta\chi_2$ is in bold, on the right, the graphs of the χ_i up to the constants c_i*

Proof of Theorem 9.19. **1.** *The case k even.* The Newtonian potential of order $n = k/2$ constructed in Theorem 9.13 satisfies

$$\delta_0 = (-\Delta)^{k/2}\xi_{k/2} = \frac{1}{\kappa_{k/2,d}}(-\Delta)^{k/2}X_k, \tag{9.34}$$

where X_k is the distribution associated by Theorem 9.5 with the function defined for $x \neq 0$ by

$$X_k(x) = \begin{cases} |x|^{k-d} & \text{if } d \text{ is odd or } d > k, \\ |x|^{k-d}\log|x| & \text{if } d \text{ is even, } d \leq k. \end{cases}$$

However, $\Delta^{k/2} = (\partial_1^2 + \cdots + \partial_d^2)^n$ is a homogeneous polynomial of degree $2n = k$ in the ∂_i, that is

$$\Delta^{k/2} = \sum_{\beta\in\mathbb{N}^d:|\beta|=k} a_\beta \partial^\beta,$$

where $a_\beta \in \mathbb{R}$. Multiplying the two terms of equality (9.34) by θ, it then follows, since $\theta\delta_0 = \delta_0$ (by Definition 5.15 of the product of a distribution with an infinitely differentiable function) and denoting $c_\beta = (-1)^{k/2}a_\beta/\kappa_{k/2,d}$, that

$$\delta_0 = \sum_{|\beta|=k} c_\beta \theta\partial^\beta X_k. \tag{9.35}$$

According to the dual Leibniz formula (Theorem A.61),

$$\theta\partial^\beta X_k = \sum_{0\leq\sigma\leq\beta} (-1)^{|\beta-\sigma|} C_\beta^\sigma \partial^\sigma (X_k\partial^{\beta-\sigma}\theta) = \partial^\beta(\theta X_k) + G_\beta, \tag{9.36}$$

where

$$G_\beta = \sum_{0\leq\sigma<\beta} (-1)^{|\beta-\sigma|} C_\beta^\sigma \partial^\sigma (X_k\partial^{\beta-\sigma}\theta).$$

In this expression, each function $\partial^{\beta-\sigma}\theta$ is zero for $|x| \le r/2$ (since then $\theta(x) = 1$) and for $|x| \ge r$ (since then $\theta(x) = 0$) and is infinitely differentiable on all of $\mathbb{R}^d$. So, the same applies to $X_k\partial^{\beta-\sigma}\theta$ and its derivatives, hence

$$G_\beta \in \mathcal{D}(\mathbb{R}^d), \quad G_\beta(x) = 0 \text{ if } |x| \le r/2 \text{ or if } |x| \ge r.$$

Thus, (9.35) and (9.36) give

$$\delta_0 = \eta_k + \sum_{|\beta|=k} \partial^\beta(c_\beta \theta X_k), \tag{9.37}$$

where $\eta_k = \sum_{|\beta|=k} c_\beta G_\beta$. This is the stated decomposition, with $\gamma_\beta = c_\beta\theta X_k$.

2. *The case k odd.* Decomposition (9.37) for $k+1$, which is even, can be written, denoting by $\mathbf{e}_i$ the multi-index whose components are all zero except the i-th which is 1,

$$\delta_0 = \eta_{k+1} + \sum_{|\beta|=k} \sum_{1\le i\le d} \partial^\beta\partial_i(c_{\beta+\mathbf{e}_i}\theta X_{k+1}).$$

With Leibniz's formula $\partial_i(\theta X_{k+1}) = \theta\partial_i X_{k+1} + X_{k+1}\partial_i\theta$ (Theorem A.60), it turns

$$\delta_0 = \eta_k + \sum_{|\beta|=k} \partial^\beta(\theta\chi_\beta),$$

where

$$\chi_\beta = \sum_{1\le i\le d} c_{\beta+\mathbf{e}_i}\partial_i X_{k+1}, \quad \eta_k = \eta_{k+1} + \sum_{|\beta|=k}\sum_{1\le i\le d} c_{\beta+\mathbf{e}_i} X_{k+1}\partial_i\theta.$$

This is the stated decomposition since, due to Lemmas 9.11 (a) and 9.14 (a),

$$\partial_i X_{k+1} = \begin{cases} (k+1-d)x_i|x|^{k-1-d} & \text{if } d \text{ is odd or } d > k, \\ x_i|x|^{k-1-d}(1+(k+1-d)\log|x|) & \text{if } d \text{ is even, } d \le k. \end{cases}$$

Observe now that η_k is an infinitely differentiable function on all of $\mathbb{R}^d$, since:
— η_{k+1} is by step 1, since $k+1$ is even;
— $X_{k+1}\partial_i\theta$ is too, since X_{k+1} is for $x \ne 0$ (Lemmas 9.11 and 9.14) and $\partial_i\theta$ is zero for $|x| \le r/2$. □

Proof of Theorem 9.20. This is a question of checking that, when $k \ge d+1$,

the distributions $\gamma_\beta = \theta\chi_\beta$ belong to $\mathcal{C}_B^{k-d-1}(\mathbb{R}^d)$.

This regularity is satisfied, because χ_β belongs to $\mathcal{C}^{k-d-1}(\mathbb{R}^d)$ since its four possible expressions (depending on the value of d and k) given by Theorem 9.19 belong to it by Lemma 9.16. Hence its product with the infinitely differentiable function θ belongs to it as well (Theorem A.60).

The condition on support is also satisfied, since the support of the product $\theta\chi_\beta$ is included in that of θ which is included in B by construction (in Theorem 9.19). □

9.8. Heaviside potential

Let us define a potential H_m solution of $\partial^\beta H_m = \delta_0$ for a suitable β, which is a primitive of the Heaviside function [5] extended to d-dimensional space.

Theorem 9.21.– *For $m \in \mathbb{N}$, $m \geq 1$, we define a function $H_m \in \mathcal{C}^{m-1}(\mathbb{R}^d)$ by*

$$H_m(x) = \begin{cases} \dfrac{1}{(m!)^d}(x_1 x_2 \dots x_d)^m & \text{if all the } x_i \text{ are} \geq 0, \\ 0 & \text{otherwise.} \end{cases}$$

It satisfies

$$(\partial_1 \partial_2 \dots \partial_d)^{m+1} H_m = \delta_0. \blacksquare$$

Dirac mass as a sum of derivatives. The Dirac mass δ_0 was decomposed in Theorem 9.18 into a sum of derivatives of order $\leq m + d + 1$ of functions in $\mathcal{C}^m(\mathbb{R}^d)$ with bounded support.

The order of the derivative obtained here to represent δ_0 as the derivative of a single function in $\mathcal{C}^m(\mathbb{R}^d)$, namely H_{m+1}, is higher since it equals $(m+2)d$. $\square$

Proof of Theorem 9.21. **1.** *Outlines.* We will calculate the derivatives in two steps. Firstly,

$$(\partial_1 \partial_2 \dots \partial_d)^m H_m = H_0, \tag{9.38}$$

where $H_0 \in \mathcal{D}'(\mathbb{R}^d)$ is defined by

$$\langle H_0, \varphi \rangle \overset{\text{def}}{=} \int_{(\mathbb{R}_+)^d} \varphi \tag{9.39}$$

and $(\mathbb{R}_+)^d \overset{\text{def}}{=} \{x \in \mathbb{R}^d : x_1 > 0, \dots, x_d > 0\}$. Next,

$$\partial_1 \partial_2 \dots \partial_d H_0 = \delta_0. \tag{9.40}$$

2. *Proof of (9.38).* For $i \in [\![0, d]\!]$ and $\varphi \in \mathcal{D}(\mathbb{R}^d)$, denote

$$\langle H_{m,i}, \varphi \rangle \overset{\text{def}}{=} c_{m,i} \int_{(\mathbb{R}_+)^d} x_1^m \dots x_i^m x_{i+1}^{m-1} \dots x_d^{m-1} \varphi(x)\, \mathrm{d}x, \tag{9.41}$$

where $c_{m,i} = 1/(m^i((m-1)!)^d)$.

5. **History of Heaviside's function.** Oliver HEAVISIDE introduced in 1894, in a treatise on symbolic calculus [41] not well justified albeit effective, the function H on $\mathbb{R}$ of value 0 for $x < 0$ and 1 for $x \geq 0$.

Observe that

$$H_{m,d} = H_m. \tag{9.42}$$

Indeed, the identification of a continuous function, here H_m, with a distribution from Theorem 3.9 is written, since the domain where the function is zero does not contribute to the integral (Theorem A.77), as

$$\langle H_m, \varphi\rangle = c_{m,d} \int_{\mathbb{R}^d} x_1^m \dots x_d^m \varphi(x)\, \mathrm{d}x =$$
$$= c_{m,d} \int_{(\mathbb{R}_+)^d} x_1^m \dots x_d^m \varphi(x)\, \mathrm{d}x = \langle H_{m,d}, \varphi\rangle.$$

Assume for the moment that, for $i \geq 1$,

$$\langle H_{m,i-1}, \varphi\rangle = -\langle H_{m,i}, \partial_i \varphi\rangle. \tag{9.43}$$

It will follow that, if $H_{m,i} \in \mathcal{D}'(\mathbb{R}^d)$, then, by Definition 5.4 of the derivative of a distribution, $H_{m,i-1} \in \mathcal{D}'(\mathbb{R}^d)$ and

$$\partial_i H_{m,i} = H_{m,i-1}.$$

Since $H_{m,d} \in \mathcal{D}'(\mathbb{R}^d)$ (because it coincides with the continuous function H_m due to (9.42)), it will inductively follow that $H_{m,i} \in \mathcal{D}'(\mathbb{R}^d)$ for each i and

$$\partial_1 \partial_2 \dots \partial_d H_{m,d} = \partial_1 \partial_2 \dots \partial_{d-1} H_{m,d-1} = \cdots = \partial_1 H_{m,1} = H_{m,0}. \tag{9.44}$$

However, from Definition (9.41), $H_{m,0} = H_{m-1,d}$ (since $c_{m,0} = c_{m-1,d}$), it will then follow that

$$\partial_1 \partial_2 \dots \partial_d H_{m,d} = H_{m-1,d}.$$

Using this equality $m - 1$ times, and then (9.44) once, we will obtain

$$(\partial_1 \partial_2 \dots \partial_d)^m H_{m,d} = \partial_1 \partial_2 \dots \partial_d H_{1,d} = H_{1,0}.$$

Which establishes (9.38), since $H_{m,d} = H_m$ (from (9.42)) and $H_{1,0} = H_0$ (from (9.41) and (9.39)). Subject to proving (9.43), which we will do in step 4.

3. *Proof of (9.40).* Let $\varphi \in \mathcal{D}(\mathbb{R}^d)$. Denote, for $i \in [\![1, d]\!]$,

$$\langle h_i, \varphi\rangle \stackrel{\text{def}}{=} \int_{(\mathbb{R}_+)^i} \varphi(x_1, \dots, x_i, 0, \dots, 0)\, \mathrm{d}x_1 \cdots \mathrm{d}x_i, \tag{9.45}$$

and, for $i = 0$,

$$\langle h_0, \varphi\rangle \stackrel{\text{def}}{=} \varphi(0, \dots, 0). \tag{9.46}$$

Assume for the moment that, for $i \geq 1$,

$$\langle h_{i-1}, \varphi\rangle = -\langle h_i, \partial_i \varphi\rangle. \tag{9.47}$$

It will follow that, if $h_i \in \mathcal{D}'(\mathbb{R}^d)$, then, again by Definition 5.4 of the derivative of a distribution, $h_{i-1} \in \mathcal{D}'(\mathbb{R}^d)$ and

$$\partial_i h_i = h_{i-1}.$$

Since $h_d = H_0$ (from their Definitions (9.45) and (9.39)), it is a distribution, and hence it will then follow inductively that

$$\partial_1 \partial_2 \dots \partial_d h_d = \partial_1 \partial_2 \dots \partial_{d-1} h_{d-1} = \dots = \partial_1 h_1 = h_0.$$

Which establishes (9.40), since $h_0 = \delta_0$ (from their Definitions, namely equality (9.46) and Definition 3.14). From which, with (9.38),

$$(\partial_1 \partial_2 \dots \partial_d)^{m+1} H_m = \partial_1 \partial_2 \dots \partial_d H_0 = \delta_0. \tag{9.48}$$

Subject to proving (9.47), which we will do in step 5.

4. *Proof of (9.43).* From Definition (9.41) of $H_{m,i}$, it is a question of showing, since $c_{m,i-1} = m c_{m,i}$, that

$$c_{m,i} \int_{(\mathbb{R}_+)^d} x_1^m \cdots x_{i-1}^m \big(m x_i^{m-1} \varphi(x) + x_i^m \partial_i \varphi(x)\big) x_{i+1}^{m-1} \cdots x_d^{m-1} \, \mathrm{d}x = 0. \tag{9.49}$$

According to Leibniz's formula (Theorem A.60),

$$m x_i^{m-1} \varphi(x) + x_i^m \partial_i \varphi(x) = \partial_i (x_i^m \varphi(x)),$$

thus (9.49) is written, after separating and permuting the variables (Theorem A.89, which applies here since the integrated function is uniformly continuous with compact support), as

$$c_{m,i} \int_{(\mathbb{R}_+)^{d-1}} x_1^m \cdots x_{i-1}^m x_{i+1}^{m-1} \cdots x_d^{m-1} \, \mathrm{d}\widehat{x}_i \int_{\mathbb{R}+} \partial_i (x_i^m \varphi(x)) \, \mathrm{d}x_i = 0, \tag{9.50}$$

where $\mathrm{d}\widehat{x}_i = \mathrm{d}x_1 \dots \mathrm{d}x_{i-1} \mathrm{d}x_{i+1} \cdots \mathrm{d}x_d$.

Let b be such that $\varphi(x) = 0$ if $x_i \geq b$. Since the domain where the function is zero does not contribute to the integral (Theorem A.77), the expression for the integral of a derivative from Theorem A.79 gives

$$\int_{\mathbb{R}+} \partial_i (x_i^m \varphi(x)) \, \mathrm{d}x_i = \int_0^b \partial_i (x_i^m \varphi(x)) \, \mathrm{d}x_i = b^m \varphi(x)|_{x_i = b} - 0^m \varphi(x)|_{x_i = 0} = 0.$$

Which implies (9.50), that is (9.49), and hence (9.43).

5. *Proof of (9.47).* From Definition (9.45) of h_i, it is a question of calculating

$$\langle h_i, \partial_i \varphi \rangle = \int_{(\mathbb{R}_+)^i} \partial_i \varphi(x_1, \dots, x_i, 0, \dots, 0) \, \mathrm{d}x_1 \cdots \mathrm{d}x_i.$$

By separating the variables (Theorem A.89, again), it follows that

$$\langle h_i, \partial_i \varphi \rangle = \int_{(\mathbb{R}_+)^{i-1}} \mathrm{d}x_1 \cdots \mathrm{d}x_{i-1} \int_{\mathbb{R}_+} \partial_i \varphi(x_1, \ldots, x_i, 0, \ldots, 0)\, \mathrm{d}x_i. \qquad (9.51)$$

The last integral becomes, by restricting it to $(0, b)$ for b such that $\varphi(x) = 0$ if $x_i \geq b$ and using the expression for the integral of a derivative (Theorem A.79, again),

$$\int_{\mathbb{R}_+} \partial_i \varphi(x_1, \ldots, x_i, 0, \ldots, 0)\, \mathrm{d}x_i = -\varphi(x_1, \ldots, x_{i-1}, 0, \ldots, 0).$$

The right-hand side of (9.51) is therefore equal to

$$-\int_{(\mathbb{R}_+)^{i-1}} \varphi(x_1, \ldots, x_{i-1}, 0, \ldots, 0)\, \mathrm{d}x_1 \cdots \mathrm{d}x_{i-1} = -\langle h_{i-1}, \varphi \rangle.$$

This proves (9.47) and hence finishes the proof of (9.48).

6. *Regularity of H_m.* For every $|\beta| \leq m - 1$, H_m possesses a partial derivative in the function sense, which equals

$$\partial^\beta H_m(x) = \begin{cases} c_\beta x_1^{m-\beta_1} \ldots x_d^{m-\beta_d} & \text{on } (\mathbb{R}_+)^d, \\ 0 & \text{outside,} \end{cases}$$

where $c_\beta - 1/((m - \beta_1!) \ldots (m - \beta_d)!)$.

These partial derivatives are continuous on all of $\mathbb{R}^d$, for every $|\beta| \leq m - 1$, which implies (Theorem A.55) that H_m is $m - 1$ times continuously differentiable, i.e.

$$H_m \in \mathcal{C}^{m-1}(\mathbb{R}^d). \ \square$$

9.9. Weighting by a singular weight

Let us show that the weighting by a weight μ with compact support which, in the neighbourhood of 0, grows less rapidly than $|x|^{-\lambda}$ with $\lambda < d$ preserves the continuity of functions.

Theorem 9.22.– *Let*

$$f \in \mathcal{C}(\Omega; E) \ \textit{and} \ \mu \in \mathcal{C}_D(\mathbb{R}^d \setminus \{0\}),$$

where Ω is an open subset of $\mathbb{R}^d$, E is a Neumann space, and D is a compact subset of $\mathbb{R}^d$. Suppose that there exist $b < \infty$ and λ such that, for every $x \neq 0$,

$$|\mu(x)| \leq \frac{b}{|x|^\lambda}, \quad \lambda < d. \qquad (9.52)$$

Then, identifying μ with a distribution in $\mathcal{D}'_D(\mathbb{R}^d)$ by Theorem 9.6,

$$f \diamond \mu \in \mathcal{C}(\Omega_D; E). \ \blacksquare$$

The proof uses the following approximation of the weight μ by continuous functions, which we will assume for a moment.

Lemma 9.23.– *Let μ satisfy the hypotheses of Theorem 9.22 and $(\alpha_n)_{n\geq 1}$ be a sequence of functions in $\mathcal{C}(\mathbb{R}^d)$ such that:*

$$0 \leq \alpha_n \leq 1, \quad \alpha_n(x) = 0 \textit{ if } |x| \leq 1/n, \quad \alpha_n(x) = 1 \textit{ if } |x| \geq 2/n.$$

Once extended to 0 by 0, the function $\alpha_n \mu$ belongs to $\mathcal{C}(\mathbb{R}^d)$ and, when $n \to \infty$,

$$\alpha_n \mu \to \mu \textit{ in } \mathcal{D}'_D(\mathbb{R}^d). \blacksquare$$

Proof of Theorem 9.22. **1.** *Notation.* To avoid any confusion, we denote by $\overline{\mu}$ the distribution associated with the singular function μ by Theorem 9.5, and we denote by $\widetilde{\alpha_n \mu}$ the extension of $\alpha_n \mu$ by 0.

2. *Method.* Let us approach $\overline{\mu}$ by the continuous functions $\widetilde{\alpha_n \mu}$ according to Lemma 9.23. Since weighting is sequentially continuous from $\mathcal{D}'(\Omega; E) \times \mathcal{D}'_D(\mathbb{R}^d)$ into $\mathcal{D}'(\Omega_D; E)$ (Theorem 7.15 (a)),

$$f \diamond \widetilde{\alpha_n \mu} \to f \diamond \overline{\mu} \text{ in } \mathcal{D}'(\Omega_D; E). \tag{9.53}$$

We are going to show that the sequence $(f \diamond \widetilde{\alpha_n \mu})_{n\in\mathbb{N}}$ is Cauchy in $\mathcal{C}(\Omega_D; E)$, which will imply that its limit $f \diamond \overline{\mu}$ belongs to this space.

3. *An equality.* By Definition 7.3 of the weighting of continuous functions, for every $x \in \Omega_D$,

$$(f \diamond \widetilde{\alpha_n \mu})(x) = \int_{\mathring{D}} f(x+y)\, \widetilde{\alpha_n \mu}(y)\, dy.$$

Let $m \geq n$. Since the domain where the function is zero does not contribute to the integral (Theorem A.77) and since $\alpha_n - \alpha_m$ is zero outside of the crown $\{y \in \mathbb{R}^d : 1/m < |y| < 2/n\}$, which we denote by $C_{1/m,2/n}$, we have

$$(f \diamond \widetilde{\alpha_n \mu} - f \diamond \widetilde{\alpha_m \mu})(x) = \int_{\mathring{D} \cap C_{1/m,2/n}} f(x+y)\, (\alpha_n - \alpha_m)(y)\, \mu(y)\, dy.$$

4. *Cauchy property.* Let $\| \ \|_{E;\nu}$ be a semi-norm of E. The upper bound of the semi-norms of the integral from Theorem 1.23 (a), the growth of the real integral (Theorem A.76 (a) and (b)) and its linearity (Theorem A.74) give, with hypothesis (9.52) and $|\alpha_n - \alpha_m| \leq 2$,

$$\begin{aligned}
&\|(f \diamond \widetilde{\alpha_n \mu} - f \diamond \widetilde{\alpha_m \mu})(x)\|_{E;\nu} \leq \\
&\qquad \leq \int_{\mathring{D} \cap C_{1/m,2/n}} \|f(x+y)\, (\alpha_n - \alpha_m)(y)\, \mu(y)\, dy\|_{E;\nu} \leq \\
&\qquad\qquad \leq \sup_{y \in \mathring{D}} \|f(x+y)\|_{E;\nu} \int_{C_{1/m,2/n}} \frac{2b}{|y|^\lambda}\, dy.
\end{aligned}$$

With the calculation of the integral of powers on a crown (Theorem A.81), it follows

$$\|(f \diamond \widetilde{\alpha_n \mu} - f \diamond \widetilde{\alpha_m \mu})(x)\|_{E;\nu} \leq \frac{2bdv_d}{d-\lambda}\left(\frac{2}{n}\right)^{d-\lambda} \sup_{y \in D} \|f(x+y)\|_{E;\nu}.$$

Definition 1.18 (a) of the semi-norms of $\mathcal{C}(\Omega; E)$ therefore gives, for every compact subset K of Ω_D,

$$\|f \diamond \widetilde{\alpha_n \mu} - f \diamond \widetilde{\alpha_m \mu}\|_{\mathcal{C}(\Omega_D;E);K,\nu} = \sup_{x \in K} \|(f \diamond \widetilde{\alpha_n \mu} - f \diamond \widetilde{\alpha_m \mu})(x)\|_{E;\nu} \leq \frac{c}{n^{d-\lambda}},$$

where

$$c = \frac{2bdv_d}{(d-\lambda)} 2^{d-\lambda} \sup_{x \in K+D} \|f(x)\|_{E;\nu} < \infty.$$

Observe that c is finite because $K + D$ is compact, as is any sum of compact subsets of $\mathbb{R}^d$ (Theorem A.27), and hence f is bounded on K, as is every continuous function on a compact set (Theorem A.32).

5. *Conclusion.* The sequence $(f \diamond \widetilde{\alpha_n \mu})_{n \in \mathbb{N}}$ is therefore Cauchy (Definition 1.3 (b)) in $\mathcal{C}(\Omega_D; E)$. It has a limit in this space, since it is sequentially complete (Theorem A.46). The limit is the same in $\mathcal{D}'(\Omega_D; E)$ (since $\mathcal{C}(\Omega_D; E)$ is topologically included in it, due to Theorem 3.8), namely $f \diamond \overline{\mu}$ from (9.53). Hence,

$$f \diamond \overline{\mu} \in \mathcal{C}(\Omega_D; E). \ \square$$

It remains to prove that $\alpha_n \mu$ converges to μ, that we had assumed to obtain (9.53).

Proof of Lemma 9.23. We again denote by $\widetilde{\alpha_n \mu}$ the extension of $\alpha_n \mu$ by 0.

1. *Convergence in $\mathcal{D}'(\mathbb{R}^d)$.* Let $\varphi \in \mathcal{D}(\mathbb{R}^d)$ and $\epsilon \leq 1/n$. Since the domain where the function is zero does not contribute to the the integral (Theorem A.77) and α_n is zero for $|x| \leq 1/n$,

$$\int_{\mathbb{R}^d} \widetilde{\alpha_n \mu} \varphi = \int_{\{x \in \mathbb{R}^d : |x| > \epsilon\}} \alpha_n \mu \varphi.$$

Therefore, again by Theorem A.77, since $\alpha_n = 1$ if $|x| \geq 2/n$,

$$\int_{\mathbb{R}^d} \widetilde{\alpha_n \mu} \varphi - \int_{\{x \in \mathbb{R}^d : |x| > \epsilon\}} \mu \varphi = \int_{\{x \in \mathbb{R}^d : |x| > \epsilon\}} (\alpha_n - 1) \mu \varphi = \int_{C_{\epsilon,2/n}} (\alpha_n - 1) \mu \varphi,$$

With hypothesis (9.52) on μ, the upper bound of the absolute value of the real integral, its growth and its linearity (Theorems 1.23 (a), A.76 (a) and A.74) give

$$\left| \int_{\mathbb{R}^d} \widetilde{\alpha_n \mu} \varphi - \int_{\{x \in \mathbb{R}^d : |x| > \epsilon\}} \mu \varphi \right| \leq b \sup_{x \in \mathbb{R}^d} |\varphi(x)| \int_{C_{\epsilon,2/n}} \frac{1}{|x|^\lambda} \, dx.$$

The calculation of the integral of powers on a crown (Theorem A.81) therefore gives

$$\left|\int_{\mathbb{R}^d} \widetilde{\alpha_n \mu} \varphi - \int_{\{x\in\mathbb{R}^d : |x|>\epsilon\}} \mu\varphi\right| \leq \frac{bdv_d}{d-\lambda} 2^{d-\lambda} \sup_{x\in\mathbb{R}^d} |\varphi(x)| \frac{1}{n^{d-\lambda}}.$$

Passing to the limit when $\epsilon \to 0$, it follows, denoting by $\overline{\widetilde{\alpha_n \mu}}$ and $\overline{\mu}$ the distributions associated with $\widetilde{\alpha_n \mu}$ and μ respectively by Theorems 3.5 and 9.5,

$$|\langle \overline{\widetilde{\alpha_n \mu}} - \overline{\mu}, \varphi \rangle| \leq \frac{bdv_d}{d-\lambda} 2^{d-\lambda} \sup_{x\in\mathbb{R}^d} |\varphi(x)| \frac{1}{n^{d-\lambda}}.$$

When $n \to \infty$, the right-hand side tends to 0. This holds for every $\varphi \in \mathcal{D}(\Omega)$, thus

$$\overline{\widetilde{\alpha_n \mu}} \to \overline{\mu} \text{ in } \mathcal{D}'(\mathbb{R}^d). \tag{9.54}$$

2. *Convergence in* $\mathcal{D}'_D(\mathbb{R}^d)$. Since the support of the distribution associated with a continuous function coincides with that of the function (Theorem 6.19), the support of $\overline{\widetilde{\alpha_n \mu}}$ is included in D, i.e.

$$\overline{\widetilde{\alpha_n \mu}} \in \mathcal{D}'_D(\mathbb{R}^d).$$

Since $\mathcal{D}'_D(\mathbb{R}^d)$ is sequentially closed in $\mathcal{D}'(\mathbb{R}^d)$ (Theorem 6.27), the limit $\overline{\mu}$ of $\overline{\widetilde{\alpha_n \mu}}$ belongs to it as well. Since $\mathcal{D}'_D(\mathbb{R}^d)$ is (by Definition 6.26) endowed with the semi-norms of $\mathcal{D}'(\mathbb{R}^d)$, convergence (9.54) therefore gives

$$\overline{\widetilde{\alpha_n \mu}} \to \overline{\mu} \text{ in } \mathcal{D}'_D(\mathbb{R}^d). \ \Box$$

A complement to Theorem 9.22. The mapping $f \mapsto f \diamond \mu$ is continuous from $\mathcal{C}(\Omega; E)$ into $\mathcal{C}_D(\Omega; E)$. This can be proven by passing to the limit when $m \to \infty$ in the last inequality in the proof of Theorem 9.22. □

An expression for $f \diamond \mu$**.** Under the hypotheses of Theorem 9.22, for every $x \in \Omega_D$,

$$(f \diamond \mu)(x) = \lim_{\epsilon \to 0} \int_{\{y\in \mathring{D}:\, |y|>\epsilon\}} f(x+y)\, \mu(y)\, dy.$$

Moreover, the mapping $y \mapsto f(x+y)\mu(y)$ is an *integrable distribution* on $\mathring{D}$, a notion that we will introduce in Volume 4 and which will allow us to write (which will generalize Definition 7.3 of the weighting of continuous functions)

$$(f \diamond \mu)(x) = \int_{\mathring{D}} f(x+y)\, \mu(y)\, dy. \tag{9.55}$$

□

Chapter 10

Line Integral of a Continuous Field

This chapter is dedicated to line integrals of continuous vector fields along a path, which is an essential tool for constructing primitives. Some classical properties for fields with values in a Banach space are extended here to values in a Neumann space.

After having defined the line integral $\oint_\Gamma q \cdot d\ell \stackrel{\text{def}}{=} \int_{t_i}^{t_e} (q \circ \Gamma) \cdot \Gamma' \, dt$ of a field $q = (q_1, \dots, q_d)$ along a path $\Gamma \in C^1([t_i, t_e]; \mathbb{R}^d)$, we give the following properties.

— Line integrals concatenate, namely $\oint_\Gamma = \sum_n \oint_{\Gamma_n}$ if $\Gamma = \vec{\cup}_n \Gamma_n$ (Theorems 10.13 and 10.15).

— Every concatenation of C^1 paths, namely every piecewise C^1 path, can be reparameterized into a C^1 path (Theorem 10.16), without changing the line integral (Theorem 10.17).

— The line integral of a gradient along a closed path Γ is zero, that is $\oint_\Gamma \nabla f \cdot d\ell = 0$ (Theorem 10.8).

— The line integral of a local gradient, namely of a field q that is of the form $q = \nabla f_B$ on every ball B, along a closed path Γ is homotopy invariant (Theorem 10.19). This is a key point in the proof of the existence of primitives on a simply connected set, given in Theorem 13.7.

We finish this chapter with some definitions and properties related to connectedness.

10.1. Line integral along a C^1 path

Since the line integral had not yet been defined for fields with values in a Neumann space, we did it in Volume 2. Let us recall the points of this construction that we will use. We begin with the definition of paths.

Definition 10.1.– *A* ***path*** *in a subset Ω of $\mathbb{R}^d$ is a mapping $\Gamma \in C([t_i, t_e]; \Omega)$, where $[t_i, t_e]$ is a closed and bounded interval in $\mathbb{R}$. We say that Γ* ***joins in*** *Ω the* ***initial point*** *$\Gamma(t_i)$ to the* ***ending point*** *$\Gamma(t_e)$. The* ***image*** *of Γ is the set $[\Gamma] \stackrel{\text{def}}{=} \{\Gamma(t) : t_i \le t \le t_e\}$.*

A ***closed path*** *is a path whose initial and ending points coincide.*

We say that a path ***is*** *C^1, or* ***of class*** *C^1, if it belongs to $C^1([t_i, t_e]; \Omega)$.* ∎

We denote

$$\mathcal{C}^1([a,b];\Omega) \stackrel{\text{def}}{=} \left\{ \begin{array}{l} \Gamma \in \mathcal{C}([a,b];\Omega) : \\ \qquad \exists \Gamma' \in \mathcal{C}([a,b];\mathbb{R}^d),\ \Gamma' = \mathrm{d}\Gamma/\mathrm{d}t \text{ in } (a,b) \end{array} \right\}. \tag{10.1}$$

In other words, $\mathcal{C}^1([a,b];\Omega)$ is the set of functions Γ from $[a,b]$ into Ω which are continuous, whose restriction is continuously differentiable from (a,b) into $\mathbb{R}^d$, and whose derivative Γ' (which is *a priori* only defined on the open interval (a,b)) has a continuous extension to $[a,b]$, again denoted by Γ'.

Geometry. The image $[\Gamma]$ of a $\mathcal{C}^1$ path is not necessarily a regular curve or a one-dimensional manifold. For example, it can be just a point, overlap itself, have angles or turning points, etc. □

Let us now define the line integral [1] of a continuous field along a $\mathcal{C}^1$ path.

Definition 10.2.– *Let $q \in \mathcal{C}(\Omega; E^d)$, where $\Omega \subset \mathbb{R}^d$ and E is a Neumann space, and $\Gamma \in \mathcal{C}^1([t_\mathrm{i}, t_\mathrm{e}];\Omega)$ be a path in Ω. We denote by Γ' the derivative of Γ.*

*The **line integral** of q along Γ is the element of E defined by*

$$\int_\Gamma q \cdot \mathrm{d}\ell \stackrel{\text{def}}{=} \int_{t_\mathrm{i}}^{t_\mathrm{e}} (q \circ \Gamma) \cdot \Gamma' \,\mathrm{d}t. \ \blacksquare$$

Inconsistent notation! To be consistent with our notation of the Cauchy integral, we should write here either $\int_{t_\mathrm{i}}^{t_\mathrm{e}} (q \circ \Gamma)(t) \cdot \Gamma'(t)\,\mathrm{d}t$ or $\int_{t_\mathrm{i}}^{t_\mathrm{e}} (q \circ \Gamma) \cdot \Gamma'$. We add $\mathrm{d}t$ to the latter to mirror the usage of $\mathrm{d}\ell$. □

Justification of Definition 10.2. For the right-hand side to be defined it is necessary, by Definition 1.22 of the Cauchy integral with values in a Neumann space, that

$$\text{the function } (q \circ \Gamma) \cdot \Gamma' \text{ is uniformly continuous on } (t_\mathrm{i}, t_\mathrm{e}). \tag{10.2}$$

1. **History of the line integral of a field along a path.** The line integral of a vector field along a path was introduced by Gaspard-Gustave DE CORIOLIS in 1829 [23] to express the work of a force, i.e. the variation of the kinetic energy of a body that moves under the action of this force.
This notion was developed as part of the theory of differential forms established around 1890–1900 by Émile CARTAN [16, Vol. II, pp. 309–396] and Henri POINCARÉ [61, Vol. III, Chap. XXII]; see, for example, [CARTAN, Henri, 17, pp. 215–219] (the son of Émile whom we mentioned above), where the results given here can be found with the field q "hidden" behind the 0-form ω and the gradient ∇f "hidden" behind the 1-form ω or $\mathrm{d}g$.

French terminology. In French, the line integral is called the *circulation*, a term that English-speakers reserve for the case where the path is closed. The French term *intégrale curviligne*, which is the word-for-word translation of "line integral," is generally reserved for the line integral of a scalar function.

However, the composition $q \circ \Gamma$ is continuous (Theorem A.30) on $[t_\mathrm{i}, t_\mathrm{e}]$, as well as Γ' (extended, by definition of $\mathcal{C}^1([t_\mathrm{i}, t_\mathrm{e}]; \Omega)$). The product $(q \circ \Gamma) \boldsymbol{\cdot} \Gamma'$ is thus continuous too (Theorem A.30, again) since it is obtained by composing with the mapping $\boldsymbol{\cdot}$ (which is continuous from $E^d \times \mathbb{R}^d$ into E by equality (1.3), p. 8). Therefore, by Heine's theorem (Theorem A.32), $(q \circ \Gamma) \boldsymbol{\cdot} \Gamma'$ is uniformly continuous on the compact set $[t_\mathrm{i}, t_\mathrm{e}]$ and, *a fortiori*, on $(t_\mathrm{i}, t_\mathrm{e})$.

It is also necessary, again to satisfy Definition 1.22 of the integral, that $(q \circ \Gamma) \boldsymbol{\cdot} \Gamma'$ has a bounded support, which is the case since $[t_\mathrm{i}, t_\mathrm{e}]$ is bounded by Definition 10.1 of a path. □

The semi-norms of E play the role classically undertaken by the norm in the Banach space case. In particular, we have the following inequality.

Theorem 10.3.– *Let $q \in \mathcal{C}(\Omega; E^d)$, where $\Omega \subset \mathbb{R}^d$ and E is a Neumann space, Γ be a $\mathcal{C}^1$ path in Ω and $\{\| \ \|_{E;\nu} : \nu \in \mathcal{N}_E\}$ the family of semi-norms of E. Then, for every $\nu \in \mathcal{N}_E$,*

$$\left\| \int_\Gamma q \boldsymbol{\cdot} \mathrm{d}\ell \right\|_{E;\nu} \leq \gamma |t_\mathrm{e} - t_\mathrm{i}| \sup_{x \in [\Gamma]} \|q(x)\|_{E^d;\nu},$$

where $[\Gamma] = \{\Gamma(t) : t_\mathrm{i} \leq t \leq t_\mathrm{e}\}$ and $\gamma = \sup_{t_\mathrm{i} < t < t_\mathrm{e}} |\Gamma'(t)| < \infty$. ▮

Proof. Definition 10.2 of the line integral, the bound of the semi-norms of the integral from Theorem 1.23 (b) and the inequality $\|u \boldsymbol{\cdot} z\|_{E;\nu} \leq \|u\|_{E^d;\nu} |z|$ established in (1.3), p. 8, successively give

$$\begin{aligned} \left\| \int_\Gamma q \boldsymbol{\cdot} \mathrm{d}\ell \right\|_{E;\nu} &= \left\| \int_{t_\mathrm{i}}^{t_\mathrm{e}} (q \circ \Gamma) \boldsymbol{\cdot} \Gamma' \, \mathrm{d}t \right\|_{E;\nu} \leq \\ &\leq |t_\mathrm{e} - t_\mathrm{i}| \sup_{t_\mathrm{i} < t < t_\mathrm{e}} \|((q \circ \Gamma) \boldsymbol{\cdot} \Gamma')(t)\|_{E;\nu} \leq \gamma |t_\mathrm{e} - t_\mathrm{i}| \sup_{x \in [\Gamma]} \|q(x)\|_{E^d;\nu}, \end{aligned} \tag{10.3}$$

where $\gamma = \sup_{t_\mathrm{i} < t < t_\mathrm{e}} |\Gamma'(t)|$. This one is finite since, by Definition 10.1 of a $\mathcal{C}^1$ path, Γ' has a continuous extension on the compact set $[t_\mathrm{i}, t_\mathrm{e}]$ and since any continuous function on a compact set is bounded (Theorem A.32). □

Let us show that the line integral of a vector field depends continuously on the field.

Theorem 10.4.– *Let $\Omega \subset \mathbb{R}^d$, E be a Neumann space and Γ a $\mathcal{C}^1$ path in Ω.*

The mapping $q \mapsto \int_\Gamma q \boldsymbol{\cdot} \mathrm{d}\ell$ is continuous, and therefore sequentially continuous, linear from $\mathcal{C}(\Omega; E^d)$ into E. ▮

Proof. Inequality (10.3) gives, with Definition 1.18 (a) of the semi-norms of $\mathcal{C}(\Omega; E^d)$,

$$\left\| \oint_\Gamma q \cdot \mathrm{d}\ell \right\|_{E;\nu} \le c \sup_{x \in [\Gamma]} \|q(x)\|_{E^d;\nu} = c\|q\|_{\mathcal{C}(\Omega;E^d);[\Gamma],\nu}.$$

Which implies the continuity of the mapping $q \mapsto \oint_\Gamma q \cdot \mathrm{d}\ell$, by the characterization of continuous linear mappings from Theorem 1.12 (a).

This mapping is therefore sequentially continuous, as is every continuous mapping (Theorem 1.10). □

Let us calculate the line integral along a path reduced to a point.

Theorem 10.5.– *Let $q \in \mathcal{C}(\Omega; E^d)$, where $\Omega \subset \mathbb{R}^d$ and E is a Neumann space. Moreover, let $a \in \Omega$ and $\Gamma_{\{a\}}$ be the path defined on $[0, 1]$ by $\Gamma_{\{a\}}(t) = a$. Then,*

$$\oint_{\Gamma_{\{a\}}} q \cdot \mathrm{d}\ell = 0_E. \ ▮$$

Proof. By Definition 10.2 of the line integral, this follows from the equality

$$\Gamma'_{\{a\}} = \frac{\mathrm{d}a}{\mathrm{d}t} = 0. \ □$$

Let us calculate the line integral along a **rectilinear path**.

Theorem 10.6.– *Let $q \in \mathcal{C}(\Omega; E^d)$, where $\Omega \subset \mathbb{R}^d$ and E is a Neumann space. Moreover, let $a \in \Omega$ and $x \in \mathbb{R}^d$ be such that $[a, x] \subset \Omega$, and $\Gamma_{\overrightarrow{a,x}}$ the path defined on $[0, 1]$ by*

$$\Gamma_{\overrightarrow{a,x}}(t) = a + t(x - a).$$

Then,

$$\oint_{\Gamma_{\overrightarrow{a,x}}} q \cdot \mathrm{d}\ell = (x - a) \cdot \int_0^1 q(a + t(x - a))\, \mathrm{d}t. \ ▮$$

Proof. By Definition 10.2 of the line integral, this follows from the equality

$$\Gamma'_{\overrightarrow{a,x}} = \frac{\mathrm{d}(a + t(x - a))}{\mathrm{d}t} = x - a. \ □$$

Let us calculate the **line integral of a gradient**.

Theorem 10.7.– *Let $f \in C^1(\Omega; E)$, where Ω is an open subset of $\mathbb{R}^d$ and E is a Neumann space, and Γ be a C^1 path in Ω with initial point a and ending point b. Then,*

$$\int_\Gamma \nabla f \cdot \mathrm{d}\ell = f(b) - f(a). \blacksquare$$

Proof. By the expression for the change of variable in a derivative from Theorem A.64 (a) with $\ell = 1$ and $\partial_i = \mathrm{d}/\mathrm{d}t$,

$$(f \circ \Gamma)' = \sum_{j=1}^{d} (\partial_j f \circ \Gamma)\, \Gamma'_j = (\nabla f \circ \Gamma) \cdot \Gamma'.$$

Definition 10.2 of the line integral therefore gives

$$\int_\Gamma \nabla f \cdot \mathrm{d}\ell = \int_{t_\mathrm{i}}^{t_\mathrm{e}} (\nabla f \circ \Gamma) \cdot \Gamma' \,\mathrm{d}t = \int_{t_\mathrm{i}}^{t_\mathrm{e}} (f \circ \Gamma)' \,\mathrm{d}t.$$

With the expression for the integral of a derivative from Theorem A.79, it follows that

$$\int_\Gamma \nabla f \cdot \mathrm{d}\ell = (f \circ \Gamma)(t_\mathrm{e}) - (f \circ \Gamma)(t_\mathrm{i}) = f(b) - f(a). \square$$

In particular, the line integral of a gradient along a closed path is zero:

Theorem 10.8.– *Let $f \in C^1(\Omega; E)$, where Ω is an open subset of $\mathbb{R}^d$ and E is a Neumann space, and Γ be a C^1 closed path in Ω. Then,*

$$\int_\Gamma \nabla f \cdot \mathrm{d}\ell = 0_E. \blacksquare$$

Proof. This follows from Theorem 10.7 since, by Definition 10.1 of a closed path, its ending point coincides with its initial point. $\square$

10.2. Change of variable in a path

Let us define the reverse path, which exchanges the initial point with the ending point.

Definition 10.9.– *Let $\Gamma \in \mathcal{C}([t_{\mathrm{i}}, t_{\mathrm{e}}]; \mathbb{R}^d)$ be a path in $\mathbb{R}^d$. The* **reverse** *path of Γ is the path $\overleftarrow{\Gamma}$ defined on $[-t_{\mathrm{e}}, -t_{\mathrm{i}}]$ by*

$$\overleftarrow{\Gamma}(t) \stackrel{\text{def}}{=} \Gamma(-t). \blacksquare$$

Let us show that the line integral changes sign when the path is reversed.

Theorem 10.10.– *Let $q \in \mathcal{C}(\Omega; E^d)$, where $\Omega \subset \mathbb{R}^d$ and E is a Neumann space, and Γ be a $\mathcal{C}^1$ path in Ω. Then,*

$$\int_{\overleftarrow{\Gamma}} q \cdot \mathrm{d}\ell = -\int_{\Gamma} q \cdot \mathrm{d}\ell. \blacksquare$$

Proof. Definitions 10.9 of the reverse path and 10.2 of the line integral give, since $\mathrm{d}(\Gamma(-t))/\mathrm{d}t = -(\mathrm{d}\Gamma/\mathrm{d}t)(-t)$,

$$\int_{\overleftarrow{\Gamma}} q \cdot \mathrm{d}\ell = \int_{-t_{\mathrm{e}}}^{-t_{\mathrm{i}}} q \circ \Gamma(-t) \cdot \frac{\mathrm{d}(\Gamma(-t))}{\mathrm{d}t}\, \mathrm{d}t = -\int_{-t_{\mathrm{e}}}^{-t_{\mathrm{i}}} \Big(q \circ \Gamma \cdot \frac{\mathrm{d}\Gamma}{\mathrm{d}t}\Big)(-t)\, \mathrm{d}t.$$

Since the integral is symmetry invariant (Theorem A.87), it follows that

$$\int_{\overleftarrow{\Gamma}} q \cdot \mathrm{d}\ell = -\int_{t_{\mathrm{i}}}^{t_{\mathrm{e}}} \Big(q \circ \Gamma \cdot \frac{\mathrm{d}\Gamma}{\mathrm{d}t}\Big)(t)\, \mathrm{d}t = -\int_{\Gamma} q \cdot \mathrm{d}\ell. \square$$

Let us show that the line integral is invariant under an increasing change of variable.

Theorem 10.11.– *Let Γ be a $\mathcal{C}^1$ path in a subset Ω of $\mathbb{R}^d$ defined on a bounded interval $[t_{\mathrm{i}}, t_{\mathrm{e}}]$, and let T be a bijection from a bounded interval $[t'_{\mathrm{i}}, t'_{\mathrm{e}}]$ onto $[t_{\mathrm{i}}, t_{\mathrm{e}}]$, such that:*

$$T \in \mathcal{C}^1([t'_{\mathrm{i}}, t'_{\mathrm{e}}]), \quad T' > 0 \text{ in } (t'_{\mathrm{i}}, t'_{\mathrm{e}}).$$

Then, $\Gamma \circ T$ is a $\mathcal{C}^1$ path in Ω and, for every $q \in \mathcal{C}(\Omega; E^d)$, where E is a Neumann space,

$$\int_{\Gamma \circ T} q \cdot \mathrm{d}\ell = \int_{\Gamma} q \cdot \mathrm{d}\ell. \blacksquare$$

Proof. **1.** *Property of* $\Gamma \circ T$. Let us check that it satisfies Definition (10.1) of

$$\Gamma \circ T \subset \mathcal{C}^1([t'_\mathrm{i}, t'_\mathrm{e}]; \mathbb{R}^d).$$

On one hand, it is necessary that $\Gamma \circ T$ is continuous from $[t'_\mathrm{i}, t'_\mathrm{e}]$ into Ω, which follows (Theorem A.30) from the continuity of both Γ and T.

On the other hand, it is necessary that the restriction of $\Gamma \circ T$ to $(t_\mathrm{i}, t_\mathrm{e})$ is differentiable, which follows from the differentiability of the restrictions of Γ and T from Theorem A.64 (b), which also gives

$$(\Gamma \circ T)' = (\Gamma' \circ T)\, T'. \tag{10.4}$$

Finally, it is necessary for $(\Gamma \circ T)'$ to have a continuous extension to $[t_\mathrm{i}, t_\mathrm{e}]$, which follows, from (10.4) and the theorem of continuous extension (Theorem A.34), from the uniform continuity of $(\Gamma' \circ T)\, T'$ established in (10.2), p. 222.

2. *Invariance of the line integral.* Due to its Definition 10.2 and equality (10.4), the line integral here equals

$$\int_{\Gamma \circ T} q \cdot \mathrm{d}\ell = \int_{t'_\mathrm{i}}^{t'_\mathrm{e}} (q \circ \Gamma \circ T) \cdot (\Gamma \circ T)' \,\mathrm{d}t = \int_{t'_\mathrm{i}}^{t'_\mathrm{e}} \big(((q \circ \Gamma) \cdot \Gamma') \circ T\big)\, T' \,\mathrm{d}t.$$

By transforming the last term with the formula of change of variable in an integral from Theorem A.85 (a), we obtain, since here $|\det[\nabla T]| = T'$ (because $\nabla T = T'$ which is positive by hypothesis),

$$\int_{\Gamma \circ T} q \cdot \mathrm{d}\ell = \int_{t_\mathrm{i}}^{t_\mathrm{e}} (q \circ \Gamma) \cdot \Gamma' \,\mathrm{d}t = \int_{\Gamma} q \cdot \mathrm{d}\ell.$$

3. *Conditions on* T. It remains to check the conditions of Theorem A.85. It suffices:

— on one hand, that $T \in \mathbf{C}^1_\mathrm{b}((t'_o, t'_\mathrm{e}))$, which follows from the hypothesis $T \in \mathcal{C}^1([t'_o, t'_\mathrm{e}])$;

— on the other hand (from Theorem A.85 (b), since $[t'_o, t'_\mathrm{e}]$ is bounded), that

$$T^{-1} \in \mathcal{C}^1((t_o, t_\mathrm{e})). \tag{10.5}$$

However, since $T' > 0$, the inverse function theorem (Theorem A.66) shows that T is invertible, T^{-1} is continuous and differentiable, and $(T^{-1})'(t) = 1/(T'(T^{-1}(t)))$, i.e.

$$(T^{-1})' = \mathcal{Q} \circ T' \circ T^{-1},$$

where $\mathcal{Q}(x) = 1/x$. Thus, $(T^{-1})'$ is continuous as is every composition of continuous functions (Theorem A.30) given that, in addition to T' and T^{-1}, $\mathcal{Q}$ is continuous from $(0, \infty)$ into $\mathbb{R}$ (Theorem A.67). Which proves (10.5). □

Independence of the parameterization. From Theorems 10.10 and 10.11, the line integral along a $\mathcal{C}^1$ path Γ depends only on its geometry (namely on its image $[\Gamma]$) and on its orientation. □

10.3. Line integral along a piecewise $\mathcal{C}^1$ path

Let us concatenate paths for which the ending point of one is the initial point of the other.

Definition 10.12.– *Let $\Gamma_1 \in \mathcal{C}([t_{i_1}, t_{e_1}]; \mathbb{R}^d)$ and $\Gamma_2 \in \mathcal{C}([t_{i_2}, t_{e_2}]; \mathbb{R}^d)$ be two paths such that*

$$\Gamma_1(t_{e_1}) = \Gamma_2(t_{i_2}).$$

Their ***concatenation*** *is the path $\Gamma_1 \vec{\cup} \Gamma_2 \in \mathcal{C}([t_{i_1}, t_{e_1} + t_{e_2} - t_{i_2}]; \mathbb{R}^d)$ defined by:*

$$(\Gamma_1 \vec{\cup} \Gamma_2)(t) \stackrel{\text{def}}{=} \begin{cases} \Gamma_1(t) & \text{for } t_{i_1} \leq t \leq t_{e_1}, \\ \Gamma_2(t + t_{i_2} - t_{e_1}) & \text{for } t_{e_1} \leq t \leq t_{e_1} + t_{e_2} - t_{i_2}. \end{cases}$$

A ***piecewise $\mathcal{C}^1$ path*** *is the concatenation of a finite number of $\mathcal{C}^1$ paths.* ▮

Let us show that the line integral along a piecewise $\mathcal{C}^1$ path is the sum of the line integrals along the pieces.

Theorem 10.13.– *Let $q \in \mathcal{C}(\Omega; E^d)$, where $\Omega \subset \mathbb{R}^d$ and E is a Neumann space, and let Γ, Γ_1, ... and Γ_N be $\mathcal{C}^1$ paths in Ω such that*

$$\Gamma = \vec{\bigcup_{1 \leq n \leq N}} \Gamma_n.$$

Then,

$$\oint_\Gamma q \cdot d\ell = \sum_{1 \leq n \leq N} \oint_{\Gamma_n} q \cdot d\ell. \text{ ▮}$$

Proof. By Definition 10.2 of the line integral, it is a question of showing that

$$\int_{t_i}^{t_e} (q \circ \Gamma) \cdot \Gamma' \, dt = \sum_{1 \leq n \leq N} \int_{t_{i_n}}^{t_{e_n}} (q \circ \Gamma) \cdot \Gamma' \, dt.$$

This follows from Chasles' relation (Theorem A.78) since, by Definition 10.12 of the concatenation, $t_i = t_{i_1} < \ldots < t_{e_n} = t_{i_{n+1}} < \ldots < t_{e_N} = t_e$. □

We now extend this property to piecewise $\mathcal{C}^1$ paths which are not necessarily $\mathcal{C}^1$ as a whole, by taking this as the definition of the line integral along such a path.

Definition 10.14.– *Let $q \in \mathcal{C}(\Omega; E^d)$, where $\Omega \subset \mathbb{R}^d$ and E is a Neumann space, and let Γ be a piecewise $\mathcal{C}^1$ path in Ω, i.e. there exist $\mathcal{C}^1$ paths Γ_n such that*

$$\Gamma = \overrightarrow{\bigcup_{1 \le n \le N}} \Gamma_n.$$

*The **line integral** of q along Γ is, here, the element of E defined by*

$$\int_\Gamma q \cdot \mathrm{d}\ell \stackrel{\text{def}}{=} \sum_{1 \le n \le N} \int_{\Gamma_n} q \cdot \mathrm{d}\ell. \blacksquare$$

Justification. The term *line integral* and the notation $\int_\Gamma$ are allowed here because, when Γ is $\mathcal{C}^1$, we recover the line integral introduced in Definition 10.2, due to the additivity property from Theorem 10.13.

This definition is allowed, although there are infinitely many splittings of Γ into $\mathcal{C}^1$ pieces, since the line integral does not depend on the splitting thanks to the following theorem. □

Let us check that the line integral along a piecewise $\mathcal{C}^1$ path is indeed independent of its splitting into $\mathcal{C}^1$ pieces.

Theorem 10.15.– *Under the hypotheses of Definition 10.14, $\sum_{1 \le n \le N} \int_{\Gamma_n} q \cdot \mathrm{d}\ell$ does not depend on the splitting of Γ into the pieces Γ_n.* $\blacksquare$

Proof. **1.** *Minimal splitting.* Define $t_1 = t_\mathrm{i}$ and then, by induction, t_{i+1} as the largest real number such that the restriction Λ_i of Γ to $[t_i, t_{i+1}]$ is a $\mathcal{C}^1$ path. Continue doing this until $t_{I+1} = t_\mathrm{e}$. Then,

$$\Gamma = \overrightarrow{\bigcup_{1 \le i \le I}} \Lambda_i.$$

This *minimal* splitting depends only on Γ, and not on the initial splitting into Γ_n.

2. *Independence of the splitting.* For every $i \in [\![1, I]\!]$, there exists n_i such that

$$\Lambda_i = \overrightarrow{\bigcup_{n_i \le n < n_{i+1}}} \Gamma_n,$$

so the additivity property from Theorem 10.13 gives

$$\sum_{1 \le n \le N} \int_{\Gamma_n} q \cdot \mathrm{d}\ell = \sum_{1 \le i \le I} \sum_{n_i \le n < n_{i+1}} \int_{\Gamma_n} q \cdot \mathrm{d}\ell = \sum_{1 \le i \le I} \int_{\Lambda_i} q \cdot \mathrm{d}\ell.$$

Therefore, every splitting $\Gamma = \overrightarrow{\bigcup}_{1 \le n \le N} \Gamma_n$ indeed produces the same sum, namely that of the minimal splitting. □

Let us show that we can **reparameterize** any piecewise C^1 path to make it into a C^1 path.

Theorem 10.16.– *Let $\Gamma = \vec{\bigcup}_{1\le n\le N} \Gamma_n$ be a piecewise C^1 path in $\mathbb{R}^d$, and $[t_\mathrm{i}, t_\mathrm{e}]$ its interval of definition.*

Then, there exists a bijection $T \in C^1([t_\mathrm{i}, t_\mathrm{e}])$ from $[t_\mathrm{i}, t_\mathrm{e}]$ onto itself such that T' is zero at the initial point and the ending point of each Γ_n, and is > 0 away from these points. For any such bijection,

$$\Gamma \circ T \textit{ is a } C^1 \textit{ path.}$$ ▌

Proof. **1.** *Reparameterization of a piece.* Let $\Gamma_n \in C^1([a, b]; \mathbb{R}^d)$ be one of the pieces of Γ and

$$T(t) \stackrel{\text{def}}{=} a + (b-a)\Big(3\Big(\frac{t-a}{b-a}\Big)^2 - 2\Big(\frac{t-a}{b-a}\Big)^3\Big).$$

Then, $T(a) = a$, $T(b) = b$, the derivative $T'(t) = 6(t-a)/(b-a) - 6((t-a)/(b-a))^2$ is continuous and > 0 on (a, b), and its extension by 0 is continuous on $[a, b]$.

According to the chain rule theorem (Theorem A.64 (b)), $\Gamma_n \circ T$ is differentiable and $(\Gamma_n \circ T)'(t) = (\Gamma_n' \circ T)(t)\, T'(t)$. When $t \to a$ or $t \to b$, $(\Gamma_n' \circ T)(t)$ remains bounded and $T'(t) \to 0$, hence

$$(\Gamma_n \circ T)'(t) \to 0.$$

2. *Global reparametrization.* By reparameterizing all the pieces of Γ, we get a function $\Gamma \circ T$ which is continuous on $[t_\mathrm{i}, t_\mathrm{e}]$, which is differentiable away from the extremities of each piece, and whose derivative tends to 0 at each of these points.

Then, $\Gamma \circ T$ satisfies Definition (10.1), p. 222, of a C^1 path, because:
— on one hand, $\Gamma \circ T$ is differentiable at the connection points of the pieces, and is continuous differentiable on $(t_\mathrm{i}, t_\mathrm{e})$ from Theorem A.57;
— on the other hand, the extension of its derivative by 0 at the points t_i and t_e is continuous on all of $[t_\mathrm{i}, t_\mathrm{e}]$. □

Let us show that the reparameterization of a piecewise C^1 path into a C^1 path does not change the line integral, neither of the path nor of its pieces.

Theorem 10.17.– *Let $q \in C(\Omega; E^d)$, where $\Omega \subset \mathbb{R}^d$ and E is a Neumann space, $\Gamma = \vec{\bigcup}_{1\le n\le N} \Gamma_n$ be a piecewise C^1 path in Ω, and T a reparameterization given by Theorem 10.16 making $\Gamma \circ T$ into a C^1 path.*

Then,

$$\int_{\Gamma \circ T} q \cdot d\ell = \int_{\Gamma} q \cdot d\ell$$

and, for every $n \in [\![1, N]\!]$,

$$\int_{\Gamma_n \circ T} q \cdot d\ell = \int_{\Gamma_n} q \cdot d\ell. \blacksquare$$

Proof. For every piece Γ_n of Γ, Theorem 10.11 on the change of variables in a line integral along a C^1 path gives, since T is C^1 on $[t_i, t_e]$ and $T' > 0$ on (t_i, t_e),

$$\int_{\Gamma_n \circ T} q \cdot d\ell = \int_{\Gamma_n} q \cdot d\ell.$$

Since $\Gamma \circ T$ is the concatenation of the $\Gamma_n \circ T$, Definition 10.14 of the line integral along a piecewise C^1 path then gives

$$\int_{\Gamma \circ T} q \cdot d\ell = \sum_{1 \leq n \leq N} \int_{\Gamma_n \circ T} q \cdot d\ell = \sum_{1 \leq n \leq N} \int_{\Gamma_n} q \cdot d\ell = \int_{\Gamma} q \cdot d\ell. \square$$

10.4. The homotopy invariance theorem

Firstly, let us define homotopic closed paths.

Definition 10.18.– *Two closed paths Γ and Γ_* in a subset Ω of $\mathbb{R}^d$, defined on the same interval $[t_i, t_e]$, are called* **homotopic** *in Ω if we can pass from one to the other via a continuous deformation. That is is say, if there exists $H \in C([t_i, t_e] \times [0, 1]; \Omega)$ such that, for every $t \in [t_i, t_e]$ and $s \in [0, 1]$,*

$$H(t, 0) = \Gamma(t), \quad H(t, 1) = \Gamma_*(t), \quad H(t_i, s) = H(t_e, s).$$

We call the **image** *of H the set $[H] = \{H(t, s) : t_i \leq t \leq t_e,\ 0 \leq s \leq 1\}$.* $\blacksquare$

Let us show that, if a field is locally a gradient, its line integral along any closed path is homotopy invariant, which we call the **homotopy invariance theorem** [2].

2. **History of the homotopy invariance theorem.** We ignore the origin of Theorem 10.19, which is classical in the theory of differential forms with values in a Banach space, see for example [CARTAN, Henri, 17, Theorem 3.7.3, p. 229], where the field q is "hidden" behind the 1-form ω, and where the existence of ∇f_B is the hypothesis "ω is closed".

Theorem 10.19.– *Let $q \in \mathcal{C}(\Omega; E^d)$, where Ω is an open subset of $\mathbb{R}^d$ and E is a Neumann space, such that, for every open ball $B \Subset \Omega$ (i.e. such that $\overline{B} \subset \Omega$), there exists $f_B \in \mathcal{C}^1(B; E)$ such that*

$$\nabla f_B = q \text{ on } B.$$

Then, if Γ and Γ_* are two homotopic $\mathcal{C}^1$ closed paths in Ω,

$$\oint_\Gamma q \cdot \mathrm{d}\ell = \oint_{\Gamma_*} q \cdot \mathrm{d}\ell. \blacksquare$$

Proof. **1.** *Intermediate closed paths.* By reparameterizing Γ and Γ_* if necessary with Theorem 10.11, we can suppose that they are defined on $[0, 1]$. Let then H be a homotopy between Γ and Γ_* in Ω, that is

$$H \in \mathcal{C}([0,1] \times [0,1]; \Omega)$$

such that, for every t and s in $[0, 1]$,

$$H(t, 0) = \Gamma(t), \quad H(t, 1) = \Gamma_*(t), \quad H(0, s) = H(1, s). \tag{10.6}$$

Let $N \in \mathbb{N}$. We define $N + 1$ closed paths $\Gamma_n \in \mathcal{C}([0, 1]; \Omega)$, where $n \in [\![0, N]\!]$, by

$$\Gamma_n(t) \stackrel{\text{def}}{=} H\Big(t, \frac{n}{N}\Big), \quad 0 \le t \le 1.$$

2. *Elementary pieces.* We split each closed path Γ_n into N pieces Γ_n^m, where $m \in [\![0, N-1]\!]$. More precisely, $\Gamma_n^m \in \mathcal{C}([m/N, (m+1)/N]; \Omega)$ is defined by

$$\Gamma_n^m(t) \stackrel{\text{def}}{=} H\Big(t, \frac{n}{N}\Big), \quad \frac{m}{N} \le t \le \frac{m+1}{N}.$$

So, see Figure 10.1 below,

$$\Gamma_n \stackrel{\text{def}}{=} \Gamma_n^0 \,\vec{\cup}\, \Gamma_n^1 \,\vec{\cup} \ldots \vec{\cup}\, \Gamma_n^{N-1}. \tag{10.7}$$

We denote by a_n^m the extremities of these pieces, that is, for n and m in $[\![0, N]\!]$,

$$a_n^m \stackrel{\text{def}}{=} H\Big(\frac{m}{N}, \frac{n}{N}\Big) \tag{10.8}$$

and we denote by T_n^m the *transversal* rectilinear path linking a_n^m to a_{n+1}^m, that is

$$T_n^m \stackrel{\text{def}}{=} \Gamma_{\overrightarrow{a_n^m, a_{n+1}^m}}.$$

3. *Choice of N.* The image

$$[H] \stackrel{\text{def}}{=} \{H(t, s) : 0 \le t \le 1,\ 0 \le s \le 1\}$$

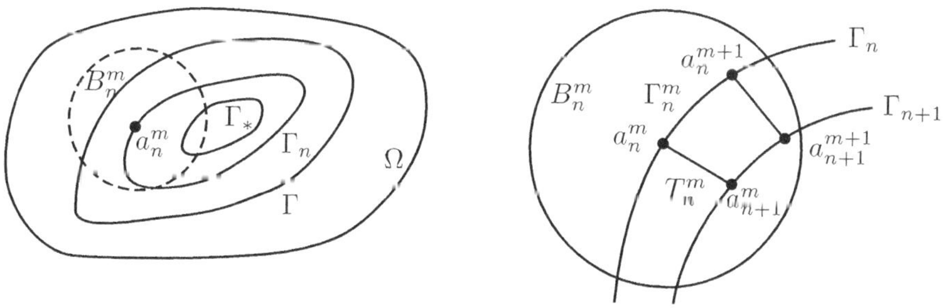

Figure 10.1. *Splitting of the intermediate closed paths into pieces*

of H is compact, as is any image of a compact set, here $[0,1] \times [0,1]$, under a continuous mapping (Theorem A.31). Thus, by the strong inclusion theorem (Theorem A.22), there exists $\delta > 0$ such that

$$[H] + B(0, \delta) \subset \Omega.$$

Choose N large enough so that $|t - t'| \le 1/N$ and $|s - s'| \le 1/N$ imply $|H(t,s) - H(t', s')| \le \delta/3$, and let B_n^m be the open ball centered at a_n^m of radius $2\delta/3$. Then,

the paths Γ_n^m, Γ_{n+1}^m, T_n^m and T_n^{m+1} are included in B_n^m.

4. *Local calculation of the line integral.* By hypothesis, there exists a function f_n^m in $C^1(B_n^m; E)$ such that

$$q = \nabla f_n^m \text{ on } B_n^m.$$

The expression for the line integral of a gradient (Theorem 10.7) then gives

$$\int_{\Gamma_n^m} q \cdot d\ell = \int_{\Gamma_n^m} \nabla f_n^m \cdot d\ell = f_n^m(a_n^{m+1}) - f_n^m(a_n^m),$$

and this equality equally holds with $n+1$ in the place of n. Likewise,

$$\int_{T_n^m} q \cdot d\ell = f_n^m(a_{n+1}^m) - f_n^m(a_n^m).$$

and this equality equally holds with $m+1$ in the place of m. Hence,

$$\int_{\Gamma_{n+1}^m} q \cdot d\ell - \int_{\Gamma_n^m} q \cdot d\ell = \int_{T_n^{m+1}} q \cdot d\ell - \int_{T_n^m} q \cdot d\ell \tag{10.9}$$

(because the two sides equal $\left(f_n^m(a_{n+1}^{m+1}) - f_n^m(a_{n+1}^m)\right) - \left(f_n^m(a_n^{m+1}) - f_n^m(a_n^m)\right)$.)

5. Invariance of the line integral along the Γ_n. Due to the additivity of the line integral with respect to paths, i.e. by Definition 10.14 of the line integral along a piecewise C^1 path, decomposition (10.7) of Γ_n gives

$$\int_{\Gamma_n} q \cdot d\ell = \int_{\Gamma_n^0} q \cdot d\ell + \cdots + \int_{\Gamma_n^{N-1}} q \cdot d\ell.$$

By summing equalities (10.9) over m from 0 to $N-1$, it then follows

$$\int_{\Gamma_{n+1}} q \cdot d\ell - \int_{\Gamma_n} q \cdot d\ell = \int_{T_n^N} q \cdot d\ell - \int_{T_n^0} q \cdot d\ell. \qquad (10.10)$$

The right-hand side is zero, since

the transversal paths T_n^0 and T_n^N coincide.

Indeed, T_n^0 links the initial points a_n^0 and a_{n+1}^0 of Γ_n and Γ_{n+1}, while T_n^N links their ending points a_n^N and a_{n+1}^N, ending points that coincide with the initial points since (by Definitions (10.8) of a_n^m and (10.6) of H)

$$a_n^0 = H\Big(0, \frac{n}{N}\Big) = H\Big(1, \frac{n}{N}\Big) = a_n^N$$

and, likewise, $a_{n+1}^0 = a_{n+1}^N$.

Equality (10.10) therefore gives

$$\int_{\Gamma_{n+1}} q \cdot d\ell = \int_{\Gamma_n} q \cdot d\ell.$$

6. Conclusion. Since this equality is true for each n,

$$\int_{\Gamma_N} q \cdot d\ell = \int_{\Gamma_0} q \cdot d\ell.$$

Which gives the stated result, since $\Gamma_0 = \Gamma$ and $\Gamma_N = \Gamma_*$. □

Stokes' formula. The homotopy invariance theorem (Theorem 10.19) is a (non-elementary!) incarnation of Stokes' formula[3]

$$\int_{\partial H} \sigma = \int_H d\sigma,$$

3. **History of Stokes' formula.** We did not find any precise reference regarding the origin of this formula. Attributed to Sir George Gabriel STOKES, it was supposedly discovered by Mikhail Vasilyevitch OSTROGRADSKY around 1820 and then rediscovered by Lord KELVIN.
In the scientific literature, it can also be found associated with the names of Carl Friedrich GAUSS or of George GREEN and in many different forms, including that in Theorem 9.2.

where σ is an exterior differential k-form and H is a $k+1$-chain, see for example [BOURBAKI, 13, § 11.3.4, p. 49] when σ takes its values in a Banach space.

Indeed, the hypothesis $\nabla f_B = q$ gives $\partial_i q_j = \partial_i \partial_j f_B = \partial_j \partial_i f_B = \partial_j q$, so the differential 1-form $\sigma = \Sigma_j q_j \, \mathrm{d}x_j$ satisfies

$$\mathrm{d}\sigma = \Sigma_{ij} \partial_i q_j \, \mathrm{d}x_i \wedge \mathrm{d}x_j = \Sigma_{i<j} (\partial_i q_j - \partial_j q_i) \, \mathrm{d}x_i \wedge \mathrm{d}x_j = 0_E.$$

Since a homotopy H between Γ and Γ_* is an oriented 2-chain with boundary $\partial H = \overrightarrow{\Gamma} \cup \overleftarrow{\Gamma}_*$, it follows that

$$\oint_\Gamma q \cdot \mathrm{d}\ell - \oint_{\Gamma_*} q \cdot \mathrm{d}\ell = \oint_{\partial H} \sigma = \oint_H \mathrm{d}\sigma = 0_E. \; \square$$

10.5. Connectedness and simply connectedness

Let us define connected sets.

Definition 10.20.– *A subset U of a separated semi-normed space is said to be* ***connected*** *if it cannot be covered by two open sets whose intersections with U are disjoint and non-empty.*

The ***connected component*** *of U generated by a point u of U is the largest connected subset of U which contains u.* ▌

Justification of the definition of a connected component. There exists indeed such a set, namely the union of the connected subsets of U containing u (this point is explained in Volume 1 [81, Justification of Definition 2.30 (b)]). $\square$

Let us show that every connected open set Ω is the union (over n) of the connected components $\Omega_{1/n}^{(a)}$ of the $\Omega_{1/n}$ (i.e. Ω with a neighborhood of its boundary of width $1/n$ removed) generated by a point a.

Theorem 10.21.– *Let Ω be a connected open subset of $\mathbb{R}^d$ and $a \in \Omega$. For $n \in \mathbb{N}^*$, let*

$$\Omega_{1/n} \stackrel{\text{def}}{=} \{x \in \mathbb{R}^d : B(x, 1/n) \subset \Omega\}$$

and $\Omega_{1/n}^{(a)}$ be the connected component of $\Omega_{1/n}$ generated by a. Then,

$$\Omega = \bigcup_{n \in \mathbb{N}^*} \Omega_{1/n}^{(a)}. \; ▌$$

Proof. Let

$$x \in \Omega.$$

Since Ω is open and connected, x is linked to a by a path Γ in Ω (Theorem A.16). Its image $[\Gamma]$ is compact (as is any image of a compact set under a continuous mapping, see Theorem A.31), therefore it is included in one of the $\Omega_{1/n}$ (since they are open sets increasing with n and covering Ω). The path Γ thus links x to a in $\Omega_{1/n}$, which implies (Theorem A.17)

$$x \in \Omega_{1/n}^{(a)}.$$

So, $\Omega \subset \bigcup_{n\in\mathbb{N}^*} \Omega_{1/n}^{(a)}$. The converse inclusion is immediate. □

Remark. The connectedness of Ω does not imply that of the $\Omega_{1/n}$ (otherwise Theorem 10.21 would be trivial). Two examples of connected open sets whose subsets $\Omega_{1/n}$ are not connected are represented in Figures 13.1, p. 266, and 13.2, p. 268. □

Let us now come to simply connected sets.

Definition 10.22.– *A subset U of a separated semi-normed space is said to be* ***simply connected*** *if every closed path in U is homotopic in U to a closed path reduced to a point.* ▮

Simply-connected versus connected. Neither of these two properties implies the other, as the following examples show.

— A pair of disjoint balls in $\mathbb{R}^d$ is simply connected but not connected.

— The crown $C = \{x \in \mathbb{R}^2 : 1 < |x| < 2\}$ is connected but not simply connected in $\mathbb{R}^2$. The product $C \times \mathbb{R}^{d-2}$ (i.e. the space between two nested infinite cylinders) has the same properties in $\mathbb{R}^d$.

For certain authors, "*simply connected*" includes "*connected*". This is obtained by replacing in Definition 10.22 "*every closed path is homotopic to* **a** *point*" with "*every closed path is homotopic to* **every** *point in U*". □

Simply-connectedness of $\mathbb{R}^d$ versus absence of "holes". If we define a *hole* as a connected and bounded component of the complement, then simply connectedness is equivalent to the absence of holes in dimension $d = 2$; in contrast, when $d \geq 3$, one does not imply or result from the other.

— In $\mathbb{R}$, every open set is simply connected, even if it has holes.

— In $\mathbb{R}^2$, an open set is simply connected if and only if it has no holes (for example, a ball in $\mathbb{R}^2$ is simply connected and has no holes; on the contrary, a crown C has a hole and is not simply connected).

— In $\mathbb{R}^d$, $d \geq 3$, a simply connected open set may have holes (this is the case for the complement of a ball). Conversely, an open set without holes is not necessarily simply connected (this is the case for the complement of an infinite cylinder or for a torus in $\mathbb{R}^3$). □

Chapter 11

Primitives of Functions

This chapter provides three results on the existence of primitives of fields $q = (q_1, \ldots, q_d)$ of functions, to which we will come back to construct primitives of distributions, in Chapter 13.

The first (Theorem 11.1) gives the existence of a primitive of every field whose line integral is zero along closed paths.

The second is the *orthogonality theorem* for functions (Theorem 11.4), which gives the existence of a primitive of every field which is *orthogonal* to test fields, namely such that $\int_\Omega q \cdot \psi = 0$ for all ψ in $\mathcal{D}(\Omega; \mathbb{R}^d)$ such that $\nabla \cdot \psi = 0$.

This result is based on the *concentration theorem* (Theorem 11.3), which says that, for any field q, the integral $\int_\Omega q \cdot \Psi$ associated with an incompressible *tubular flow* Ψ with support in a tube of axis Γ is equal to the integral $\oint_\Gamma q \diamond \rho \cdot d\ell$ concentrated on the closed path Γ.

The third is *Poincaré's theorem* (Theorem 11.5), which gives the existence of a primitive on a ball of every field which satisfies *Poincaré's condition* $\partial_i q_j = \partial_j q_i$.

11.1. Primitive of a function field with a zero line integral

Let us show that every continuous vector field $q = (q_1, \ldots, q_d)$ whose line integral along every closed path is zero has a primitive.

Theorem 11.1.– *Let $q \in \mathcal{C}(\Omega; E^d)$, where Ω is an open subset of $\mathbb{R}^d$ and E is a Neumann space, be such that, for every C^1 closed path Γ in Ω,*

$$\oint_\Gamma q \cdot d\ell = 0_E. \tag{11.1}$$

Then, there exists $f \in \mathcal{C}^1(\Omega; E)$ such that

$$\nabla f = q. \blacksquare$$

Proof. **1.** *Definition of f.* In each connected component Ω_m of Ω, we choose a point a_m. Since each Ω_m is connected and open (Theorem A.14, since Ω is open), each of its points x is linked (Theorem A.16) to the point a_m by a C^1 path $\Gamma(a_m, x)$ in Ω_m, and hence in Ω. We will check in three steps that a suitable function is defined by

$$f(x) \stackrel{\text{def}}{=} \int_{\Gamma(a_m,x)} q \cdot \mathrm{d}\ell.$$

2. *Obtaining a function.* We first check that $\int_{\Gamma(a_m,x)} q \cdot \mathrm{d}\ell$ does not depend on the path connecting a_m to x. So, let Γ and Γ_* be two such paths.

The concatenation $\Gamma \vec{\cup} \overleftarrow{\Gamma_*}$ of Γ and of the reverse path of Γ_* is a piecewise C^1 closed path. Due to the additivity of the line integral with respect to paths, i.e. by Definition 10.14 of the line integral along a concatenation, and since the sign of the line integral changes along the reverse path by Theorem 10.10,

$$\int_{\Gamma \vec{\cup} \overleftarrow{\Gamma_*}} q \cdot \mathrm{d}\ell = \int_{\Gamma} q \cdot \mathrm{d}\ell + \int_{\overleftarrow{\Gamma_*}} q \cdot \mathrm{d}\ell = \int_{\Gamma} q \cdot \mathrm{d}\ell - \int_{\Gamma_*} q \cdot \mathrm{d}\ell.$$

In reparameterizing $\Gamma \vec{\cup} \overleftarrow{\Gamma_*}$ via Theorem 10.16, we obtain (Theorem 10.17) a C^1 closed path with the same line integral, which is then zero by hypothesis. Therefore, we indeed have

$$\int_{\Gamma} q \cdot \mathrm{d}\ell = \int_{\Gamma_*} q \cdot \mathrm{d}\ell. \tag{11.2}$$

3. *Obtaining a primitive.* We now check that f has partial derivatives which satisfy

$$\partial_i f = q_i.$$

Let $x \in \Omega$, $\eta > 0$ be such that the ball $\{y \in \mathbb{R}^d : |y - x| \le \eta\}$ is included in Ω, and s be a non-zero real number such that $|s| \le \eta$.

Again from Definition 10.14 and Theorem 10.10,

$$f(x + s\mathbf{e}_i) - f(x) = \int_{\Gamma(a,x+s\mathbf{e}_i)} q \cdot \mathrm{d}\ell - \int_{\Gamma(a,x)} q \cdot \mathrm{d}\ell = \int_{\Lambda} q \cdot \mathrm{d}\ell, \tag{11.3}$$

where $\Lambda = \overleftarrow{\Gamma(a,x)} \vec{\cup} \Gamma(a, x + s\mathbf{e}_i)$.

This path Λ connects x to $x + s\mathbf{e}_i$, and the line integral is independent of the path in Ω connecting these two points from (11.2), hence equality (11.3) holds when Λ is the rectilinear path $\Gamma_{\overrightarrow{x,x+s\mathbf{e}_i}}$. Using the expression for a line integral along such a path from Theorem 10.6, equality (11.3) then gives

$$f(x + s\mathbf{e}_i) - f(x) = s\mathbf{e}_i \cdot \int_0^1 q(x + ts\mathbf{e}_i)\,\mathrm{d}t = s \int_0^1 q_i(x + ts\mathbf{e}_i)\,\mathrm{d}t.$$

So,

$$f(x+s\mathbf{e}_i)-f(x)-sq_i(x)=s\int_0^1 (q_i(x+ts\mathbf{e}_i)-q_i(x))\,\mathrm{d}t.$$

For every semi-norm $\| \ \|_{E;\nu}$ of E, the upper bound of the semi-norms of the integral from Theorem 1.23 (b) gives

$$\|f(x+s\mathbf{e}_i)-f(x)-sq_i(x)\|_{E;\nu}\leq |s| \sup_{0\leq r\leq s} \|q_i(x+r\mathbf{e}_i)-q_i(x)\|_{E;\nu}.$$

For every $\epsilon>0$, we can (by Definition 1.9 (a) of continuity) choose η such that the right-hand side is $\leq \epsilon|s|$, since q_i is continuous. Therefore, by characterization (1.5) from Definition 1.16 of the partial derivatives of a function,

$$\partial_i f(x)=q_i(x).$$

4. *Regularity of f.* The partial derivatives $\partial_i f$ being continuous since the q_i are by hypothesis, Theorem A.55 shows that f is continuous and continuously differentiable, i.e.

$$f\in C^1(\Omega;E).\ \Box$$

Connected components of an open subset of $\mathbb{R}^d$. Observe that the number of points a_m to be fixed in the proof of Theorem 11.1 is countable (possibly finite), since every open subset of $\mathbb{R}^d$ has a countable number of connected components [Vol. 2, Property (9.2), p. 193]. □

11.2. Tubular flows and concentration theorem

Let us construct a divergence-free (i.e. with zero divergence) test field with support in a tubular neighborhood of a path [1]. The **divergence** [2] of a field $\psi=(\psi_1,\ldots,\psi_d)$ i

$$\nabla\cdot\psi \stackrel{\text{def}}{=} \partial_1\psi_1+\cdots\partial_d\psi_d.$$

Theorem 11.2.– *Let $\Gamma\in C^1([t_{\mathrm{i}},t_{\mathrm{e}}];\mathbb{R}^d)$ be a closed path in $\mathbb{R}^d$, B a compact subset of $\mathbb{R}^d$ and*

$$\mathcal{T}\stackrel{\text{def}}{=}[\Gamma]+B$$

---

1. **History of the tubular flow construction.** The divergence-free field Ψ from Theorem 11.2 was introduced by Jacques SIMON in 1993 [77, Lemma, p. 1170] by constructing a concentrated incompressible flow $\vec{\delta}_\Gamma$ and then regularizing it, as explained in the comment *Underlying idea: the concentrated flow*, on the following page.
The concentrated incompressible field was also constructed by Stanislav Konstantinovitch SMIRNOV in 1993 [85, p. 842] to conversely decompose any incompressible field ψ as an integral $\psi=\int_\mu \vec{\delta}_{\Gamma_\mu}\mathrm{d}\mu$ of concentrated fields.

2. **History of divergence.** The term *divergence* was introduced by William Kingdon CLIFFORD, in 1878 [21].

a **tube**, *where* $[\Gamma] = \{\Gamma(t) : t_i \le t \le t_e\}$. *Moreover, let* $\rho \in C_B^\infty(\mathbb{R}^d)$.

We define a **tubular flow** $\Psi \in C_{\mathcal{T}}^\infty(\mathbb{R}^d; \mathbb{R}^d)$ *by*

$$\Psi(x) \stackrel{\text{def}}{=} \int_{t_i}^{t_e} \rho(x - \Gamma(t))\, \Gamma'(t)\, dt.$$

It satisfies

$$\nabla \cdot \Psi = 0. \blacksquare$$

Terminology. We say **tubular flow** because Ψ is the velocity field of an incompressible flow (since its divergence is zero) with support in the tube $\mathcal{T}$ of axis $[\Gamma]$, see the figure below. Its velocity is zero outside of $\mathcal{T}$ and its flux through any section S of $\mathcal{T}$ oriented as Γ is equal to 1. □

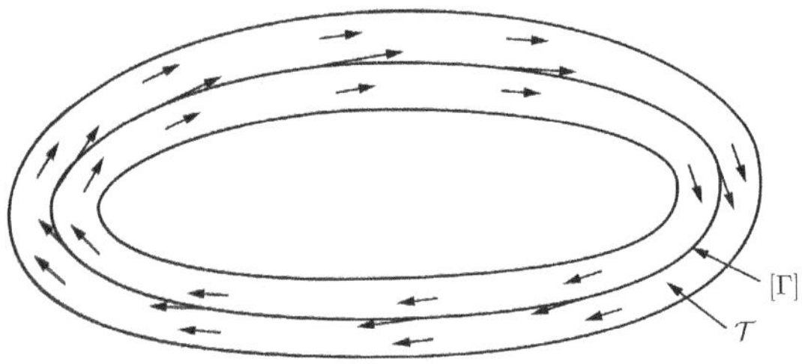

Figure 11.1. *Tubular flow* Ψ*, in the tube* $\mathcal{T}$ *of axis* $[\Gamma]$

Utility of a tubular flow. The construction of the tubular flow is a key step, via the *concentration theorem* (Theorem 11.3) in our construction of primitives, see the comment *Utility of the concentration theorem*, p. 242. □

Underlying idea: the concentrated flow. We call **concentrated flow** on $[\Gamma]$ the distribution $\vec{\delta}_\Gamma$ in $\mathcal{D}'(\mathbb{R}^d; \mathbb{R}^d)$, with support $[\Gamma]$, defined, for every test function $\varphi \in \mathcal{D}(\mathbb{R}^d)$, by

$$\langle \vec{\delta}_\Gamma, \varphi \rangle \stackrel{\text{def}}{=} \int_{t_i}^{t_e} (\varphi \circ \Gamma)\, \Gamma'\, dt.$$

The tubular flow Ψ is (according to Definition 7.1 of the weighting by a regular function) a regular approximation of the concentrated flow $\vec{\delta}_\Gamma$:

$$\Psi = \vec{\delta}_\Gamma \diamond \rho.$$

If $[\Gamma]$ is a regular curve, $\vec{\delta}_\Gamma$ is, at every point $\Gamma(t)$, a "vectorial mass" equal to the tangent vector oriented as Γ. Observe that, for every test field $\phi \in \mathcal{D}(\mathbb{R}^d; \mathbb{R}^d)$,

$$\langle \vec{\delta}_\Gamma, \phi \rangle = \sum_{i=1}^d \langle (\vec{\delta}_\Gamma)_i, \phi_i \rangle = \int_{t_i}^{t_e} (\phi \circ \Gamma) \cdot \Gamma'\, dt = \oint_\Gamma \phi \cdot d\ell.$$

The concentrated flow $\vec{\delta}_\Gamma$ is *incompressible*, meaning divergence-free, since the line integral of a gradient along a closed path is always zero (Theorem 10.8) whence

$$\langle \nabla \cdot \vec{\delta}_\Gamma, \varphi \rangle = -\langle \vec{\delta}_\Gamma, \nabla\varphi \rangle = -\oint_\Gamma \nabla\varphi \cdot d\ell = 0.$$

The divergence of the tubular flow Ψ is also zero, since

$$\nabla \cdot \Psi = \nabla \cdot (\vec{\delta}_\Gamma \diamond \rho) = (\nabla \cdot \vec{\delta}_\Gamma) \diamond \rho = 0. \ \square$$

Proof of Theorem 11.2. **1.** *Regularity of Ψ.* Its definition can be written as

$$\Psi(x) = L(R(x)),$$

where, for every function $g \in \mathcal{C}(\mathbb{R}^d)$,

$$L(g) \stackrel{\text{def}}{=} \int_{t_i}^{t_e} g(-\Gamma(t))\,\Gamma'(t)\,\mathrm{d}t,$$

and where $R(x)(y) = \rho(x+y)$, i.e. $R(x) = \tau_{-x}\rho$ where τ_x is the translation.

The bound of the semi-norms (here, the absolute value) of the integral from Theorem 1.23 (b) and Definition 1.18 (b) of the semi-norms (reduced here to a norm) of $\mathcal{C}_b(\mathbb{R}^d)$ give

$$|L(g)| \le |t_e - t_i| \sup_{y\in\mathbb{R}^d} |g(y)| \sup_{t_i\le t\le t_e} |\Gamma'(t)| = \gamma \|g\|_{\mathcal{C}_b(\mathbb{R}^d)},$$

where γ only depends on Γ. Hence, by the characterization of continuous linear mappings from Theorem 1.12 (a),

$$L \text{ is continuous from } \mathcal{C}_b(\mathbb{R}^d) \text{ into } \mathbb{R}^d.$$

However, $R \in \mathcal{C}^\infty(\mathbb{R}^d; \mathcal{C}_b(\mathbb{R}^d))$ due to the property of differentiation of the translation from Theorem 7.7 (since $\rho \in \mathcal{D}(\mathbb{R}^d)$, by hypothesis). The composite mapping $L \circ R$, i.e. Ψ, therefore satisfies (Theorem A.59)

$$\Psi \in \mathcal{C}^\infty(\mathbb{R}^d; \mathbb{R}^d).$$

2. *Support of Ψ.* If $x \notin [\Gamma] + B$, then, for every $t \in [t_i, t_e]$, we have $x - \Gamma(t) \notin B$, hence $\rho(x - \Gamma(t)) = 0$, and therefore $\Psi(x) = 0$. The support of Ψ is thus included in the tube $\mathcal{T} = [\Gamma] + B$. Since this is compact (as is any sum of compact subsets of $\mathbb{R}^d$, see Theorem A.27),

$$\Psi \in \mathcal{C}^\infty_{\mathcal{T}}(\mathbb{R}^d; \mathbb{R}^d).$$

3. *Divergence of Ψ.* Let $x \in \mathbb{R}^d$. Since each component L_i is continuous linear from $\mathcal{C}_b(\mathbb{R}^d)$ into $\mathbb{R}$, it commutes with the partial derivatives ∂_i by Theorem A.59, thus

$$(\nabla \cdot \Psi)(x) = \sum_{i=1}^d \partial_i \Psi_i(x) = \sum_{i=1}^d \partial_i (L_i R)(x) = \sum_{i=1}^d L_i(\partial_i R)(x).$$

Using the definitions of L and R, this can be written, denoting $r(y) = -\rho(x-y)$, as

$$(\nabla \cdot \Psi)(x) = \int_{t_i}^{t_e} \sum_{i=1}^d \partial_i \rho(x - \Gamma(t))\,\Gamma_i'(t)\,\mathrm{d}t = \int_{t_i}^{t_e} \nabla r(\Gamma(t)) \cdot \Gamma'(t)\,\mathrm{d}t.$$

The last term is Definition 10.2 of the line integral of ∇r on Γ, which is zero as any line integral of a gradient along a closed path (Theorem 10.8). Therefore,

$$(\nabla \cdot \Psi)(x) = 0. \ \square$$

Let us show that, for any field q, the integral $\int_\Omega q \cdot \Psi$ is equal to the integral "concentrated" over Γ, $⨍_\Gamma q \diamond \rho \cdot \mathrm{d}\ell$. We call this the **concentration theorem** [3].

Theorem 11.3.– *Let $q \in \mathcal{C}(\Omega; E^d)$, where Ω is an open subset of $\mathbb{R}^d$ and E is a Neumann space.*

Let $\mathcal{T} = [\Gamma] + B$ be a tube included in Ω, where Γ is a $\mathcal{C}^1$ closed path in Ω and B is a compact subset of $\mathbb{R}^d$, $\rho \in \mathcal{C}_B^\infty(\mathbb{R}^d)$ and $\Psi \in \mathcal{C}_\mathcal{T}^\infty(\mathbb{R}^d; \mathbb{R}^d)$ the tubular flow given by Theorem 11.2. Then,

$$\int_\Omega q(x) \cdot \Psi(x)\,\mathrm{d}x = ⨍_\Gamma q \diamond \rho \cdot \mathrm{d}\ell. \ \blacksquare$$

Utility of the concentration theorem. Theorem 11.3 is a *key point* in our proof of the *orthogonality theorem for functions* (Theorem 11.4). Indeed, it shows (see equality (11.5), p. 244) that the orthogonality condition $\int_\Omega q \cdot \psi = 0_E$ implies the condition $⨍_\Gamma q \cdot \mathrm{d}\ell = 0_E$ for any closed path Γ, a condition with which we know how to explicitly construct a primitive (see the proof of Theorem 11.1). $\square$

Proof of Theorem 11.3. The definition of Ψ gives,

$$\int_\Omega q(x) \cdot \Psi(x)\,\mathrm{d}x = \int_\Omega \sum_{i=1}^d q_i(x) \Big(\int_{t_i}^{t_e} \rho(x - \Gamma(t))\, \Gamma_i'(t)\,\mathrm{d}t \Big)\,\mathrm{d}x.$$

By permuting the order of the integration via Theorem A.91, we obtain

$$\int_\Omega q(x) \cdot \Psi(x)\,\mathrm{d}x = \int_{t_i}^{t_e} \sum_{i=1}^d \Big(\int_\Omega q_i(x) \rho(x - \Gamma(t))\,\mathrm{d}x \Big) \Gamma_i'(t)\,\mathrm{d}t.$$

Due to the expression for the weighting of a continuous function from Theorem 7.23,

$$\int_\Omega q_i(x) \rho(x - \Gamma(t))\,\mathrm{d}x = (q_i \diamond \rho)(\Gamma(t)).$$

It then becomes, with Definition 10.2 of the line integral,

$$\int_\Omega q(x) \cdot \Psi(x)\,\mathrm{d}x = \int_{t_i}^{t_e} (q \diamond \rho)(\Gamma(t)) \cdot \Gamma'(t)\,\mathrm{d}t = ⨍_\Gamma q \diamond \rho \cdot \mathrm{d}\ell. \ \square$$

3. **History of the concentration theorem.** Theorem 11.3 had been established, for E a Banach space, by Jacques SIMON in 1993 [79, p. 207, last equality].

11.3. The orthogonality theorem for functions

Let us show that a function field $q = (q_1, \dots, q_d)$ has a primitive f whenever it is "orthogonal" to every divergence-free test field $\psi = (\psi_1, \dots, \psi_d)$. This is the **orthogonality theorem**[4] for functions, which is a step in the proof of the general orthogonality theorem (Theorem 13.5).

Theorem 11.4.– *Let $q \in \mathcal{C}(\Omega; E^d)$, where Ω is an open subset of $\mathbb{R}^d$ and E is a Neumann space, be such that:*

$$\int_\Omega q \cdot \psi = 0_E, \quad \forall \psi \in \mathcal{D}(\Omega; \mathbb{R}^d) \text{ such that } \nabla \cdot \psi = 0. \tag{11.4}$$

Then there exists $f \in \mathcal{C}^1(\Omega; E)$ such that

$$\nabla f = q. \blacksquare$$

We denote by $\mathcal{D}(\Omega; \mathbb{R}^d)$ the space of **test fields**, i.e. of infinitely differentiable functions from Ω into $\mathbb{R}^d$ with compact support.

Orthogonality. By generalizing the notion of orthogonality with respect to a scalar product, we can say that a continuous field q satisfying (11.4) is *orthogonal* to the space

$$\mathcal{D}_{\mathrm{div}}(\Omega; \mathbb{R}^d) \stackrel{\mathrm{def}}{=} \{\psi \in \mathcal{D}(\Omega; \mathbb{R}^d) : \nabla \cdot \psi = 0\},$$

with respect to the bilinear mapping $(q, \psi) \mapsto \int_\Omega q \cdot \psi$ from $\mathcal{C}(\Omega; E^d) \times \mathcal{D}(\Omega; \mathbb{R}^d)$ into E.

Condition (11.4) is also necessary for q to be a gradient, therefore $\mathcal{D}_{\mathrm{div}}(\Omega; \mathbb{R}^d)$ is *the orthogonal complement* of the space $\mathcal{C}_\nabla(\Omega; E^d)$ of continuous fields which are gradients, which we can denote

$$\mathcal{C}_\nabla(\Omega; E^d) = (\mathcal{D}_{\mathrm{div}}(\Omega; \mathbb{R}^d))^\perp. \square$$

Proof of Theorem 11.4. Let Γ be a $\mathcal{C}^1$ closed path in Ω. Since its image $[\Gamma]$ is compact, the strong inclusion theorem (Theorem A.22) provides $r > 0$ such that the tube $\mathcal{T} = [\Gamma] + B(0, r)$ is included in Ω. Let n_Γ be an integer $\geq 1/r$.

For $n \geq n_\Gamma$, let $\Psi_n \in \mathcal{C}^\infty_{\mathcal{T}_n}(\mathbb{R}^d; \mathbb{R}^d)$, where $\mathcal{T}_n = [\Gamma] + B(0, 1/n)$, be the tubular flow given by Theorem 11.2 for a regularizing function ρ_n given by Definition 8.1.

The restriction of Ψ_n belongs to $\mathcal{D}(\Omega; \mathbb{R}^d)$ and $\nabla \cdot \Psi_n = 0$, therefore hypothesis (11.4) gives

$$\int_\Omega q \cdot \Psi_n = 0_E.$$

4. **History of the orthogonality theorem.** Olga LADYZHENSKAYA proved in 1963 [48, Theorem 1, p. 27] Theorem 11.4 for $q \in L^2(\Omega; \mathbb{R}^d)$, i.e. for $E = \mathbb{R}$ and q square-integrable.

The concentration theorem (Theorem 11.3) then gives

$$\oint_\Gamma q \diamond \rho_n \cdot d\ell = \int_\Omega q \cdot \Psi_n = 0_E. \tag{11.5}$$

Due to the properties of regularization of continuous functions from Theorem 8.7 (a),

$$q \diamond \rho_n \to q \ \text{ in } \mathcal{C}(\Omega_{B(0,1/n_\Gamma)}; E).$$

The line integral depending sequentially continuously on q (Theorem 10.4 for the domain $\Omega_{B(0,1/n_\Gamma)}$), it follows that

$$\oint_\Gamma q \diamond \rho_n \cdot d\ell \to \oint_\Gamma q \cdot d\ell.$$

In the limit, (11.5) gives

$$\oint_\Gamma q \cdot d\ell = 0_E.$$

Which implies the existence of f such that $\nabla f = q$ according to Theorem 11.1. □

11.4. Poincaré's theorem

Let us show that every continuously differentiable vector field q on a ball which satisfies Poincaré's condition $\partial_i q_j = \partial_j q_i$ has a primitive. It is a special case of **Poincaré's theorem** [5], and a step in its generalization to distributions which will be done in Theorem 13.7.

Theorem 11.5.– *Let $q \in \mathcal{C}^1(B; E^d)$, where B is an open ball in $\mathbb{R}^d$ and E is a Neumann space, be such that, for every i and j in $[\![1, d]\!]$,*

$$\partial_i q_j = \partial_j q_i.$$

Then, there exists $f \in \mathcal{C}^2(B; E)$ such that

$$\nabla f = q. \ \blacksquare$$

The case of a star shaped open set. Theorem 11.5 and its proof given below are more generally valid for a star shaped open subset of $\mathbb{R}^d$ instead of a ball B, see Theorem 9.5 in Volume 2 [82]. □

5. **History of Poincaré's theorem.** Theorem 11.5 had been established by Henri POINCARÉ, in 1899 [61, p. 10], for real values.

Proof of Theorem 11.5. **1.** *Construction of f.* Denote by a the center of B. We will verify that a primitive is given, for every $x \in B$, by

$$f(x) \stackrel{\text{def}}{=} (x-a) \cdot \int_0^1 q(a+t(x-a))\,\mathrm{d}t.$$

2. *Differentiating under the integral sign.* Denote, for $j \in [\![1, d]\!]$,

$$p_j(x) \stackrel{\text{def}}{=} \int_0^1 q_j(a+t(x-a))\,\mathrm{d}t$$

and assume for the moment that we can differentiate this integral under the integral sign, i.e. that $p_j \in \mathcal{C}^1(B; E)$ and

$$\partial_i p_j(x) = \int_0^1 t\partial_i q_j(a+t(x-a))\,\mathrm{d}t. \tag{11.6}$$

Then, $f(x) = \sum_{j=1}^d (x_j - a_j) p_j(x)$, hence Theorem A.60 on differentiation of a product and its Leibniz's formula give

$$f \in \mathcal{C}^1(B; E) \tag{11.7}$$

and

$$\begin{aligned} \partial_i f(x) &= p_i(x) + \sum_{j=1}^d (x_j - a_j)\partial_i p_j(x) = \\ &= \int_0^1 q_i(a+t(x-a)) + t\sum_{j=1}^d (x-a)_j \partial_i q_j(a+t(x-a))\,\mathrm{d}t. \end{aligned} \tag{11.8}$$

3. *Use of Poincaré's hypothesis.* According to the expression for the change of variables in a derivative (Theorem A.64 (a) for $T(t) = a + t(x-a)$), we have, since $\mathrm{d}T_j(t)/\mathrm{d}t = (x-a)_j$,

$$\frac{\mathrm{d}}{\mathrm{d}t}\big(q_i(a+t(x-a))\big) = \sum_{j=1}^d \partial_j q_i(a+t(x-a))(x-a)_j,$$

With this equality and the hypothesis $\partial_i q_j = \partial_j q_i$, (11.8) gives

$$\partial_i f(x) = \int_0^1 q_i(a+t(x-a)) + t\frac{\mathrm{d}}{\mathrm{d}t}\big(q_i(a+t(x-a))\big)\,\mathrm{d}t.$$

Again using Leibniz's formula and then the expression for the integral of a derivative from Theorem A.79, we finally obtain

$$\partial_i f(x) = \int_0^1 \frac{\mathrm{d}}{\mathrm{d}t}\big(tq_i(a+t(x-a))\big)\,\mathrm{d}t = q_i(x),$$

and hence

$$\nabla f = q.$$

4. *Regularity of f.* Since $f \in \mathcal{C}^1(B; E)$ from (11.7), and since $\partial_i f = q_i$ which is continuously differentiable by hypothesis,

$$f \in \mathcal{C}^2(B; E).$$

5. *Verification of (11.6).* This equality is given by Theorem A.92 on differentiation under the integral sign for the functions defined on the product $B \times [0, 1]$ by

$$h(x, t) = q_j(a + t(x - a)) \text{ and } g_i(x, t) = t\partial_i q_j(a + t(x - a)).$$

More precisely, we apply the said theorem to $B' \times (0, 1)$, where B' is an open ball centered at a, containing x and strictly included in B. Let us check the hypotheses of the said theorem.

— The functions h and g_i are uniformly continuous and bounded on $B' \times (0, 1)$, since they are so on the compact set $\overline{B'} \times [0, 1]$ by Heine's theorem (Theorem A.32), because they are continuous there.

— The mapping $x \mapsto h(x, t)$ is differentiable and $\partial h/\partial x_i(x, t) = g_i(x, t)$, for every fixed t. This is straightforward and finishes the proof of (11.6), and therefore that of Theorem 11.5. □

Chapter 12

Properties of Primitives of Distributions

This chapter is dedicated to the properties of distributions f which are primitives of a field q, i.e. such that $\nabla f = q$. In other words, we deduce some properties of f from those of its gradient. The existence of primitives will be the topic of the following chapter; here, we focus on the case where they exist.

We show the following properties.

— A distribution whose derivatives are zero is a locally constant function, i.e. constant on each connected component of its domain of definition (Theorem 12.3), which is far from being elementary. A field q which has a primitive therefore has infinitely many, each differing by a constant on each connected component (Theorem 12.6).

— A distribution with continuous derivatives is a continuously differentiable function (Theorem 12.4).

— A harmonic distribution is an infinitely differentiable function (Theorem 12.10).

— A distribution with a continuous Laplacian is a continuously differentiable function (Theorem 12.11).

We complete them as follows.

— We provide a locally explicit formulation of a primitive (Theorem 12.7).

— We construct a mapping $q \mapsto f$, from the space of gradients $\mathcal{D}'_\nabla(\Omega; E^d)$ into $\mathcal{D}'(\Omega; E)$, which is sequentially continuous (Theorem 12.9).

— We show that a sequence of distributions whose gradients converge is itself convergent, once conveniently normalized (Theorem 12.8).

This is based on the representation formulae for a distribution by its derivatives

$$f = f \diamond \eta_n + \sum_{i=1}^d \partial_i f \diamond \partial_i \gamma_n, \quad (f \diamond \eta_n)(x) = (f \diamond \eta_n)(a) + \int_{\Gamma(a,x)} \nabla f \diamond \gamma_n \cdot \mathrm{d}\ell,$$

where γ_n is the localized elementary potential and η_n is a correction term. These formulae, established in Theorems 12.1 and 12.2, take place respectively on $\Omega_{1/n}$ (which is Ω with a neighborhood of width $1/n$ of its boundary removed) and in its connected subsets, $\Gamma(a, x)$ being an arbitrary path linking a to x in $\Omega_{1/n}$.

12.1. Representation by derivatives

We begin with two representation formulae for distributions via weighting of its derivatives or of its Laplacian, *modulo* a regular correction term [1].

1. **History of the represention of a distribution in terms of its gradient. Real distributions.** Vladimir Ivanovitch KRYLOFF established an analogous formula to (12.1) for real functions on all of $\mathbb{R}^d$ in 1947 [47].

Theorem 12.1.– *Let $f \in \mathcal{D}'(\Omega; E)$, where Ω is an open subset of $\mathbb{R}^d$ and E is a Neumann space. Moreover let γ_n be the localized elementary potential and η_n be the correction term introduced in Theorem 9.17 for $r = 1/n$, where $n \in \mathbb{N}^*$.*

Then, on the open set $\Omega_{1/n} \stackrel{\text{def}}{=} \{x \in \Omega : B(x, 1/n) \in \Omega\}$,

$$f = f \diamond \eta_n + \sum_{i=1}^{d} \partial_i f \diamond \partial_i \gamma_n \qquad (12.1)$$

and

$$f = f \diamond \eta_n - \Delta f \diamond \gamma_n. \qquad (12.2)$$

Moreover,

$$f \diamond \eta_n \in \mathcal{C}^\infty(\Omega_{1/n}; E),$$

and $f \diamond \eta_n \in \mathbf{C}^\infty_{\mathrm{b}}(\omega; E)$ for every bounded open subset ω of $\Omega_{1/n}$. ∎

Recall that we denote by $B(x, r)$ the closed ball with center x and radius r.

Global representations on $\mathbb{R}^d$. For $\Omega = \mathbb{R}^d$ these representations and those from Theorem 12.2 are global, because $\mathbb{R}^d_{1/n} = \mathbb{R}^d$. □

Representations up to the boundary. When Ω is "Lipschitz", i.e. when its boundary is locally the graph of a Lipschitz function, local representations up to the boundary can be obtained by replacing the ball $B(0, 1/n)$ by a cone of vertex 0; this is detailed in the comment *Explicit primitive up to the boundary*, p. 256. □

Utility of the representation formulae. These formulae and those of Theorem 12.2 are the basis for all results relating to primitives of distributions: explicit expression, continuous dependence, uniqueness and others in this chapter, and construction of primitives in the following chapter. □

Proof of Theorem 12.1. **1.** Formula (12.1). Recall that, according to Theorem 9.17, $\gamma_n \in \mathcal{D}'_B(\mathbb{R}^d)$ and $\eta_n \in \mathcal{C}^\infty_B(\mathbb{R}^d)$, where $B = B(0, 1/n)$, so that, by Definition 7.12 of weighting,

$$f \diamond \gamma_n \in \mathcal{D}'(\Omega_{;1/n} E), \quad f \diamond \eta_n \in \mathcal{D}'(\Omega_{1/n}; E).$$

According to Theorem 9.17 (c),

$$\delta_0 = \eta_n - \Delta \gamma_n.$$

Laurent SCHWARTZ extended Kryloff's formula to distributions in $\mathcal{D}'(\mathbb{R}^d)$, in 1950 [69, Chap. VI, § 6, No. 8, p. 183].

Vector values. Jacques SIMON established the formulae of Theorems 12.1 and 12.2 for E a Banach space in 1993 [79, Lemma 2, p. 204].

Since the Dirac mass is the neutral element of weighting (Theorem 7.19) it follows that, on $\Omega_{1/n}$,

$$f = f \diamond \delta_0 = f \diamond \eta_n - f \diamond \Delta\gamma_n.$$

Whence, since $f \diamond \partial_i g = -\partial_i f \diamond g$ (Theorem 7.17),

$$f = f \diamond \eta_n + \sum_{i=1}^{d} \partial_i f \diamond \partial_i \gamma_n.$$

2. *Formula (12.2).* Formula (12.1) implies, since $\partial_i f \diamond \partial_i g = -\partial_i(\partial_i f) \diamond g$, that

$$f = f \diamond \eta_n + \sum_{i=1}^{d} \partial_i f \diamond \partial_i \gamma_n = f \diamond \eta_n - \Delta f \diamond \gamma_n.$$

3. *Regularity of $f \diamond \eta_n$.* Since $\eta_n \in \mathcal{D}_B(\mathbb{R}^d)$ (Theorem 9.17 (b)), the regularizing property of the weighting by a regular function (Theorem 7.5) gives

$$f \diamond \eta_n \in \mathcal{C}^\infty(\Omega_{1/n}; E).$$

If ω is a bounded open subset of $\Omega_{1/n}$, then $\omega + B$ is a bounded open subset of Ω, and the support of η_n is included in B, again by Theorem 9.17 (b). The regularizing property of weighting from Theorem 7.8 thus gives $f \diamond \eta_n \in \mathbf{C}^\infty_{\mathrm{b}}((\omega + B)_B; E)$. That is (since, by Definition 7.1, $(\omega + B)_B = \omega$),

$$f \diamond \eta_n \in \mathbf{C}^\infty_{\mathrm{b}}(\omega; E). \ \square$$

Let us now give a representation formula for a distribution solely in terms of its derivatives, up to a constant, on connected sets.

Theorem 12.2.– *Let $f \in \mathcal{D}'(\Omega; E)$, where Ω is an open subset of $\mathbb{R}^d$ and E is a Neumann space. Moreover let γ_n be the localized elementary potential and η_n the correction term introduced in Theorem 9.17 for $r = 1/n$, where $n \in \mathbb{N}^*$, and $\Omega_{1/n} = \{x \in \Omega : B(x, 1/n) \in \Omega\}$. Finally, let ω be a connected open subset of $\Omega_{1/n}$ and $a \in \omega$.*

Then, on ω,

$$f = (f \diamond \eta_n)(a) + (\nabla f \diamond \eta_n)^* + \sum_{i=1}^{d} \partial_i f \diamond \partial_i \gamma_n,$$

where $(\nabla f \diamond \eta_n)^* \in \mathcal{C}^\infty(\omega; E)$ is defined, for every $x \in \omega$, by

$$(\nabla f \diamond \eta_n)^*(x) \overset{\text{def}}{=} \int_{\Gamma(a,x)} \nabla f \diamond \eta_n \cdot \mathrm{d}\ell,$$

where $\Gamma(a, x)$ is an arbitrary C^1 path linking a to x in ω and $\nabla f \diamond \eta_n \in \mathcal{C}^\infty(\omega; E^d)$. ∎

Proof. ***1. Preliminaries.*** Theorem 12.1 gives, among others,

$$f \diamond \eta_n \in \mathcal{C}^\infty(\omega; E).$$

Applying this to ∇f, instead of f, we obtain

$$\nabla f \diamond \eta_n \in \mathcal{C}^\infty(\omega; E^d).$$

2. *Representation formula.* Since ω is open and connected, there exists a path $\Gamma(a, x)$ of class $\mathcal{C}^1$ linking a to x in ω (Theorem A.16). The expression for the line integral of the gradient of a $\mathcal{C}^1$ function from Theorem 10.7 then gives

$$(f \diamond \eta_n)(x) = (f \diamond \eta_n)(a) + \fint_{\Gamma(a,x)} \nabla(f \diamond \eta_n) \cdot \mathrm{d}\ell.$$

Since $\nabla(f \diamond \eta_n) = (\nabla f) \diamond \eta_n$ according to Theorem 7.17, this can be written

$$f \diamond \eta_n = (f \diamond \eta_n)(a) + (\nabla f \diamond \eta_n)^*. \qquad (12.3)$$

The representation formula (12.1) from Theorem 12.1 hence gives the stated formula.

3. *Independence of* $(\nabla f \diamond \eta_n)^*$ *on* Γ. Equality (12.3) shows that $(\nabla f \diamond \eta_n)^*(x)$, which is $\fint_{\Gamma(a,x)} \nabla f \diamond \eta_n \cdot \mathrm{d}\ell$, does not depend on the path Γ linking a to x in ω.

4. *Regularity of* $(\nabla f \diamond \eta_n)^*$. So, $(\nabla f \diamond \eta_n)^*$ is a function equal to $f \diamond \eta_n$ up to the addition of $(f \diamond \eta_n)(a)$, hence it is infinitely differentiable too. □

Representations without a correction term. Every distribution $f \in \mathcal{D}'(\mathbb{R}^d)$ **with compact support** satisfies

$$f = \sum_{i=1}^{d} \partial_i \xi \diamond \partial_i \check{f}, \qquad (12.4)$$

where ξ is the elementary potential defined in Theorem 9.10 and $\check{f}$ is the image of f under the symmetry $x \mapsto -x$. Indeed, applying Theorems 7.19, 8.25, 9.10 and 7.17 successively, we obtain

$$f = f \diamond \delta_0 = \check{\delta}_0 \diamond \check{f} = \delta_0 \diamond \check{f} = -\Delta\xi \diamond \check{f} = \sum_{i=1}^{d} \partial_i \xi \diamond \partial_i \check{f}.$$

Observe that here we can also use the convolution, i.e., see (7.15), p. 155,

$$f = \sum_{i=1}^{d} \partial_i \xi \star \partial_i f.$$

These formulae do not extend to any distribution on $\mathbb{R}^d$, because the right-hand side in (12.4) is not defined if the support of f is not compact. Nor do they extend to any distribution on an open set $\Omega \neq \mathbb{R}^d$, since the support of f is not compact unless f is zero in a whole neighborhood of the boundary of Ω.

It is to overcome these difficulties that:

— we have replaced the potential ξ with the localized potential $\gamma_n = \theta_n \xi$; this brings out the additional term $f \diamond \eta_n$, which is regular, in the representation formula of Theorem 12.1;

— we have introduced the weighting (of a distribution on Ω), in place of the convolution (of a distribution on $\mathbb{R}^d$). □

12.2. Distribution whose derivatives are zero or continuous

Let us show that a distribution whose derivatives are zero is a **locally constant** function [2].

Theorem 12.3.– *Let $f \in \mathcal{D}'(\Omega; E)$, where Ω is an open subset of $\mathbb{R}^d$ and E is a Neumann space, be such that*

$$\nabla f = 0.$$

Then, f is a constant function on each connected component of Ω.

If in addition Ω is connected, then f is a constant function on all of Ω. ▮

Proof. **1.** *General case.* Let $a \in \Omega$, $n \in \mathbb{N}^*$ be such that the open set $\Omega_{1/n}$ (which is $\{x \in \Omega : B(x, 1/n) \subset \Omega\}$) contains a, and ω be an open ball centered at a included in $\Omega_{1/n}$. Since $\nabla f = 0$, the identity from Theorem 12.2 reduces to

$$f = (f \diamond \eta)(a) \text{ on } \omega.$$

Hence,

$$f \text{ is a constant function on a ball around each point in } \Omega.$$

Given $c \in E$, the set $X_c \stackrel{\text{def}}{=} \{x \in \Omega : f(x) = c\}$ is therefore open. The set $Y_c \stackrel{\text{def}}{=} \{x \in \Omega : f(x) \neq c\}$ is too since it is the union of the open sets X_d for $d \neq c$.

Let Λ be a connected component of Ω and $b \in \Lambda$. The disjoint open sets $X_{f(b)}$ and $Y_{f(b)}$ cover Λ, hence one covers it and the other is empty, by Definition 10.20 of connectedness. Since $X_{f(b)}$ is not empty, it is the one that covers Λ. Hence,

$$f \text{ is constant on } \Lambda, \text{ equal to } f(b).$$

2. *Connected case.* If Ω is connected, then it is its only connected component, thus f is constant on it. □

Elementary, my dear Watson? That a distribution whith a zero gradient be a constant function is not as straightforward to prove as it is to state. Our proof is based, via the representation formula (12.1), on weighting, localized potential, etc.

Laurent SCHWARTZ, for his part, obtained this property after constructing primitives on parallelepipeds under Poincaré's condition, which is not simple either [69, Chap. II, § 6, Theorem VI]. □

---

2. **History of Theorem 12.3.** Laurent SCHWARTZ stated in 1945 [68, Theorem, p. 64] and proved in 1950 [69, Chap. II, § 6, Theorem VI, p. 59] that a distribution in $\mathcal{D}'(\mathbb{R}^d)$ with zero first-order partial derivatives is a constant function.

Let us show that a distribution whose derivatives are continuous functions is itself a continuous function [3].

Theorem 12.4.– *Let $f \in \mathcal{D}'(\Omega; E)$, where Ω is an open subset of $\mathbb{R}^d$ and E is a Neumann space, be such that, for every $i \in [\![1, d]\!]$,*

$$\partial_i f \in \mathcal{C}(\Omega; E).$$

Then,

$$f \in \mathcal{C}^1(\Omega; E)$$

and the $\partial_i f$ are the derivatives of f also in the function sense. ▮

Proof. **1.** *Continuity of f.* Let us focus on the last term in the representation formula (12.1) from Theorem 12.1, namely the following equality on the open set $\Omega_{1/n} = \{x \in \Omega : B(x, 1/n) \subset \Omega\}$,

$$f = f \diamond \eta_n + \sum_{i=1}^{d} \partial_i f \diamond \partial_i \gamma_n. \tag{12.5}$$

The localized elementary potential γ_n is, according to Theorem 9.17 (a), infinitely differentiable away from 0, its support is included in the ball $B(0, 1/n)$, and

$$|\partial_i \gamma_n(x)| \leq b|x|^{1-d}.$$

Since the weighting by such a singular weight preserves continuity (Theorem 9.22), the hypothesis $\partial_i f \in \mathcal{C}(\Omega; E)$ implies

$$\partial_i f \diamond \partial_i \gamma_n \in \mathcal{C}(\Omega_{1/n}; E).$$

Moreover, $f \diamond \eta_n \in \mathcal{C}^\infty(\Omega_{1/n}; E)$ (Theorem 12.1, again), thus identity (12.5) gives

$$f \in \mathcal{C}(\Omega_{1/n}; E).$$

The $\Omega_{1/n}$ being open and covering Ω, f is in particular continuous on a ball around any point in Ω, hence

$$f \in \mathcal{C}(\Omega; E).$$

3. **History of Theorem 12.4.** Laurent SCHWARTZ proved in 1950 [69, Chap. II, § 6, Theorem VII, p. 61] that a distribution of $\mathcal{D}'(\mathbb{R}^d)$ whose derivatives are continuous is a continuously differentiable function, whose derivatives in the function sense coincide with those in the distribution sense.

2. *Differentiability of f.* Let $(\rho_n)_{n\geq 1}$ be a regularizing sequence (Definition 8.1). Since f and $\partial_i f$ are continuous, the properties of regularization of continuous functions from Theorem 8.7 (a) give, for every $k \in \mathbb{N}^*$ and $n \geq k$, when $n \to \infty$,

$$f \diamond \rho_n \to f \ \text{ in } \mathcal{C}(\Omega_{1/k}; E), \tag{12.6}$$

$$(\partial_i f) \diamond \rho_n \to \partial_i f \ \text{ in } \mathcal{C}(\Omega_{1/k}; E). \tag{12.7}$$

Since $(\partial_i f) \diamond \rho_n = \partial_i (f \diamond \rho_n)$ (Theorem 7.17), convergence (12.7) gives

$$\partial_i (f \diamond \rho_n) \to \partial_i f \ \text{ in } \mathcal{C}(\Omega_{1/k}; E). \tag{12.8}$$

So far, $\partial_i f$ and $\partial_i (f \diamond \rho_n)$ were derivatives in the distribution sense. Since $f \diamond \rho_n$ is, by Definition 8.2, a regular function, $\partial_i (f \diamond \rho_n)$ coincides with its derivative in the function sense by Theorem 5.5.

Convergences (12.6) and (12.8) then imply, according to Theorem A.56, that

$$f \in \mathcal{C}^1(\Omega_{1/k}; E).$$

Since the $(\Omega_{1/k})_{k\geq 1}$ cover Ω,

$$f \in \mathcal{C}^1(\Omega; E).$$

3. *Derivatives in the function sense.* Now Theorem 5.5 shows that $\partial_i f$ is equally the derivative in the function sense. (Theorem A.56 asserts this too.) □

Uniform continuity. If Ω is "Lipschitz", which corresponds to a "uniform cone" property, and if the derivatives $\partial_i f$ of a distribution f are continuous and bounded, then f is uniformly continuous. □

12.3. Uniqueness of a primitive

Let us introduce the subspace of fields that have a primitive.

Definition 12.5.– *Let Ω be an open subset of $\mathbb{R}^d$ and E a Neumann space. We denote*

$$\mathcal{D}'_\nabla(\Omega; E^d) \overset{\text{def}}{=} \{q \in \mathcal{D}'(\Omega; E^d) : \exists f \in \mathcal{D}'(\Omega; E) \text{ such that } \nabla f = q\},$$

a vector space that we endow with the semi-norms of $\mathcal{D}'(\Omega; E^d)$. ▮

Let us show the uniqueness of a primitive, provided it exists, up to the addition of a constant on each connected component of Ω.

Theorem 12.6.– *Let $q \in \mathcal{D}'_{\nabla}(\Omega; E^d)$, where Ω is an open subset of $\mathbb{R}^d$ and E is a Neumann space. Then:*

(a) *The field q has infinitely many primitives, i.e. distributions $f \in \mathcal{D}'(\Omega; E)$ such that $\nabla f = q$.*

(b) *Every primitive comes from a given one by adding an arbitrary constant on each connected component Ω_m of Ω.*

(c) *Given, for each connected component Ω_m of Ω, $c_m \in E$ and*

$$\phi_m \in \mathcal{D}(\Omega_m) \text{ such that } \int_{\Omega_m} \phi_m \neq 0,$$

there exists a single primitive f such that, for each m,

$$\langle f, \phi_m \rangle_{\Omega_m} = c_m. \blacksquare$$

Abuse of notation. We denote $\langle f, \phi_m \rangle_{\Omega_m}$ in place of $\langle f|_{\Omega_m}, \phi_m \rangle_{\Omega_m}$ in order to be more legible. □

A simpler case. If Ω is connected, its only connected component is Ω itself, which simplifies Statements (b) and (c). □

Proof of Theorem 12.6. Let g be a primitive, namely $\nabla g = q$.

1. *Property (a).* Every distribution $f = g + c$ is another primitive, provided that c is constant on each Ω_m.

2. *Property (b).* Every other primitive f satisfies $\nabla(f - g) = 0_{E^d}$, therefore $f - g$ is constant on each connected component Ω_m according to Theorem 12.3.

3. *Property (c).* Due to (b), any other primitive is of the form $f = g + e_m$ on Ω_m where $e_m \in E$, thus the condition $\langle f, \phi_m \rangle_{\Omega_m} = c_m$ can be written (with Theorem 3.9)

$$\langle g, \phi_m \rangle_{\Omega_m} + e_m \int_{\Omega_m} \phi_m = c_m,$$

which uniquely determines the value of e_m. □

12.4. Locally explicit primitive

Let us give an explicit expression [4], on any connected subset of the domain without an arbitrary neighborhood of its boundary, of a primitive of a distribution field, provided it exists.

4. **History of obtaining an explicit primitive.** Laurent SCHWARTZ showed in 1950 [69, Chap. VI, § 6, Formula (VI,6;5), p. 183] that if $q \in \mathcal{D}'_{\nabla}(\mathbb{R}^d; \mathbb{R}^d)$ has a compact support, then $f = \sum_{i=1}^d q_i \star \partial_i \xi$, where ξ is the elementary Newtonian potential, is a primitive of it.

Theorem 12.7.– *Let*

$$q \in \mathcal{D}'_{\nabla}(\Omega; E^d),$$

where Ω *is an open subset of* $\mathbb{R}^d$ *and* E *is a Neumann space.*

Let γ_n *be the localized elementary potential and* η_n *the correction term introduced in Theorem 9.17 for* $r = 1/n$*, and* $\Omega_{1/n} \stackrel{\text{def}}{=} \{x \in \mathbb{R}^d : B(x, 1/n) \subset \Omega\}$*, where* $n \in \mathbb{N}^*$*. Finally, let* ω *be a connected open subset of* $\Omega_{1/n}$ *and* $a \in \omega$*.*

Then, we define $f_\omega \in \mathcal{D}'(\omega, E)$ *such that*

$$\nabla f_\omega = q \ \ on\ \omega$$

by

$$f_\omega \stackrel{\text{def}}{=} (q \diamond \eta_n)^* + \sum_{i=1}^{d} q_i \diamond \partial_i \gamma_n,$$

where $(q \diamond \eta_n)^* \in \mathcal{C}^\infty(\omega; E)$ *is defined, for every* $x \in \omega$*, by*

$$(q \diamond \eta_n)^*(x) \stackrel{\text{def}}{=} \int_{\Gamma(a,x)} q \diamond \eta_n \cdot \mathrm{d}\ell,$$

where $\Gamma(a, x)$ *is an arbitrary* C^1 *path linking* a *to* x *in* ω*.* ▮

Proof. By Definition 12.5 of $\mathcal{D}'_{\nabla}(\Omega; E^d)$, there exists $f \in \mathcal{D}'(\Omega, E)$ such that

$$\nabla f = q \ \ \text{on}\ \Omega.$$

The representation formula for f by its derivatives from Theorem 12.2 is written, by replacing ∇f by q, as

$$f = (f \diamond \eta_n)(a) + (q \diamond \eta_n)^* + \sum_{i=1}^{d} q_i \diamond \partial_i \gamma_n \ \ \text{on}\ \omega.$$

That is

$$f = (f \diamond \eta_n)(a) + f_\omega.$$

Hence,

$$\nabla f_\omega = \nabla f = q \ \ \text{on}\ \omega. \ \square$$

Sergei L'vovich SOBOLEV would have previously established a formula $f = \sum_{i=1}^{d} q_i \star \zeta_i$, for an unbounded set Ω which has the *cone property*, where the ζ_i have their support in an infinite cone.
Jacques SIMON established the representation of Theorem 12.7 in 1993 [79, Theorem 10, p. 208] for an arbitrary Ω and E a Banach space. (When Ω is bounded, formulae of Schwartz and Sobolev are impossible since the supports of ξ and ζ_i are unbounded; they need to be adjusted by a *correction term*, here $(q \diamond \eta_n)^*$).

Complement. Observe that, under the hypotheses of Theorem 12.7, for any primitive f of q, there exists $c_\omega \in E$ such that

$$f = f_\omega + c_\omega \ \text{ on } \omega.$$

Indeed, by hypothesis, ω is connected and $\nabla(f - f_\omega)$ is zero on it, so $f - f_\omega$ is constant there according to Theorem 12.3. □

Explicit primitive up to the boundary. The explicit expression for the primitives of a distribution from Theorem 12.7 is not valid up to the boundary $\partial\Omega$ of Ω, since the points of ω are situated at a distance $\geq 1/n$ from $\partial\Omega$.

A primitive up to a regular section of the boundary is obtained by replacing the ball $B = B(0, 1/n)$ with a conic section K of B, in which case Ω_K goes up to a portion of the boundary as in Figure 7.2, p. 143.

More precisely [SIMON, 79 or 84], we obtain a representation on the domain Ω_K by completing the radial localization by θ done in Theorem 9.17 with a localization in the cone K by a function of the form $\sigma(x/|x|)$ where $\sigma \in \mathcal{C}^\infty(\partial B)$ is zero on $\partial B \setminus K$. We thus obtain analogous representation formulae to those in Theorems 12.1, 12.2 and 12.7, with functions $\zeta_{i,n} \in L^1(\mathbb{R}^d)$ having their support in K instead of the $\partial_i \gamma_n$ and with another correction term η_n.

For example, if Ω is bounded with a $\mathcal{C}^1$ boundary, there exist a finite number of cones K_m such that the domains Ω_{K_m} cover Ω, and thus such that an explicit primitives on all of Ω may be obtained by connecting the explicit primitives on each of them. □

Explicit primitive of a continuous field. If $q \in \mathcal{C}(\Omega; E^d)$ and Ω is star-shaped with respect to a point a, the expression for the primitive from Theorem 12.7 can be written, by choosing for $\Gamma(a, x)$ the rectilinear path linking a to x and by using expression (9.55), p. 220, of the weighting of a continuous function by a singular weight, on $\Omega_{1/n}$,

$$f(x) = (x - a) \cdot \int_0^1 (q \diamond \eta_n)(a + t(x - a))\, \mathrm{d}t + \sum_{i=1}^d (q_i \diamond \partial_i \gamma_n)(x) =$$

$$= (x - a) \cdot \int_0^1 \int_{|y|<1/n} q(a + t(x - a) + y)\, \eta_n(y)\, \mathrm{d}y\, \mathrm{d}t + \int_{|y|<1/n} q(x + y) \cdot \nabla \gamma_n(y)\, \mathrm{d}y.$$

This formula is more complicated than that obtained for functions in Volume 2 [82, Theorem 9.1 (c)], namely

$$f(x) = c + (x - a) \cdot \int_0^1 q(a + t(x - a))\, \mathrm{d}t,$$

which applies in all of Ω. But it has the advantage that it extends to distributions by replacing the integrals with respect to y with the corresponding weightings. Note that these two formulae do not generally define the same primitive (they differ by a constant). □

12.5. Continuous primitive mapping

We begin with the convergence of distributions whose gradients converge [5].

5. **History of the convergence of distributions whose gradients converge.** Theorems 12.8 and 12.9 are new, included for real distributions.

Theorem 12.8.– *Let Ω be an open subset of $\mathbb{R}^d$, E a Neumann space and, for each connected component Ω_m of Ω, let*

$$\phi_m \in \mathcal{D}(\Omega_m) \text{ be such that } \int_{\Omega_m} \phi_m \neq 0.$$

Let $(f_n)_{n\in\mathbb{N}}$ be a sequence and f an element of $\mathcal{D}'(\Omega; E)$ such that

$$\nabla f_n \to \nabla f \text{ in } \mathcal{D}'(\Omega; E^d)$$

and, for each m,

$$\langle f_n, \phi_m \rangle_{\Omega_m} \to \langle f, \phi_m \rangle_{\Omega_m} \text{ in } E.$$

Then,

$$f_n \to f \text{ in } \mathcal{D}'(\Omega; E). \blacksquare$$

Before proving this result, let us use it to construct a sequentially continuous linear primitive mapping.

Theorem 12.9.– *Let Ω be an open subset of $\mathbb{R}^d$, E a Neumann space and, for each connected component Ω_m of Ω, let*

$$\phi_m \in \mathcal{D}(\Omega_m) \text{ be such that } \int_{\Omega_m} \phi_m \neq 0.$$

For every $q \in \mathcal{D}'_\nabla(\Omega; E^d)$, let $f \in \mathcal{D}'(\Omega; E)$ be the unique distribution (given by Theorem 12.6 (c)) such that

$$\nabla f = q, \quad \langle f, \phi_m \rangle_{\Omega_m} = 0_E, \ \forall m. \tag{12.9}$$

Then, the mapping $q \mapsto f$ is sequentially continuous linear from $\mathcal{D}'_\nabla(\Omega; E^d)$ into $\mathcal{D}'(\Omega; E)$. $\blacksquare$

Topology. Recall that the space $\mathcal{D}'_\nabla(\Omega; E^d)$ of distribution fields with a primitive is, by Definition 12.5, endowed with the semi-norms of $\mathcal{D}'(\Omega; E^d)$. $\square$

Notation. We could denote by ∇^{-1} the mapping constructed in Theorem 12.8, i.e. $\nabla^{-1}q \stackrel{\text{def}}{=} f$, and call it **an inverse mapping of the gradient**. It is defined from $\mathcal{D}'_\nabla(\Omega; E^d)$ into $\mathcal{D}'(\Omega; E)$. $\square$

A simpler case. If Ω is connected, its only connected component is Ω itself, which simplifies the statements of Theorems 12.8 and 12.9. $\square$

Proof of Theorem 12.9. **1.** *Linearity.* It obviously follows from (12.9).

2. *Continuity.* Let $(q_n)_{n\in\mathbb{N}}$ and q be such that

$$q_n \to q \ \text{ in } \mathcal{D}'_\nabla(\Omega; E^d). \tag{12.10}$$

According to theorem 12.6 (c), there exists a unique distribution $f \in \mathcal{D}'(\Omega; E)$ satisfying (12.9) and, for each $n \in \mathbb{N}$, a unique distribution $f_n \in \mathcal{D}'(\Omega; E)$ such that

$$\nabla f_n = q_n, \quad \langle f_n, \phi_m \rangle_{\Omega_m} = 0_E, \ \forall m.$$

Since $\mathcal{D}'_\nabla(\Omega; E^d)$ is, by Definition 12.5, endowed with the semi-norms of $\mathcal{D}'(\Omega; E^d)$, the convergence (12.10) takes place in the latter. So,

$$\nabla f_n \to \nabla f \ \text{ in } \mathcal{D}'(\Omega; E^d), \quad \langle f_n, \phi_m \rangle_{\Omega_m} = \langle f, \phi_m \rangle_{\Omega_m}, \ \forall m.$$

Theorem 12.8 then gives

$$f_n \to f \ \text{ in } \mathcal{D}'(\Omega; E). \ \Box$$

Let us come to the convergence of distributions whose gradients converge.

Proof of Theorem 12.8. **1. The case where Ω is connected.** Here, there is a single connected component, Ω itself, and a single function ϕ_m, that we denote by ϕ. Assume that

$$\nabla f_n \to 0 \ \text{ in } \mathcal{D}'(\Omega; E^d), \quad \langle f_n, \phi \rangle_\Omega \to 0_E. \tag{12.11}$$

We are going to show in seven steps that

$$\langle f_n, \varphi \rangle_\Omega \to 0_E. \tag{12.12}$$

1.a. *Localization.* Let $a \in \Omega$, $k \in \mathbb{N}^*$, $\Omega_{1/k} = \{x \in \mathbb{R}^d : B(x, 1/k) \subset \Omega\}$ and

$$\Omega^{(a)}_{1/k} \text{ be the connected component of } \Omega_{1/k} \text{ containing } a.$$

The $(\Omega^{(a)}_{1/k})_{k\in\mathbb{N}^*}$ form an open cover of Ω (Theorem 10.21), thus there exists a finite subcover of the compact set $\operatorname{supp}\varphi \cup \operatorname{supp}\phi$. Since the $\Omega^{(a)}_{1/k}$ grow with k, one of them covers Ω. Denote by ω this set $\Omega^{(a)}_{1/k}$; so

$$\omega \text{ is connected, and it contains } a \text{ and the supports of } \varphi \text{ and } \phi.$$

Then, $\varphi|_\omega \in \mathcal{D}(\omega)$ and, by Definition 6.1 of the restriction, $\langle f_n|_\omega, \varphi|_\omega \rangle_\omega = \langle f_n, \varphi \rangle_\Omega$. It suffices therefore to show, in omitting the restriction symbol, that

$$f_n \to 0 \ \text{ in } \mathcal{D}'(\omega; E). \tag{12.13}$$

***1.b.** Decomposition.* According to the representation formula of a distribution by its derivatives from Theorem 12.2,

$$f_n = h_n + \sum_{i=1}^{d} \partial_i f_n \diamond \partial_i \gamma_k + c_n \ \text{ on } \omega, \tag{12.14}$$

where $c_n \subset E$ and, for every $x \in \omega$,

$$h_n(x) = \int_{\Gamma(a,x)} \nabla f_n \diamond \eta_k \cdot \mathrm{d}\ell, \tag{12.15}$$

$\Gamma(a, x)$ being an arbitrary C^1 path linking a to x in ω (such a path exists due to Theorem A.16, since ω is open and connected).

Let us show that each term in the right-hand side of (12.14) goes to 0, as $n \to \infty$.

***1.c.** Convergence of h_n.* The weighting by a regular function η_k is sequentially continuous from $\mathcal{D}'(\Omega; E^d)$ into $\mathcal{C}^\infty(\Omega_{1/k}; E^d)$ (Theorem 7.11 (a)), and thus into $\mathcal{C}(\omega; E^d)$. Convergence (12.11) of ∇f_n therefore gives

$$\nabla f_n \diamond \eta_k \to 0 \ \text{ in } \mathcal{C}(\omega; E^d). \tag{12.16}$$

Denote by $\{\| \, \|_{E;\nu} : \nu \in \mathcal{N}_E\}$ the family of semi-norms of E and assume for a moment that, for every compact subset K of ω and every $\nu \in \mathcal{N}_E$, there exists a compact subset Q of Ω and $c \in \mathbb{R}$ such that, for every n,

$$\|h_n\|_{\mathcal{C}(\omega;E);K,\nu} \leq c \|\nabla f_n \diamond \eta_k\|_{\mathcal{C}(\omega;E^d);Q,\nu}. \tag{12.17}$$

Then, convergence (12.16) will imply that of h_n in $\mathcal{C}(\omega; E)$, and therefore, since the identity is sequentially continuous from $\mathcal{C}(\omega; E)$ into $\mathcal{D}'(\omega; E)$ (Theorem 3.8),

$$h_n \to 0 \ \text{ in } \mathcal{D}'(\omega; E). \tag{12.18}$$

***1.d.** Verification of (12.17).* By Definition 1.18 (a) of the semi-norms of $\mathcal{C}(\omega; E)$, it is a question of showing that

$$\sup_{x \in K} \|h_n(x)\|_{E;\nu} \leq c \sup_{z \in Q} \|(\nabla f_n \diamond \eta_k)(z)\|_{E^d;\nu}. \tag{12.19}$$

For this, we use the bound of the semi-norms of the line integral from Theorem 10.3, namely

$$\left\| \int_\Lambda \nabla f_n \diamond \eta_k \cdot \mathrm{d}\ell \right\|_{E;\nu} \leq \gamma_\Lambda \sup_{z \in [\Lambda]} \|(\nabla f_n \diamond \eta_k)(z)\|_{E^d;\nu}, \tag{12.20}$$

where $\gamma_\Lambda = |t_\mathrm{e} - t_\mathrm{i}| \sup_{t_\mathrm{i} \leq t \leq t_\mathrm{e}} |\Lambda'(t)|$ and $[\Lambda] = \{\Lambda(t) : t_\mathrm{i} \leq t \leq t_\mathrm{e}\}$.

If we knew how to bound $\gamma_{\Gamma(a,x)} \leq c$ and impose that $\Gamma(a,x)$ remains in Q when x spans the compact set K, we would obtain (12.19). Since we do not know how to do this for all these x, we will do it when x remains in a ball $\mathring{B}(y,\epsilon)$ and then we will reduce to a finite number of such balls covering K.

So let $B(y,\epsilon)$ be a closed ball included in ω and Λ be a path linking a to y. For each $x \in \mathring{B}(y,\epsilon)$, let Σ be the rectilinear path linking y to x. Then, $\Lambda \overrightarrow{\cup} \Sigma$ links a to x and the additivity of the line integral with respect to the paths, that is Definition 10.14 of the line integral along a concatenation, gives, with (12.15),

$$h_n(x) = \int_\Lambda \nabla f_n \diamond \eta_k \cdot \mathrm{d}\ell + \int_\Sigma \nabla f_n \diamond \eta_k \cdot \mathrm{d}\ell.$$

Using inequality (12.20) for Λ and for Σ, we obtain (since Σ is defined on $[0,1]$ by $\Sigma(t) = y + t(x-y)$ and so $\gamma_\Sigma = |x-y| \leq \epsilon$)

$$\sup_{x\in\mathring{B}(y,\epsilon)} \|h_n(x)\|_{E;\nu} \leq (\gamma_\Lambda + \epsilon) \sup_{z\in[\Lambda]\cup B(y,\epsilon)} \|(\nabla f_n \diamond \eta_k)(z)\|_{E^d;\nu}. \tag{12.21}$$

The open balls $\mathring{B}(y,\epsilon)$ cover K, thus there exists a finite subcover. Denote by y_j, ϵ_j and Λ_j, where $j \in [\![1,J]\!]$, the corresponding elements. Then, (12.21) gives the desired inequality (12.19) with

$$c = \sup_{1\leq j\leq J} \gamma_{\Lambda_j} + \epsilon_j, \quad Q = \bigcup_{1\leq j\leq J} [\Lambda_j] \cup B(y_j,\epsilon_j).$$

(The set Q is compact, because it is closed and bounded as is any finite union of closed bounded sets (Theorem A.11) and hence is compact in $\mathbb{R}^d$ due to the Borel–Lebesgue theorem (Theorem A.26 (b)).)

1.e. *Convergence of the second term in the right-hand side of (12.14).* The weighting by $\partial_i\gamma_k$ is sequentially continuous from $\mathcal{D}'(\Omega;E)$ into $\mathcal{D}'(\Omega_{1/k};E)$ (Theorem 7.15 (b)) and the restriction is sequentially continuous from $\mathcal{D}'(\Omega_{1/k};E)$ into $\mathcal{D}'(\omega;E)$ (Theorem 6.3), therefore convergence (12.11) of ∇f_n gives

$$\partial_i f_n \diamond \partial_i\gamma_k \to 0 \text{ in } \mathcal{D}'(\omega;E). \tag{12.22}$$

1.f. *Convergence of c_n.* Since ω contains the support of ϕ, this one belongs to $\mathcal{D}(\omega)$ and decomposition (12.14) gives

$$\langle f_n,\phi\rangle_\omega = \langle h_n,\phi\rangle_\omega + \sum_{i=1}^d \langle \partial_i f_n \diamond \partial_i\gamma_k, \phi\rangle_\omega + c_n \int_\omega \phi.$$

Let us examine these terms.

— Firstly, $\langle f_n,\phi\rangle_\omega$ goes to 0_E by hypothesis (12.11).

— On the other hand, convergences (12.18) and (12.22) imply the convergence of $\langle h_n, \phi\rangle_\omega$ and $\langle \partial_i f_n \diamond \partial_i \gamma_k, \phi\rangle_\omega$ to 0_E.

— Finally, $\int_\omega \phi$ is equal to $\int_\Omega \phi$ which is not zero, by hypothesis.

Therefore,

$$c_n \to 0_E. \tag{12.23}$$

1.g. *Conclusion.* With decomposition (12.14) of f_n, convergences (12.18), (12.22) and (12.23) give convergence (12.13), which itself gives (12.12).

2. The case of a general Ω. We return to the case where Ω has several connected components Ω_m and where

$$\nabla f_n \to \nabla f \text{ in } \mathcal{D}'(\Omega; E^d), \quad \langle f_n, \phi_m\rangle_{\Omega_m} \to \langle f, \phi_m\rangle_{\Omega_m}, \forall m.$$

Let $\varphi \in \mathcal{D}(\Omega)$. According to Theorem 6.8 on the additivity with respect to disjoint open sets (which is the case for connected components due to Theorem A.14), we have $\varphi|_{\Omega_m} \in \mathcal{D}(\Omega_m)$, only a finite number of these restrictions are not zero and, denoting by M the finite set of these m,

$$\langle f_n - f, \varphi\rangle_\Omega = \sum_{m\in M} \langle (f_n - f)|_{\Omega_m}, \varphi|_{\Omega_m}\rangle_{\Omega_m}.$$

Step 1 applied to the restriction of $f_n - f$ to Ω_m gives, for each m,

$$\langle (f_n - f)|_{\Omega_m}, \varphi|_{\Omega_m}\rangle_{\Omega_m} \to 0_E.$$

Therefore, $\langle f_n - f, \varphi\rangle_\Omega \to 0_E$. This holds for every $\varphi \in \mathcal{D}(\Omega)$, which proves that

$$f_n \to f \text{ in } \mathcal{D}'(\Omega; E). \ \square$$

12.6. Harmonic distributions, distributions with a continuous Laplacian

Let us show that a harmonic distribution is an infinitely differentiable function.

Theorem 12.10.– *Let $f \in \mathcal{D}'(\Omega; E)$, where Ω is an open subset of $\mathbb{R}^d$ and E is a Neumann space, be a **harmonic** distribution, i.e.*

$$\Delta f = 0.$$

Then,

$$f \in \mathcal{C}^\infty(\Omega; E). \ \blacksquare$$

Proof. The representation formula (12.2) from Theorem 12.1 reduces here to

$$f = f \diamond \eta_n \ \text{ in } \Omega_{1/n}.$$

Theorem 12.1 gives in addition $f \diamond \eta_n \in \mathcal{C}^\infty(\Omega_{1/n}; E)$. So, f is an infinitely differentiable function on each $\Omega_{1/n}$, and therefore on their union Ω. □

Uniform continuity. In dimension $d = 1$, harmonicity is written as $f'' = 0$. Hence, if Ω in an interval, $f(x) = c_0 + c_1 x$. So, f is uniformly continuous and it is bounded if Ω is bounded.

In dimension $d \geq 2$, even if Ω is "regular" and bounded, harmonicity does not imply that f is uniformly continuous, or is bounded, on all of Ω. □

Let us show that a distribution with a continuous Laplacian is a differentiable function.

Theorem 12.11.– *Let $f \in \mathcal{D}'(\Omega; E)$, where Ω is an open subset of $\mathbb{R}^d$ and E is a Neumann space, be such that*

$$\Delta f \in \mathcal{C}(\Omega; E).$$

Then,

$$f \in \mathcal{C}^1(\Omega; E). \blacksquare$$

Proof. According to the representation formula (12.2) from Theorem 12.1,

$$f = f \diamond \eta_n - \Delta f \diamond \gamma_n \ \text{ on } \Omega_{1/n}. \tag{12.24}$$

The localized elementary potential γ_n is, according to Theorem 9.17 (a), continuous away from 0, its support is included in the ball $B(0, 1/n)$, and

$$|\gamma_n(x)| \leq b|x|^{1-d}.$$

Since the weighting by such a singular weight preserves continuity (Theorem 9.22), hypothesis $\Delta f \in \mathcal{C}(\Omega; E)$ implies

$$\Delta f \diamond \gamma_n \in \mathcal{C}(\Omega_{1/n}; E).$$

Theorem 9.17 (a) gives in addition

$$|\partial_i \gamma_n(x)| \leq b|x|^{1-d},$$

hence, again by Theorem 9.22,

$$\Delta f \diamond \partial_i \gamma_n \in \mathcal{C}(\Omega_{1/n}; E).$$

However, $\partial_i(\Delta f \diamond \gamma_n) = -\Delta f \diamond \partial_i \gamma_n$ (according to Theorem 7.17), so

$$\partial_i(\Delta f \diamond \gamma_n) \in \mathcal{C}(\Omega_{1/n}; E).$$

Therefore,

$$\Delta f \diamond \gamma_n \in \mathcal{C}^1(\Omega_{1/n}; E).$$

Since $f \diamond \eta_n \in \mathcal{C}^\infty(\Omega_{1/n}; E)$ (again by Theorem 12.1), equality (12.24) gives

$$f \in \mathcal{C}^1(\Omega_{1/n}; E).$$

So, f is a continuously differentiable function on each $\Omega_{1/n}$, and thus on their union Ω.

□

Second derivatives. If Δf is continuous, the second derivatives of f are locally L^p for every finite p [AGMON–DOUGLIS–NIRENBERG, 1], but not necessarily for $p = \infty$. *A fortiori*, f is not necessarily twice differentiable and its Laplacian is not necessarily defined in the function sense. In particular:

there exists $f \in \mathcal{C}^1(\Omega)$ *with compact support such that* $\Delta f \in \mathcal{C}(\Omega)$ *but* $f \notin \mathcal{C}^2(\Omega)$. (12.25)

□

An example of the utility of distributions. There exist continuous functions, $g \in \mathcal{C}(\Omega)$, for which the Poisson equation

$$-\Delta u = g \text{ on } \Omega, \quad u = 0 \text{ on } \partial\Omega,$$

does not have a solution $u \in \mathcal{C}^2(\Omega)$, i.e. in the *classical* sense of twice continuously differentiable functions. For example, this is the case for $g = \Delta f$, where f satisfies (12.25).

In contrast, as we will see in Volume 7, the Poisson equation has a unique solution u in the distribution space $H^1(\Omega)$, the solution here being $u = -f$.

This illustrates the decisive utility of distributions for solving partial differential equations. Furthermore, even when a *classical* solution exists, it is harder to obtain (it is often obtained by showing that the distribution solution is regular). □

Chapter 13

Existence of Primitives

This chapter is dedicated to obtaining conditions for a distributions field $q = (q_1, \ldots, q_d)$ to have a primitive f, i.e. for $\nabla f = q$. The main conditions are the following.

— On an arbitrary open set Ω, it suffices that q is orthogonal to divergence-free test fields, namely that $\langle q, \psi \rangle = 0$ for every ψ such that $\nabla \cdot \psi = 0$. This is the *orthogonality theorem* for distributions (Theorem 13.5).

— When Ω is simply connected, it suffices that q satisfies Poincaré's condition $\partial_i q_j = \partial_j q_i$ for all i and j. This is *Poincaré's theorem* generalized to distributions (Theorem 13.7).

These conditions are necessary and sufficient.

We prove these results as follows.

— First of all, we reduce the existence of a primitive on Ω to the existence of a primitive on each of its subsets $\Omega_{1/n} = \{x : B(x, 1/n) \subset \Omega\}$, i.e. on Ω with a neighborhood of its boundary of width $1/n$ removed, by a method of gluing. This is the *peripheral gluing theorem* (Theorem 13.1).

— Next, we reduce the existence of a primitive of a distribution field q to the existence of a primitive of a function field, namely $q \diamond \eta_n$ on $\Omega_{1/n}$, when q satisfies Poincaré's condition. This is the *theorem on reducing to the function case* (Theorem 13.2).

— The orthogonality theorem for distributions then easily follows from the orthogonality theorem for functions (Theorem 11.4).

— Obtaining Poincaré's theorem for distributions is more delicate, because the $\Omega_{1/n}$ are not necessarily simply connected, so we cannot directly apply Poincaré's theorem for the functions (Theorem 11.5). We overcome this difficulty by juggling with n and line integrals along homotopic closed paths.

We also show that every field that is a limit of gradients has a primitive (Theorem 13.13).

We add the following.

— In dimension two, every divergence-free distribution field on a simply connected set is derived from a "current function", which here is a scalar distribution. This is *Haar's lemma* (Theorem 13.8).

— If Ω is simply connected, the existence of local primitives implies the existence of a global primitive (Theorem 13.9). We give counterexamples in the converse case (Theorems 13.10 and 13.11).

— We compare the various conditions of existence of a primitive, in § 13.7.

13.1. Peripheral gluing

Let us show that, if a distribution field is a gradient on each $\Omega_{1/n}$, then it is a gradient on all of Ω. We call this result the **peripheral gluing theorem** [1].

Theorem 13.1.– *Let $q \in \mathcal{D}'(\Omega; E^d)$, where Ω is an open subset of $\mathbb{R}^d$ and E is a Neumann space, and $\Omega_{1/n} \stackrel{\text{def}}{=} \{x \in \mathbb{R}^d : B(x, 1/n) \subset \Omega\}$.*

Suppose that, for every $n \in \mathbb{N}^*$, there exists $f_n \in \mathcal{D}'(\Omega_{1/n}; E)$ such that

$$\nabla f_n = q \text{ in } \Omega_{1/n}.$$

Then, there exists $f \in \mathcal{D}'(\Omega, E)$ such that

$$\nabla f = q. \blacksquare$$

Caution. The primitives f_n **do not necessarily glue together**, since they may differ by a constant. Even after subtracting the said constant, they may not always glue; for example if $\Omega_{1/n}$ is not connected yet $\Omega_{1/(n+1)}$ is.

This is the case if Ω is, like in Figure 13.1, a "dumbbell" made of two balls of radius 1 connected by a cylinder of radius $1/2$; then $\Omega_{1/2}$ is made of two disjoint "pointed" balls of radius $1/2$ while $\Omega_{1/3}$ is a connected dumbbell. Then, a primitive of $q = 0$ is given on $\Omega_{1/2}$ by $f_2 = 0$ on the left-hand ball and $f_2 = 1$ on the right-hand ball; it cannot be glued with the primitive given on $\Omega_{1/3}$ by $f_3 = 0$, even after subtracting a constant from the latter.

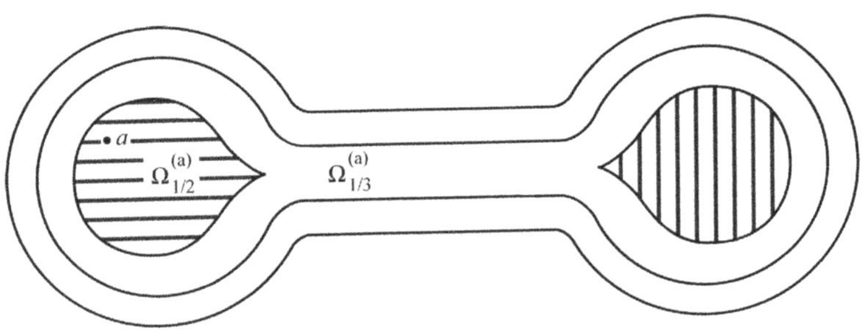

Figure 13.1. *Connected component $\Omega^{(a)}_{1/n}$ of $\Omega_{1/n}$ containing a. $\Omega^{(a)}_{1/2}$ is hatched horizontally, $\Omega_{1/2}$ is the union of the two hatched areas and $\Omega^{(a)}_{1/3}$ is the inner dumbbell; it coincides with $\Omega_{1/3}$* $\square$

1. **History of the peripheral gluing.** Theorem 13.1 has its seeds in the constructive proof of the existence of a primitive given in [77, § 5, p. 1171–1172].

Terminology. We call the gluing in Theorem 13.1 **peripheral** to distinguish it from the *gluing of local primitives* (Theorem 13.9), in which we glue on all of Ω, and not only along its periphery, primitives which are only given on balls.

We speak of the gluing **of** primitives, not **of the** primitives, to underline that, as we explained, the primitives f_n cannot necessarily be glued together, but that there exist some that do. □

Utility of Theorem 13.1. The peripheral gluing is a step in the *reduction to the function case* (Theorem 13.2) which is itself a step in the proof of the *orthogonality theorem* (Theorem 13.5) which is the main result of existence of primitives. □

Proof of Theorem 13.1. **1.** *Case where Ω is connected: method.* Let $a \in \Omega$ and

$$\Omega_{1/n}^{(a)} \text{ be the connected component of } \Omega_{1/n} \text{ containing } a.$$

We construct primitives g_n on $\Omega_{1/n}^{(a)}$ which glue together by subtracting a suitable constant from each f_n. More precisely, for each $n \geq 1$, $g_n \in \mathcal{D}'(\Omega_{1/n}^{(a)}; E)$ and

$$\nabla g_n = q \text{ on } \Omega_{1/n}^{(a)}, \tag{13.1}$$

and, for each $n \geq 2$,

$$g_n = g_{n-1} \text{ on } \Omega_{1/(n-1)}^{(a)}. \tag{13.2}$$

This makes sense because $\Omega_{1/(n-1)}^{(a)} \subset \Omega_{1/n}^{(a)}$.

2. *Constructing the g_n.* Let us choose for g_1 the restriction of f_1 to $\Omega_1^{(a)}$, and then proceed by induction on n, assuming that g_{n-1} has been determined. Since the restriction of the gradient is the gradient of the restriction (Theorem 6.4),

$$\nabla(f_n - g_{n-1}) = q - q = 0_E \text{ on } \Omega_{1/(n-1)}^{(a)}.$$

Since this set is connected (by construction), $f_n - g_{n-1}$ is a constant on it according to Theorem 12.3. By subtracting this constant from f_n, we obtain a (the only) distribution $g_n \in \mathcal{D}'(\Omega_{1/n}^{(a)}; E)$ satisfying (13.1) and (13.2).

3. *Gluing the g_n.* Since it is connected (for the moment), Ω is the union of the $\Omega_{1/n}^{(a)}$ according to Theorem 10.21. Moreover (13.2) implies that $g_n = g_i$ on $\Omega_{1/n}^{(a)} \cap \Omega_{1/i}^{(a)}$ for every n and i. The gluing theorem for distributions (Theorem 6.16) then provides $f \in \mathcal{D}'(\Omega; E)$ such that

$$f = g_n \text{ on each } \Omega_{1/n}^{(a)}.$$

With (13.1), it follows that

$$\nabla f = \nabla g_n = q \text{ on each } \Omega_{1/n}^{(a)},$$

hence the gluing theorem for equalities (Theorem 6.10) shows that

$$\nabla f = q \text{ on } \Omega.$$

4. General case. Now Ω is an arbitrary open subset of $\mathbb{R}^d$. The case of a connected open set provides a primitive f_m on each connected component Ω_m of Ω. For each m and m',

$$f_m = f_{m'} \text{ on } \Omega \cap \Omega_{m'},$$

since this intersection is empty. So, again due to the gluing theorem for distributions, the f_m may be glued together into a distribution $f \in \mathcal{D}'(\Omega; E)$.

Finally, since $\nabla f = \nabla f_m = q$ on Ω_m, the gluing theorem for equalities again gives

$$\nabla f = q \text{ on } \Omega. \square$$

Examples of connected sets Ω such that $\Omega_{1/n}$ is not connected. If Ω is connected and bounded and if its boundary is regular (for example, if it is a C^1 manifold), then the $\Omega_{1/n}$ are connected after a certain rank n, so it would be enough to start from this rank for the f_n to glue together up to a constant. It is the case of the "dumbbell" in Figure 13.1, p. 266.

But there are connected bounded sets where none of the $\Omega_{1/n}$ are connected. This is the case if, as in Figure 13.2, Ω is made up from a sequence of balls with radius 2^{2-m}, $m \geq 1$, linked together by cylinders of radius 2^{-m} and of this length. Indeed, $\Omega_{1/n}$ always has two connected components, the first is common to the balls linked by the cylinders of radius $2^{-m} > 1/n$, and the second is included in the subsequent ball.

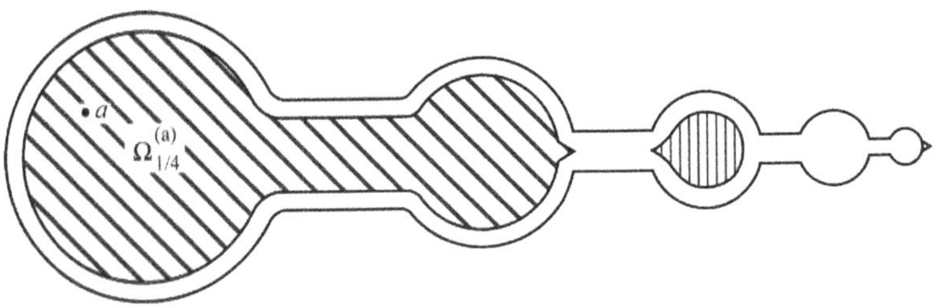

Figure 13.2. *A connected set where none of the $\Omega_{1/n}$ are connected. $\Omega^{(a)}_{1/4}$ is cross-hatched, $\Omega_{1/4}$ is the union of the two hatched regions* $\square$

13.2. Reduction to the function case

Let us reduce the search for primitives of a distribution field that satisfies **Poincaré's condition**

$$\partial_i q_j = \partial_j q_i$$

to a search for primitives of a function field. This will be used to establish the orthogonality and Poincaré's theorems for distributions (Theorems 13.5 and 13.7). We call this result the **reduction to the function case theorem**[2].

Theorem 13.2.– *Let $q \in \mathcal{D}'(\Omega; E^d)$, where Ω is an open subset of $\mathbb{R}^d$ and E is a Neumann space, be such that, for every i and j in $[\![1, d]\!]$,*

$$\partial_i q_j = \partial_j q_i.$$

Let $\eta_n \in \mathcal{C}^\infty_{B(0,1/n)}(\mathbb{R}^d)$ be the correction term given by Theorem 9.17 for $r = 1/n$, and $\Omega_{1/n} = \{x \in \mathbb{R}^d : B(x, 1/n) \subset \Omega\}$.

Suppose that the field $q \diamond \eta_n$, which belongs to $\mathcal{C}^\infty(\Omega_{1/n}; E^d)$, has for each $n \in \mathbb{N}^*$ a primitive in either the distribution or function sense, i.e. that there exists h_n belonging either to $\mathcal{D}'(\Omega_{1/n}; E)$ or to $\mathcal{C}^1(\Omega_{1/n}; E)$ such that

$$\nabla h_n = q \diamond \eta_n.$$

Then, there exists $f \in \mathcal{D}'(\Omega; E)$ such that

$$\nabla f = q. \blacksquare$$

Caution. The weighted distribution $q \diamond \eta_n$ is regular since η_n is, but **it is not a regular approximation of** q because $(\eta_n)_{n\in\mathbb{N}}$ is not a regularizing sequence. □

Utility of Theorem 13.2. The reduction to the function case is a step in the proof of the *orthogonality theorem* (Theorem 13.5) which is the main result of existence of primitives. □

The proof will use the following representation formula, which will also be used to establish Poincaré's theorem for distributions.

Theorem 13.3.– *Let $q \in \mathcal{D}'(\Omega; E^d)$, where Ω is an open subset of $\mathbb{R}^d$ and E is a Neumann space, be such that, for every i and j in $[\![1, d]\!]$,*

$$\partial_i q_j = \partial_j q_i.$$

Let γ_n be the localized elementary potential and η_n be the correction term introduced in Theorem 9.17 for $r = 1/n$ and $\Omega_{1/n} \stackrel{\text{def}}{=} \{x \in \mathbb{R}^d : B(x, 1/n) \subset \Omega\}$, where $n \in \mathbb{N}^*$.

Then,

$$q = q \diamond \eta_n + \nabla k_n \quad on\ \Omega_{1/n},$$

where $k_n \in \mathcal{D}'(\Omega_{1/n}; E)$ is defined by $k_n \stackrel{\text{def}}{=} \sum_{i=1}^d q_i \diamond \partial_i \gamma_n$. ■

2. **History of the reduction to the function case.** Theorem 13.2 has its seeds in the constructive proof of the existence of a primitive given in [77, § 5, p. 1171–1172].

Proof. By the representation formula (12.1) from Theorem 12.1 for q_j, we have on $\Omega_{1/n}$

$$q_j = q_j \diamond \eta_n + \sum_{i=1}^{d} \partial_i q_j \diamond \partial_i \gamma_n.$$

With Poincaré's hypothesis $\partial_i q_j = \partial_j q_i$, this gives

$$q_j = q_j \diamond \eta_n + \sum_{i=1}^{d} \partial_j q_i \diamond \partial_i \gamma_n.$$

However, $\partial_j q_i \diamond \partial_i \gamma_n = \partial_j (q_i \diamond \partial_i \gamma_n)$ according to the first expression for the derivative of a weighted distribution from Theorem 7.17, so we finally have

$$q_j = q_j \diamond \eta_n + \partial_j \Big(\sum_{i=1}^{d} q_i \diamond \partial_i \gamma_n \Big).$$

Which is the stated formula by definition of k_n. □

We are now able to prove the reduction to the function case theorem.

Proof of Theorem 13.2. Theorem 12.1 gives $q \diamond \eta_n \in \mathcal{C}^\infty(\Omega_{1/n}; E^d)$.

By hypothesis, there exists a distribution (it may be a function) h_n such that

$$\nabla h_n = q \diamond \eta_n.$$

The representation formula from Theorem 13.3, which is satisfied thanks to Poincaré's hypothesis, then gives

$$q = \nabla(h_n + k_n) \text{ on } \Omega_{1/n}.$$

Since this holds for every $n \in \mathbb{N}^*$, the peripheral gluing theorem (Theorem 13.1) provides $f \in \mathcal{D}'(\Omega; E)$ such that

$$\nabla f = q. \ \square$$

13.3. The orthogonality theorem

Before we come to the orthogonality theorem, let us show that every distribution field $q = (q_1, \dots, q_d)$ "orthogonal" to divergence-free test fields $\psi = (\psi_1, \dots, \psi_d)$ satisfies Poincaré's condition $\partial_i q_j = \partial_j q_i$.

Theorem 13.4.– *Let $q \in \mathcal{D}'(\Omega; E^d)$, where Ω is an open subset of $\mathbb{R}^d$ and E is a Neumann space, be such that*

$$\langle q, \psi \rangle = 0_E, \text{ for every } \psi \in \mathcal{D}(\Omega; \mathbb{R}^d) \text{ such that } \nabla \cdot \psi = 0. \tag{13.3}$$

Then, for every i and j in $[\![1, d]\!]$,

$$\partial_i q_j = \partial_j q_i. \blacksquare$$

We denote by $\mathcal{D}(\Omega; \mathbb{R}^d)$ the space of **test fields**, i.e. of infinitely differentiable functions from Ω into $\mathbb{R}^d$ with compact support. We do not endow it with a topology.

If q is a distribution field, we denote

$$\langle q, \psi \rangle \stackrel{\text{def}}{=} \sum_{i=1}^{d} \langle q_i, \psi_i \rangle.$$

Orthogonality. By generalizing the notion of orthogonality with respect to a scalar product, we can say that a field q satisfying condition (13.3) is *orthogonal to* the set

$$\mathcal{D}_{\text{div}}(\Omega; \mathbb{R}^d) \stackrel{\text{def}}{=} \{\psi \in \mathcal{D}(\Omega; \mathbb{R}^d) : \nabla \cdot \psi = 0\}$$

with respect to the bilinear mapping $\langle\ ,\ \rangle$ from $\mathcal{D}'(\Omega; E^d) \times \mathcal{D}(\Omega; \mathbb{R}^d)$ into E. □

Proof of Theorem 13.4. Assume $i \neq j$, since otherwise the conclusion is obvious. Given $\varphi \in \mathcal{D}(\Omega)$, we then define $\psi \in \mathcal{D}(\Omega; \mathbb{R}^d)$ by:

$$\psi_i = \partial_j \varphi, \quad \psi_j = -\partial_i \varphi, \quad \psi_k = 0 \text{ otherwise (i.e. if } k \neq i \text{ and } k \neq j).$$

Since the derivatives in the function sense commute according to Schwarz's theorem (Theorem 1.17),

$$\nabla \cdot \psi = \partial_i(\partial_j \varphi) + \partial_j(-\partial_i \varphi) = 0.$$

The orthogonality hypothesis (13.3) then implies, with Definition 5.4 of the derivative of a distribution,

$$\langle \partial_i q_j - \partial_j q_i, \varphi \rangle = \langle q_i, \partial_j \varphi \rangle + \langle q_j, -\partial_i \varphi \rangle = \langle q, \psi \rangle = 0_E.$$

This holds for any test function $\varphi \in \mathcal{D}(\Omega)$, therefore

$$\partial_i q_j - \partial_j q_i = 0_E. \ \square$$

Let us show that a distribution field q has a primitive f whenever it is orthogonal to every divergence-free test field. This is the **orthogonality theorem** [3] for distributions, which extends *de Rham's duality theorem* to vector-valued distributions.

Theorem 13.5.– *Let $q \in \mathcal{D}'(\Omega; E^d)$, where Ω is an open subset of $\mathbb{R}^d$ and E is a Neumann space, be such that:*

$$\langle q, \psi \rangle = 0_E, \ \textit{for every } \psi \in \mathcal{D}(\Omega; \mathbb{R}^d) \textit{ such that } \nabla \cdot \psi = 0. \tag{13.4}$$

Then, there exists $f \in \mathcal{D}'(\Omega, E)$ such that

$$\nabla f = q. \blacksquare$$

Detailed formulation of Theorem 13.5. In other words, if elements $q_1, \ldots, q_d$ of $\mathcal{D}'(\Omega; E)$ satisfy

$$\sum_{i=1}^{d} \langle q_i, \varphi_i \rangle = 0_E, \quad \text{for all of the } \varphi_1, \ldots, \varphi_d \text{ of } \mathcal{D}(\Omega) \text{ such that } \sum_{i=1}^{d} \partial_i \varphi_i = 0,$$

then there exists $f \in \mathcal{D}'(\Omega; E)$ such that

$$\partial_i f = q_i, \text{ for every } i \in [\![1, d]\!]. \ \square$$

Optimality of Theorem 13.5. The orthogonality condition (13.4) is necessary and sufficient for q to have a primitive because, if $q = \nabla f$, then for every $\psi \in \mathcal{D}(\Omega; \mathbb{R}^d)$ such that $\nabla \cdot \psi = 0$,

$$\langle q, \psi \rangle = \sum_{i=1}^{d} \langle \partial_i f, \psi_i \rangle = -\sum_{i=1}^{d} \langle f, \partial_i \psi_i \rangle = -\langle f, \nabla \cdot \psi \rangle = 0_E. \tag{13.5}$$

$\square$

3. **History of the orthogonality theorem (Theorem 13.5). Real currents.** Georges de RHAM proved in 1955 [63, Theorem 17', p. 114] that a **current** T is homological to 0 if and only if $T[\psi] = 0$ for every differential form ψ which is $\mathcal{C}^\infty$, closed and with compact support (a current generalizes a differential form on a manifold just as a distribution generalizes a function; for a differential form, this result states that *every closed differential form is exact*). This is **de Rham's duality theorem**.

Real distributions. Jacques-Louis LIONS observed in 1969 [54, p. 69] that Theorem 13.5 for real values follows from considering the current $T = q_1 \mathrm{d}x_1 + \cdots + q_d \mathrm{d}x_d$ (the passage from differential forms to primitives is well explained, for functions, in [RUDIN, 66, § 10.42 and 10.43, p. 262–264]).
Due to the importance of this result for solving the Navier–Stokes equations, many more direct and elementary proofs were given for particular real-valued distributions: by Olga LADYZHENSKAYA in 1963 [48, Theorem 1, p. 28] for $q \in (L^2(\Omega))^d$; by Luc TARTAR in 1978 [88] for $q \in (H^{-1}(\Omega))^d$; by Jacques SIMON in 1993 [77] for $q \in (\mathcal{D}'(\Omega))^d$.

Vector values. Jacques SIMON proved Theorem 13.5 for E a Banach space in 1993 [78, Theorem 5 (ii), p. 4] (or [79, Theorem 13 p. 210], when Ω is Lipschitz), by the virtue of a constructive method (the proof by Georges DE RHAM [63] does not seem to extend to this case, because it uses reflexivity properties of the space of currents).
The method for the present generalization to E a Neumann space is inspired from that of [78].

Formulation in terms of orthogonality. Theorem 13.5 and its converse that we have just seen can be expressed in the form

$$\mathcal{D}'_\nabla(\Omega; E^d) = (\mathcal{D}_{\text{div}}(\Omega; \mathbb{R}^d))^\perp,$$

where $\mathcal{D}'_\nabla(\Omega; E^d)$ designates the set of distribution fields which are gradients, $\mathcal{D}_{\text{div}}(\Omega; \mathbb{R}^d)$ designates the set of divergence-free test fields, and $\perp$ is the orthogonality with respect to the bilinear mapping $\langle\, ,\, \rangle$ from $\mathcal{D}'(\Omega; E^d) \times \mathcal{D}(\Omega; \mathbb{R}^d)$ into E. □

Proof of Theorem 13.5. **1.** *Reducing to the function case.* According to the reduction to the function case theorem (Theorem 13.2), it suffices to check two conditions.

— The first is Poincaré's condition $\partial_i q_j = \partial_j q_i$, which follows from the orthogonality hypothesis (13.4) according to Theorem 13.4.

— The second, which remains to be checked, is the existence of a primitive of the function field $q \diamond \eta_n$, i.e. of $h_n \in C^1(\Omega_{1/n}; E)$ such that

$$\nabla h_n = q \diamond \eta_n, \tag{13.6}$$

where $\eta_n \in C^\infty_{B(0,1/n)}(\mathbb{R}^d)$ is the correction term introduced in Theorem 9.17 for $r = 1/n$ and $\Omega_{1/n} = \{x \in \mathbb{R}^d : B(x, 1/n) \subset \Omega\}$.

2. *Obtaining a function h_n satisfying (13.6).* According to the orthogonality theorem for functions (Theorem 11.4), it suffices to check that the field $q \diamond \eta_n$ is orthogonal to every divergence-free test field. Thus, let

$$\psi \in \mathcal{D}(\Omega_{1/n}; \mathbb{R}^d) \text{ be such that } \nabla \cdot \psi = 0.$$

By Definition 7.12 of weighting,

$$\langle q \diamond \eta_n, \psi \rangle_{\Omega_{1/n}} = \sum_{i=1}^d \langle q_i \diamond \eta_n, \psi_i \rangle_{\Omega_{1/n}} = \sum_{i=1}^d \langle q_i, \eta_n \diamond \check{\psi}_i \rangle_\Omega = \langle q, \eta_n \diamond \check{\psi} \rangle_\Omega, \tag{13.7}$$

where $\check{\psi} \in \mathcal{D}(\mathbb{R}^d; \mathbb{R}^d)$ is defined by

$$\check{\psi}(x) = \psi(-x) \text{ if } -x \in \Omega_{1/n}, \quad \check{\psi}(x) = 0 \text{ otherwise.}$$

However, according to the equality $\partial_i(g \diamond \mu) = -g \diamond \partial_i \mu$ from Theorem 7.17,

$$\nabla \cdot (\eta_n \diamond \check{\psi}) = \sum_{i=1}^d \partial_i(\eta_n \diamond \check{\psi}_i) = -\sum_{i=1}^d \eta_n \diamond \partial_i \check{\psi}_i = -\eta_n \diamond (\nabla \cdot \check{\psi}).$$

Since $\nabla \cdot \psi = 0$, we have $\nabla \cdot \check{\psi} = -(\nabla \cdot \psi)\check{} = 0$, hence

$$\nabla \cdot (\eta_n \diamond \check{\psi}) = 0.$$

The orthogonality hypothesis (13.4) then gives $\langle q, \eta_n \diamond \check{\psi} \rangle_\Omega = 0_E$, i.e. with (13.7),

$$\langle q \diamond \eta_n, \psi \rangle_{\Omega_{1/n}} = 0_E.$$

Since $q \diamond \eta_n$ is (Theorem 13.2) a continuous field, this can be written as (Theorem 3.9)

$$\int_{\Omega_{1/n}} q \diamond \eta_n \cdot \psi = 0_E.$$

This is the orthogonality condition (11.4) for functions of Theorem 11.4 that we are looking for. The said theorem hence provides a primitive $h_n \in \mathcal{C}^1(\Omega_{1/n}; E)$ of $q \diamond \rho_n$.

3. *Conclusion.* Since condition (13.6) is thus satisfied, we can apply the reduction to the function case theorem (Theorem 13.2), which provides a primitive f of q. □

Let us show that, in dimension one, every distribution has a primitive [4].

Theorem 13.6.– *Let $q \in \mathcal{D}'(\Omega; E)$, where Ω is an open subset of $\mathbb{R}$ and E is a Neumann space. Then, there exists $f \in \mathcal{D}'(\Omega, E)$ such that*

$$\frac{\mathrm{d}f}{\mathrm{d}t} = q. \blacksquare$$

Proof. According to the orthogonality theorem (Theorem 13.5), it suffices to check that $\langle q, \psi \rangle$ is zero for every $\psi \in \mathcal{D}(\Omega)$ such that $\nabla \cdot \psi = 0$, i.e. here such that $\psi' = 0$.

This is the case since such a function ψ is zero, because it is constant (for example by Theorem 12.3) and is zero outside of a compact set (by definition of $\mathcal{D}(\Omega)$). □

13.4. Poincaré's generalized theorem

Let us show that every distribution field $q = (q_1, \ldots, q_d)$ on a simply connected set satisfying Poincaré's condition $\partial_i q_j = \partial_j q_i$ has a primitive. This is **Poincaré's theorem** [5], generalized to distributions.

---

4. **History of Theorem 13.6.** Laurent SCHWARTZ proved in 1950 [69, Chap. II, § 4, Theorem I, p. 51] that every distribution in $\mathcal{D}'(\mathbb{R})$ has a primitive, unique up to a constant.

5. **History of Poincaré's theorem. For real functions.** It is Theorem 11.5 established by POINCARÉ in 1899 [61, p. 10].

Generalization to distributions. Laurent SCHWARTZ proved in 1950 [69, Chap. II, § 6, Theorem VI, p. 59] Theorem 13.7 for distributions in $\mathcal{D}'(\mathbb{R}^d)$. He had stated this result for dimension 2 in 1945 [68, p. 64]. His method, which uses induction on dimension d, does not seem to extend to the case of a simply connected set.

Generalization to vector values. Jacques SIMON proved in 1993 [78, Theorem 5, p. 4] Theorem 13.7 for E a Banach space, thanks to a constructive method. The present generalization to E a Neumann space follows the method of [78].

Theorem 13.7.– *Let $q \in \mathcal{D}'(\Omega; E^d)$, where E is a Neumann space and*

$$\Omega \text{ is a simply connected open subset of } \mathbb{R}^d,$$

be such that, for every i and j in $[\![1, d]\!]$,

$$\partial_i q_j = \partial_j q_i.$$

Then, there exists $f \in \mathcal{D}'(\Omega; E)$ such that

$$\nabla f = q. \blacksquare$$

Optimality of Theorem 13.7. The condition $\partial_i q_j = \partial_j q_i$ is necessary for a field q to have a primitive because, if $q = \nabla f$ then, since derivatives commute (Theorem 5.7),

$$\partial_i q_j = \partial_i \partial_j f = \partial_j \partial_i f = \partial_j q_i. \tag{13.8}$$

When Ω is simply connected, it is therefore both necessary and sufficient.

For an arbitrary open set Ω, it is necessary but not always sufficient: an example of a field satisfying $\partial_i q_j = \partial_j q_i$ on a crown, but which does not have a primitive, is given in Theorem 13.10 for dimension $d = 2$ and in Theorem 13.11 for $d \geq 3$. $\square$

Proof of Theorem 13.7. **1.** *Reduction of the problem.* According to the reduction to the function case theorem (Theorem 13.2), it suffices, since Poincaré's condition $\partial_i q_j = \partial_j q_i$ is satisfied by hypothesis here, to check that the function field $q \diamond \eta_n$ has a primitive, i.e. that there exists $h_n \in \mathcal{C}^1(\Omega_{1/n}; E)$ such that

$$\nabla h_n = q \diamond \eta_n,$$

where $\eta_n \in \mathcal{C}^\infty_{B(0,1/n)}(\mathbb{R}^d)$ is the correction term given by Theorem 9.17 (b) for $r = 1/n$ and $\Omega_{1/n} = \{x \in \mathbb{R}^d : B(x, 1/n) \subset \Omega\}$.

According to Theorem 11.1 on the existence of primitives of a function field, it suffices that, for every closed path Γ in $\Omega_{1/n}$,

$$\int_\Gamma q \diamond \eta_n \cdot \mathrm{d}\ell = 0_E. \tag{13.9}$$

Let us check it in three steps.

2. *Local primitive.* Due to the equality $\partial_i(g \diamond \mu) = \partial_i g \diamond \mu$ from Theorem 7.17 and the hypothesis $\partial_i q_j = \partial_j q_i$,

$$\partial_i (q_j \diamond \eta_n) = \partial_i q_j \diamond \eta_n = \partial_j q_i \diamond \eta_n = \partial_j (q_i \diamond \eta_n).$$

So, the function field $q \diamond \eta_n$ satisfies Poincaré's hypothesis, therefore it has a local primitive by Poincaré's theorem for functions (Theorem 11.5). That is to say, for every open ball B included in $\Omega_{1/n}$, there exists $h_B \in \mathcal{C}^1(B; E)$ such that

$$\nabla h_B = q \diamond \eta_n \ \text{ in } B. \tag{13.10}$$

3. *Invariance of the line integral: an easy case.* Let Γ be a closed path in $\Omega_{1/n}$. Since Ω is simply connected (Definition 10.22), Γ is homotopic (Definition 10.18) in Ω to a closed path Γ_* reduced to a point. If this homotopy H takes place in $\Omega_{1/n}$, i.e. if its image $[H]$ (Definition 10.18) satisfies

$$[H] \subset \Omega_{1/n},$$

the homotopy invariance of the line integral of a local gradient (Theorem 10.19) gives, since the line integral along a path reduced to a point is zero (Theorem 10.5),

$$\int_\Gamma q \diamond \eta_n \cdot \mathrm{d}\ell = \int_{\Gamma_*} q \diamond \eta_n \cdot \mathrm{d}\ell = 0_E. \tag{13.11}$$

But $[H]$ does not necessarily belong to $\Omega_{1/n}$. Let us thus prove (13.11) without this hypothesis.

4. *Invariance of the line integral when* $[H] \not\subset \Omega_{1/n}$. Since the image $[H]$ is compact (as image of the compact set $[t_\mathrm{i}, t_\mathrm{e}] \times [0,1]$ under the continuous mapping H, see Theorem A.31) and the $\Omega_{1/n}$ cover Ω, there exists $m \geq n$ such that

$$[H] \subset \Omega_{1/m}.$$

Theorem 10.19 therefore gives, in the place of (13.11),

$$\int_\Gamma q \diamond \eta_m \cdot \mathrm{d}\ell = 0_E. \tag{13.12}$$

On the other hand, the representation formula from Theorem 13.3 gives, for every n,

$$q = q \diamond \eta_n + \nabla k_n \ \text{ on } \Omega_{1/n},$$

where $k_n \in \mathcal{D}'(\Omega_{1/n}; E)$. By subtracting the formula for m from that for n, it follows, since $\Omega_{1/m}$ contains $\Omega_{1/n}$, that

$$q \diamond (\eta_n - \eta_m) = \nabla(k_n - k_m) \ \text{ on } \Omega_{1/n}.$$

Since $q \diamond \eta_n$ and $q \diamond \eta_m$ are regular fields (Theorem 13.2), $\nabla(k_n - k_m)$ is continuous so, due to Theorem 12.4,

$$k_n - k_m \in \mathcal{C}^1(\Omega_{1/n}; E).$$

Since the line integral along a closed path of the gradient of a C^1 field is zero due to Theorem 10.8, it follows that

$$\int_\Gamma q \diamond (\eta_n - \eta_m) \cdot \mathrm{d}\ell = 0_E.$$

Hence, with (13.12),

$$\int_\Gamma q \diamond \eta_n \cdot \mathrm{d}\ell = 0_E.$$

This establishes (13.9) and thus finishes the proof of Theorem 13.7. □

A simple case. If all of the $\Omega_{1/n}$ are simply connected, the second step in the proof of Theorem 13.7 is sufficient, since then (13.10) implies the existence of a primitive h_n of $q \diamond \eta_n$ by the gluing theorem of local primitives of continuous fields on a simply connected set established in Volume 2 [82, Theorem 9.4].

This is the case if Ω is star-shaped or, more generally if it is retractable, i.e. if it is homotopic "in itself" to one of its points. But $\Omega_{1/n}$ is not necessarily simply connected, as the following examples show. □

Simply-connected sets whose $\Omega_{1/n}$ are not all simply connected. Let us give two examples.

Example 1: one of the $\Omega_{1/n}$ is not simply connected. A torus with small radius 1 completed by a disc of thickness $1/2$, and more precisely the domain Ω in $\mathbb{R}^3$ generated by the rotation around its vertical axis of symmetry of the "dumbbell" in $\mathbb{R}^2$ represented in Figure 13.1, p. 266, is simply connected, since the closed paths in the torus can retract into the disc.

In contrast, $\Omega^{(a)}_{1/2}$ is not simply connected, since the disc has disappeared but not the torus. However, $\Omega^{(a)}_{1/3}$ and the following ones are simply connected.

Example 2: none of the $\Omega_{1/n}$ are simply connected. The domain Ω in $\mathbb{R}^3$ generated by the rotation around the vertical axis of the largest ball of the domain ω in $\mathbb{R}^2$ represented in Figure 13.2, p. 268 (the sequence of balls connected by cylinders), is simply connected.

But none of the $\Omega_{1/n}$ are connected since, as we have seen p. 268, $\omega_{1/n}$ always has two connected components and the rotation of the component disconnected from the largest ball generates a torus. □

13.5. Current of an incompressible two dimensional field

Let us show that, on a simply connected set in $\mathbb{R}^2$, every divergence-free distribution field $v = (v_1, v_2)$ can be derived from a "current function", which here is a distribution. This is **Haar's lemma** [6], generalized to distributions.

6. **History of Haar's lemma.** Alfréd HAAR proved the existence of a current function for a divergence-free field $v \in C^1(\Omega)$ between 1926 [37] and 1929 [38].

Theorem 13.8.– *Let $v \in \mathcal{D}'(\Omega; E^2)$, where Ω is a simply connected open subset of $\mathbb{R}^2$ and E is a Neumann space, be such that*

$$\partial_1 v_1 + \partial_2 v_2 = 0.$$

*Then, there exists a "**current function**" $f \in \mathcal{D}'(\Omega; E)$ such that*

$$v_1 = \partial_2 f, \quad v_2 = -\partial_1 f. \tag{13.13}$$

∎

Proof. Let us consider the field $q \in \mathcal{D}'(\Omega; E^2)$ given by

$$q \overset{\text{def}}{=} (-v_2, v_1). \tag{13.14}$$

It satisfies Poincaré's condition, which reduces here to $\partial_1 q_2 = \partial_2 q_1$, since

$$\partial_1 q_2 - \partial_2 q_1 = \partial_1 v_1 + \partial_2 v_2 = 0.$$

Therefore, due to the generalized Poincaré's theorem (Theorem 13.7), there exists a distribution f such that

$$\partial_1 f = q_1 = -v_2, \quad \partial_2 f = q_2 = v_1. \ \square$$

Precise definition of q. Definition (13.14) is intentionally abusive, to gain clarity. Indeed, the left-hand term q belongs to $\mathcal{D}'(\Omega; E^2)$, while the right-hand term $(-v_2, v_1)$ belongs to $(\mathcal{D}'(\Omega; E))^2$, since v_1 and v_2 belong to $\mathcal{D}'(\Omega; E)$.

To be more precise, it would be necessary to replace (13.14) with:

q is the field in $\mathcal{D}'(\Omega; E^2)$ whose components are $q_1 = -v_2$ and $q_2 = v_1$.

Namely, q is the image of $(-v_2, v_1)$ under the isomorphism from $(\mathcal{D}'(\Omega; E))^2$ onto $\mathcal{D}'(\Omega; E^2)$ obtained in Theorem 5.3. In perfectly rigorous terms (albeit again less clear!), q is defined for every $\varphi \in \mathcal{D}(\Omega)$ by

$$\langle q, \varphi \rangle \overset{\text{def}}{=} (-(\langle v, \varphi \rangle)_2, (\langle v, \varphi \rangle)_1). \ \square$$

Short form of Theorem 13.8. Denoting $^\perp$ the rotation by $\pi/2$ in the direct sense, and so

$$\nabla^\perp \overset{\text{def}}{=} (\partial_2, -\partial_1),$$

this result can be expressed as:

if Ω is simply connected and $\nabla \cdot v = 0$, then there exists f such that $v = \nabla^\perp f$. (13.15)

Observe that

$$\Delta f = \nabla^\perp \cdot \nabla^\perp f = \nabla^\perp \cdot v = \partial_2 v_1 - \partial_1 v_2. \ \square$$

Uniqueness. The distribution f obtained in Theorem 13.8 is unique up to a the addition of a constant on each connected component of Ω, due to Theorem 12.3. In particular, if Ω is connected, f is unique up to the addition of a single constant. □

Current versus current function. These two concepts have nothing to do with one another.
— A *current* generalizes a differential form like a distribution generalizes a function, see Note 3, p. 272.
— A *current function* is a scalar function (or a distribution) from which an incompressible two-dimensional field derives by relation (13.13). □

13.6. Global versus local primitives

Let us first observe that, on a simply connected set, if a field has local primitives, it has a global primitive. We name this result the **gluing theorem for local primitives**[7].

Theorem 13.9.– *Let $q \in \mathcal{D}'(\Omega; E^d)$, where E is a Neumann space and*

$$\Omega \textit{ is a simply connected open subset of } \mathbb{R}^d,$$

be such that, for every open ball $B \Subset \Omega$, there exists $f_B \in \mathcal{D}'(B; E)$ such that

$$\nabla f_B = q \textit{ on } B.$$

Then, there exists $f \in \mathcal{D}'(\Omega; E)$ such that

$$\nabla f = q. \blacksquare$$

Proof. Since the partial derivatives commute (Theorem 5.7), the hypothesis $\nabla f_B = q$ implies, for every i and j, on B,

$$\partial_i q_j = \partial_i \partial_j f_B = \partial_j \partial_i f_B = \partial_j q_i. \tag{13.16}$$

This equality being true on each ball B, it is true on their union Ω according to the gluing theorem for equalities (Theorem 6.10), i.e., for every i and j,

$$\partial_i q_j = \partial_j q_i \text{ on } \Omega.$$

Since Ω is simply connected, the generalized Poincaré's theorem (Theorem 13.7) thus provides a distribution f such that

$$\nabla f = q \text{ on } \Omega. \square$$

Generalization. In Theorem 13.9, we can replace "for every open ball $B \Subset \Omega$" by:

"for every $x \in \Omega$, for at least one open ball B centered at x". □

Gluing of local primitives versus peripheral gluing. Theorem 13.9 is complementary to the *peripheral gluing theorem* (Theorem 13.1), in which we only glue at the periphery of Ω primitives that are given excepted on a neighborhood of the boundary, but this for a set Ω not necessarily simply connected. □

We now show that there exist sets where the existence of local primitives does not ensure the existence of a global one. Let us begin with an example in dimension $d = 2$ with real values.

7. **History of the gluing theorem for local primitives.** This easy result does not seem to have been stated for distributions.

Theorem 13.10.– *Let $\Omega = \{x \in \mathbb{R}^2 : |x| > 1\}$ and $q \in \mathcal{C}^\infty(\Omega; \mathbb{R}^2)$ be the field defined, for every $x \in \Omega$, by*

$$q(x) = \Big(-\frac{x_2}{|x|^2}, \frac{x_1}{|x|^2}\Big).$$

For every open ball B included in Ω, there exists $f_B \in \mathcal{C}^\infty(B)$ such that $\nabla f_B = q$ on B, and yet no distribution $f \in \mathcal{D}'(\Omega)$ exists such that $\nabla f = q$ on all of Ω. ▌

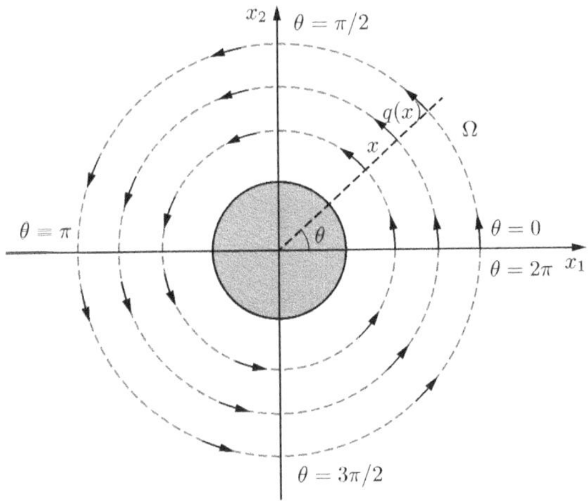

Figure 13.3. *Field q with θ for local but not global primitive. Ω is the exterior of the grey disc*

Proof of Theorem 13.10. The field q is the "rotating" field represented on Figure 13.3. In polar coordinates,

$$q(\theta, r) \stackrel{\text{def}}{=} \frac{e_\theta}{r}.$$

Since $\nabla = e_r \partial_r + (e_\theta/r)\partial_\theta$,

$$\nabla\theta = q \text{ except at } \theta = 0. \tag{13.17}$$

Indeed, θ is discontinuous on the half-line $D = \{(r, \theta) : \theta = 0\}$: it equals 0 on one side, 2π on the other.

The field q does not have a primitive, else this would be a continuous function as is every distribution with continuous derivatives (Theorem 12.4) and its restriction to $\Omega \setminus D$ would be of the form $\theta + c$ by Theorem 12.3 which would contradict its continuity.

However $\nabla\theta = q$ on every ball B, since q is equally a gradient on the balls that intersect D, as it can be verified by choosing another half-line for the origin of θ. □

Let us now show that there are such sets in every dimension greater than two.

Theorem 13.11.– *If $d \geq 2$ and E is a Neumann space not reduced to $\{0_E\}$, there exist an open subset Ω of $\mathbb{R}^d$ and a field $q \in \mathcal{C}^\infty(\Omega; E^d)$ such that:*
— for every ball B included in Ω, there is a function $f_B \in \mathcal{C}^\infty(B; E)$ such that $\nabla f_B = q$ on B;
— there is no distribution $f \in \mathcal{D}'(\Omega; E)$ such that $\nabla f = q$ on all of Ω. ▮

Proof. Let $\underline{\Omega}$ be the open subset of $\mathbb{R}^2$ and $\underline{q}$ the field in $\mathcal{C}^\infty(\underline{\Omega}; \mathbb{R}^2)$ constructed in Theorem 13.10, $\underline{f}_B$ a primitive of $\underline{q}$ on B, and $u \in E$, $u \neq 0_E$.

In dimension $d = 2$, a convenient field on $\underline{\Omega}$ is defined by $q(x) = \underline{q}(x)u$ because:
— the function defined by $f_B(x) = \underline{f}_B(x)u$ is a primitive of q on B;
— the field q does not have a global primitive f, otherwise it would be a continuous function as is every distribution with continuous derivatives (Theorem 12.4); and, for $\theta \neq 0$, since $\nabla(\theta u) = q$ from (13.17), it would be of the form $f = \theta u + c$ where $c \in E$ (due to Theorem 12.3), which contradicts its continuity on the half-line $\theta = 0$.

In higher dimensions, a convenient field q on $\underline{\Omega} \times \mathbb{R}^{d-2}$ is defined by

$$q(x_1, \ldots, x_d) = (\underline{q}_1(x_1, x_2), \underline{q}_2(x_1, x_2), 0, \ldots, 0)\, u. \ \square$$

Necessity of simple-connectedness for gluing local primitives. Simple-connectedness is sufficient for every field with local primitives to have a global primitive, due to the gluing of local primitives theorem (Theorem 13.9), and thus for any field satisfying Poincaré's condition $\partial_i q_j = \partial_j q_i$ to have a primitive.

It is necessary if $d = 1$ or 2, but no longer is if $d \geq 3$, although it is for $d = 3$ when the set is regular.

These results, which were communicated to me by Pierre DREYFUSS and Nicolas DEPAUW, appeal to difficult considerations of algebraic topology presented in [DREYFUSS, 27]. Let us give an overview.

Case $d = 1$. Every open subset of $\mathbb{R}$ is simply connected, and therefore has the gluing property for local primitives.

Case $d = 2$. Every not simply connected open subset Ω of $\mathbb{R}^2$ has at least one *hole*, i.e. there is a point $z \notin \Omega$ which is circled by a closed path Γ in Ω. It therefore does not have the gluing property for local primitives, since the field q introduced in Theorem 13.10, after a translation so that z is at the origin, is locally a gradient on Ω, but is not one globally.

Case $d = 3$. The exterior of the *Alexander horned sphere* (which is represented and studied on p. 171 in [HATCHER, 39]) possesses the gluing property for local primitives, although it is not simply connected.

On the contrary, for a bounded open subset of $\mathbb{R}^3$ that is locally on one side of the graph of a continuous function, the gluing property for local primitives implies simply connectedness.

Case $d \geq 4$. The gluing property for local primitives on an open subset of $\mathbb{R}^d$ does not imply its simply connectedness, even if it is locally on one side of the graph of a continuous function. $\square$

13.7. Comparison of the existence conditions of a primitive

Let us compare the conditions of existence of a primitive of the previous sections [8].

Theorem 13.12.– *Let $q \in \mathcal{D}'(\Omega; E^d)$, where Ω is an open subset of $\mathbb{R}^d$ and E is a Neumann space. Then:*

(a) $q \in \mathcal{D}'_\nabla(\Omega; E^d) \quad \Leftrightarrow \quad \exists f \in \mathcal{D}'(\Omega; E)$ *such that* $\nabla f = q$

$\Leftrightarrow \quad \langle q, \psi \rangle = 0_E, \ \forall \psi \in \mathcal{D}(\Omega; \mathbb{R}^d)$ *such that* $\nabla \cdot \psi = 0.$

(b) $q \in \mathcal{D}'_\nabla(\Omega; E^d) \quad \Rightarrow \quad \partial_i q_j = \partial_j q_i, \ \forall i, \ \forall j.$

(c) $\partial_i q_j = \partial_j q_i, \ \forall i, \ \forall j \quad \Leftrightarrow \quad \forall$ *ball* $B \Subset \Omega, \ \exists f_B \in \mathcal{D}'(B; E), \ \nabla f_B = q$ *on* B.

(d) *If Ω is simply connected:*

$$
\begin{aligned}
q \in \mathcal{D}'_\nabla(\Omega; E^d) \quad &\Leftrightarrow \quad \exists f \in \mathcal{D}'(\Omega; E) \text{ such that } \nabla f = q \\
&\Leftrightarrow \quad \langle q, \psi \rangle = 0_E, \ \forall \psi \in \mathcal{D}(\Omega; \mathbb{R}^d) \text{ such that } \nabla \cdot \psi = 0 \\
&\Leftrightarrow \quad \partial_i q_j = \partial_j q_i, \ \forall i, \ \forall j \\
&\Leftrightarrow \quad \forall \text{ ball } B \Subset \Omega, \ \exists f_B \in \mathcal{D}'(B; E) \text{ s.t. } \nabla f_B = q \text{ on } B.
\end{aligned}
$$

(e) *When $d \geq 2$, there exists Ω such that:*

$$\partial_i q_j = \partial_j q_i, \ \forall i, \ \forall j \quad \not\Rightarrow \quad q \in \mathcal{D}'_\nabla(\Omega; E^d). \ \blacksquare$$

Proof. **1.** *Equivalences (a).* The first equivalence is Definition 12.5 of $\mathcal{D}'_\nabla(\Omega; E^d)$.

The direct part of the second equivalence is given by equality (13.5), p. 272, the converse is the orthogonality theorem (Theorem 13.5).

2. *Implication (b).* This implication is given by equality (13.8), p. 275.

3. *Equivalence (c).* The direct part is given by Poincaré's theorem (Theorem 13.7), since every ball is simply connected. The converse follows from equality (13.16), p. 279.

---

8. **History of Theorem 13.12.** The reader will refer to Notes 3, p. 272, and 5, p. 274, for the history of the direct parts, and in particular for the contributions of Georges DE RHAM and Laurent SCHWARTZ.

4. *Equivalences (d).* These equivalences follow from equivalences (a) and (c), and from the fact that, if Ω is simply connected, every local gradient is a gradient according to the gluing of local primitives theorem (Theorem 13.9), and, evidently, conversely.

5. *Property (e).* Such an Ω is given by Theorem 13.11, taking equivalence (c) into account. □

13.8. Limits of gradients

Let us show that every limit of gradients is a gradient[9] and that, up to constants, the primitives converge.

Theorem 13.13.– *Let $(f_n)_{n\in\mathbb{N}}$ be a sequence in $\mathcal{D}'(\Omega; E)$, where Ω is an open subset of $\mathbb{R}^d$ and E is a Neumann space, and $q \in \mathcal{D}'(\Omega; E^d)$ such that*

$$\nabla f_n \to q \ \text{ in } \mathcal{D}'(\Omega; E^d). \tag{13.18}$$

Then, there exists $f \in \mathcal{D}'(\Omega; E)$ such that

$$\nabla f = q$$

and, for each connected component Ω_m of Ω and for every $n \in \mathbb{N}$, there exists $c_{m,n} \in E$ such that the locally constant function σ_n which equals $c_{m,n}$ in Ω_m for each m satisfies

$$f_n - \sigma_n \to f \ \text{ in } \mathcal{D}'(\Omega; E). \ \blacksquare$$

A simpler case. If Ω is connected, its only connected component is Ω itself and thus σ_n is constant. □

Proof of Theorem 13.13. **1.** Existence of f. Let

$$\psi \in \mathcal{D}'(\Omega; \mathbb{R}^d) \ \text{ be such that } \ \nabla \cdot \psi = 0.$$

By Definition 5.4 of derivatives,

$$\langle \nabla f_n, \psi \rangle = \sum_{i=1}^{d} \langle \partial_i f_n, \psi_i \rangle = -\sum_{i=1}^{d} \langle f_n, \partial_i \psi_i \rangle = -\langle f_n, \nabla \cdot \psi \rangle = 0_E.$$

9. **History of a primitive of limits of gradients.** Theorems 13.13 and 13.14 are new to the knowledge of the author, included for real distributions.

With Hypothesis (13.18) on the convergence of gradients, it follows

$$\langle q, \psi \rangle = 0_E.$$

This holds for every divergence-free ψ, thus the orthogonality theorem (Theorem 13.5) provides $f \in \mathcal{D}'(\Omega; E)$ such that

$$\nabla f = q.$$

2. *Existence of the* $c_{m,n}$. For every connected component Ω_m, let

$$\phi_m \in \mathcal{D}(\Omega_m) \text{ be such that } \int_{\Omega_m} \phi_m \neq 0$$

and let

$$c_{m,n} \stackrel{\text{def}}{=} \langle f_n - f, \phi_m \rangle_{\Omega_m}.$$

Then,

$$\nabla(f_n - \sigma_n) \to \nabla f \text{ in } \mathcal{D}'(\Omega; E^d)$$

and, for every $n \in \mathbb{N}$ and every m,

$$\langle f_n - \sigma_n, \phi_m \rangle_{\Omega_m} = \langle f, \phi_m \rangle_{\Omega_m}.$$

Therefore, $f_n - \sigma_n$ depends continuously on its gradient from Theorem 12.8, namely

$$f_n - \sigma_n \to f \text{ in } \mathcal{D}'(\Omega; E). \ \square$$

Observe that the set of gradients is sequentially closed.

Theorem 13.14.– *The space $\mathcal{D}'_\nabla(\Omega; E^d)$, where Ω is an open subset of $\mathbb{R}^d$ and E is a Neumann space, is sequentially closed in $\mathcal{D}'(\Omega; E^d)$.* ∎

Proof. Let $(q_n)_{n\in\mathbb{N}}$ be a sequence in $\mathcal{D}'_\nabla(\Omega; E^d)$ and $q \in \mathcal{D}'(\Omega; E^d)$ such that

$$q_n \to q \text{ in } \mathcal{D}'(\Omega; E^d).$$

By Definition 12.5 of $\mathcal{D}'_\nabla(\Omega, E^d)$, for every n, there exists $f_n \in \mathcal{D}'(\Omega; E)$ such that $\nabla f_n = q_n$. Theorem 13.13 then provides $f \in \mathcal{D}'(\Omega; E)$ such that $\nabla f = q$. So,

$$q \in \mathcal{D}'_\nabla(\Omega; E^d). \ \square$$

Chapter 14

Distributions of Distributions

This chapter is dedicated to the space $\mathcal{D}'(\Omega_1; \mathcal{D}'(\Omega_2; E))$ of distributions on Ω_1 with values in the space $\mathcal{D}'(\Omega_2; E)$ of distributions on Ω_2.

— We characterize these distributions, their bounded sets and their convergent sequences (§ 14.1 to 14.3).

— We characterize the sets in which every bounded sequence has a convergent subsequence (§ 14.4). In particular, we show that every bounded sequence in $\mathcal{D}'(\Omega_1; \mathcal{D}'(\Omega_2))$ has a convergent subsequence.

— We give conditions for a bounded sequence in $\mathcal{D}'(\Omega_1; \mathcal{D}'(\Omega_2; E))$ to have a converging subsequence in $\mathcal{D}'(\Omega_1; \mathcal{D}'(\Omega_2; F))$ (§ 14.5) or in $\mathcal{D}'(\Omega_1; \mathcal{D}'(\Omega_2; E\text{-weak}))$ (§ 14.6).

14.1. Characterization

The space $\mathcal{D}'(\Omega_1; \mathcal{D}'(\Omega_2; E))$ is well-defined (by Definition 3.1 of distributions) when E is a Neumann space, since then $\mathcal{D}'(\Omega_2; E)$ is a Neumann space according to Theorem 4.5.

It is, by Definition 3.1, provided with the following family of semi-norms.

Definition 14.1.– *Let Ω_1 be an open subset of $\mathbb{R}^{d_1}$, Ω_2 an open subset of $\mathbb{R}^{d_2}$ and E a Neumann space whose family of semi-norms is denoted by $\{\| \,\|_{E;\nu} : \nu \in \mathcal{N}_E\}$.*

The space $\mathcal{D}'(\Omega_1; \mathcal{D}'(\Omega_2; E))$ of ***distributions of distributions*** *is endowed with the semi-norms, indexed by $\varphi_1 \in \mathcal{D}(\Omega_1)$, $\varphi_2 \in \mathcal{D}(\Omega_2)$ and $\nu \in \mathcal{N}_E$,*

$$\|f\|_{\mathcal{D}'(\Omega_1;\mathcal{D}'(\Omega_2;E));\varphi_1,\varphi_2,\nu} \stackrel{\text{def}}{=} \|\langle\langle f, \varphi_1\rangle_{\Omega_1}, \varphi_2\rangle_{\Omega_2}\|_{E;\nu}. \;\blacksquare \qquad (14.1)$$

Utility of distributions of distributions. Distributions of distributions are particularly useful for the study of the so-called *evolution* equations governing quantities depending on time t and position x that are represented, after separating the variables via the kernel theorem (Theorem 15.10), by an element of $\mathcal{D}'((0,T); \mathcal{D}'(\Omega))$. □

Let us characterize distributions of distributions, denoting by $\mathrm{Lin}(E; F)$ the vector space of linear mappings from E into F.

Theorem 14.2.– *Let*

$$f \in \mathrm{Lin}(\mathcal{D}(\Omega_1); \mathrm{Lin}(\mathcal{D}(\Omega_2); E)),$$

where Ω_1 is an open subset of $\mathbb{R}^{d_1}$, Ω_2 is an open subset of $\mathbb{R}^{d_2}$ and E is a Neumann space whose family of semi-norms is denoted by $\{\| \ \|_{E;\nu} : \nu \in \mathcal{N}_E\}$.

Then, each of the following three properties is equivalent to

$$f \in \mathcal{D}'(\Omega_1; \mathcal{D}'(\Omega_2; E)).$$

(a) *For every $\varphi_1 \in \mathcal{D}(\Omega_1)$ and $\nu \in \mathcal{N}_E$, there exist $p_2 \in \mathcal{C}^+(\Omega_2)$ and $c_2 \in \mathbb{R}$ such that: for every $\varphi_2 \in \mathcal{D}(\Omega_2)$,*

$$\|\langle\langle f, \varphi_1\rangle_{\Omega_1}, \varphi_2\rangle_{\Omega_2}\|_{E;\nu} \leq c_2 \|\varphi_2\|_{\mathcal{D}(\Omega_2);p_2}. \tag{14.2}$$

And, for every $\varphi_2 \in \mathcal{D}(\Omega_2)$ and $\nu \in \mathcal{N}_E$, there exist $p_1 \in \mathcal{C}^+(\Omega_1)$ and $c_1 \in \mathbb{R}$ such that: for every $\varphi_1 \in \mathcal{D}(\Omega_1)$,

$$\|\langle\langle f, \varphi_1\rangle_{\Omega_1}, \varphi_2\rangle_{\Omega_2}\|_{E;\nu} \leq c_1 \|\varphi_1\|_{\mathcal{D}(\Omega_1);p_1}. \tag{14.3}$$

(b) *For every $\varphi_1 \in \mathcal{D}(\Omega_1)$, the mapping $\varphi_2 \mapsto \langle\langle f, \varphi_1\rangle_{\Omega_1}, \varphi_2\rangle_{\Omega_2}$ is sequentially continuous from $\mathcal{D}(\Omega_2)$ into E.*

And, for every $\varphi_2 \in \mathcal{D}(\Omega_2)$, the mapping $\varphi_1 \mapsto \langle\langle f, \varphi_1\rangle_{\Omega_1}, \varphi_2\rangle_{\Omega_2}$ is sequentially continuous from $\mathcal{D}(\Omega_1)$ into E.

(c) *The mapping $(\varphi_1, \varphi_2) \mapsto \langle\langle f, \varphi_1\rangle_{\Omega_1}, \varphi_2\rangle_{\Omega_2}$ is sequentially continuous from $\mathcal{D}(\Omega_1) \times \mathcal{D}(\Omega_2)$ into E.* ▌

Proof. **1.** Characterization *(a)*. According to the characterization of distributions from Theorem 3.3, $f \in \mathcal{D}'(\Omega_1; \mathcal{D}'(\Omega_2; E))$ is equivalent to the two following properties:

— for every $\varphi_1 \in \mathcal{D}(\Omega_1)$, $\langle f, \varphi_1\rangle_{\Omega_1} \in \mathcal{D}'(\Omega_2; E)$, which is characterized by (14.2);

— for every semi-norm of $\mathcal{D}'(\Omega_2; E)$, i.e. for every $\varphi_2 \in \mathcal{D}(\Omega_2)$ and $\nu \in \mathcal{N}_E$, there exist $p_1 \in \mathcal{C}^+(\Omega_1)$ and $c_1 \in \mathbb{R}$ such that, for every $\varphi_1 \in \mathcal{D}(\Omega_1)$,

$$\|\langle f, \varphi_1\rangle_{\Omega_1}\|_{\mathcal{D}'(\Omega_2;E);\varphi_2,\nu} = \|\langle\langle f, \varphi_1\rangle_{\Omega_1}, \varphi_2\rangle_{\Omega_2}\|_{E;\nu} \leq c_1 \|\varphi_1\|_{\mathcal{D}(\Omega_1);p_1},$$

that is (14.3).

2. Characterization *(b)*. For every $\varphi_1 \in \mathcal{D}(\Omega_1)$, Property (14.2) is equivalent, as we have seen, to $\langle f, \varphi_1\rangle_{\Omega_1} \in \mathcal{D}'(\Omega_2; E)$ and so, according to the characterization of distributions from Theorem 3.4 (b), to the first sequential continuity property of (b).

Similarly, (14.3) is equivalent to the second property of (b).

3. *Characterization (c).* Let $f \in \mathcal{D}'(\Omega_1; \mathcal{D}'(\Omega_2; E))$. If $\varphi_{1,n} \to \varphi_1$ in $\mathcal{D}(\Omega_1)$, then, due to (b),

$$\langle f, \varphi_{1,n} \rangle_{\Omega_1} \to \langle f, \varphi_1 \rangle_{\Omega_1} \text{ in } \mathcal{D}'(\Omega_2; E).$$

If moreover $\varphi_{2,n} \to \varphi_2$ in $\mathcal{D}(\Omega_2)$, then, since (Theorem 4.4 (c)) the mapping $\langle \ , \ \rangle_{\Omega_2}$ is sequentially continuous from $\mathcal{D}'(\Omega_2; E) \times \mathcal{D}(\Omega_2)$ into E,

$$\langle \langle f, \varphi_{1,n} \rangle_{\Omega_1}, \varphi_{2,n} \rangle_{\Omega_2} \to \langle \langle f, \varphi_1 \rangle_{\Omega_1}, \varphi_2 \rangle_{\Omega_2} \text{ in } E.$$

Conversely, this property implies properties (b) of sequentially continuity, and hence $f \in \mathcal{D}'(\Omega_1; \mathcal{D}'(\Omega_2; E))$. □

Slightly abusive notation. The notation $\langle f, \varphi_1 \rangle_{\Omega_1}$ in the place of $f(\varphi_1)$ and $\langle \langle f, \varphi_1 \rangle_{\Omega_1}, \varphi_2 \rangle_{\Omega_2}$ in the place of $(f(\varphi_1))(\varphi_2)$ anticipates the fact that f is a distribution of distributions. □

Non-continuity. The sequential continuity cannot be replaced with continuity in characterization (c) of Theorem 14.2, even if $E = \mathbb{R}$. Indeed, if Ω_1 and Ω_2 are not empty and $E \neq \{0\}$:

$$\begin{cases} \text{There exists } f \in \mathcal{D}'(\Omega_1; \mathcal{D}'(\Omega_2; E)) \text{ such that the mapping} \\ (\varphi_1, \varphi_2) \mapsto \langle \langle f, \varphi_1 \rangle_{\Omega_1}, \varphi_2 \rangle_{\Omega_2} \text{ is not continuous from } \mathcal{D}(\Omega_1) \times \mathcal{D}(\Omega_2) \text{ into } E. \ \square \end{cases} \tag{14.4}$$

Proof. We define such a distribution of distributions by

$$\langle \langle f, \varphi_1 \rangle_{\Omega_1}, \varphi_2 \rangle_{\Omega_2} = e \sum_{n \in \mathbb{N}} (\partial_1^n \varphi_1)(a)\, \varphi_2(b_n),$$

where $e \in E$, $e \neq 0$, $a \in \Omega_1$ and the sequence $(b_n)_{n \in \mathbb{N}}$ is made up of distinct points in Ω_2 converging to a point outside of Ω_2 (if $\Omega_2 = \mathbb{R}^{d_2}$, we replace this condition with $|b_n| = n$). Let a semi-norm of E be such that $\|e\|_{E;\nu} = \alpha \neq 0$.

If the mapping $(\varphi_1, \varphi_2) \mapsto \langle \langle f, \varphi_1 \rangle_{\Omega_1}, \varphi_2 \rangle_{\Omega_2}$ were continuous, the characterization of multilinear continuous mappings from Theorem A.38 (b) would provide (since the family of semi-norms of $\mathcal{D}(\Omega)$ is filtering by Theorem 2.7) the existence of $p_1 \in \mathcal{C}^+(\Omega_1)$, $p_2 \in \mathcal{C}^+(\Omega_2)$ and $c \in \mathbb{R}$ such that

$$\alpha \Big| \sum_{n \in \mathbb{N}} (\partial_1^n \varphi_1)(a)\, \varphi_2(b_n) \Big| \leq c \|\varphi_1\|_{\mathcal{D}(\Omega_1);p_1} \|\varphi_2\|_{\mathcal{D}(\Omega_2);p_2}. \tag{14.5}$$

Let $B(a,r)$ be a closed ball included in Ω_1 and m be an integer bounding p_1 from above on $B(a,r)$. Choose φ_1 with support in $B(a,r)$ and φ_2 such that $\varphi_2(b_{m+1}) = 1$ and $\varphi_2(b_n) = 0$ if $n \neq m+1$. Then, (14.5) would give, denoting $b = c \|\varphi_2\|_{\mathcal{D}(\Omega_2);p_2} / \alpha$,

$$|(\partial_1^{m+1} \varphi_1)(a)| \leq b \|\varphi_1\|_{\mathcal{D}(\Omega_1);p_1} \leq bm \sup_{|\beta| \leq m} \sup_{|x_1 - a| \leq r} |\partial^\beta \varphi_1(x_1)|.$$

However, there exists a function φ_1 which does not satisfy this inequality (we will verify this in Lemma 16.13), which contradicts (14.5) and so proves (14.4). □

Schwartz's topology. The space $\mathcal{D}'(\Omega_1; \mathcal{D}'(\Omega_2; E))$ does not change when we use the uniform topology, i.e. that of the uniform convergence on the bounded subsets of the $\mathcal{D}(\Omega_i)$, instead of the simple convergence on the $\mathcal{D}(\Omega_i)$ used here, although it depends *a priori* on the topology with which $\mathcal{D}'(\Omega_2; E)$ is endowed. That is to say,

$$\mathcal{D}'(\Omega_1; \mathcal{D}'(\Omega_2; E)\text{-unif})\text{-unif} = \mathcal{D}'(\Omega_1; \mathcal{D}'(\Omega_2; E)).$$

Its elements remain the same, as well as its convergent sequences and its bounded subsets. On the contrary, its uniform topology is, as always, stronger than the simple topology, that is to say,

$$\mathcal{D}'(\Omega_1; \mathcal{D}'(\Omega_2; E)\text{-unif})\text{-unif} \subsetneq \mathcal{D}'(\Omega_1; \mathcal{D}'(\Omega_2; E)).$$

These properties are proven in [SIMON, 84], in progress. □

14.2. Bounded sets

Let us characterize the **bounded sets of distributions of distributions**.

Theorem 14.3.– *Let $\mathcal{F} \subset \mathcal{D}'(\Omega_1; \mathcal{D}'(\Omega_2; E))$, where Ω_1 is an open subset of $\mathbb{R}^{d_1}$, Ω_2 is an open subset of $\mathbb{R}^{d_2}$ and E is a Neumann space.*

Then, each of the following properties is equivalent to:

$$\mathcal{F} \text{ is bounded in } \mathcal{D}'(\Omega_1; \mathcal{D}'(\Omega_2; E)).$$

(a) *For every $\varphi_1 \in \mathcal{D}(\Omega_1)$, every $\varphi_2 \in \mathcal{D}(\Omega_2)$ and every semi-norm $\| \ \|_{E;\nu}$ of E,*

$$\sup_{f \in \mathcal{F}} \| \langle \langle f, \varphi_1 \rangle_{\Omega_1}, \varphi_2 \rangle_{\Omega_2} \|_{E;\nu} < \infty.$$

(b) *For every $\varphi_1 \in \mathcal{D}(\Omega_1)$ and $\varphi_2 \in \mathcal{D}(\Omega_2)$,*

$$\{ \langle \langle f, \varphi_1 \rangle_{\Omega_1}, \varphi_2 \rangle_{\Omega_2} : f \in \mathcal{F} \} \text{ is bounded in } E. \ ▮$$

Proof. According to Definitions 1.2 of a bounded set in a semi-normed space and 14.1 of the semi-norms of $\mathcal{D}'(\Omega_1; \mathcal{D}'(\Omega_2; E))$, $\mathcal{F}$ is bounded in $\mathcal{D}'(\Omega_1; \mathcal{D}'(\Omega_2; E))$ if and only if: for all $\varphi_1 \in \mathcal{D}(\Omega_1)$, $\varphi_2 \in \mathcal{D}(\Omega_2)$ and $\nu \in \mathcal{N}_E$,

$$\sup_{f \in \mathcal{F}} \|f\|_{\mathcal{D}'(\Omega_1;\mathcal{D}'(\Omega_2;E));\varphi_1,\varphi_2,\nu} = \sup_{f \in \mathcal{F}} \| \langle \langle f, \varphi_1 \rangle_{\Omega_1}, \varphi_2 \rangle_{\Omega_2} \|_{E;\nu} < \infty.$$

That is, again from Definition 1.2, equivalent to Property (b). □

Let us show that every bounded set of distributions of distributions is equicontinuous on the product spaces $\mathcal{C}^\infty_{K_1}(\Omega_1) \times \mathcal{C}^\infty_{K_2}(\Omega_2)$.

Theorem 14.4.– *Let*

$$\mathcal{F} \text{ be a bounded subset of } \mathcal{D}'(\Omega_1; \mathcal{D}'(\Omega_2; E)),$$

where Ω_1 is an open subset of $\mathbb{R}^{d_1}$, Ω_2 is an open subset of $\mathbb{R}^{d_2}$ and E is a Neumann space. And let $\| \ \|_{E;\nu}$ be a semi-norm of E, K_1 a compact subset of Ω_1 and K_2 a compact subset of Ω_2.

Then, there exist $m \in \mathbb{N}$ and $c \in \mathbb{R}$ such that, for every $\varphi_1 \in \mathcal{C}^\infty_{K_1}(\Omega_1)$ and $\varphi_2 \in \mathcal{C}^\infty_{K_2}(\Omega_2)$,

$$\sup_{f \in \mathcal{F}} \| \langle \langle f, \varphi_1 \rangle_{\Omega_1}, \varphi_2 \rangle_{\Omega_2} \|_{E;\nu} \leq c \|\varphi_1\|_{\mathcal{C}^m_b(\Omega_1)} \|\varphi_2\|_{\mathcal{C}^m_b(\Omega_2)}. \ ▮$$

Utility of Theorem 14.4. The equicontinuity of the bounded sets of distributions of distributions will be used to prove the surjectivity of the separation of variables in distributions (Theorem 15.10). □

Proof of Theorem 14.4. Let us use the analog of the Banach–Steinhaus for bilinear mappings recalled in Theorem A.40, for the family of mappings T_f from $\mathcal{C}^\infty_{K_1}(\Omega_1) \times \mathcal{C}^\infty_{K_2}(\Omega_2)$ into E defined, for $f \in \mathcal{F}$, by

$$T_f(\varphi_1, \varphi_2) \stackrel{\text{def}}{=} \langle\langle f, \varphi_1\rangle_{\Omega_1}, \varphi_2\rangle_{\Omega_2}.$$

Les us check the hypotheses of the said theorem.

— The spaces $\mathcal{C}^\infty_{K_1}(\Omega_1)$ and $\mathcal{C}^\infty_{K_2}(\Omega_2)$ are Fréchet (Theorem A.53).

— Each mapping T_f is continuous. Indeed, it is sequentially continuous on $\mathcal{D}(\Omega_1) \times \mathcal{D}(\Omega_2)$ (Theorem 14.2 (c)), and therefore (Theorem A.29, because $\mathcal{C}^\infty_{K_i}(\Omega_i) \subsetneq \mathcal{D}(\Omega_i)$, see Theorem 2.10) on $\mathcal{C}^\infty_{K_1}(\Omega_1) \times \mathcal{C}^\infty_{K_2}(\Omega_2)$; since this is a Fréchet space (Theorem A.25), T_f is continuous there (Theorem 1.11).

— For every semi-norm $\| \;\|_{E;\nu}$ of E, Definition 14.1 of the semi-norms of $\mathcal{D}'(\Omega_1; \mathcal{D}'(\Omega_2; E))$ gives

$$\sup_{f\in\mathcal{F}} \|T_f(\varphi_1, \varphi_2)\|_{E;\nu} = \sup_{f\in\mathcal{F}} \|f\|_{\mathcal{D}'(\Omega_1;\mathcal{D}'(\Omega_2;E));\varphi_1,\varphi_2,\nu} < \infty.$$

Theorem A.40 therefore gives the existence of $c > 0$ and of two finite sets M_1 and M_2 of integers such that, for all $\varphi_1 \in \mathcal{C}^\infty_{K_1}(\Omega_1)$ and $\varphi_2 \in \mathcal{C}^\infty_{K_2}(\Omega_2)$,

$$\sup_{f\in\mathcal{F}} \|\langle\langle f, \varphi_1\rangle_{\Omega_1}, \varphi_2\rangle_{\Omega_2}\|_{E;\nu} \le c \sup_{m_1\in M_1} \|\varphi_1\|_{\mathcal{C}^\infty_b(\Omega_1);m_1} \sup_{m_2\in M_2} \|\varphi_2\|_{\mathcal{C}^\infty_b(\Omega_2);m_2}.$$

The stated inequality is then satisfied with $m = \sup\{M_1 \cup M_2\}$ since, from equality (1.6), p. 13, $\| \;\|_{\mathcal{C}^\infty_b(\Omega);m} = \| \;\|_{\mathcal{C}^m_b(\Omega)}$. □

14.3. Convergent sequences

Let us characterize the **convergent sequences of distributions of distributions**.

Theorem 14.5.– *Let $(f_n)_{n\in\mathbb{N}}$ be a sequence of $\mathcal{D}'(\Omega_1; \mathcal{D}'(\Omega_2; E))$, where Ω_1 is an open subset of $\mathbb{R}^{d_1}$, Ω_2 is an open subset of $\mathbb{R}^{d_2}$ and E is a Neumann space, and $f \in \mathcal{D}'(\Omega_1; \mathcal{D}'(\Omega_2; E))$.*

Then, each of the following properties is equivalent to

$$f_n \to f \;\; in\; \mathcal{D}'(\Omega_1; \mathcal{D}'(\Omega_2; E)).$$

(a) *For every $\varphi_1 \in \mathcal{D}(\Omega_1)$, every $\varphi_2 \in \mathcal{D}(\Omega_2)$ and every semi-norm $\| \;\|_{E;\nu}$ of E,*

$$\|\langle\langle f_n - f, \varphi_1\rangle_{\Omega_1}, \varphi_2\rangle_{\Omega_2}\|_{E;\nu} \to 0.$$

(b) *For every $\varphi_1 \in \mathcal{D}(\Omega_1)$ and $\varphi_2 \in \mathcal{D}(\Omega_2)$,*

$$\langle\langle f_n, \varphi_1\rangle_{\Omega_1}, \varphi_2\rangle_{\Omega_2} \to \langle\langle f, \varphi_1\rangle_{\Omega_1}, \varphi_2\rangle_{\Omega_2} \;\; in\; E. \;\blacksquare$$

Proof. According to Definitions 1.3 of a convergent sequence in a semi-normed space and 14.1 of the semi-norms of $\mathcal{D}'(\Omega_1; \mathcal{D}'(\Omega_2; E))$, the convergence of f_n to f in $\mathcal{D}'(\Omega_1; \mathcal{D}'(\Omega_2; E))$ is equivalent to: for every $\varphi_1 \in \mathcal{D}(\Omega_1)$, $\varphi_2 \in \mathcal{D}(\Omega_2)$ and $\nu \in \mathcal{N}_E$,

$$\|f_n - f\|_{\mathcal{D}'(\Omega_1;\mathcal{D}'(\Omega_2;E));\varphi_1,\varphi_2,\nu} = \|\langle\langle f_n - f, \varphi_1\rangle_{\Omega_1}, \varphi_2\rangle_{\Omega_2}\|_{E;\nu} \to 0.$$

That is equivalent, again from Definition 1.3, to $\langle\langle f_n - f, \varphi_1\rangle_{\Omega_1}, \varphi_2\rangle_{\Omega_2} \to 0_E$ for every $\varphi_1 \in \mathcal{D}(\Omega_1)$ and $\varphi_2 \in \mathcal{D}(\Omega_2)$. □

Let us give another characterization of these sequences providing a convergence of $\langle\langle f_n - f, \varphi_1\rangle_{\Omega_1}, \varphi_2\rangle_{\Omega_2}$ that is uniform [1] with respect to φ_1 and φ_2.

Theorem 14.6.– *Let $(f_n)_{n\in\mathbb{N}}$ be a sequence of $\mathcal{D}'(\Omega_1; \mathcal{D}'(\Omega_2; E))$, where Ω_1 is an open subset of $\mathbb{R}^{d_1}$, Ω_2 is an open subset of $\mathbb{R}^{d_2}$ and E is a Neumann space, and $f \in \mathcal{D}'(\Omega_1; \mathcal{D}'(\Omega_2; E))$. Then,*

$$f_n \to f \text{ in } \mathcal{D}'(\Omega_1; \mathcal{D}'(\Omega_2; E))$$

if and only if, for every semi-norm $\|\ \|_{E;\nu}$ of E, every compact subset K_1 of Ω_1 and every compact subset K_2 of Ω_2, there exist $m \in \mathbb{N}$ and a decreasing real sequence $(c_n)_{n\in\mathbb{N}}$, $c_n \to 0$, such that: for every $\varphi_1 \in C^\infty_{K_1}(\Omega_1)$, $\varphi_2 \in C^\infty_{K_2}(\Omega_2)$ and $n \in \mathbb{N}$,

$$\|\langle\langle f_n - f, \varphi_1\rangle_{\Omega_1}, \varphi_2\rangle_{\Omega_2}\|_{E;\nu} \le c_n \|\varphi_1\|_{C^m_b(\Omega_1)} \|\varphi_2\|_{C^m_b(\Omega_2)}. \tag{14.6}$$

▮

Utility of Theorem 14.6. Inequality (14.6) is one of the two inequalities by the virtue of which we prove the kernel theorem without resorting, like Laurent SCHWARTZ did, to delicate topological properties, see the comment *Originality of the proof*, p. 317. More precisely, this inequality provides inequalities (15.28) and (15.30), which are respectively used to prove that the separation of variables is surjective and that its inverse mapping is sequentially continuous. □

Principle of the proof of Theorem 14.6. It is the extension to distributions of distributions of the proof of Theorem 8.27, which is the analogous property for $\mathcal{D}'(\Omega; E)$. □

Proof of Theorem 14.6. Inequality (14.6) implies the convergence of f_n to f due to the characterization of convergent sequences of distributions of distributions from Theorem 14.5 (a), so it remains to prove the converse.

1. **History of the uniform convergence with respect to test functions.** The uniform convergence given in Theorem 14.6 is new, including for real values. Its relation with the uniform convergence on the bounded subsets of $\mathcal{D}(\Omega)$ is explained, in the simpler case of $\mathcal{D}'(\Omega; E)$, in the comment *Comparison of Theorems 8.27 and 8.28*, p. 190.

Suppose thus that

$$f_n \to f \text{ in } \mathcal{D}'(\Omega_1; \mathcal{D}'(\Omega_2; E)), \tag{14.7}$$

and let $\nu \in \mathcal{N}_E$, K_1 be a compact subset of Ω_1, and K_2 a compact subset of Ω_2. Let us establish (14.6) by contradiction, in five steps.

1. Choice of m. Due to the strong inclusion theorem (Theorem A.22), there are $r_1 > 0$ and $r_2 > 0$ such that $K_1 + B_1(0, r_1) \subset \Omega_1$ and $K_2 + B_2(0, r_2) \subset \Omega_2$, where B_i denotes a closed ball in $\mathbb{R}^{d_i}$. Let $r = \inf\{r_1, r_2\}/2$ and

$$\omega_1 \stackrel{\text{def}}{=} K_1 + \mathring{B}_1(0, r), \;\; \omega_2 \stackrel{\text{def}}{=} K_2 + \mathring{B}_2(0, r), \;\; Q_1 \stackrel{\text{def}}{=} \overline{\omega_1}, \;\; Q_2 \stackrel{\text{def}}{=} \overline{\omega_2}.$$

These are respectively two open sets and two compact sets.

The sequence $(f_n - f)_{n \in \mathbb{N}}$ is bounded in $\mathcal{D}'(\Omega_1; \mathcal{D}'(\Omega_2; E))$, as is every convergent sequence (Theorem A.5). It is therefore equicontinuous on $\mathcal{C}^\infty_{Q_1}(\Omega) \times \mathcal{C}^\infty_{Q_2}(\Omega)$ according to Theorem 14.4. More precisely, this shows that there exist $m \geq 1$ and $b \in \mathbb{R}$ such that: for every $\varphi_1 \in \mathcal{C}^\infty_{Q_1}(\Omega_1)$ and $\varphi_2 \in \mathcal{C}^\infty_{Q_2}(\Omega_2)$,

$$\|\langle\langle f_n - f, \varphi_1 \rangle_{\Omega_1}, \varphi_2 \rangle_{\Omega_2}\|_{E;\nu} \leq b \|\varphi_1\|_{\mathcal{C}^{m-1}_{\mathrm{b}}(\Omega_1)} \|\varphi_2\|_{\mathcal{C}^{m-1}_{\mathrm{b}}(\Omega_2)}. \tag{14.8}$$

2. Extraction of subsequences. If Property (14.6) was not satisfied, there would exist a subsequence, that we again denote by $(f_n)_{n \in \mathbb{N}}$, for which there would exist $a > 0$ and sequences $(\phi^1_n)_{n \in \mathbb{N}}$ of $\mathcal{C}^\infty_{K_1}(\Omega_1)$ and $(\phi^2_n)_{n \in \mathbb{N}}$ of $\mathcal{C}^\infty_{K_2}(\Omega_2)$ such that, for every n,

$$\|\phi^1_n\|_{\mathcal{C}^m_{\mathrm{b}}(\Omega_1)} \leq 1, \quad \|\phi^2_n\|_{\mathcal{C}^m_{\mathrm{b}}(\Omega_2)} \leq 1, \tag{14.9}$$

$$\|\langle\langle f_n - f, \phi^1_n \rangle_{\Omega_1}, \phi^2_n \rangle_{\Omega_2}\|_{E;\nu} \geq a. \tag{14.10}$$

Since every bounded sequence in $\mathcal{C}^m_K(\Omega)$ has a subsequence converging in $\mathcal{C}^{m-1}_K(\Omega)$ (Lemma 2.15), we could extract a pair of subsequences such that, still denoting by n their current index,

$$\phi^1_n \to \phi^1 \text{ in } \mathcal{C}^{m-1}_{K_1}(\Omega_1), \quad \phi^2_n \to \phi^2 \text{ in } \mathcal{C}^{m-1}_{K_2}(\Omega_2). \tag{14.11}$$

3. Contradiction of (14.10). Let us assume for the moment that there exist $\varphi_1 \in \mathcal{C}^\infty_{Q_1}(\Omega_1)$ and $\varphi_2 \in \mathcal{C}^\infty_{Q_2}(\Omega_2)$ (we will obtain them by regularizing ϕ^1 and ϕ^2) such that

$$\|\varphi_1 - \phi^1\|_{\mathcal{C}^{m-1}_{\mathrm{b}}(\Omega_1)} \leq \frac{a}{4b}, \; \|\varphi_2 - \phi^2\|_{\mathcal{C}^{m-1}_{\mathrm{b}}(\Omega_2)} \leq \frac{a}{4b}, \; \|\varphi_2\|_{\mathcal{C}^{m-1}_{\mathrm{b}}(\Omega_2)} \leq 2. \tag{14.12}$$

Decompose

$$\langle\langle f_n - f, \phi^1_n \rangle_{\Omega_1}, \phi^2_n \rangle_{\Omega_2} = X + Y + \langle\langle f_n - f, \varphi_1 \rangle_{\Omega_1}, \varphi_2 \rangle_{\Omega_2}, \tag{14.13}$$

where

$$X \stackrel{\text{def}}{=} \langle\langle f_n - f, \phi^1_n \rangle_{\Omega_1}, \phi^2_n - \varphi_2 \rangle_{\Omega_2}, \quad Y \stackrel{\text{def}}{=} \langle\langle f_n - f, \phi^1_n - \varphi_1 \rangle_{\Omega_1}, \varphi_2 \rangle_{\Omega_2}.$$

Inequality (14.8) would give

$$\|X\|_{E;\nu} \leq b\|\phi_n^1\|_{\mathcal{C}_b^{m-1}(\Omega_1)}\|\phi_n^2 - \varphi_2\|_{\mathcal{C}_b^{m-1}(\Omega_2)}.$$

By decomposing $\phi_n^2 - \varphi_2 = (\phi_n^2 - \phi^2) + (\phi^2 - \varphi_2)$ and using (14.9) and (14.12), we would obtain

$$\|X\|_{E;\nu} \leq \frac{a}{4} + b\|\phi_n^2 - \phi^2\|_{\mathcal{C}_b^{m-1}(\Omega_2)}.$$

Similarly, (14.8), (14.9) and (14.12) would give

$$\|Y\|_{E;\nu} \leq b\|\phi_n^1 - \varphi_1\|_{\mathcal{C}_b^{m-1}(\Omega_1)}\|\varphi_2\|_{\mathcal{C}_b^{m-1}(\Omega_2)} \leq \frac{a}{2} + 2b\|\phi_n^1 - \phi^1\|_{\mathcal{C}_b^{m-1}(\Omega_1)}.$$

With these inequalities, decomposition (14.13) would give

$$\|\langle\langle f_n - f, \phi_n^1\rangle_{\Omega_1}, \phi_n^2\rangle_{\Omega_2}\|_{E;\nu} \leq \frac{3a}{4} + \epsilon_n,$$

where

$$\epsilon_n = b\|\phi_n^2 - \phi^2\|_{\mathcal{C}_b^{m-1}(\Omega_2)} + 2b\|\phi_n^1 - \phi^1\|_{\mathcal{C}_b^{m-1}(\Omega_1)} + \|\langle\langle f_n - f, \varphi_1\rangle_{\Omega_1}, \varphi_2\rangle_{\Omega_2}\|_{E;\nu}.$$

According to convergences (14.7) and (14.11), ϵ_n would tend to 0, thus, for n large enough, inequality (14.10) would not be satisfied. This is the desired contradiction.

This proves the stated inequality (14.6), subject to proving (14.12).

4. Verification of (14.12). The extension $\widetilde{\phi}^1$ of ϕ^1 by 0 belongs to $\mathbf{C}_b^{m-1}(\mathbb{R}^{d_1})$, hence its regularized $\widetilde{\phi^1} \diamond \rho_k$ belongs to $\mathbf{C}_b^\infty(\mathbb{R}^{d_1})$ and satisfies (Theorem 8.7 (b))

$$\widetilde{\phi^1} \diamond \rho_k \to \widetilde{\phi^1} \text{ in } \mathbf{C}_b^{m-1}(\mathbb{R}^{d_1}).$$

Since this space is, by Definition 1.20, endowed with the semi-norms of $\mathcal{C}_b^{m-1}(\mathbb{R}^d)$, the restriction $\varphi_1 \stackrel{\text{def}}{=} (\widetilde{\phi^1} \diamond \rho_k)|_{\Omega_1}$ satisfies, for k large enough,

$$\|\varphi_1 - \phi^1\|_{\mathcal{C}_b^{m-1}(\Omega_1)} \leq \frac{a}{4b}.$$

In addition, φ_1 belongs to $\mathcal{C}_{Q_1}^\infty(\Omega_1)$ as soon as $k \geq 1/r$, because then the inclusion of the support of a weighted distribution from Theorem 7.10 gives

$$\operatorname{supp}(\widetilde{\phi^1} \diamond \rho_k) \subset \operatorname{supp}\widetilde{\phi^1} - \operatorname{supp}\rho_k \subset K_1 + B(0,r) = Q_1.$$

Similarly, the restriction $\varphi_2 \stackrel{\text{def}}{=} (\widetilde{\phi^2} \diamond \rho_k)|_{\Omega_2}$ belongs to $\mathcal{C}_{Q_2}^\infty(\Omega_2)$ and satisfies (since $\|\phi^2\|_{\mathcal{C}_b^{m-1}(\Omega_2)} \leq 1$ from (14.9) and (14.11)), for k large enough,

$$\|\varphi_2 - \phi^2\|_{\mathcal{C}_b^{m-1}(\Omega_2)} \leq \frac{a}{4b}, \quad \|\varphi_2\|_{\mathcal{C}_b^{m-1}(\Omega_2)} \leq 2.$$

5. Conclusion. This proves Property (14.6), except for the decreasing of the sequence $(c_n)_{n\in\mathbb{N}}$. This is obtained by replacing c_n by $c'_n = \sup_{m\geq n} c_m$. □

14.4. Extraction of convergent subsequences

Let us characterize sets of distributions of distributions that are relatively sequentially compact, namely in which every sequence has a convergent subsequence[2].

Theorem 14.7.– *Let $\mathcal{F} \subset \mathcal{D}'(\Omega_1; \mathcal{D}'(\Omega_2; E))$, where Ω_1 is an open subset of $\mathbb{R}^{d_1}$, Ω_2 is an open subset of $\mathbb{R}^{d_2}$ and E is a Neumann space. Then:*

every sequence in $\mathcal{F}$ has a subsequence which converges in $\mathcal{D}'(\Omega_1; \mathcal{D}'(\Omega_2; E))$

if, and only if, for every $\varphi_1 \in \mathcal{D}(\Omega_1)$ and $\varphi_2 \in \mathcal{D}(\Omega_2)$,

$$\begin{cases} \text{every sequence in } \{\langle\langle f, \varphi_1\rangle_{\Omega_1}, \varphi_2\rangle_{\Omega_2} : f \in \mathcal{F}\} \\ \text{has a subsequence which converges in } E. \end{cases}$$ ▌

Proof. According to the characterization of relatively sequentially compact sets of distributions from Theorem 4.6, every sequence in $\mathcal{F}$ has a subsequence which converges in $\mathcal{D}'(\Omega_1; \mathcal{D}'(\Omega_2; E))$ if and only if every sequence in the set $\{\langle f, \varphi_1\rangle_{\Omega_1} : f \in \mathcal{F}\}$ has a subsequence which converges in $\mathcal{D}'(\Omega_2; E)$.

Which, again due to Theorem 4.6, is equivalent to the fact that every sequence in the set $\{\langle\langle f, \varphi_1\rangle_{\Omega_1}, \varphi_2\rangle_{\Omega_2} : f \in \mathcal{F}\}$ has a subsequence which converges in E. □

Observe that, in particular, bounded sets of real distributions of distributions are relatively sequentially compact.

Theorem 14.8.– *Given open subsets Ω_1 of $\mathbb{R}^{d_1}$ and Ω_2 of $\mathbb{R}^{d_2}$,*

every bounded sequence in $\mathcal{D}'(\Omega_1; \mathcal{D}'(\Omega_2))$ has a convergent subsequence. ▌

Proof. Let $(f_n)_{n\in\mathbb{N}}$ be a bounded sequence in $\mathcal{D}'(\Omega_1; \mathcal{D}'(\Omega_2))$. For it to have a convergent subsequence, it suffices according to Theorem 14.7 to check that, for every $\varphi_1 \in \mathcal{D}(\Omega_1)$ and every $\varphi_2 \in \mathcal{D}(\Omega_2)$,

$$\{\langle\langle f_n, \varphi_1\rangle_{\Omega_1}, \varphi_2\rangle_{\Omega_2}\}_{n\in\mathbb{N}} \text{ has a convergent subsequence in } \mathbb{R}.$$

This is the case because this set is bounded in $\mathbb{R}$ by the characterization of bounded sets of distributions of distributions from Theorem 14.3 (b), and thus each of its sequences has a convergent subsequence due to the Bolzano–Weierstrass theorem (Theorem A.26 (c)). □

2. **History of the characterization of relatively sequentially compact sets.** The characterization given in Theorem 14.7 is new.

14.5. Change of the space of values

We show here that, if the space E of values is included in a "larger" space, the same holds for the corresponding spaces of distributions of distributions. Let us begin with the case of a topological inclusion.

Theorem 14.9.– *Let Ω_1 be an open subset of $\mathbb{R}^{d_1}$, Ω_2 an open subset of $\mathbb{R}^{d_2}$, and E and F two Neumann spaces such that*

$$E \underset{\rightarrow}{\subseteq} F.$$

Then,

$$\mathcal{D}'(\Omega_1; \mathcal{D}'(\Omega_2; E)) \underset{\rightarrow}{\subseteq} \mathcal{D}'(\Omega_1; \mathcal{D}'(\Omega_2; F)). \blacksquare$$

Proof. Since $E \underset{\rightarrow}{\subseteq} F$, Theorem 4.8 gives $\mathcal{D}'(\Omega_2; E) \underset{\rightarrow}{\subseteq} \mathcal{D}'(\Omega_2; F)$, from which it gives the stated topological inclusion. □

Let us come to the case of an inclusion that is only "sequentially continuous".

Theorem 14.10.– *Let Ω_1 be an open subset of $\mathbb{R}^{d_1}$, Ω_2 an open subset of $\mathbb{R}^{d_2}$, and E and F two Neumann spaces such that E is a vector subspace of F and*

the identity is sequentially continuous from E into F.

Then, $\mathcal{D}'(\Omega_1; \mathcal{D}'(\Omega_2; E))$ is a vector subspace of $\mathcal{D}'(\Omega_1; \mathcal{D}'(\Omega_2; F))$ and

$$\begin{cases} \textit{the identity is sequentially continuous from} \\ \mathcal{D}'(\Omega_1; \mathcal{D}'(\Omega_2; E)) \textit{ into } \mathcal{D}'(\Omega_1; \mathcal{D}'(\Omega_2; F)). \end{cases} \blacksquare$$

Proof. Since the identity from E into F is sequentially continuous, Theorem 4.9 on changing the space of values of a distribution shows that $\mathcal{D}'(\Omega_2; E)$ is a vector subspace of $\mathcal{D}'(\Omega_2; F)$ and that the identity is sequentially continuous from $\mathcal{D}'(\Omega_2; E)$ into $\mathcal{D}'(\Omega_2; F)$.

The same Theorem 4.9 now for $E_2 = \mathcal{D}'(\Omega_2; E)$ and $F_2 = \mathcal{D}'(\Omega_2; F)$ then gives the stated properties. □

Let us finish with the case of a "sequentially compacting" topological inclusion, namely such that the bounded subsets of E are relatively sequentially compact[3] in F.

3. **History of sequential compactness property.** The sequential compactness in $\mathcal{D}'(\Omega_1; \mathcal{D}'(\Omega_2; F))$ given in Theorem 14.11 seems to be new, as well as those in $\mathcal{D}'(\Omega_1; \mathcal{D}'(\Omega_2; E\text{-weak}))$ given in Theorems 14.12 and 14.13.

Theorem 14.11.– *Let Ω_1 be an open subset of $\mathbb{R}^{d_1}$, Ω_2 an open subset of $\mathbb{R}^{d_2}$, and E and F two Neumann spaces such that $E \underset{\rightarrow}{\subseteq} F$ and*

every bounded sequence in E has a subsequence which converges in F.

Then, $\mathcal{D}'(\Omega_1; \mathcal{D}'(\Omega_2; E)) \underset{\rightarrow}{\subseteq} \mathcal{D}'(\Omega_1; \mathcal{D}'(\Omega_2; F))$ and

$$\begin{cases} \text{every bounded sequence in } \mathcal{D}'(\Omega_1; \mathcal{D}'(\Omega_2; E)) \text{ has a} \\ \text{subsequence which converges in } \mathcal{D}'(\Omega_1; \mathcal{D}'(\Omega_2; F)). \end{cases} ∎$$

Proof. Theorem 4.10 on compactness shows that $\mathcal{D}'(\Omega_2; E) \underset{\rightarrow}{\subseteq} \mathcal{D}'(\Omega_2; F)$ and that every bounded sequence in $\mathcal{D}'(\Omega_2; E)$ has a subsequence which converges in $\mathcal{D}'(\Omega_2; F)$.

The same Theorem 4.10 now for $E_2 = \mathcal{D}'(\Omega_2; E)$ and $F_2 = \mathcal{D}'(\Omega_2; F)$ then gives the stated result. □

14.6. Distributions of distributions with values in E-weak

Let us show that every distribution of distributions with values in E-weak is a distribution of distributions with values in E.

Theorem 14.12.– *Let Ω_1 be an open subset of $\mathbb{R}^{d_1}$, Ω_2 an open subset of $\mathbb{R}^{d_2}$ and E a separated semi-normed space such that*

E-weak is a Neumann space.

Then, E is a Neumann space,

$$\mathcal{D}'(\Omega_1; \mathcal{D}'(\Omega_2; E)) = \mathcal{D}'(\Omega_1; \mathcal{D}'(\Omega_2; E\text{-weak})),$$

these two spaces have the same bounded subsets, and

$$\mathcal{D}'(\Omega_1; \mathcal{D}'(\Omega_2; E)) \underset{\rightarrow}{\subseteq} \mathcal{D}'(\Omega_1; \mathcal{D}'(\Omega_2; E\text{-weak})). ∎$$

Proof. **1.** *Completion of E.* Since E-weak is a Neumann space, E is as well due to Theorem 4.13, which allows to define $\mathcal{D}'(\Omega_1; \mathcal{D}'(\Omega_2; E))$.

2. *Algebraic equality.* An element of $\mathcal{D}'(\Omega_1; \mathcal{D}'(\Omega_2; E))$ is, by the characterization of distributions from Theorem 3.4 (e), a linear mapping from $\mathcal{D}(\Omega_1)$ into $\mathcal{D}'(\Omega_2; E)$ which, for every compact subset K of Ω_1, transforms the bounded subsets of $\mathcal{C}_K^\infty(\Omega_1)$ into bounded subsets of $\mathcal{D}'(\Omega_2; E)$.

Since $\mathcal{D}'(\Omega_2; E)$ and $\mathcal{D}'(\Omega_2; E\text{-weak})$ here are equal and have the same bounded subsets due to Theorem 4.14, we can exchange E with E-weak which proves that

$$\mathcal{D}'(\Omega_1; \mathcal{D}'(\Omega_2; E)) = \mathcal{D}'(\Omega_1; \mathcal{D}'(\Omega_2; E\text{-weak})).$$

3. *Topological inclusion.* Since $E \subsetneq E$-weak (Theorem 4.12), Theorem 4.8 on the change of the space of values of distributions gives $\mathcal{D}'(\Omega_2; E) \subsetneq \mathcal{D}'(\Omega_2; E\text{-weak})$; with this, again Theorem 4.8 gives

$$\mathcal{D}'(\Omega_1; \mathcal{D}'(\Omega_2; E)) \subsetneq \mathcal{D}'(\Omega_1; \mathcal{D}'(\Omega_2; E\text{-weak})).$$

4. *Equality of the bounded sets.* According to the characterization of bounded sets of distributions of distributions from Theorem 14.3 (b), a bounded subset of $\mathcal{D}'(\Omega_1; \mathcal{D}'(\Omega_2; E))$ is a subset $\mathcal{F}$ such that, for every $\varphi_1 \in \mathcal{D}(\Omega_1)$ and $\varphi_2 \in \mathcal{D}(\Omega_2)$,

$$\{\langle\langle f, \varphi_1\rangle_{\Omega_1}, \varphi_2\rangle_{\Omega_2} : f \in \mathcal{F}\} \text{ is bounded in } E.$$

Here, we can replace E with E-weak since they have the same bounded subsets due to the Banach–Mackey theorem (Theorem A.41). Which proves that

$$\mathcal{D}'(\Omega_1; \mathcal{D}'(\Omega_2; E)) \text{ has the same bounded subsets as } \mathcal{D}'(\Omega_1; \mathcal{D}'(\Omega_2; E\text{-weak})). \ \square$$

Let us show that, if E is *extractable*, i.e. if every bounded sequence in E has a subsequence which converges in E-weak, then every bounded sequence in $\mathcal{D}'(\Omega_1; \mathcal{D}'(\Omega_2; E))$ has a subsequence which converges in $\mathcal{D}'(\Omega_1; \mathcal{D}'(\Omega_2; E\text{-weak}))$.

Theorem 14.13.– *Let Ω_1 be an open subset of $\mathbb{R}^{d_1}$, Ω_2 an open subset of $\mathbb{R}^{d_2}$ and E a separated semi-normed space such that*

$$\begin{cases} \textit{every bounded sequence in } E \textit{ has a subsequence} \\ \textit{which converges in } E\text{-weak}. \end{cases} \tag{14.14}$$

Then, E and E-weak are Neumann spaces, and

$$\begin{cases} \textit{every bounded sequence in } \mathcal{D}'(\Omega_1; \mathcal{D}'(\Omega_2; E)) \textit{ has a subsequence} \\ \textit{which converges in } \mathcal{D}'(\Omega_1; \mathcal{D}'(\Omega_2; E\text{-weak})). \end{cases} \blacksquare$$

Proof. Hypothesis (14.14) means that E is extractable (Definition A.43), which implies (Theorem A.44) that E-weak is a Neumann space. Then (Theorem 4.13) E also is a Neumann space.

Since $E \subsetneq E$-weak (Theorem 4.12), this allows us to apply Theorem 14.11 with $F = E$-weak, which gives the stated property. □

Equivalent formulation. Theorem 14.13 can be expressed as follows: if the bounded subsets of E are relatively sequentially compact in E-weak, then the bounded subsets of $\mathcal{D}'(\Omega; E)$ are relatively sequentially compact in $\mathcal{D}'(\Omega; E\text{-weak})$. □

Topological equality. The spaces $\mathcal{D}'(\Omega_1; \mathcal{D}'(\Omega_2; E))$ and $\mathcal{D}'(\Omega_1; \mathcal{D}'(\Omega_2; E\text{-weak}))$ are topologically equal only if E and E-weak are. □

Chapter 15

Separation of Variables

This chapter is essentially dedicated to show (Theorem 15.10) that the separation of variables is a continuous bijection from $\mathcal{D}'(\Omega_1 \times \Omega_2; E)$ onto $\mathcal{D}'(\Omega_1; \mathcal{D}'(\Omega_2; E))$ whose inverse mapping, i.e. the regrouping of variables, is sequentially continuous.

This is the diabolical *vector-valued kernel theorem* of Laurent SCHWARTZ, but for E sequentially complete, which is more general than the quasi-complete he uses, and with a weaker topology on $\mathcal{D}'(\Omega_2; E)$ which explains why the inverse mapping, that he gets continuous, is only sequentially continuous here.

We prove this by reducing by regularization to the function case and by using the following tools.

— The uniform convergence with respect to test functions of the sequences of distributions of distributions form the previous chapter (Theorem 14.6).

— The "control" of test functions on $\Omega_1 \otimes \Omega_2$ by tensor products of test functions (Theorems 15.5 and 15.7).

We finish with the regrouping of variables and their permutation (§ 15.6 and 15.7).

15.1. Tensor products of test functions

This section, like the following two, is dedicated to some technical results which will be used to justify Definition 15.8 of the separation of variables and to show that it is bijective in Theorem 15.10.

Let us first give an expression of the supremum (of the semi-norms) of a function in terms of **tensor product** of test functions. This is defined by

$$(\varphi_1 \otimes \varphi_2)(x_1, x_2) \stackrel{\text{def}}{=} \varphi_1(x_1)\varphi_2(x_2).$$

Theorem 15.1.– *Let $f \in \mathcal{C}(\Omega_1 \times \Omega_2; E)$, where Ω_1 is an open subset of $\mathbb{R}^{d_1}$, Ω_2 is an open subset of $\mathbb{R}^{d_2}$ and E is a Neumann space. For every semi-norm $\| \ \|_{E;\nu}$ of E,*

$$\sup_{\Omega_1 \times \Omega_2} \|f\|_{E;\nu} = \sup_{\substack{\varphi_1 \in \mathcal{D}(\Omega_1) \\ \varphi_1 \neq 0}} \ \sup_{\substack{\varphi_2 \in \mathcal{D}(\Omega_2) \\ \varphi_2 \neq 0}} \frac{\| \int_{\Omega_1 \times \Omega_2} f \varphi_1 \otimes \varphi_2 \|_{E;\nu}}{\int_{\Omega_1} |\varphi_1| \int_{\Omega_2} |\varphi_2|}. \ \blacksquare$$

Proof. Denote by X the value of the right-hand side of the stated inequality, and let $y_1 \in \Omega_1$, $y_2 \in \Omega_2$ and $\varepsilon > 0$.

1. *Expression for* $f(y_1, y_2)$. By Definition 1.9 (a) of continuity, there exists $r > 0$ such that

$$\|f - f(y_1, y_2)\|_{E;\nu} < \varepsilon \text{ on the ball } B((y_1, y_2), r). \tag{15.1}$$

Let ω_1 and ω_2 be two open balls in $\mathbb{R}^{d_1}$ and $\mathbb{R}^{d_2}$ of radius $r/\sqrt{2}$ respectively centered at y_1 and y_2. Then, $\omega_1 \times \omega_2 \subset B((y_1, y_2), r)$.

Let $\varphi_1 \in \mathcal{D}(\Omega_1)$ and $\varphi_2 \in \mathcal{D}(\Omega_2)$ be such that (such functions are obtained by translating the regularizing functions of Definition 8.1)

$$\varphi_1 \geq 0,\ \varphi_1 = 0 \text{ outside of } \omega_1,\ \int_{\omega_1} \varphi_1 = 1,$$

$$\varphi_2 \geq 0,\ \varphi_2 = 0 \text{ outside of } \omega_2,\ \int_{\omega_2} \varphi_2 = 1.$$

The expression for the integral of a tensor product (Theorem A.90) gives

$$f(y_1, y_2) = \int_{\omega_1 \times \omega_2} f(y_1, y_2)\, \varphi_1(x_1)\, \varphi_2(x_2)\, \mathrm{d}x_1 \mathrm{d}x_2.$$

We decompose the right-hand side into the sum $I + J$, where

$$I \stackrel{\text{def}}{=} \int_{\omega_1 \times \omega_2} f(x_1, x_2)\, \varphi_1(x_1)\, \varphi_2(x_2)\, \mathrm{d}x_1 \mathrm{d}x_2,$$

$$J \stackrel{\text{def}}{=} \int_{\omega_1 \times \omega_2} (f(y_1, y_2) - f(x_1, x_2))\, \varphi_1(x_1)\, \varphi_2(x_2)\, \mathrm{d}x_1 \mathrm{d}x_2.$$

2. *Upper bound of* $\|f(y_1, y_2)\|_{E;\nu}$. Since the domain where the function is zero does not contribute to the integral (Theorem A.77) and $\varphi_1 \otimes \varphi_2$ is zero outside of $\omega_1 \times \omega_2$,

$$\|I\|_{E;\nu} = \left\| \int_{\Omega_1 \times \Omega_2} f \varphi_1 \otimes \varphi_2 \right\|_{E;\nu} = \frac{\| \int_{\Omega_1 \times \Omega_2} f \varphi_1 \otimes \varphi_2 \|_{E;\nu}}{\int_{\Omega_1} |\varphi_1| \int_{\Omega_2} |\varphi_2|} \leq X.$$

On the other hand, inequality (15.1) implies, with the bound of the semi-norms of the integral from Theorem 1.23 (a), the growth of the real integral (Theorem A.76 (a)) and again the expression for the integral of a tensor product (Theorem A.90),

$$\|J\|_{E;\nu} \leq \epsilon \int_{\omega_1 \times \omega_2} |\varphi_1 \otimes \varphi_2| = \epsilon \int_{\omega_1} |\varphi_1| \int_{\omega_2} |\varphi_2| = \epsilon.$$

Hence,

$$\|f(y_1, y_2)\|_{E;\nu} = \|I + J\|_{E;\nu} \leq X + \varepsilon.$$

Since this inequality is true for every $\epsilon > 0$, it is true for $\epsilon = 0$. Therefore,

$$\sup_{(y_1,y_2)\in\Omega_1\times\Omega_2} \|f(y_1,y_2)\|_{E,\nu} \le X.$$

3. *Converse inequality.* The upper bound of the absolute value of the real integral, its growth and its linearity (Theorems 1.23 (a), A.76 (a) and A.74) and then the expression for the integral of a tensor product (Theorem A.90) give

$$\Big\| \int_{\Omega_1\times\Omega_2} f\varphi_1\otimes\varphi_2 \Big\|_{E;\nu} \le \sup_{\Omega_1\times\Omega_2} \|f\|_{E;\nu} \int_{\Omega_1\times\Omega_2} |\varphi_1|\otimes|\varphi_2|$$
$$= \sup_{\Omega_1\times\Omega_2} \|f\|_{E;\nu} \int_{\Omega_1} |\varphi_1| \int_{\Omega_2} |\varphi_2|.$$

Therefore, X is bounded by $\sup_{\Omega_1\times\Omega_2} \|f\|_{E;\nu}$, and thus is equal to it. □

Let us now bound the semi-norms in $\mathcal{D}(\Omega_1\times\Omega_2)$ of a tensor product of test functions.

Theorem 15.2.– *Let $\varphi_1 \in \mathcal{D}(\Omega_1)$ and $p \in \mathcal{C}^+(\Omega_1\times\Omega_2)$, where Ω_1 is an open subset of $\mathbb{R}^{d_1}$ and Ω_2 is an open subset of $\mathbb{R}^{d_2}$.*

Then, there exists $p_2 \in \mathcal{C}^+(\Omega_2)$ such that: for every $\varphi_2 \in \mathcal{D}(\Omega_2)$,

$$\|\varphi_1\otimes\varphi_2\|_{\mathcal{D}(\Omega_1\times\Omega_2);p} \le \|\varphi_2\|_{\mathcal{D}(\Omega_2);p_2}. \ \blacksquare$$

Proof. **1.** *Construction of p_2.* Denote by K_1 the support of φ_1 and, for every $x_2 \in \Omega_2$,

$$r(x_2) \stackrel{\text{def}}{=} \sup_{x_1\in K_1} \sup_{|\beta_1|\le p(x_1,x_2)} p(x_1,x_2)(1+|\partial^{\beta_1}\varphi_1(x_1)|).$$

The function r so defined is bounded on every compact subset K_2 of Ω_2. Indeed, denoting $m = \sup_{(x_1,x_2)\in K_1\times K_2} p(x_1,x_2)$, which is finite since every continuous function on a compact set is bounded (Theorem A.32), we have

$$\sup_{x_2\in K_2} r(x_2) \le \sup_{|\beta_1|\le m} \sup_{(x_1,x_2)\in K_1\times K_2} p(x_1,x_2)(1+|\partial^{\beta_1}\varphi_1(x_1)|)$$

which is finite too, since $K_1\times K_2$ is compact in $\mathbb{R}^{d_1+d_2}$ and the functions p and $p|\partial^{\beta_1}\varphi_1|$ are continuous.

A function which is bounded on every compact set can be bounded by a continuous function, as we will see in Theorem 15.3. Thus, there exists $p_2 \in \mathcal{C}^+(\Omega_2)$ such that

$$p_2 \ge r \ \text{ on } \Omega_2.$$

2. Inequality. By Definition 2.5 of the semi-norms of $\mathcal{D}(\Omega)$,

$$\|\varphi_1 \otimes \varphi_2\|_{\mathcal{D}(\Omega_1 \times \Omega_2);p} =$$

$$= \sup_{x_1 \in \Omega_1,\, x_2 \in \Omega_2} \sup_{|\beta_1|+|\beta_2| \leq p(x_1,x_2)} p(x_1, x_2)|\partial^{\beta_1}\varphi_1(x_1)||\partial^{\beta_2}\varphi_2(x_2)|.$$

In this supremum, we can replace Ω_1 by K_1 (since $\partial^{\beta_1}\varphi_1$ is zero outside of it), and replace the condition $|\beta_1| + |\beta_2| \leq p(x_1, x_2)$ by the two conditions $|\beta_1| \leq p(x_1, x_2)$ and $|\beta_2| \leq p(x_1, x_2)$. This gives

$$\|\varphi_1 \otimes \varphi_2\|_{\mathcal{D}(\Omega_1 \times \Omega_2);p} \leq$$

$$\leq \sup_{x_2 \in \Omega_2} \sup_{x_1 \in K_1} \sup_{|\beta_2| \leq p(x_1,x_2)} \sup_{|\beta_1| \leq p(x_1,x_2)} p(x_1, x_2)|\partial^{\beta_1}\varphi_1(x_1)||\partial^{\beta_2}\varphi_2(x_2)|.$$

However, for every $x_1 \in K_1$,

$$p(x_1, x_2) \leq r(x_2) \leq p_2(x_2)$$

and

$$\sup_{|\beta_1| \leq p(x_1,x_2)} p(x_1, x_2)|\partial^{\beta_1}\varphi_1(x_1)| \leq r(x_2) \leq p_2(x_2).$$

Whence, again with Definition 2.5,

$$\|\varphi_1 \otimes \varphi_2\|_{\mathcal{D}(\Omega_1 \times \Omega_2);p} \leq \sup_{x_2 \in \Omega_2} \sup_{|\beta_2| \leq p_2(x_2)} p_2(x_2)|\partial^{\beta_2}\varphi_2(x_2)| = \|\varphi_2\|_{\mathcal{D}(\Omega_2);p_2}.$$

This proves Theorem 15.2 subject to check the following property that we used to obtain p_2 in step 1. □

It remains to bound a function that is finite on any compact set by a continuous function.

Theorem 15.3.– *If a real function f on an open subset Ω of $\mathbb{R}^d$ has a finite upper bound on every compact subset of Ω, there exists $p \in C^+(\Omega)$ such that*

$$p \geq f \text{ on } \Omega. \blacksquare$$

Proof. ***1.*** *Construction of p.* Let $\{\kappa_k\}_{n\in\mathbb{N}}$ be the cover of Ω by crown-shaped sets (Definition 2.18), and $\{\alpha_k\}_{n\in\mathbb{N}}$ a subordinate partition of unity (their properties used here are recalled in Theorem 2.19 and Definition 2.20).

We define p on Ω by

$$p(x) \stackrel{\text{def}}{=} \sum_{k \in \mathbb{N}} c_k \alpha_k(x),$$

where, denoting $f^+(y) \stackrel{\text{def}}{=} \sup\{f(y), 0\}$ and $\kappa_{-1} = \emptyset$,

$$c_k \stackrel{\text{def}}{=} \sup_{y \in \kappa_{k-1} \cup \kappa_k \cup \kappa_{k+1}} f^+(y).$$

This sum makes sense and is finite because:

— On one hand, only a finite number of these terms are non-zero. Indeed, each point x of Ω belongs to (at least) one crown-shaped set κ_k, this one only intersects κ_{k-1} and κ_{k+1}, and each α_k is zero outside of κ_k.

— On the other hand, each c_k is finite. Indeed, $\overline{\kappa_k}$ is a compact subset of Ω, thus f is bounded from above on it by hypothesis.

2. *Continuity of* p. If $x \in \kappa_k$, due to what we have just seen, the definition of p reduces to

$$p(x) = c_{k-1}\alpha_{k-1}(x) + c_k \alpha_k(x) + c_{k+1}\alpha_{k+1}(x).$$

Therefore, p is continuous on each crown-shaped set κ_k, and hence on their union Ω.

3. *Inequality.* If $x \in \kappa_k$, then c_{k-1}, c_k and c_{k+1} are greater than $f^+(x)$, thus

$$p(x) \geq f^+(x)(\alpha_{k-1} + \alpha_k + \alpha_{k+1})(x) = f^+(x) \sum_{n \in \mathbb{N}} \alpha_n(x) = f^+(x),$$

since $\sum_{n \in \mathbb{N}} \alpha_n = 1$. So, $p \geq f$ on each κ_k, and therefore on Ω. □

15.2. Decomposition of test functions on a product of sets

We again give a technical result, here on the decomposition of a test function on a product of sets into a sum of its derivatives weighted by tensor products of $\mathcal{C}^m$ functions with bounded support.

Theorem 15.4.– *Let $\phi \in \mathcal{D}(\mathbb{R}^{d_1+d_2})$, m be an integer ≥ 1 and r a real number > 0. Then, on $\mathbb{R}^{d_1+d_2}$,*

$$\phi = \sum_{\beta_1 \in \mathbb{N}^{d_1} : |\beta_1| \leq m+d_1+1} \; \sum_{\beta_2 \in \mathbb{N}^{d_2} : |\beta_2| \leq m+d_2+1} (-1)^{|\beta_1|+|\beta_2|} \partial_{x_1}^{\beta_1} \partial_{x_2}^{\beta_2} \phi \diamond (\zeta_{\beta_1} \otimes \zeta_{\beta_2}),$$

where, for $i = 1$ or 2, ζ_{β_i} satisfies the decomposition of the Dirac mass $\delta_0^{d_i}$ of $\mathbb{R}^{d_i}$ of Theorem 9.18 and thus where, denoting $B_i = \{x_i \in \mathbb{R}^{d_i} : |x_i| \leq r\}$,

$$\zeta_{\beta_i} \in \mathcal{C}^m_{B_i}(\mathbb{R}^{d_i}). \blacksquare$$

Proof. **1.** *Recall.* The decomposition of Theorem 9.18 is written here

$$\delta_0^{d_i} = \sum_{\beta_i \in \mathbb{N}^{d_i}:|\beta_i| \le m+d_i+1} \partial_{x_i}^{\beta_i} \zeta_{\beta_i} \text{ on } \mathbb{R}^{d_i}.$$

2. *Decomposition with respect to one of the two variables.* Let $\varphi_i \in \mathcal{D}(\mathbb{R}^{d_i})$. By weighting it by $\delta_0^{d_i}$, which is the neutral element of weighting (Theorem 7.19), we obtain, with the equality $\partial^\beta f \diamond \mu = (-1)^{|\beta|} f \diamond \partial^\beta \mu$ from Theorem 7.17,

$$\varphi_i = \varphi_i \underset{x_i}{\diamond} \delta_0^{d_i} = \sum_{|\beta_i| \le m+d_i+1} \varphi_i \underset{x_i}{\diamond} \partial_{x_i}^{\beta_i} \zeta_{\beta_i} = \sum_{|\beta_i| \le m+d_i+1} (-1)^{|\beta_i|} \partial_{x_i}^{\beta_i} \varphi_i \underset{x_i}{\diamond} \zeta_{\beta_i}.$$

So, with the expression for the weighting of continuous functions from Theorem 7.14, for every $x_i \in \mathbb{R}^{d_i}$,

$$\varphi_i(x_i) = \sum_{|\beta_i| \le m+d_i+1} (-1)^{|\beta_i|} \int_{\mathbb{R}^{d_i}} \partial_{x_i}^{\beta_i} \varphi_i(x_i + y_i)\, \zeta_{\beta_i}(y_i)\, dy_i. \tag{15.2}$$

3. *Decomposition of ϕ.* Given $x_1 \in \mathbb{R}^{d_1}$, equality (15.2) is in particular satisfied by $\varphi_2(x_2) = \phi(x_1, x_2)$. That is written

$$\phi(x_1, x_2) = \sum_{|\beta_2| \le m+d_2+1} (-1)^{|\beta_2|} g_{\beta_2}(x_1, x_2), \tag{15.3}$$

where

$$g_{\beta_2}(x_1, x_2) \overset{\text{def}}{=} \int_{\mathbb{R}^{d_2}} \partial_{x_2}^{\beta_2} \phi(x_1, x_2 + y_2)\, \zeta_{\beta_2}(y_2)\, dy_2. \tag{15.4}$$

Now using (15.2) for $\varphi_1(x_1) = g_{\beta_2}(x_1, x_2)$ with x_2 fixed, (15.3) gives

$$\phi(x_1, x_2) = \sum_{|\beta_1| \le m+d_1+1} \sum_{|\beta_2| \le m+d_2+1} (-1)^{|\beta_1|+|\beta_2|} g_{\beta_1,\beta_2}(x_1, x_2), \tag{15.5}$$

where

$$g_{\beta_1,\beta_2}(x_1, x_2) \overset{\text{def}}{=} \int_{\mathbb{R}^{d_1}} \partial_{x_1}^{\beta_1} g_{\beta_2}(x_1 + y_1, x_2)\, \zeta_{\beta_1}(y_1)\, dy_1. \tag{15.6}$$

4. *An expression for $\partial_{x_1}^{\beta_1} g_{\beta_2}$.* Due to the theorem on differentiation of a Cauchy integral under the integral sign (Theorem A.92), the function $x_1 \mapsto g_{\beta_2}(x_1, x_2)$ is differentiable and, if $\partial_{x_1}^{\beta_1}$ is a first-order partial derivative, i.e. if $|\beta_1| = 1$, equality (15.4) gives

$$\partial_{x_1}^{\beta_1} g_{\beta_2}(x_1, x_2) = \int_{\mathbb{R}^{d_2}} \partial_{x_1}^{\beta_1} \partial_{x_2}^{\beta_2} \phi(x_1, x_2 + y_2)\, \zeta_{\beta_2}(y_2)\, dy_2. \tag{15.7}$$

In fact, Theorem A.92 assumes that we are integrating over a bounded set, but we can return to this case by replacing $\mathbb{R}^{d_2}$ with the ball $\mathring{B}_2$ in (15.4) and (15.7), which does not change the value of the integrals according to Theorem A.77 since ζ_{β_2}, and hence the integrand, is zero outside it.

By induction, equality (15.7) is true for all $\beta_1 \in \mathbb{N}^{d_1}$. Indeed, if it is true for all β_1 such that $|\beta_1| \leq k$, the equality for $|\beta_1| = 1$ we have just established implies that for $|\beta_1| \leq k+1$.

5. Conclusion. With (15.7), equality (15.6) gives

$$g_{\beta_1,\beta_2}(x_1,x_2) =$$
$$= \int_{\mathbb{R}^{d_1}} \Big(\int_{\mathbb{R}^{d_2}} \partial_{x_1}^{\beta_1} \partial_{x_2}^{\beta_2} \phi(x_1+y_1, x_2+y_2)\, \zeta_{\beta_2}(y_2)\, \mathrm{d}y_2 \Big) \zeta_{\beta_1}(y_1)\, \mathrm{d}y_1 .$$

By grouping the variables with Theorem A.89 (which applies here since the integrand is uniformly continuous with compact support), we get

$$g_{\beta_1,\beta_2}(x_1,x_2) = \int_{\mathbb{R}^{d_1+d_2}} \partial_{x_1}^{\beta_1} \partial_{x_2}^{\beta_2} \phi(x_1+y_1, x_2+y_2)\, \zeta_{\beta_1}(y_1)\, \zeta_{\beta_2}(y_2)\, \mathrm{d}(y_1,y_2).$$

That is, again according to the expression for the weighting of continuous functions from Theorem 7.14,

$$g_{\beta_1,\beta_2} = \partial_{x_1}^{\beta_1} \partial_{x_2}^{\beta_2} \phi \diamond (\zeta_{\beta_1} \otimes \zeta_{\beta_2}).$$

Which, with (15.5), gives the stated equality. □

15.3. The tensorial control theorem

This section is once again dedicated to a technical result, the *tensorial control theorem*, which we will use to prove the kernel theorem. It provides an estimate of the value $\langle f, \phi \rangle$ of a distribution f on a product $\Omega_1 \times \Omega_2$, valid for any test function ϕ from an estimate for test functions only of the tensorial form $\varphi_1 \otimes \varphi_2$.

We begin with the case where f is a function [1].

Theorem 15.5.– *Let Ω_1 be an open subset of $\mathbb{R}^{d_1}$, Ω_2 an open subset of $\mathbb{R}^{d_2}$, E a Neumann space, $\| \;\|_{E;\nu}$ one of its semi-norms, $m \in \mathbb{N}$, $r > 0$ and, for $i = 1$ or 2,*

$$\omega_i \stackrel{\text{def}}{=} \{x_i \in \mathbb{R}^{d_i} : B_i(x_i, r) \subset \Omega_i\},$$

1. **History of the tensorial control theorem.** Theorem 15.5 is new including in the case where $E = \mathbb{R}$, as well as its extension to the distribution case in Theorem 15.7. It avoids the use of the properties of strict approximation of Laurent SCHWARTZ and of nuclearity of Alexandre GROTHENDIECK in the proof of the kernel theorem, see the comment *Originality of the proof*, p. 317.

where $B_i(x_i, r) \stackrel{\text{def}}{=} \{y_i \in \mathbb{R}^{d_i} : |y_i - x_i| \le r\}$. *In other words,* ω_i *is* Ω_i *without a neighborhood of width* r *of its boundary.*

Then, there exists $b \in \mathbb{R}$, *depending only on* m, r, d_1 *and* d_2, *such that: if* $f \in \mathcal{C}(\Omega_1 \times \Omega_2; E)$ *and* $c \in \mathbb{R}$ *satisfy, for every* $\varphi_1 \in \mathcal{D}(\Omega_1)$ *and* $\varphi_2 \in \mathcal{D}(\Omega_2)$,

$$\Big\| \int_{\Omega_1 \times \Omega_2} f\, \varphi_1 \otimes \varphi_2 \Big\|_{E;\nu} \le c \|\varphi_1\|_{\mathcal{C}_b^m(\Omega_1)} \|\varphi_2\|_{\mathcal{C}_b^m(\Omega_2)}, \tag{15.8}$$

then, for every $\phi \in \mathcal{D}(\omega_1 \times \omega_2)$, *its extension by* 0 *(also denoted by* ϕ*) satisfies*

$$\Big\| \int_{\Omega_1 \times \Omega_2} f\phi \Big\|_{E;\nu} \le cb\sigma_\phi \|\phi\|_{\mathcal{C}_b^{2m+d_1+d_2+2}(\omega_1 \times \omega_2)},$$

where $\sigma_\phi \stackrel{\text{def}}{=} |\{(x_1, x_2) \in \omega_1 \times \omega_2 : \phi(x_1, x_2) \ne 0\}|$, *i.e.* σ_ϕ *is the measure of the interior of the support of* ϕ. ▮

Proof. **1.** Decomposition of $\int f\phi$. Let

$$f \in \mathcal{C}(\Omega_1 \times \Omega_2; E), \quad \phi \in \mathcal{D}(\omega_1 \times \omega_2).$$

According to the decomposition given in Theorem 15.4 of the test function ϕ, or more precisely of its extension by 0 which we also denote by ϕ,

$$\int_{\Omega_1 \times \Omega_2} f\phi = \sum_{|\beta_1| \le m+d_1+1} \sum_{|\beta_2| \le m+d_2+1} (-1)^{|\beta_1|+|\beta_2|} \times \\ \times \int_{\Omega_1 \times \Omega_2} f\left(\partial_{x_1}^{\beta_1} \partial_{x_2}^{\beta_2} \phi \diamond (\zeta_{\beta_1} \otimes \zeta_{\beta_2})\right), \tag{15.9}$$

where $\zeta_{\beta_i} \in \mathcal{C}_{B_i}^m(\mathbb{R}^{d_i})$.

2. *A sufficient condition.* Assume for a moment that there exist two real numbers b_{β_1} and b_{β_2}, independent of f and ϕ, such that for every $\phi \in \mathcal{D}(\omega_1 \times \omega_2)$,

$$\Big\| \int_{\Omega_1 \times \Omega_2} f\left(\phi \diamond (\zeta_{\beta_1} \otimes \zeta_{\beta_2})\right) \Big\|_{E;\nu} \le cb_{\beta_1} b_{\beta_2} \sigma_\phi \sup_{\omega_1 \times \omega_2} |\phi|, \tag{15.10}$$

where $\sigma_\phi = |\mathcal{O}|$ and

$$\mathcal{O} \stackrel{\text{def}}{=} \{(x_1, x_2) \in \omega_1 \times \omega_2 : \phi(x_1, x_2) \ne 0\}.$$

Since $\partial_{x_1}^{\beta_1} \partial_{x_2}^{\beta_2} \phi$ also belongs to $\mathcal{D}(\omega_1 \times \omega_2)$ from Theorem 2.6, this inequality gives

$$\Big\| \int_{\Omega_1 \times \Omega_2} f\left(\partial_{x_1}^{\beta_1} \partial_{x_2}^{\beta_2} \phi \diamond (\zeta_{\beta_1} \otimes \zeta_{\beta_2})\right) \Big\|_{E;\nu} \le cb_{\beta_1} b_{\beta_2} \sigma_\phi \sup_{\omega_1 \times \omega_2} |\partial_{x_1}^{\beta_1} \partial_{x_2}^{\beta_2} \phi|.$$

Decomposition (15.9) therefore gives

$$\Big\| \int_{\Omega_1\times\Omega_2} f\phi \Big\|_{E;\nu} \le c \sum_{|\beta_1|\le m+d_1+1} \sum_{|\beta_2|\le m+d_2+1} b_{\beta_1} b_{\beta_2} \sigma_\phi \sup_{\omega_1\times\omega_2} |\partial_{x_1}^{\beta_1}\partial_{x_2}^{\beta_2}\phi|,$$

and so, denoting $b = \sum_{|\beta_1|\le m+d_1+1}\sum_{|\beta_2|\le m+d_2+1} b_{\beta_1} b_{\beta_2}$,

$$\Big\| \int_{\Omega_1\times\Omega_2} f\phi \Big\|_{E;\nu} \le cb\sigma_\phi \sup_{|(\beta_1,\beta_2)|\le 2m+d_1+d_2+2} \sup_{\omega_1\times\omega_2} |\partial_{x_1}^{\beta_1}\partial_{x_2}^{\beta_2}\phi|.$$

This is the desired inequality, according to Definition 1.18 (b) of the norm of $\mathcal{C}_{\mathrm{b}}^m(\Omega)$.

It therefore remains to check (15.10), which we will do in step 4 with the help of the following inequalities.

3. *Two equalities.* Let us use the equality of Theorem 7.24, namely

$$\int_\Omega f(\phi \diamond \check{\mu}) = \int_{\Omega_D} (f\diamond\mu)\phi,$$

where $\check{\mu}(x) = \mu(-x)$, for $\Omega = \Omega_1\times\Omega_2$, $D = B_1(0,r)\times B_2(0,r)$ and $\mu = \check{\zeta}_{\beta_1}\otimes\check{\zeta}_{\beta_2}$. Then, $\Omega_D = \omega_1\times\omega_2$ and $\check{\mu} = \zeta_{\beta_1}\otimes\zeta_{\beta_2}$, which gives

$$\int_{\Omega_1\times\Omega_2} f(\phi\diamond(\zeta_{\beta_1}\otimes\zeta_{\beta_2})) = \int_{\omega_1\times\omega_2} (f\diamond(\check{\zeta}_{\beta_1}\otimes\check{\zeta}_{\beta_2}))\phi. \tag{15.11}$$

Setting $\phi = \varphi_1\otimes\varphi_2$ and observing that weighting and tensor product can be permuted in the left-hand side, as we will check in Lemma 15.6, we obtain, for every $\varphi_1\in\mathcal{D}(\omega_1)$ and $\varphi_2\in\mathcal{D}(\omega_2)$,

$$\int_{\Omega_1\times\Omega_2} f((\varphi_1 \underset{x_1}{\diamond} \zeta_{\beta_1})\otimes(\varphi_2 \underset{x_2}{\diamond} \zeta_{\beta_1})) = \int_{\omega_1\times\omega_2} (f\diamond(\check{\zeta}_{\beta_1}\otimes\check{\zeta}_{\beta_2}))(\varphi_1\otimes\varphi_2). \tag{15.12}$$

4. *Proof of (15.10).* Let

$$\varphi_1\in\mathcal{D}(\omega_1),\ \ \varphi_2\in\mathcal{D}(\omega_2).$$

Each function $\varphi_i \underset{x_i}{\diamond} \zeta_{\beta_i}$, where $i=1$ or 2, belongs to $\mathcal{D}(\Omega_i)$ since, from Theorem 7.18, its support is included in $\operatorname{supp}\varphi_i + \operatorname{supp}\rho_n^i$ and therefore in Ω_i. Hypothesis (15.8) then gives

$$\Big\| \int_{\Omega_1\times\Omega_2} f((\varphi_1 \underset{x_1}{\diamond} \zeta_{\beta_1})\otimes(\varphi_2 \underset{x_2}{\diamond} \zeta_{\beta_1})) \Big\|_{E;\nu} \le c\|\varphi_1 \underset{x_1}{\diamond} \zeta_{\beta_1}\|_{\mathcal{C}_{\mathrm{b}}^m(\Omega_1)} \|\varphi_2 \underset{x_1}{\diamond} \zeta_{\beta_2}\|_{\mathcal{C}_{\mathrm{b}}^m(\Omega_2)}.$$

According to Lemma 7.25,

$$\|\varphi_i \underset{x_i}{\diamond} \zeta_{\beta_i}\|_{\mathcal{C}_{\mathrm{b}}^m(\mathbb{R}^{d_i})} \le b_{\beta_i} \int_{\mathbb{R}^{d_i}} |\varphi_i|,$$

where $b_{\beta_i} \overset{\text{def}}{=} \|\zeta_{\beta_i}\|_{C_b^m(\mathbb{R}^{d_i})}$. Transforming the left-hand side with equality (15.12) therefore gives

$$\Big\|\int_{\omega_1\times\omega_2} (f \diamond (\check{\zeta}_{\beta_1} \otimes \check{\zeta}_{\beta_2}))(\varphi_1 \otimes \varphi_2)\Big\|_{E;\nu} \leq c b_{\beta_1} b_{\beta_2} \int_{\mathbb{R}^{d_1}} |\varphi_1| \int_{\mathbb{R}^{d_2}} |\varphi_2|.$$

The expression of the supremum (of the semi-norms) of a function in terms of tensor products of test functions from Theorem 15.1 then gives

$$\sup_{\omega_1\times\omega_2} \|f \diamond (\check{\zeta}_{\beta_1} \otimes \check{\zeta}_{\beta_2})\|_{E;\nu} \leq c b_{\beta_1} b_{\beta_2}. \tag{15.13}$$

The bound of the semi-norms of an integral from Theorem 1.23 (b) then gives, since $|\mathcal{O}| = \sigma_\phi$,

$$\Big\|\int_{\mathcal{O}} f \diamond (\check{\zeta}_{\beta_1} \otimes \check{\zeta}_{\beta_2})\, \phi\Big\|_{E;\nu} \leq c b_{\beta_1} b_{\beta_2} \sigma_\phi \sup_{\omega_1\times\omega_2} |\phi|. \tag{15.14}$$

Here, we can replace $\mathcal{O}$ with $\omega_1 \times \omega_2$ (since $\phi = 0$ outside of $\mathcal{O}$ and the set where the function is zero does not contribute to the integral, see Theorem A.77), thus the left-hand side is equal to that of (15.10) due to equality (15.11).

This proves (15.10), and hence Theorem 15.5, subject to proving Lemma 15.6. □

It remains to check the following property of commutation of weighting with tensor product, which was used to establish (15.12).

Lemma 15.6.– *Let $\varphi_1 \in \mathcal{C}(\mathbb{R}^{d_1})$, $\varphi_2 \in \mathcal{C}(\mathbb{R}^{d_2})$, $\zeta_1 \in \mathcal{C}_{D_1}(\mathbb{R}^{d_1})$ and $\zeta_2 \in \mathcal{C}_{D_2}(\mathbb{R}^{d_2})$, where D_1 and D_2 are compact sets. Then,*

$$(\varphi_1 \otimes \varphi_2) \diamond (\zeta_{\beta_1} \otimes \zeta_{\beta_2}) = (\varphi_1 \underset{x_1}{\diamond} \zeta_{\beta_1}) \otimes (\varphi_2 \underset{x_2}{\diamond} \zeta_{\beta_1}). \blacksquare$$

Proof. The expression for the weighting of continuous functions from Theorem 7.14, here on $\mathbb{R}^{d_1+d_2}$, gives for every $x_1 \in \mathbb{R}^{d_1}$ and $x_2 \in \mathbb{R}^{d_2}$,

$$\begin{aligned}((\varphi_1 \otimes \varphi_2) \diamond (\zeta_{\beta_1} \otimes \zeta_{\beta_2}))(x_1, x_2) &= \\ &= \int_{\mathbb{R}^{d_1+d_2}} \varphi_1(x_1 + y_1)\varphi_2(x_2 + y_2)\zeta_{\beta_1}(y_1)\zeta_{\beta_2}(y_2)\, \mathrm{d}(y_1, y_2).\end{aligned}$$

The right-hand side equals, according to the expression for the integral of a tensor product from Theorem A.90,

$$\Big(\int_{\mathbb{R}^{d_1}} \varphi_1(x_1 + y_1)\zeta_{\beta_1}(y_1)\, \mathrm{d}y_1\Big)\Big(\int_{\mathbb{R}^{d_2}} \varphi_2(x_2 + y_2)\zeta_{\beta_2}(y_2)\, \mathrm{d}y_2\Big).$$

Hence, again with Theorem 7.14 now on $\mathbb{R}^{d_1}$ and on $\mathbb{R}^{d_2}$,

$$(\varphi_1 \underset{x_1}{\diamond} \zeta_{\beta_1})(x_1)(\varphi_2 \underset{x_2}{\diamond} \zeta_{\beta_1})(x_2) = ((\varphi_1 \underset{x_1}{\diamond} \zeta_{\beta_1}) \otimes (\varphi_2 \underset{x_2}{\diamond} \zeta_{\beta_1}))(x_1, x_2). \ \square$$

We come to the general **tensorial control theorem**, where f is a distribution, by reducing to the function case via regularization.

Theorem 15.7.– *Let Ω_1 be an open subset of $\mathbb{R}^{d_1}$, Ω_2 an open subset of $\mathbb{R}^{d_2}$, E a Neumann space, $\| \|_{E;\nu}$ one of its semi-norms, $m \in \mathbb{N}$, $r > 0$ and, for $i = 1$ or 2,*

$$\omega_i \overset{\text{def}}{=} \{x_i \in \mathbb{R}^{d_i} : B_i(x_i, r) \subset \Omega_i\}.$$

Then, there exists $b \in \mathbb{R}$, depending only on m, r, d_1 and d_2, such that: if $f \in \mathcal{D}'(\Omega_1 \times \Omega_2; E)$ and $c \in \mathbb{R}$ satisfy, for every $\varphi_1 \in \mathcal{D}(\Omega_1)$ and $\varphi_2 \in \mathcal{D}(\Omega_2)$,

$$\|\langle f, \varphi_1 \otimes \varphi_2 \rangle_{\Omega_1 \times \Omega_2}\|_{E;\nu} \leq c \|\varphi_1\|_{\mathcal{C}_b^m(\Omega_1)} \|\varphi_2\|_{\mathcal{C}_b^m(\Omega_2)}, \tag{15.15}$$

then, for every $\phi \in \mathcal{D}(\omega_1 \times \omega_2)$, its extension by 0 (also denoted by ϕ) satisfies

$$\|\langle f, \phi \rangle_{\Omega_1 \times \Omega_2}\|_{E;\nu} \leq cb\sigma_\phi \|\phi\|_{\mathcal{C}_b^{2m+d_1+d_2+2}(\omega_1 \times \omega_2)},$$

where σ_ϕ is the measure of the interior of the support of ϕ. ▮

Proof. Let us come back to the case of a function, and more precisely to Theorem 15.5 for the regular approximation $f \diamond (\rho_n^1 \otimes \rho_n^2)$.

1. *Regularization.* For $i = 1$ or 2, let

$$\omega_i' \overset{\text{def}}{=} \{x_i \in \mathbb{R}^{d_i} : B_i(x_i, r/2) \subset \Omega_i\}$$

and ρ_n^i be a regularizing function (Definition 8.1) on $\mathbb{R}^{d_i}$, where $n > 2/r$.

Then, $\rho_n^i \in \mathcal{D}_{B_i(0,r/2)}(\mathbb{R}^{d_i})$ and

$$\rho_n^1 \otimes \rho_n^2 \in \mathcal{D}_D(\mathbb{R}^{d_1+d_2}),$$

where $D = B_1(0, r/2) \times B_2(0, r/2)$. Moreover,

$$(\Omega_1 \times \Omega_2)_D = \omega_1' \times \omega_2'$$

and, from the regularizing property of weighting (Theorem 7.5),

$$f \diamond (\rho_n^1 \otimes \rho_n^2) \in \mathcal{C}(\omega_1' \times \omega_2'; E).$$

Let $\varphi_1' \in \mathcal{D}(\omega_1')$ and $\varphi_2' \in \mathcal{D}(\omega_2')$. Their extensions by 0 (also denoted by φ_1' and φ_2') satisfy

$$\varphi_1' \underset{x_1}{\diamond} \check{\rho}_n^1 \in \mathcal{D}(\Omega_1), \quad \varphi_2' \underset{x_2}{\diamond} \check{\rho}_n^2 \in \mathcal{D}(\Omega_2),$$

where $\check{\rho}^i(x) \stackrel{\text{def}}{=} \rho^i(-x)$. Indeed, due to the inclusion of the support of a weighted distribution from Theorem 7.18,

$$\mathrm{supp}(\varphi'_i \underset{x_i}{\diamond} \check{\rho}^i_n) \subset \mathrm{supp}\, \varphi'_i - \mathrm{supp}\, \check{\rho}^i_n \subset \Omega_i.$$

2. An equality. Note that Definition 7.12 of weighting and its commutativity up to a symmetry (Theorem 8.25) give, for $g \in \mathcal{D}'(\Omega; E)$, $\mu \in \mathcal{D}'_D(\mathbb{R}^d)$ and $\varphi \in \mathcal{D}(\Omega_D)$,

$$\langle g \diamond \mu, \varphi \rangle_{\Omega_D} = \langle g, \mu \diamond \check{\varphi} \rangle_\Omega = \langle g, \varphi \diamond \check{\mu} \rangle_\Omega.$$

For $g = f$, $\mu = \rho^1_n \otimes \rho^2_n$ and $\phi = \varphi'_1 \otimes \varphi'_2$, this equality is written

$$\langle f \diamond (\rho^1_n \otimes \rho^2_n), \varphi'_1 \otimes \varphi'_2 \rangle_{\omega'_1 \times \omega'_2} = \langle f, (\varphi'_1 \otimes \varphi'_2) \diamond (\check{\rho}^1_n \otimes \check{\rho}^2_n) \rangle_{\Omega_1 \times \Omega_2}.$$

By permuting the weighting and the tensor product on the right-hand side using Lemma 15.6, it becomes

$$\langle f \diamond (\rho^1_n \otimes \rho^2_n), \varphi'_1 \otimes \varphi'_2 \rangle_{\omega'_1 \times \omega'_2} = \langle f, (\varphi'_1 \underset{x_1}{\diamond} \check{\rho}^1_n) \otimes (\varphi'_2 \underset{x_2}{\diamond} \check{\rho}^2_n) \rangle_{\Omega_1 \times \Omega_2}. \tag{15.16}$$

3. Obtaining hypothesis (15.8) of Theorem 15.5. Hypothesis (15.15) gives, using equality (15.16),

$$\| \langle f \diamond (\rho^1_n \otimes \rho^2_n), \varphi'_1 \otimes \varphi'_2 \rangle_{\omega'_1 \times \omega'_2} \|_{E;\nu} \le c \| \varphi'_1 \underset{x_1}{\diamond} \check{\rho}^1_n \|_{\mathcal{C}^m_b(\Omega_1)} \| \varphi'_2 \underset{x_2}{\diamond} \check{\rho}^2_n \|_{\mathcal{C}^m_b(\Omega_2)}.$$

According to the estimate of weighting from Lemma 7.25,

$$\| \varphi'_i \underset{x_i}{\diamond} \check{\rho}^i_n \|_{\mathcal{C}^m_b(\mathbb{R}^{d_i})} \le \| \varphi'_i \|_{\mathcal{C}^m_b(\mathbb{R}^{d_i})} \int_{\mathbb{R}^{d_i}} |\check{\rho}^i_n| = \| \varphi'_i \|_{\mathcal{C}^m_b(\omega'_i)}.$$

By transforming the left-hand side of the previous inequality with the identification of continuous functions with distributions (Theorem 3.9), it then becomes

$$\left\| \int_{\omega'_1 \times \omega'_2} (f \diamond (\rho^1_n \otimes \rho^2_n))(\varphi'_1 \otimes \varphi'_2) \right\|_{E;\nu} \le c \| \varphi'_1 \|_{\mathcal{C}^m_b(\omega'_1)} \| \varphi'_2 \|_{\mathcal{C}^m_b(\omega'_2)}. \tag{15.17}$$

4. Use of Theorem 15.5. According to Theorem 15.5, i.e. to the function case, inequality (15.17) implies the existence of b such that, for every $\phi \in \mathcal{D}(\omega_1 \times \omega_2)$,

$$\left\| \int_{\omega'_1 \times \omega'_2} f \diamond (\rho^1_n \otimes \rho^2_n) \phi \right\|_{E;\nu} \le c b \sigma_\phi \| \phi \|_{\mathcal{C}^{2m+d_1+d_2+2}_b(\omega_1 \times \omega_2)}.$$

Again according to the identification of continuous functions with distributions from Theorem 3.9, this inequality is written

$$\| \langle f \diamond (\rho^1_n \otimes \rho^2_n), \phi \rangle_{\omega'_1 \times \omega'_2} \|_{E;\nu} \le c b \sigma_\phi \| \phi \|_{\mathcal{C}^{2m+d_1+d_2+2}_b(\omega_1 \times \omega_2)}.$$

5. Passing to the limit. It only remains to observe (we will justify it in step 6) that

$$f \diamond (\rho_n^1 \otimes \rho_n^2) \to f \text{ in } \mathcal{D}'(\omega_1' \times \omega_2'; E) \tag{15.18}$$

to obtain, in the limit,

$$\|\langle f, \phi \rangle_{\omega_1' \times \omega_2'}\|_{E;\nu} \leq cb\sigma_\phi \|\phi\|_{\mathcal{C}_b^{2m+d_1+d_2+2}(\omega_1 \times \omega_2)}.$$

Which proves Theorem 15.7 since, by Definition 6.1 of restriction,

$$\langle f, \phi \rangle_{\omega_1' \times \omega_2'} = \langle f, \phi \rangle_{\Omega_1 \times \Omega_2}.$$

6. Justification of (15.18). The convergence would result from the local convergence of regular approximations from Theorem 8.4 if $\rho_n^1 \otimes \rho_n^2$ satisfied Definition 8.1 of a regularizing sequence.

This is almost the case since, from the expression for the integral of a tensor product (Theorem A.90),

$$\int_{\mathbb{R}^{d_1+d_2}} \rho_n^1 \otimes \rho_n^2 = \int_{\mathbb{R}^{d_1}} \rho_n^1 \int_{\mathbb{R}^{d_2}} \rho_n^2 = 1.$$

On the contrary, the property of support is not satisfied; but it is by $\rho_{2n}^1 \otimes \rho_{2n}^2$ since

$$(\rho_{2n}^1 \otimes \rho_{2n}^2)(x) = 0 \quad \text{if} \quad |x| = (|x_1|^2 + |x_2|^2)^{1/2} \geq \frac{1}{n},$$

because then $|x_1|$ or $|x_2|$ is $\geq 1/(2n)$ and thus $(\rho_{2n}^1)(x_1)$ or $(\rho_{2n}^2)(x_2)$ is zero.

The sequence $(\rho_{2n}^1 \otimes \rho_{2n}^2)_{n \in \mathbb{N}}$ hence satisfies Definition 8.1, thus Theorem 8.4 gives convergence (15.18) for even n, which suffices for step 5. □

15.4. Separation of variables

We finally come to the **separation of variables of a distribution** defined on a product $\Omega_1 \times \Omega_2$, which is the main focus of this chapter.

Definition 15.8.– *Let*

$$f \in \mathcal{D}'(\Omega_1 \times \Omega_2; E),$$

where Ω_1 is an open subset of $\mathbb{R}^{d_1}$, Ω_2 is an open subset of $\mathbb{R}^{d_2}$ and E is a Neumann space. We define

$$\underline{f} \in \mathcal{D}'(\Omega_1; \mathcal{D}'(\Omega_2; E))$$

by: for every $\varphi_1 \in \mathcal{D}(\Omega_1)$ and $\varphi_2 \in \mathcal{D}(\Omega_2)$,

$$\langle \langle \underline{f}, \varphi_1 \rangle_{\Omega_1}, \varphi_2 \rangle_{\Omega_2} \stackrel{\text{def}}{=} \langle f, \varphi_1 \otimes \varphi_2 \rangle_{\Omega_1 \times \Omega_2}. \blacksquare$$

Justification. It is necessary to verify that, as we claimed,

$$\underline{f} \in \mathcal{D}'(\Omega_1; \mathcal{D}'(\Omega_2; E)).$$

According to the characterization of distributions of distributions from Theorem 14.2 (a), it suffices to check the following three criteria.

1. *First criterion.* This is the following obvious linearity property,

$$\underline{f} \in \text{Lin}(\mathcal{D}(\Omega_1); \text{Lin}(\mathcal{D}(\Omega_2); E)).$$

2. *Second criterion.* This is the existence, for every $\varphi_1 \in \mathcal{D}(\Omega_1)$ and $\nu \in \mathcal{N}_E$, of $c \in \mathbb{R}$ and $p_2 \in \mathcal{C}^+(\Omega_2)$ such that, for every $\varphi_2 \in \mathcal{D}(\Omega_2)$,

$$\|\langle\langle \underline{f}, \varphi_1 \rangle_{\Omega_1}, \varphi_2 \rangle_{\Omega_2}\|_{E;\nu} \leq c \|\varphi_2\|_{\mathcal{D}(\Omega_2);p_2}. \tag{15.19}$$

Let us check this. From the characterization of distributions from Theorem 3.3, there exist a function $p \in \mathcal{C}^+(\Omega_1 \times \Omega_2)$ and $c \in \mathbb{R}$ such that, for every $\phi \in \mathcal{D}(\Omega_1 \times \Omega_2)$,

$$\|\langle f, \phi \rangle_{\Omega_1 \times \Omega_2}\|_{E;\nu} \leq c \|\phi\|_{\mathcal{D}(\Omega_1 \times \Omega_2);p}. \tag{15.20}$$

Once p and φ_1 are fixed, Theorem 15.2 provides $p_2 \in \mathcal{C}^+(\Omega_2)$ such that, for every $\varphi_2 \in \mathcal{D}(\Omega_2)$,

$$\|\varphi_1 \otimes \varphi_2\|_{\mathcal{D}(\Omega_1 \times \Omega_2);p} \leq \|\varphi_2\|_{\mathcal{D}(\Omega_2);p_2}.$$

Inequality (15.20) for $\phi = \varphi_1 \otimes \varphi_2$ joined with the definition of $\underline{f}$ then gives (15.19).

3. *Third criterion.* This is the existence, for every $\varphi_2 \in \mathcal{D}(\Omega_2)$ and $\nu \in \mathcal{N}_E$, of $c \in \mathbb{R}$ and $p_1 \in \mathcal{C}^+(\Omega_1)$ such that, for every $\varphi_1 \in \mathcal{D}(\Omega_1)$,

$$\|\langle\langle \underline{f}, \varphi_1 \rangle_{\Omega_1}, \varphi_2 \rangle_{\Omega_2}\|_{E;\nu} \leq c \|\varphi_1\|_{\mathcal{D}(\Omega_1);p_1}.$$

This inequality is obtained by swapping the roles of Ω_1 and Ω_2 in the proof of (15.19), and hence in Theorem 15.2. □

Let us check that we recover the separation of variables of functions.

Theorem 15.9.– *If $f \in \mathcal{C}(\Omega_1 \times \Omega_2; E)$, the separation of variables in the distribution sense from Definition 15.8 coincides with that in the function sense.*

Which is to say that, with identification (3.6), p. 52, of continuous functions with distributions, Definition 15.8 also gives $\underline{f} \in \mathcal{C}(\Omega_1; \mathcal{C}(\Omega_2; E))$ and, for every $x_1 \in \Omega_1$ and $x_2 \in \Omega_2$,

$$(\underline{f}(x_1))(x_2) = f(x_1, x_2). \blacksquare$$

Proof. Let $\bar{f}$ be the distribution associated with the function f by Theorem 3.5, and $\underline{\bar{f}}$ be the distribution obtained by separating its variables via Definition 15.8. It is a question of showing that

$$\underline{\bar{f}} = \overline{\overline{\underset{\sim}{f}}^{1}}^{2}, \tag{15.21}$$

where $\underset{\sim}{f}$ is the function obtained by separating the variables of f, and $^{-1}$ and $^{-2}$ are the mappings associating (by Theorem 3.5) a distribution with a continuous function, respectively on Ω_1 and Ω_2.

Let $\varphi_1 \in \mathcal{D}(\Omega_1)$ and $\varphi_2 \in \mathcal{D}(\Omega_2)$. Definition 15.8 of the separation of variables (of $\bar{f}$), and then the definition of $\bar{f}$ from Theorem 3.5 and that of the tensor product, successively give

$$\langle\langle \underline{\bar{f}}, \varphi_1 \rangle_{\Omega_1}, \varphi_2 \rangle_{\Omega_2} = \langle\langle \bar{f}, \varphi_1 \otimes \varphi_2 \rangle_{\Omega_1 \times \Omega_2} =$$
$$= \int_{\Omega_1 \times \Omega_2} f \varphi_1 \otimes \varphi_2 = \int_{\Omega_1 \times \Omega_2} f(x_1, x_2) \varphi_1(x_1) \varphi_2(x_2)\, \mathrm{d}(x_1, x_2).$$

By separating the variables in the integral (Theorem A.89) and then in the function f, it comes

$$\langle\langle \underline{\bar{f}}, \varphi_1 \rangle_{\Omega_1}, \varphi_2 \rangle_{\Omega_2} = \int_{\Omega_2} \Big(\int_{\Omega_1} f(x_1, x_2) \varphi_1(x_1) \varphi_2(x_2)\, \mathrm{d}x_1 \Big)\, \mathrm{d}x_2 =$$
$$= \int_{\Omega_2} \Big(\int_{\Omega_1} (\underset{\sim}{f}(x_1))(x_2) \varphi_1(x_1)\, \mathrm{d}x_1 \Big) \varphi_2(x_2)\, \mathrm{d}x_2.$$

Using the definitions of $^{-1}$ and $^{-2}$ from Theorem 3.5, this gives

$$\langle\langle \underline{\bar{f}}, \varphi_1 \rangle_{\Omega_1}, \varphi_2 \rangle_{\Omega_2} = \int_{\Omega_2} \langle \overline{\underset{\sim}{f}}^{1}, \varphi_1 \rangle_{\Omega_1}(x_2)\, \varphi_2(x_2)\, \mathrm{d}x_2 = \Big\langle \overline{\langle \overline{\underset{\sim}{f}}^{1}, \varphi_1 \rangle_{\Omega_1}}^{2}, \varphi_2 \Big\rangle_{\Omega_2}.$$

This proves (15.21), i.e. we obtain the same distribution when we separate the variables in the distribution sense as that when we separate them in the function sense.

The above use of Theorem A.89 is allowed since the function $f\varphi_1 \otimes \varphi_2$ belongs to $\mathbf{C}_{K_1 \times K_2}(\Omega_1 \times \Omega_2; E)$, where K_i is the support of φ_i. □

15.5. The kernel theorem

Let us show that the **separation of variables** of a distribution is a continuous bijection with a sequentially continuous inverse. This is the **kernel theorem** [2].

2. **History of the kernel theorem.** Laurent SCHWARTZ showed in 1957 [72, p. 125] that, when E is quasi-complete, which is stronger than the sequentially completeness used here, the separation of variables

Theorem 15.10.– *Let Ω_1 be an open subset of $\mathbb{R}^{d_1}$, Ω_2 an open subset of $\mathbb{R}^{d_2}$ and E a Neumann space.*

The separation of variables is a continuous, and hence sequentially continuous, linear bijection from $\mathcal{D}'(\Omega_1 \times \Omega_2; E)$ onto $\mathcal{D}'(\Omega_1; \mathcal{D}'(\Omega_2; E))$, and its inverse mapping is sequentially continuous linear. ∎

Interest in the separation of variables and the kernel theorem. The separation of variables plays an essential role in the study of *evolution equations*, which govern quantities depending on time t and position x. Such a quantity is represented by a distribution in (t, x), namely an element of $\mathcal{D}'((0,T)\times\Omega)$ if it is real-valued.

The kernel theorem allows its identification with an element of $\mathcal{D}'((0,T); \mathcal{D}'(\Omega))$, and so the use of properties specific to one-dimensional distributions, such as compactness or regularity properties.

This also reveals the interest of distributions with values in a general Neumann space E, including for real-valued equations: real values may be *exchanged* for reduction to dimension one. □

Why "kernel"? Laurent SCHWARTZ called a **kernel** any distribution of $\mathcal{D}'(\mathbb{R}^d \times \mathbb{R}^d)$, by analogy with integral kernels [69, p. 138–142]. Similarly, he called **vector-valued kernel** any distribution in $\mathcal{D}'(\mathbb{R}^{d_1} \times \mathbb{R}^{d_2}; E)$ [72, p. 124]. □

Proof of Theorem 15.10. Denote by $\{\| \ \|_{E;\nu} : \nu \in \mathcal{N}_E\}$ the family of semi-norms of E.

Since the separation of variables, i.e. the mapping $f \mapsto \underline{f}$ of Definition 15.8, is evidently linear, we focus on the other properties.

1. Continuity. Let $f \in \mathcal{D}'(\Omega_1 \times \Omega_2; E)$. Definitions 14.1 of the semi-norms of distributions of distributions, 15.8 of $\underline{f}$ and 3.1 of the semi-norms of distributions, give successively, for every $\varphi_1 \in \mathcal{D}(\Omega_1)$, $\varphi_2 \in \mathcal{D}(\Omega_2)$ and $\nu \in \mathcal{N}_E$,

$$\begin{aligned} \|\underline{f}\|_{\mathcal{D}'(\Omega_1;\mathcal{D}'(\Omega_2;E));\varphi_1,\varphi_2,\nu} &= \|\langle\langle \underline{f}, \varphi_1\rangle_{\Omega_1}, \varphi_2\rangle_{\Omega_2}\|_{E;\nu} = \\ &= \|\langle f, \varphi_1 \otimes \varphi_2\rangle_{\Omega_1\times\Omega_2}\|_{E;\nu} = \|f\|_{\mathcal{D}'(\Omega_1\times\Omega_2;E);\varphi_1\otimes\varphi_2,\nu}. \end{aligned} \tag{15.22}$$

This implies the continuity of the mapping $f \mapsto \underline{f}$, according to the characterization of linear continuous mappings from Theorem 1.12 (a).

This mapping is thus sequentially continuous, as is every continuous mapping (Theorem 1.10).

is an isomorphism from $\mathcal{D}'(\mathbb{R}^{d_1} \times \mathbb{R}^{d_2}; E)$-unif onto $\mathcal{D}'(\mathbb{R}^{d_1}; \mathcal{D}'(\mathbb{R}^{d_2}; E)$-unif)-unif, i.e. when these spaces are endowed with the uniform convergence topology on the bounded subsets of $\mathcal{D}(\mathbb{R}^{d_i})$.
These properties are new for the simple topology and for a Neumann space E considered here, but then the inverse mapping is only sequentially continuous. Our method of proof is also new, see the comment *Originality of the proof*, p. 317.

2. Injectivity. Let $f \in \mathcal{D}'(\Omega_1 \times \Omega_2; E)$ be such that

$$\underline{f} = 0.$$

From (15.22), for every $\varphi_1 \in \mathcal{D}(\Omega_1)$, $\varphi_2 \in \mathcal{D}(\Omega_2)$ and $\nu \in \mathcal{N}_E$,

$$\|\underline{f}\|_{\mathcal{D}'(\Omega_1;\mathcal{D}'(\Omega_2;E));\varphi_1,\varphi_2,\nu} = \|\langle f, \varphi_1 \otimes \varphi_2 \rangle_{\Omega_1 \times \Omega_2}\|_{E;\nu} = 0.$$

Hypothesis (15.15) of the tensorial control theorem (Theorem 15.7) is thus satisfied with $c = 0$, hence this theorem gives

$$\|\langle f, \phi \rangle_{\Omega_1 \times \Omega_2}\|_{E;\nu} = 0,$$

for every $\phi \in \mathcal{D}(\omega_1 \times \omega_2)$, where $\omega_i = \{x_i \in \mathbb{R}^{d_i} : B_i(x_i, r) \subset \Omega_i\}$ and r is any > 0 real number.

For every $\phi \in \mathcal{D}(\Omega_1 \times \Omega_2)$, there is $r > 0$ such that this is satisfied, therefore

$$f = 0.$$

Which proves that the mapping $f \mapsto \underline{f}$ is injective.

3. Surjectivity. Let

$$g \in \mathcal{D}'(\Omega_1; \mathcal{D}'(\Omega_2; E)).$$

It is a question of showing that there exists $f \in \mathcal{D}'(\Omega_1 \times \Omega_2; E)$ such that

$$\underline{f} = g. \tag{15.23}$$

We will obtain such a distribution f as the limit of functions f_n obtained by regrouping the variables in regular approximations g_n of g, in eight steps.

3.a. *Regularization of g*. Let R_n^1 and R_n^2 be the regularization operators, respectively on Ω_1 and Ω_2, given by Definition 8.13 and let

$$g_n \stackrel{\text{def}}{=} R_n^2 R_n^1 g.$$

Since R_n^2 is sequentially continuous linear from $\mathcal{D}'(\Omega_2; E)$ into $\mathbf{C}_\text{b}(\Omega_2; E)$ according to Theorem 8.23, the image of g under this mapping is (Definition 5.10)

$$R_n^2 g \in \mathcal{D}'(\Omega_1; \mathbf{C}_\text{b}(\Omega_2; E)).$$

Again Theorem 8.23, now for R_n^1, then gives

$$R_n^1 R_n^2 g \in \mathbf{C}_\text{b}(\Omega_1; \mathbf{C}_\text{b}(\Omega_2; E)).$$

Since the regularization operators, here R_n^1, commute (Theorem 8.20) with linear mappings, here R_n^2, this function is equal to $R_n^2 R_n^1 g$, that is g_n. So,

$$g_n \in \mathbf{C}_b(\Omega_1; \mathbf{C}_b(\Omega_2; E)). \tag{15.24}$$

3.b. *Convergence of the* g_n. Due to the convergence of global approximations (Theorem 8.16), when $n \to \infty$,

$$R_n^1 g \to g \text{ in } \mathcal{D}'(\Omega_1; \mathcal{D}'(\Omega_2; E)).$$

For every $\varphi_1 \in \mathcal{D}(\Omega_1)$, we therefore have

$$\langle R_n^1 g, \varphi_1 \rangle_{\Omega_1} \to \langle g, \varphi_1 \rangle_{\Omega_1} \text{ in } \mathcal{D}'(\Omega_2; E).$$

The convergence of global approximations of approximations (Theorem 8.17) then gives

$$R_n^2(\langle R_n^1 g, \varphi_1 \rangle_{\Omega_1}) \to \langle g, \varphi_1 \rangle_{\Omega_1} \text{ in } \mathcal{D}'(\Omega_2; E).$$

Since the image of an image is the image of the composition (Theorem 5.13), the left-hand side is equal to $\langle R_n^2 R_n^1 g, \varphi_1 \rangle_{\Omega_1}$, that is $\langle g_n, \varphi_1 \rangle_{\Omega_1}$. Therefore,

$$g_n \to g \text{ in } \mathcal{D}'(\Omega_1; \mathcal{D}'(\Omega_2; E)). \tag{15.25}$$

3.c. *Some parameters.* Let

$$\phi \in \mathcal{D}(\Omega_1 \times \Omega_2).$$

According to the strong inclusion theorem (Theorem A.22), there exists $r > 0$ such that $\operatorname{supp} \phi + B_{d_1+d_2}(0, 2r) \subset \Omega_1 \times \Omega_2$, where $B_d(x, r)$ designates the closed ball in $\mathbb{R}^d$. For $i = 1$ and 2, let

$$D_i \stackrel{\text{def}}{=} \{x_i \in \mathbb{R}^{d_i} : (x_1, x_2) \in \operatorname{supp} \phi\}.$$

It is a compact subset of $\mathbb{R}^{d_i}$ (because it is closed and bounded, as is the support of ϕ),

$$\operatorname{supp} \phi \subset D_1 \times D_2 \tag{15.26}$$

and $D_i + B(0, 3r) \subset \Omega_i$. Finally, let

$$\omega_i \stackrel{\text{def}}{=} D_i + \mathring{B}_i(0, r), \quad \Lambda_i \stackrel{\text{def}}{=} D_i + \mathring{B}_i(0, 2r), \quad K_i \stackrel{\text{def}}{=} D_i + B_i(0, 2r).$$

So,

$$D_i \Subset \omega_i \Subset \Lambda_i \subset \overline{\Lambda_i} = K_i \Subset \Omega_i. \tag{15.27}$$

3.d. *Uniform convergence of the* $g_n - g_{n'}$. Convergent sequences of distributions of distributions have a property of uniform convergence with respect to φ_1 and φ_2, according to Theorem 14.6. More precisely, due to this theorem, convergence (15.25) implies, for every $\nu \in \mathcal{N}_E$, the existence of $m \in \mathbb{N}$ and of a decreasing sequence $(c_n)_{n\in\mathbb{N}}$, $c_n \to 0$, such that: for every $\varphi_1 \in \mathcal{C}^\infty_{K_1}(\Omega_1)$ and $\varphi_2 \in \mathcal{C}^\infty_{K_2}(\Omega_2)$,

$$\|\langle\langle g_n - g, \varphi_1\rangle_{\Omega_1}, \varphi_2\rangle_{\Omega_2}\|_{E;\nu} \le c_n \|\varphi_1\|_{\mathcal{C}^m_{\mathrm{b}}(\Omega_1)} \|\varphi_2\|_{\mathcal{C}^m_{\mathrm{b}}(\Omega_2)}.$$

Then, for all $n' \ge n$,

$$\|\langle\langle g_n - g_{n'}, \varphi_1\rangle_{\Omega_1}, \varphi_2\rangle_{\Omega_2}\|_{E;\nu} \le 2c_n \|\varphi_1\|_{\mathcal{C}^m_{\mathrm{b}}(\Omega_1)} \|\varphi_2\|_{\mathcal{C}^m_{\mathrm{b}}(\Omega_2)}. \tag{15.28}$$

3.e. *Regrouping the variables.* By regrouping the variables in the function g_n, we obtain, thanks to the continuity Property (15.24), a function such that

$$f_n \in \mathbf{C}_{\mathrm{b}}(\Omega_1 \times \Omega_2; E).$$

Indeed, regrouping the variables is an isomorphism from $\mathbf{C}_{D_1}(\Omega_1; \mathbf{C}_{D_2}(\Omega_2; E))$ onto $\mathbf{C}_{D_1\times D_2}(\Omega_1 \times \Omega_2; E)$, according to Theorem A.50 since E is a Neumann space.

The separation of variables of f_n gives g_n in the function sense and therefore (Theorem 15.9) in the distribution sense. Then, Definition 15.8 of this separation and the identification of f_n with a distribution (Theorem 3.9) successively give

$$\langle\langle g_n, \varphi_1\rangle_{\Omega_1}, \varphi_2\rangle_{\Omega_2} = \langle f_n, \varphi_1 \otimes \varphi_2\rangle_{\Omega_1\times\Omega_2} = \int_{\Omega_1\times\Omega_2} f_n \varphi_1 \otimes \varphi_2.$$

3.f. *Use of the tensorial control theorem.* Inequality (15.28) can thus be written as

$$\Big\|\int_{\Omega_1\times\Omega_2} (f_n - f_{n'}) \varphi_1 \otimes \varphi_2\Big\|_{E;\nu} \le 2c_n \|\varphi_1\|_{\mathcal{C}^m_{\mathrm{b}}(\Omega_1)} \|\varphi_2\|_{\mathcal{C}^m_{\mathrm{b}}(\Omega_2)}.$$

This is true for every $\varphi_i \in \mathcal{C}^\infty_{K_i}(\Omega_i)$ and, *a fortiori*, for every $\varphi_i \in \mathcal{D}(\Lambda_i)$ according to (15.27). Then, Theorem 15.5 (i.e. the function case of the tensorial control theorem) shows, since the support of ϕ is included in $\omega_1 \times \omega_2$ by (15.26), that there exists $b = b(m, r)$ such that

$$\Big\|\int_{\Lambda_1\times\Lambda_2} (f_n - f_{n'}) \phi\Big\|_{E;\nu} \le 2c_n b \|\phi\|_{\mathcal{C}^{2m+d_1+d_2+2}_{\mathrm{b}}(\omega_1\times\omega_2)}. \tag{15.29}$$

3.g. *Cauchy property of the f_n.* On the left-hand side, we can replace $\Lambda_1 \times \Lambda_2$ with $\Omega_1 \times \Omega_2$, since the domain where the function is zero does not contribute to the integral (Theorem A.77) and since ϕ is zero outside of $\Lambda_1 \times \Lambda_2$ according to (15.26) and (15.27).

Also using the expression for the semi-norms of $\mathcal{D}'(\Omega_1 \times \Omega_2; E)$ for a continuous function from Theorem 3.9 and denoting $e_\phi = 2b\|\phi\|_{\mathcal{C}^{2m+d_1+d_2+2}_{\mathrm{b}}(\omega_1\times\omega_2)}$, inequality (15.29) is written as

$$\|f_n - f_{n'}\|_{\mathcal{D}'(\Omega_1\times\Omega_2;E);\phi,\nu} \le c_n e_\phi.$$

According to step 3.d, this inequality is true for every $n' \geq n$, and $c_n \to 0$, hence

$$\sup_{n' \geq n} \|f_n - f_{n'}\|_{\mathcal{D}'(\Omega_1 \times \Omega_2;E);\phi,\nu} \to 0.$$

This is the case for every $\phi \in \mathcal{D}(\Omega_1 \times \Omega_2)$ and $\nu \in \mathcal{N}_E$.

3.h. *Obtaining* f. The sequence $(f_n)_{n\in\mathbb{N}}$ is therefore Cauchy in $\mathcal{D}'(\Omega_1 \times \Omega_2; E)$, which is sequentially complete (Theorem 4.5). It therefore has a limit f there.

The sequential continuity of the separation of variables, i.e. of the mapping $f \mapsto \underline{f}$, established in step 1 gives $\underline{f_n} \to \underline{f}$. Since $\underline{f_n} = g_n$ and $g_n \to g$ from steps 3.e and 3.b,

$$\underline{f} = g.$$

This proves (15.23), and hence the surjectivity of the mapping $f \mapsto \underline{f}$.

4. Inverse mapping. It remains to prove that the mapping $\underline{f} \mapsto f$ is sequentially continuous. Let thus f_n and f belong to $\mathcal{D}'(\Omega_1 \times \Omega_2; E)$ and be such that

$$\underline{f_n} \to \underline{f} \text{ in } \mathcal{D}'(\Omega_1; \mathcal{D}'(\Omega_2; E)).$$

Let $\phi \in \mathcal{D}(\Omega_1 \times \Omega_2)$ and ω_i, Λ_i and K_i be the domains introduced in step 3.c.

The property of uniform convergence with respect to φ_1 and φ_2 of convergent sequences of distributions of distributions from Theorem 14.6 gives, for every $\nu \in \mathcal{N}_E$, the existence of $m \in \mathbb{N}$ and a decreasing sequence $(c_n)_{n\in\mathbb{N}}$, $c_n \to 0$, such that: for every $\varphi_1 \in C^\infty_{K_1}(\Omega_1)$ and $\varphi_2 \in C^\infty_{K_2}(\Omega_2)$,

$$\|\langle\langle \underline{f_n} - \underline{f}, \varphi_1\rangle_{\Omega_1}, \varphi_2\rangle_{\Omega_2}\|_{E;\nu} \leq c_n \|\varphi_1\|_{C^m_b(\Omega_1)} \|\varphi_2\|_{C^m_b(\Omega_2)}. \tag{15.30}$$

That is, by Definition 15.8 of the separation of variables,

$$\|\langle f_n - f, \varphi_1 \otimes \varphi_2\rangle_{\Omega_1\times\Omega_2}\|_{E;\nu} \leq c_n \|\varphi_1\|_{C^m_b(\Omega_1)} \|\varphi_2\|_{C^m_b(\Omega_2)}.$$

Then, Theorem 15.7 (i.e. now the distribution case of the tensorial control theorem) shows that there exists $b = b(m, r)$ such that

$$\|\langle f_n - f, \phi\rangle_{\Omega_1\times\Omega_2}\|_{E;\nu} \leq c_n b \|\phi\|_{C^{2m+d_1+d_2+2}_b(\omega_1\times\omega_2)}.$$

This holds for every $\phi \in \mathcal{D}(\Omega_1 \times \Omega_2)$ and $\nu \in \mathcal{N}_E$, therefore (Theorem 4.3 (b))

$$f_n \to f \text{ in } \mathcal{D}'(\Omega_1 \times \Omega_2; E).$$

Which proves that

the mapping $\underline{f} \mapsto f$ is sequentially continuous.

It is linear as is any inverse mapping of a linear bijection [Vol. 1, Theorem 7.17 (b)]. □

Originality of the proof of Theorem 15.10. Laurent SCHWARTZ proved that, when E is quasi-complete, which is stronger than the sequential completeness used here, the separation of variables is an isomorphism from $\mathcal{D}'(\mathbb{R}^{d_1} \times \mathbb{R}^{d_2}; E)$-unif onto $\mathcal{D}'(\mathbb{R}^{d_1}; \mathcal{D}'(\mathbb{R}^{d_2}; E)\text{-unif})$-unif [72, p. 125]. For that, he used some fine and difficult topological properties, including:
— the strict approximation property of $\mathcal{C}^\infty(\mathbb{R}^d)$, $\mathcal{D}(\mathbb{R}^d)$ and $\mathcal{D}'(\mathbb{R}^d)$ [72, p. 5 and 10];
— the real-valued kernel theorem [72, p. 93], which itself relies on the topological tensor products of Alexandre GROTHENDIECK and on the nuclear character of $\mathcal{C}^\infty(\mathbb{R}^d)$ and $\mathcal{D}'(\mathbb{R}^d)$ [36].

In contrast, we proved here that the separation of variables is a continuous bijection from $\mathcal{D}'(\Omega_1 \times \Omega_2; E)$ onto $\mathcal{D}'(\Omega_1; \mathcal{D}'(\Omega_2; E))$ whose inverse mapping is sequentially continuous, by using new inequalities that are elementary, albeit laborious to establish, that provide:
— a uniform convergence with respect to test functions of sequences of distributions of distributions (Theorem 14.6);
— a "control" of test functions on the product of open sets by tensor products of test functions (Theorems 15.5 and 15.7). □

Identification. The (not recommended) identification of f with $\underline{f}$ in Theorem 15.10 would give:

$$\mathcal{D}'(\Omega_1 \times \Omega_2; E) \simeq \mathcal{D}'(\Omega_1; \mathcal{D}'(\Omega_2; E)),$$

$$\mathcal{D}'(\Omega_1 \times \Omega_2; E) \underset{\rightleftarrows}{\subset} \mathcal{D}'(\Omega_1; \mathcal{D}'(\Omega_2; E)). \ \square$$

15.6. Regrouping of variables

The variables of a distribution of distributions may be **regrouped**, as follows.

Theorem 15.11.– *Let*

$$f \in \mathcal{D}'(\Omega_1; \mathcal{D}'(\Omega_2; E)),$$

where Ω_1 is an open subset of $\mathbb{R}^{d_1}$, Ω_2 is an open subset of $\mathbb{R}^{d_2}$ and E is a Neumann space.

Then, there exists a unique distribution

$$\bar{f} \in \mathcal{D}'(\Omega_1 \times \Omega_2; E)$$

such that, after separating its variables, we obtain f, i.e. such that, for every $\varphi_1 \in \mathcal{D}(\Omega_1)$ and $\varphi_2 \in \mathcal{D}(\Omega_2)$,

$$\langle \bar{f}, \varphi_1 \otimes \varphi_2 \rangle_{\Omega_1 \times \Omega_2} = \langle \langle f, \varphi_1 \rangle_{\Omega_1}, \varphi_2 \rangle_{\Omega_2}. \ \blacksquare$$

Proof. Since the separation of variables is a bijection (Theorem 15.10), there exists a unique distribution $\bar{f} \in \mathcal{D}'(\Omega_1 \times \Omega_2; E)$ such that, after separating its variables, we obtain f.

Definition 15.8 of the separation of variables then gives the stated equality. □

Let us show that the regrouping of variables is a sequentially continuous bijection with continuous inverse.

Theorem 15.12.– *Let Ω_1 be an open subset of $\mathbb{R}^{d_1}$, Ω_2 an open subset of $\mathbb{R}^{d_2}$ and E a Neumann space.*

The regrouping of variables is a sequentially continuous linear bijection from $\mathcal{D}'(\Omega_1; \mathcal{D}'(\Omega_2; E))$ onto $\mathcal{D}'(\Omega_1 \times \Omega_2; E)$, whose inverse mapping is continuous, and hence sequentially continuous, linear. ▮

Proof. This is a just rewriting of the properties of the separation of variables given in Theorem 15.10, since the regrouping of variables is its inverse mapping. □

15.7. Permutation of variables

Let us define the **permutation of variables** of a distribution of distributions.

Definition 15.13.– *Let $f \in \mathcal{D}'(\Omega_1; \mathcal{D}'(\Omega_2; E))$, where Ω_1 is an open subset of $\mathbb{R}^{d_1}$, Ω_2 is an open subset of $\mathbb{R}^{d_2}$ and E is a Neumann space.*

We define $\breve{f} \in \mathcal{D}'(\Omega_2; \mathcal{D}'(\Omega_1; E))$ by: for every $\varphi_1 \in \mathcal{D}(\Omega_1)$ and $\varphi_2 \in \mathcal{D}(\Omega_2)$,

$$\langle\langle \breve{f}, \varphi_2 \rangle_{\Omega_2}, \varphi_1 \rangle_{\Omega_1} \stackrel{\text{def}}{=} \langle\langle f, \varphi_1 \rangle_{\Omega_1}, \varphi_2 \rangle_{\Omega_2}. \ ▮$$

Justification. The mapping $\breve{f}$ is a distribution of distributions according to any of the characterizations in Theorem 14.2, since Ω_1 and Ω_2 play the same role there. □

Let us check that we recover the permutation of variables of a function.

Theorem 15.14.– *If $f \in \mathcal{C}(\Omega_1; \mathcal{C}(\Omega_2; E))$, permuting its variables in the distribution sense given by Definition 15.13 coincides with that in the function sense.*

That is to say that, with identification (3.6), p. 52, of a continuous function with a distribution, Definition 15.13 also gives $\breve{f} \in \mathcal{C}(\Omega_2; \mathcal{C}(\Omega_1; E))$ and, for every $x_1 \in \Omega_1$ and $x_2 \in \Omega_2$,

$$(\breve{f}(x_2))(x_1) = f(x_1))(x_2). \ ▮$$

Proof. **1.** *Objective.* Denoting $\overline{}^{1}$ and $\overline{}^{2}$ the mappings associating (by Theorem 3.5) a distribution to a continuous function, respectively on Ω_1 and Ω_2, and denoting by $\overline{}$ their composition, it is a question of checking that

$$\overline{\breve{f}} = \breve{\overline{f}}. \tag{15.31}$$

2. *Proof of (15.31).* Let $\varphi_1 \in \mathcal{D}(\Omega_1)$ and $\varphi_2 \in \mathcal{D}(\Omega_2)$. Using Theorem 3.5 twice, we get

$$\begin{aligned}\langle\langle \overline{f}, \varphi_1\rangle_{\Omega_1}, \varphi_2\rangle_{\Omega_2} &= \Big\langle \overline{\langle \overline{f}^{1}, \varphi_1\rangle_{\Omega_1}}^{2}, \varphi_2\Big\rangle_{\Omega_2} = \int_{\Omega_2} \langle \overline{f}^{1}, \varphi_1\rangle_{\Omega_1}(x_2)\,\varphi_2(x_2)\,\mathrm{d}x_2 = \\ &= \int_{\Omega_2}\Big(\int_{\Omega_1} (f(x_1))(x_2)\varphi_1(x_1)\,\mathrm{d}x_1\Big)\varphi_2(x_2)\,\mathrm{d}x_2.\end{aligned}$$

That is

$$\langle\langle \overline{f}, \varphi_1\rangle_{\Omega_1}, \varphi_2\rangle_{\Omega_2} = \int_{\Omega_2}\int_{\Omega_1} \Phi, \tag{15.32}$$

where

$$(\Phi(x_1))(x_2) \stackrel{\text{def}}{=} (f(x_1))(x_2)\varphi_1(x_1)\varphi_2(x_2).$$

Definition 15.13 of the permutation of variables for the distribution $\overline{f}$, equality (15.32) and the permutation of variables in the Cauchy integral (Theorem A.91) successively give

$$\langle\langle \breve{\overline{f}}, \varphi_2\rangle_{\Omega_2}, \varphi_1\rangle_{\Omega_1} = \langle\langle \overline{f}, \varphi_1\rangle_{\Omega_1}, \varphi_2\rangle_{\Omega_2} = \int_{\Omega_2}\int_{\Omega_1} \Phi = \int_{\Omega_1}\int_{\Omega_2} \breve{\Phi}.$$

With equality (15.32) for the function $\breve{f}$ (to which corresponds $\breve{\Phi}$), this gives

$$\langle\langle \breve{\overline{f}}, \varphi_2\rangle_{\Omega_2}, \varphi_1\rangle_{\Omega_1} = \langle\langle \overline{\breve{f}}, \varphi_2\rangle_{\Omega_2}, \varphi_1\rangle_{\Omega_1}.$$

This proves (15.31), except that the above use of Theorem A.91 requires

$$\Phi \in \mathbf{C}_\mathrm{b}(\Omega_1; \mathbf{C}_\mathrm{b}(\Omega_2; E)). \tag{15.33}$$

3. *Verification of (15.33).* We proceed in two steps.

— The product by φ_2, i.e. the mapping $g \mapsto g\varphi_2$, is continuous linear from $\mathcal{C}(\Omega_2; E)$ into $\mathbf{C}_\mathrm{b}(\Omega_2; E)$ (Theorem A.49). The image of f under this mapping, which we denote by $f\varphi_2$, therefore belongs to $\mathcal{C}(\Omega_1; \mathbf{C}_\mathrm{b}(\Omega_2; E))$, according to Theorem A.48.

— Theorem A.49, now for the space of values $\mathbf{C}_\mathrm{b}(\Omega_2; E)$, then shows that the product of $f\varphi_2$ with φ_1, i.e. Φ, belongs to $\mathbf{C}_\mathrm{b}(\Omega_1; \mathbf{C}_\mathrm{b}(\Omega_2; E))$. □

Let us show that the permutation of variables is an isomorphism [3].

Theorem 15.15.– *Let Ω_1 be an open subset of $\mathbb{R}^{d_1}$, Ω_2 an open subset of $\mathbb{R}^{d_2}$ and E a Neumann space.*

The permutation of variables is an isomorphism from $\mathcal{D}'(\Omega_1; \mathcal{D}'(\Omega_2; E))$ onto $\mathcal{D}'(\Omega_2; \mathcal{D}'(\Omega_1; E))$. ▮

Proof. Let $\{\| \ \|_{E;\nu} : \nu \in \mathcal{N}_E\}$ be the family of semi-norms of E.

Definition 15.13 of the mapping $\check{}$ implies that it is linear, injective (since $\check{f} = 0$ implies $f = 0$) and surjective (since $\check{\check{f}} = f$), and thus bijective.

From Definition 14.1 of the semi-norms of distributions of distributions, for every $\varphi_1 \in \mathcal{D}(\Omega_1)$, $\varphi_2 \in \mathcal{D}(\Omega_2)$ and $\nu \in \mathcal{N}_E$,

$$\|\check{f}\|_{\mathcal{D}'(\Omega_2;\mathcal{D}'(\Omega_1;E));\varphi_2,\varphi_1,\nu} = \|\langle\check{f}, \varphi_2\rangle_{\Omega_2}, \varphi_1\rangle_{\Omega_1}\|_{E;\nu} =$$
$$= \|\langle\langle f, \varphi_1\rangle_{\Omega_1}, \varphi_2\rangle_{\Omega_2}\|_{E;\nu} = \|f\|_{\mathcal{D}'(\Omega_1;\mathcal{D}'(\Omega_2;E));\varphi_1,\varphi_2,\nu}.$$

Which proves that $\check{}$ is bicontinuous, according to the characterization of continuous linear mappings from Theorem 1.12 (a). It is therefore an isomorphism. □

Observe that, even without permuting the variables, the order of test functions does not matter.

Theorem 15.16.– *Let $f \in \mathcal{D}'(\Omega_1; \mathcal{D}'(\Omega_2; E))$, where Ω_1 is an open subset of $\mathbb{R}^{d_1}$, Ω_2 is an open subset of $\mathbb{R}^{d_2}$ and E is a Neumann space.*

Then, for every $\varphi_1 \in \mathcal{D}(\Omega_1)$ and $\varphi_2 \in \mathcal{D}(\Omega_2)$,

$$\langle\langle f, \varphi_1\rangle_{\Omega_1}, \varphi_2\rangle_{\Omega_2} = \langle\langle f, \varphi_2\rangle_{\Omega_2}, \varphi_1\rangle_{\Omega_1}. \ ▮$$

3. **History of the properties of the permutation of variables.** In 1957, Laurent SCHWARTZ showed [72, p. 125] that when E is quasi-complete, which is more restrictive than the sequential completeness used here, the permutation of variables is an isomorphism from the space $\mathcal{D}'(\mathbb{R}^{d_1}; \mathcal{D}'(\mathbb{R}^{d_2}; E)\text{-unif})\text{-unif}$ onto $\mathcal{D}'(\mathbb{R}^{d_2}; \mathcal{D}'(\mathbb{R}^{d_1}; E)\text{-unif})\text{-unif}$, i.e. when these spaces are endowed with the topology of uniform convergence on the bounded subsets of $\mathcal{D}(\mathbb{R}^{d_i})$.
These properties are new for the simple topology and when E is Neumann as considered here. Our method of proof is equally new (Laurent SCHWARTZ used his method of separation of variables discussed in the comment *Originality of the proof*, p. 317).

Meaning of the terms in Theorem 15.16.

— In the left-hand side, $\langle f, \varphi_1\rangle_{\Omega_1}$ is the value of f at φ_1; it belongs to $\mathcal{D}'(\Omega_2; E)$. Subsequently, $\langle\langle f, \varphi_1\rangle_{\Omega_1}, \varphi_2\rangle_{\Omega_2}$ is its value at φ_2; it belongs to E.

— In the right-hand side, $\langle f, \varphi_2\rangle_{\Omega_2}$ is the image of f under the mapping $\langle\,\cdot\,, \varphi_2\rangle_{\Omega_2}$, which is continuous linear from $\mathcal{D}'(\Omega_2; E)$ into E; it belongs to $\mathcal{D}'(\Omega_1; E)$. And $\langle\langle f, \varphi_2\rangle_{\Omega_2}, \varphi_1\rangle_{\Omega_1}$ is its value under φ_1; it belongs to E, like the left-hand side. □

Proof of Theorem 15.16. The stated equality is a particular case of Definition 5.10 of the image of a distribution, here f, under a continuous linear mapping, here $g \mapsto \langle g\,\cdot\,, \varphi_2\rangle_{\Omega_2}$ which is continuous linear from $\mathcal{D}'(\Omega_2; E)$ into E according to Theorem 4.4 (a). □

Let us show that the permutation of variables does not change the effect of test functions.

Theorem 15.17.– *Let $f \in \mathcal{D}'(\Omega_1; \mathcal{D}'(\Omega_2; E))$, where Ω_1 is an open subset of $\mathbb{R}^{d_1}$, Ω_2 is an open subset of $\mathbb{R}^{d_2}$ and E is a Neumann space, and $\breve{f} \in \mathcal{D}'(\Omega_2; \mathcal{D}'(\Omega_1; E))$ be the distribution obtained by permuting its variables (Definition 15.13).*

Then, for every $\varphi_1 \in \mathcal{D}(\Omega_1)$ and $\varphi_2 \in \mathcal{D}(\Omega_2)$:

(a) $$\langle \breve{f}, \varphi_1\rangle_{\Omega_1} = \langle f, \varphi_1\rangle_{\Omega_1}.$$

(b) $$\langle\langle \breve{f}, \varphi_1\rangle_{\Omega_1}, \varphi_2\rangle_{\Omega_2} = \langle\langle \breve{f}, \varphi_2\rangle_{\Omega_2}, \varphi_1\rangle_{\Omega_1} = \langle\langle f, \varphi_1\rangle_{\Omega_1}, \varphi_2\rangle_{\Omega_2} = \langle\langle f, \varphi_2\rangle_{\Omega_2}, \varphi_1\rangle_{\Omega_1}.$$ ∎

Meaning of the terms in equality (a).

— The left-hand side $\langle \breve{f}, \varphi_1\rangle_{\Omega_1}$ is the image of $\breve{f}$ under the mapping $\langle\,\cdot\,, \varphi_1\rangle_{\Omega_1}$ which is continuous linear from $\mathcal{D}'(\Omega_1; E)$ into E; it belongs to $\mathcal{D}'(\Omega_2; E)$.

— The right-hand side $\langle f, \varphi_1\rangle_{\Omega_1}$ is the value of f under the test function φ_1; it belongs to $\mathcal{D}'(\Omega_2; E)$, just like the left-hand side. □

Proof of Theorem 15.17. **1.** *Equalities (b).* The first equality of (b) is Theorem 15.16 applied to $\breve{f}$, the second is the definition of $\breve{f}$, and the third is Theorem 15.16.

2. *Equality (a).* This equality follows from $\langle\langle \breve{f}, \varphi_1\rangle_{\Omega_1}, \varphi_2\rangle_{\Omega_2} = \langle\langle f, \varphi_1\rangle_{\Omega_1}, \varphi_2\rangle_{\Omega_2}$ which is included in (b). □

Identification. The (not recommended) identification of f with $\breve{f}$ in Theorem 15.10 would give:

$$\mathcal{D}'(\Omega_1; \mathcal{D}'(\Omega_2; E)) \approxeq \mathcal{D}'(\Omega_2; \mathcal{D}'(\Omega_1; E)).\ \square$$

Another permutation of variables. Definition 15.13 of the permution of variables of a distribution of distributions is analogous to the permution of variables of a distribution on $\Omega_1 \times \Omega_2$ introduced in Theorem 5.30. This is why we denote both by $\breve{f}$, although it is a little abusive since they refer to a distribution f belonging to different spaces. □

Chapter 16

Banach Space Valued Distributions

This chapter is dedicated to the specifics of distributions with values in a Banach space.
— Such a distribution is a sum $\sum_{n\in\mathbb{N}} \partial^{\beta_n} g_n$ of derivatives of continuous functions g_n with compact support which, on each compact subset K of Ω, are all zero from a finite rank n_K (Theorem 16.9). Locally, it is even the derivative of a single continuous function with compact support (Theorem 16.11).
— Such a distribution on an open set Ω is extendable to all of $\mathbb{R}^d$ if and only if, on every bounded subset ω of Ω, it is the derivative $\partial^\beta g$ of a uniformly continuous function g (Theorem 16.16).
— These properties are no longer true in some Fréchet spaces (Theorems 16.12 and 16.17).

These properties follow from the fact that every distribution with values in a Banach space is locally of finite order, a property that we begin by studying, in § 16.1 to 16.4.

16.1. Finite order distributions

Let us define finite order distributions [1].

Definition 16.1.– *Let $f \in \mathcal{D}'(\Omega; E)$, where Ω is an open subset of $\mathbb{R}^d$ and E is a Neumann space, and let $\{\| \; \|_{E;\nu} : \nu \in \mathcal{N}_E\}$ be the family of semi-norms of E.*

(a) *We say that f is **of finite order** if there exists $m \in \mathbb{N}$ such that, for every $\nu \in \mathcal{N}_E$, there exists $c_\nu \in \mathbb{R}$ such that: for every $\varphi \in \mathcal{D}(\Omega)$,*

$$\|\langle f, \varphi\rangle\|_{E;\nu} \leq c_\nu \|\varphi\|_{\mathcal{C}_b^m(\Omega)}.$$

We say then that f is of order m, and we denote

$$|||f|||_{m,\nu} \stackrel{\text{def}}{=} \sup_{\varphi\in\mathcal{D}(\Omega),\, \varphi\neq 0} \frac{\|\langle f, \varphi\rangle\|_{E;\nu}}{\|\varphi\|_{\mathcal{C}_b^m(\Omega)}}.$$

1. **History of finite order distributions.** Laurent SCHWARTZ introduced this notion for real-valued distributions in 1950 [69, Chap. I, § 2, p. 26]. He extended it to distributions with values in a locally convex separated topological vector space in 1957 [72, Chap. I, § 2, Definition, p. 49].

The preceding inequality is hence verified with $c_\nu = |||f|||_{m,\nu}$.

(b) *We say that* f *is* **locally of finite order** *if, for every compact subset* K *of* Ω, *there exists* $m_K \in \mathbb{N}$ *such that, for every* $\nu \in \mathcal{N}_E$, *there exists* $c_{K,\nu} \in \mathbb{R}$ *such that: for every* $\varphi \in \mathcal{C}_K^\infty(\Omega)$,

$$\|\langle f, \varphi\rangle\|_{E;\nu} \leq c_{K,\nu} \|\varphi\|_{\mathcal{C}_b^{m_K}(\Omega)}.$$ ▮

Let us show that every distribution with values in a Banach space is locally of finite order.

Theorem 16.2.– *Every distribution on an open subset of* $\mathbb{R}^d$ *with values in a Banach space is locally of finite order.* ▮

Proof. Let $f \in \mathcal{D}'(\Omega; E)$, where Ω is an open subset of $\mathbb{R}^d$ and E is a Banach space, and K be a compact subset of Ω. By the characterization of distributions from Theorem 3.4 (d), f is continuous linear from $\mathcal{C}_K^\infty(\Omega)$ into E.

The space $\mathcal{C}_K^\infty(\Omega)$ is, by Definition 2.3 (a), endowed with the family of semi-norms of $\mathcal{C}_b^\infty(\Omega)$ which is, by Definition 1.19 (b), indexed by $m \in \mathbb{N}$, and which is filtering (Definition 1.8) according to Theorem A.52.

Since the family of semi-norms of E here reduces to its norm, the characterization of continuous linear mappings from Theorem 1.12 (b) shows that there exist $m_K \in \mathbb{N}$ and $c_K \in \mathbb{R}$ such that, for every $\varphi \in \mathcal{C}_K^\infty(\Omega)$,

$$\|\langle f, \varphi\rangle\|_E \leq c_K \|\varphi\|_{\mathcal{C}_b^\infty(\Omega);m_K}. \tag{16.1}$$

Therefore, f satisfies Definition 16.1 (b) of a distribution which is locally of finite order, since by equality (1.6), p. 13, $\|\varphi\|_{\mathcal{C}_b^\infty(\Omega);m_K} = \|\varphi\|_{\mathcal{C}_b^{m_K}(\Omega)}$. □

Let us now show that every distribution with values in a Banach space with compact support, is globally of finite order.

Theorem 16.3.– *Every distribution on an open subset of* $\mathbb{R}^d$ *with values in a Banach space whose support is compact is of finite order.* ▮

Proof. Let Ω be an open subset of $\mathbb{R}^d$, E a Banach space and $f \in \mathcal{D}'(\Omega; E)$ such that

$$\operatorname{supp} f \text{ is compact.}$$

Since $\operatorname{supp} f \subset \Omega$, the strong inclusion theorem (Theorem A.22) provides $r > 0$ such that $\operatorname{supp} f + B(0,r) \subset \Omega$. Then, $\omega = \operatorname{supp} f + \mathring{B}(0,r)$ is an open set such that

$$\operatorname{supp} f \Subset \omega \Subset \Omega.$$

Let

$$\mathcal{O} \stackrel{\text{def}}{=} \operatorname{nihil} f$$

be the annihilation domain of f. By Definition 6.18 of support, $\operatorname{supp} f = \Omega \setminus \mathcal{O}$, thus

$$\omega \cup \mathcal{O} = \Omega.$$

Let $(\alpha_\omega, \alpha_\mathcal{O})$ be a partition of unity subordinate to this cover of Ω, given by Theorem 2.21. By Definition 2.20 of such a partition, α_ω and $\alpha_\mathcal{O}$ belong to $\mathcal{C}^\infty(\Omega)$ and $\alpha_\omega + \alpha_\mathcal{O} = 1$. For every $\varphi \in \mathcal{D}(\Omega)$, the products $\alpha_\omega \varphi$ and $\alpha_\mathcal{O}\varphi$ belong to $\mathcal{D}(\Omega)$ by property (5.9), p. 95. Therefore,

$$\langle f, \varphi \rangle = \langle f, \alpha_\omega \varphi \rangle + \langle f, \alpha_\mathcal{O} \varphi \rangle.$$

Moreover, $\operatorname{supp}(\alpha_\omega \varphi) \subset \operatorname{supp} \alpha_\omega \subset \omega$, thus inequality (16.1) for $K = \overline{\omega}$ and then the bound of the norm of a product of functions from Theorem 2.23 give successively

$$\|\langle f, \alpha_\omega \varphi \rangle\|_E \le c_K \|\alpha_\omega \varphi\|_{\mathcal{C}_b^{m_K}(\Omega)} \le c_K 2^{m_K} \|\alpha_\omega\|_{\mathcal{C}_b^{m_K}(\Omega)} \|\varphi\|_{\mathcal{C}_b^{m_K}(\Omega)}. \tag{16.2}$$

On the other hand, $\operatorname{supp}(\alpha_\mathcal{O}\varphi) \subset \mathcal{O}$ so, by Definition 6.17 of the cancellation domain,

$$\langle f, \alpha_\mathcal{O} \varphi \rangle = 0_E.$$

The first term of (16.2) is thus equal to $\|\langle f, \varphi\rangle\|_E$. Therefore, f is of order m_K. □

The case of continuous functions. Every continuous function f with values in a Neumann space E is locally a distribution of order 0. Indeed, denoting $c = \int_{\mathring{K}} \|f(x)\|_{E;\nu}$, we have, for every $\varphi \in \mathcal{C}_K^0(\Omega)$,

$$\|\langle f, \varphi \rangle\|_{E;\nu} = \Big\| \int_\Omega f\varphi \Big\|_{E;\nu} \le c \sup_{x \in \Omega} |\varphi(x)| = c\|\varphi\|_{\mathcal{C}_b^0(\Omega)}.$$

In order for a continuous functions with values in a Neumann space to be globally of order 0, it is necessary that it be *integrable*, a notion that we will see in Volume 4. □

The case of measures. Any measure f with values in a Neumann space E is locally a distribution of order 0. Indeed, according to the characterization of measures from Theorem 3.13, there exist $p \in \mathcal{C}^+(\Omega)$ and $c \in \mathbb{R}$ such that, for every $\varphi \in \mathcal{K}(\Omega)$,

$$\|\langle f, \varphi \rangle\|_{E;\nu} \le c \sup_{x \in \Omega} p(x)|\varphi(x)|,$$

and so, denoting $c_K = \sup_{x \in K} p(x)$, we have for every $\varphi \in \mathcal{C}_K^0(\Omega)$,

$$\|\langle f, \varphi \rangle\|_{E;\nu} = c c_K \|\varphi\|_{\mathcal{C}_b^0(\Omega)}.$$

In particular, any positive distribution is locally of order 0, because it is a measure (Theorem 5.34).

For a measure with values in a Neumann space to be globally of order 0, it is necessary that it is *of a finite mass*. This is the case for measures with compact support, and in particular of the Dirac mass. □

16.2. Weighting of a finite order distribution

Let us show that the weighting of a distribution of order m by a $\mathcal{C}^m$ weight produces a bounded continuous function.

Theorem 16.4.– *Let*

$$f \in \mathcal{D}'(\Omega; E) \text{ be a distribution of order } m \text{ and } \mu \in \mathcal{C}^m_D(\mathbb{R}^d),$$

where Ω is an open subset of $\mathbb{R}^d$, E is a Neumann space, $m \in \mathbb{N}$ and D is a compact subset of $\mathbb{R}^d$. Then,

$$f \diamond \mu \in \mathcal{C}_b(\Omega_D; E). \blacksquare$$

We supplemente this result with an estimate.

Theorem 16.5.– *Suppose that the hypotheses of Theorem 16.4 are satisfied and let $\{\| \; \|_{E;\nu} : \nu \in \mathcal{N}_E\}$ be the family of semi-norms of E. Then, for every $\nu \in \mathcal{N}_E$,*

$$\|f \diamond \mu\|_{\mathcal{C}_b(\Omega_D;E);\nu} \leq |||f|||_{m,\nu} \|\mu\|_{\mathcal{C}^m_b(\mathbb{R}^d)},$$

where $|||f|||_{m,\nu} \stackrel{\text{def}}{=} \sup_{\varphi \in \mathcal{D}(\Omega),\, \varphi \neq 0} \|\langle f, \varphi \rangle\|_{E;\nu} / \|\varphi\|_{\mathcal{C}^m_b(\Omega)}$. $\blacksquare$

Proofs of Theorems 16.4 and 16.5. We are going to establish the continuity of $f \diamond \mu$ by showing that the functions $f \diamond \mu_n$, where μ_n is a regular approximation of μ, locally form a Cauchy sequence of continuous functions.

1. *Estimation of $f \diamond \mu_n$.* Let $(\rho_n)_{n \geq 1}$ be a regularizing sequence (Definition 8.1) and

$$\mu_n \stackrel{\text{def}}{=} \mu \diamond \rho_n.$$

Let $k \geq 1$ and

$$B \stackrel{\text{def}}{=} \{x \in \mathbb{R}^d : |x| \leq 1/k\}.$$

For $n \geq k$,

$$\mu_n \in \mathcal{C}^\infty_{D+B}(\mathbb{R}^d). \tag{16.3}$$

Indeed, μ_n is an infinitely differentiable function thanks to the regularizing property of weighting by a regular function (Theorem 7.5), and its support is included in $\operatorname{supp} \mu - \operatorname{supp} \rho_n$, and hence in $D + B$, by Theorem 7.18.

It follows, again according to Theorem 7.5, that

$$f \diamond \mu_n \in \mathcal{C}^\infty(\Omega_{D+B}; E).$$

By Definition 7.1 of the weighting by a regular function (which coincides with a general weighting according to Theorem 7.13), for every $x \in \Omega_{D+B}$, we have $\tau_x \mu_n \in \mathcal{D}(\Omega)$ and

$$(f \diamond \mu_n)(x) = \langle f, \tau_x \mu_n \rangle_\Omega.$$

By hypothesis, f is of order m so, by bounding the semi-norm $\| \ \|_{E;\nu}$ of the right-hand side using Definition 16.1 (a) of this property, we obtain, for every $\nu \in \mathcal{N}_E$,

$$\|(f \diamond \mu_n)(x)\|_{E;\nu} \leq |||f|||_{m,\nu} \|\tau_x \mu_n\|_{\mathcal{C}_b^m(\Omega)},$$

By taking the supremum with respect to x, we obtain, by Definition 1.18 (b) of the semi-norms of $\mathcal{C}_b^m(\Omega; E)$, since $\|\tau_x \mu_n\|_{\mathcal{C}_b^m(\Omega)} \leq \|\mu_n\|_{\mathcal{C}_b^m(\mathbb{R}^d)}$,

$$\|f \diamond \mu_n\|_{\mathcal{C}_b(\Omega_{D+B};E);\nu} \leq |||f|||_{m,\nu} \|\mu_n\|_{\mathcal{C}_b^m(\mathbb{R}^d)}. \tag{16.4}$$

2. Convergence of $f \diamond \mu_n$. Let $n' \geq n$. Due to (16.3), $\mu_{n'} - \mu_n \in \mathcal{C}_{D+B}^\infty(\mathbb{R}^d)$, thus inequality (16.4) applied to this function gives

$$\|f \diamond \mu_{n'} - f \diamond \mu_n\|_{\mathcal{C}_b(\Omega_{D+B};E);\nu} \leq |||f|||_{m,\nu} \|\mu_{n'} - \mu_n\|_{\mathcal{C}_b^m(\mathbb{R}^d)}. \tag{16.5}$$

The hypothesis $\mu \in \mathcal{C}_b^m(\mathbb{R}^d)$ implies, according to the local convergence property (which is global here, since $\mathbb{R}^d_{1/k} = \mathbb{R}^d$) of regular approximations of a continuous function from Theorem 8.7 (b), when $n \to \infty$,

$$\mu_n \to \mu \ \text{ in } \mathcal{C}_b^m(\mathbb{R}^d). \tag{16.6}$$

This result requires $\mu \in \mathbf{C}_b^m(\mathbb{R}^d)$, i.e. that μ and its derivatives of order $\leq m$ are uniformly continuous, which is the case due to a corollary (Theorem A.33) of Heine's theorem since they are continuous with compact support.

With inequality (16.5), it follows that the sequence $(f \diamond \mu_n)_{n \geq 1/r}$ is Cauchy in $\mathcal{C}_b(\Omega_{D+B}; E)$. Given that this space is sequentially complete (Theorem A.46, since E is), this sequence has a limit ℓ in it, i.e.

$$f \diamond \mu_n \to \ell \ \text{ in } \mathcal{C}_b(\Omega_{D+B}; E). \tag{16.7}$$

This limit is

$$\ell = f \diamond \mu. \tag{16.8}$$

Indeed, on one hand this convergence takes place in $\mathcal{D}'(\Omega_{D+B}; E)$ since, according to Theorem 3.8, $\mathcal{C}(\Omega_{D+B}; E) \subsetneq \mathcal{D}'(\Omega_{D+B}; E)$. And on the other hand,

$$f \diamond \mu_n \to f \diamond \mu \text{ in } \mathcal{D}'(\Omega_{D+B}; E),$$

since $\mu_n \to \mu$ in $\mathcal{D}'_{D+B}(\mathbb{R}^d)$ due to Theorem 8.6 (b) and since the mapping $\mu \mapsto f \diamond \mu$ is sequentially continuous from this space into $\mathcal{D}'(\Omega_{D+B}; E)$ due to Theorem 7.15 (a).

3. *Continuity of $f \diamond \mu$.* By (16.7) and (16.8),

$$f \diamond \mu \in \mathcal{C}_{\mathrm{b}}(\Omega_{D+B}; E).$$

It follows that $f \diamond \mu$ is continuous at every point of Ω_D, i.e.

$$f \diamond \mu \in \mathcal{C}(\Omega_D; E),$$

since

$$\Omega_D \text{ is the union of the } \Omega_{D+B} \text{ for } k \geq 1. \tag{16.9}$$

Indeed, if $x \in \Omega_D$, then $B(x, 1/k) \subset \Omega_D$ for k large enough (since Ω_D is open, see Theorem 7.2), that is $B(x, 1/k) + D = x + D + B \subset \Omega$, and thus $x \in \Omega_{D+B}$.

4. *Boundedness and estimation of $f \diamond \mu$.* By passing to the limit in inequality (16.4) with convergences (16.6) and (16.7), we obtain

$$\|f \diamond \mu\|_{\mathcal{C}_{\mathrm{b}}(\Omega_{D+B}; E); \nu} \leq |||f|||_{m,\nu} \|\mu\|_{\mathcal{C}^m_{\mathrm{b}}(\mathbb{R}^d)}.$$

That is, by Definition 1.18 (b) of the semi-norms of $\mathcal{C}_{\mathrm{b}}(\Omega; E)$,

$$\sup_{x \in \Omega_{D+B}} \|(f \diamond \mu)(x)\|_{E;\nu} \leq |||f|||_{m,\nu} \|\mu\|_{\mathcal{C}^m_{\mathrm{b}}(\mathbb{R}^d)}.$$

By taking the supremum for $k \geq 1$, it follows, with (16.9) and continuity obtained in step 3, that

$$f \diamond \mu \in \mathcal{C}_{\mathrm{b}}(\Omega_D; E),$$

and, again by Definition 1.18 (b),

$$\|f \diamond \mu\|_{\mathcal{C}_{\mathrm{b}}(\Omega_D; E); \nu} \leq |||f|||_{m,\nu} \|\mu\|_{\mathcal{C}^m_{\mathrm{b}}(\mathbb{R}^d)}. \quad \square$$

16.3. Finite order distribution as derivatives of continuous functions

Let us represent every finite order distribution as a sum of derivatives of continuous functions [2].

---

2. **History of decomposition of finite order distributions.** Laurent SCHWARTZ proved in 1957 [72, Proposition 24, p. 86] a more general result than Theorem 16.6, namely the stated decomposition for any semi-normed space E, i.e. the decomposition of any continuous linear mapping f from $\mathcal{K}^m(\Omega)$ into E (recall that it is not desirable to call such a mapping a distribution, since it does not have the expected properties, see Section 3.4, *The case where E is not a Neumann space*, p. 53).

Theorem 16.6.– *Let*

$$f \in \mathcal{D}'(\Omega; E) \text{ be a distribution of order } m,$$

where Ω is an open subset of $\mathbb{R}^d$, E is a Neumann space and $m \in \mathbb{N}$.

Then, there exist functions $g_\beta \in \mathcal{C}(\Omega; E)$ such that

$$f = \sum_{0 \le |\beta| \le m+d+1} \partial^\beta g_\beta. \tag{16.10}$$

▌

We can choose these functions g_β depending linearly and continuously on f, as follows.

Theorem 16.7.– *Let Ω be an open subset of $\mathbb{R}^d$, E a Neumann space, $m \in \mathbb{N}$, $r > 0$ and $\{\| \ \|_{E;\nu} : \nu \in \mathcal{N}_E\}$ the family of semi-norms of E.*

For every distribution $f \in \mathcal{D}'(\Omega; E)$ of order m, there exists a family $(g_\beta)_{|\beta| \le m+d+1}$ of functions in $\mathcal{C}(\Omega; E)$ satisfying decomposition (16.10) such that:

$$\text{the mapping } f \mapsto (g_\beta)_{|\beta| \le m+d+1} \text{ is linear,}$$

$$\operatorname{supp} g_\beta \subset \operatorname{supp} f + B_r,$$

where $B_r \stackrel{\text{def}}{=} \{x \in \mathbb{R}^d : |x| \le r\}$ and, such that, for every compact subset K of Ω and every $\nu \in \mathcal{N}_E$,

$$\|g_\beta\|_{\mathcal{C}(\Omega;E);K,\nu} \le c_{m,K,\Omega} |||f|||_{m,\nu},$$

where $|||f|||_{m,\nu} \stackrel{\text{def}}{=} \sup_{\varphi \in \mathcal{D}(\Omega),\, \varphi \ne 0} \|\langle f, \varphi\rangle\|_{E;\nu} / \|\varphi\|_{\mathcal{C}_b^m(\Omega)}$ and $c_{m,K,\Omega}$ depends neither on f nor on ν. ▌

Maximum order of the derivatives. In Theorems 16.6 and 16.7, we cannot replace $m+d+1$ by $m+d-1$, see [SCHWARTZ, 72, Vol. I, p. 89]. If $d = 1$, we cannot even replace $m+d+1$ by $m+d$. We do not know, for $d \ge 2$, if we can replace $m+d+1$ by $m+d$. □

Proof of Theorem 16.6. We are going to proceed in three steps: representation on Ω_B; localization; gluing.

1. *Representation on* Ω_B. Let $r > 0$ and $B_r \stackrel{\text{def}}{=} \{x \in \mathbb{R}^d : |x| \le r\}$.

According to the decomposition of the Dirac mass from Theorem 9.18, there exist functions $\zeta_\beta \in \mathcal{C}^m_{B_r}(\mathbb{R}^d)$ such that

$$\delta_0 = \sum_{\beta \in \mathbb{N}^d : 0 \le |\beta| \le m+d+1} \partial^\beta \zeta_\beta.$$

Since the Dirac mass is the neutral element of weighting (Theorem 7.19), we have, with the equality $\partial^\beta(f \diamond \mu) = (-1)^{|\beta|} f \diamond \partial^\beta \mu$ from Theorem 7.17, on Ω_{B_r},

$$f = f \diamond \delta_0 = \sum_{|\beta| \leq m+d+1} f \diamond \partial^\beta \zeta_\beta = \sum_{|\beta| \leq m+d+1} (-1)^{|\beta|} \partial^\beta (f \diamond \zeta_\beta). \tag{16.11}$$

Since f is a distribution of order m and the ζ_β are $\mathcal{C}^m$, Theorem 16.4 gives

$$f \diamond \zeta_\beta \in \mathcal{C}_\mathrm{b}(\Omega_{B_r}; E). \tag{16.12}$$

So, (16.11) is the desired decomposition, but only on Ω_{B_r}. To obtain it on all of Ω, we are going to decompose f into an (infinite) sum of distributions f_k whose respective supports are compact, then we will glue the representations on $\mathbb{R}^d$ of their extensions by 0_E.

2. Localization. Let $(\kappa_k)_{k \in \mathbb{N}}$ be the cover of Ω by crown-shaped sets given by Definition 2.18 and $(\alpha_k)_{k \in \mathbb{N}}$ a subordinate partition of unity (Definition 2.20; it exists by Theorem 2.21). We define

$$f_k \in \mathcal{D}'(\mathbb{R}^d; E),$$

by: for every $\phi \in \mathcal{D}(\mathbb{R}^d)$,

$$\langle f_k, \phi \rangle_{\mathbb{R}^d} \overset{\text{def}}{=} \langle f, \alpha_k \phi|_\Omega \rangle_\Omega.$$

Indeed, f_k is the localized extension $\widetilde{\alpha_k f}$ from Definition 6.12, which applies here since $\widetilde{\alpha_k} \in \mathcal{C}^\infty(\mathbb{R}^d)$ and $\operatorname{supp} \widetilde{\alpha_k} \subset \Omega$ (since $\widetilde{\alpha_k} = 0$ outside of κ_k, and thus $\operatorname{supp} \widetilde{\alpha_k} \subset \overline{\kappa_k} \subset \Omega$).

By hypothesis, f is of order m, so Definition 16.1 of this property gives, for every $\phi \in \mathcal{D}(\mathbb{R}^d)$ and every semi-norm $\| \ \|_{E;\nu}$ of E,

$$\|\langle f_k, \phi \rangle_{\mathbb{R}^d}\|_{E;\nu} = \|\langle f, \alpha_k \phi \rangle_\Omega\|_{E;\nu} \leq |||f|||_{m,\nu} \|\alpha_k \phi\|_{\mathcal{C}_\mathrm{b}^m(\Omega)}.$$

Hence, with the bound of the norm of a product of $\mathcal{C}^m$ functions from Theorem 2.23,

$$\|\langle f_k, \phi \rangle_{\mathbb{R}^d}\|_{E;\nu} \leq 2^m |||f|||_{m,\nu} \|\alpha_k\|_{\mathcal{C}_\mathrm{b}^m(\Omega)} \|\phi\|_{\mathcal{C}_\mathrm{b}^m(\mathbb{R}^d)}. \tag{16.13}$$

So, f_k is a distribution on $\mathbb{R}^d$ of order m (Definition 16.1).

Decomposition (16.11) for $\Omega = \mathbb{R}^d$ and $r = r_k$ then gives, on all of $\mathbb{R}^d$ (because $\mathbb{R}^d_{B_r} = \mathbb{R}^d$),

$$f_k = \sum_{0 \leq |\beta| \leq m+d+1} (-1)^{|\beta|} \partial^\beta (f_k \diamond \zeta_{\beta,k}), \tag{16.14}$$

where, due to (16.12),

$$f_k \diamond \zeta_{\beta,k} \in \mathcal{C}_b(\mathbb{R}^d; E).$$

By the expression for the localized extension from Theorem 6.13,

$$f_k = \widetilde{\alpha_k f} = \alpha_k f \text{ in } \Omega. \tag{16.15}$$

Since the support of a product is included in the intersection of the supports due to Theorem 6.24, the support of f_k is included in that of α_k, which is included in κ_k by definition of a partition of unity. Therefore, according to the inclusion of the support of a weighted distribution from Theorem 7.18,

$$\operatorname{supp}(f_k \diamond \zeta_{\beta,k}) \subset \operatorname{supp} f_k - \operatorname{supp} \zeta_{\beta,k} \subset \kappa_k + B_{r_k}.$$

3. Gluing. Choose

$$r_k \leq \frac{1}{k-1} - \frac{1}{k}. \tag{16.16}$$

Then, by Definition 2.18 of crown-shaped sets, $\kappa_k + B_{r_k}$ is included in the set $\kappa_{k-1} \cup \kappa_k \cup \kappa_{k+1}$, thus

$$\operatorname{supp}(f_k \diamond \zeta_{\beta,k}) \subset \kappa_{k-1} \cup \kappa_k \cup \kappa_{k+1}.$$

For each β, we define g_β on Ω by

$$g_\beta \stackrel{\text{def}}{=} (-1)^{|\beta|} \sum_{k \geq 0} f_k \diamond \zeta_{\beta,k}.$$

This sum makes sense since, at each point, it only has a finite number of non-zero terms. More precisely,

$$g_\beta = (-1)^{|\beta|}(f_{k-1} \diamond \zeta_{\beta,k-1} + f_k \diamond \zeta_{\beta,k} + f_{k+1} \diamond \zeta_{\beta,k+1}) \text{ on } \kappa_k, \tag{16.17}$$

since (Theorem 2.19 (b)) κ_k only intersects κ_{k-1} and κ_{k+1}, and hence the other functions $f_i \diamond \zeta_{\beta,i}$ are zero there. This equality shows moreover that g_β is continuous on each κ_k, and hence on their union Ω, i.e.

$$g_\beta \in \mathcal{C}(\Omega; E).$$

On the other hand, by Definition 2.20 of a partition of unity, $\sum_{k \geq 0} \alpha_k = 1$. On κ_k, at most α_{k-1}, α_k and α_{k+1} are not zero, therefore, with (16.15),

$$f = (\alpha_{k-1} + \alpha_k + \alpha_{k+1}) f = f_{k-1} + f_k + f_{k+1} \text{ on } \kappa_k.$$

Summing equalities (16.14) relative to $k-1$, k and $k+1$ and using (16.17), we therefore obtain

$$f = \sum_{0 \leq |\beta| \leq m+d+1} \partial^\beta g_\beta$$

on each κ_k. This equality is thus true on the union of the κ_k, i.e. on Ω, according to the gluing theorem for equalities (Theorem 6.10). □

Now let us show that the g_β depend linearly and continuously on f.

Proof of Theorem 16.7. Consider the functions g_β constructed in the proof of Theorem 16.6.

1. *Linearity.* From (16.17) and (16.15), on κ_k,

$$g_\beta = (-1)^{|\beta|}\big((\alpha_{k-1}f) \diamond \zeta_{\beta,k-1} + (\alpha_k f) \diamond \zeta_{\beta,k} + (\alpha_{k+1}f) \diamond \zeta_{\beta,k+1}\big). \quad (16.18)$$

Since the weighting is bilinear (Theorem 7.15 (a)) and so is the product, the equalities characterizing the linearity of the mapping $f \mapsto g_\beta$ are therefore satisfied on each κ_k, and hence on their union Ω.

2. *Support of g_β.* Let r' be such that $0 < r' < r$. We can choose the r_k such that, in addition to (16.16),

$$r_k \le r'.$$

The inclusion of the support of a weighted distribution from Theorem 7.18 then gives, since $\operatorname{supp}(\alpha_k f) \subset \operatorname{supp} f$ due to Theorem 6.24,

$$\operatorname{supp}((\alpha_k f) \diamond \zeta_{\beta,k}) \subset \operatorname{supp}(\alpha_k f) - \operatorname{supp} \zeta_{\beta,k} \subset \operatorname{supp} f + B_{r'}.$$

Since the support in the distribution sense coincides with that in the function sense due to Theorem 6.19, Definition 2.1 of the support of a function shows that the function $(\alpha_k f) \diamond \zeta_{\beta,k}$ is zero on the set

$$\mathcal{O} = \Omega \setminus (\operatorname{supp} f + B_{r'}).$$

Equality (16.18) then shows that the function g_β is zero on each $\mathcal{O} \cap \kappa_k$, and hence on their union $\mathcal{O}$. Hence, again by Definition 2.1 of the support of a function and since $r' < r$,

$$\operatorname{supp} g_\beta \subset \overline{\operatorname{supp} f + B_{r'}} \cap \Omega \subset \operatorname{supp} f + B_r.$$

3. *Semi-norm of g_β.* According to inequality (16.13), f_k is a distribution on $\mathbb{R}^d$ of order m. Thus the estimate for the semi-norms of a weight from Theorem 16.5 gives

$$\|f_k \diamond \zeta_{\beta,k}\|_{\mathcal{C}_b(\mathbb{R}^d;E);\nu} \le |||f_k|||_{m,\nu} \|\zeta_{\beta,k}\|_{\mathcal{C}_b^m(\mathbb{R}^d)}.$$

On the other hand, (16.13) gives, with Definition 16.1 of $|||\ |||_{m;\nu}$,

$$|||f_k|||_{m,\nu} \le 2^m |||f|||_{m,\nu} \|\alpha_k\|_{\mathcal{C}_b^m(\Omega)}.$$

Hence, with Definition 1.18 (b) of the semi-norms of $\mathcal{C}_b(\mathbb{R}^d; E)$,

$$\sup_{x\in\mathbb{R}^d} \|(f_k \diamond \zeta_{\beta,k})(x)\|_{E;\nu} \le 2^m \|\alpha_k\|_{\mathcal{C}_b^m(\Omega)} \|\zeta_{\beta,k}\|_{\mathcal{C}_b^m(\mathbb{R}^d)} |||f|||_{m,\nu}.$$

With equality (16.17), if follows then, for every $x \in \kappa_k$,

$$\|g_\beta(x)\|_{E,\nu} \leq c_{m,k,\Omega} |||f|||_{m,\nu},$$

where

$$c_{m,k,\Omega} = 2^m \sum_{i=k-1}^{k+1} \|\alpha_i\|_{\mathcal{C}_b^m(\Omega)} \|\zeta_{\beta,i}\|_{\mathcal{C}_b^m(\mathbb{R}^d)}.$$

Now, let K be a compact subset of Ω. It only intersects a finite number of the κ_k (Theorem 2.19 (c)), so, by Definition 1.18 (a) of the semi-norms of $\mathcal{C}(\Omega; E)$,

$$\|g_\beta\|_{\mathcal{C}(\Omega;E);K,\nu} = \sup_{x \in K} \|g_\beta(x)\|_{E;\nu} \leq c_{m,K,\Omega} |||f|||_{m,\nu},$$

where $c_{m,K,\Omega} = \sup_{k \in N_K} c_{m,k,\Omega}$ and N_K is the set of the k such that $K \cap \kappa_k \neq \emptyset$. □

16.4. Finite order distribution as derivative of a single function

Let us locally represent any finite order distribution as the derivative of a single bounded continuous function [3].

Theorem 16.8.– *Let*

$$f \in \mathcal{D}'(\Omega; E) \textit{ be a distribution of order } m,$$

where Ω is an open subset of $\mathbb{R}^d$, E is a Neumann space and $m \in \mathbb{N}$.

Then, for every open set $\omega \Subset \Omega$, there exists $g \in \mathcal{C}_b(\Omega; E)$ such that

$$f = (\partial_1 \ldots \partial_d)^{m+2} g \textit{ on } \omega. \blacksquare$$

Proof. We are going to show that the stated properties are satisfied by the restriction to Ω of the function defined on all of $\mathbb{R}^d$ by

$$g \stackrel{\text{def}}{=} \epsilon \widetilde{\alpha f} \diamond \zeta H_{m+1},$$

3. **History of the representation by a single derivative.** Laurent SCHWARTZ proved Theorem 16.8 for $E = \mathbb{R}^d$ in 1950 [69, Chap. III, § 6, p. 83–84]. His proof is based on the Hahn–Banach theorem. Jérôme LEMOINE and Jacques SIMON proved it for E a Banach space in 1996 [52, p. 33–34], using the method expressed here.

where $\widetilde{\alpha f} \in \mathcal{D}'(\mathbb{R}^d; E)$ is the localized extension given by Definition 6.12, H_{m+1} is the Heaviside potential introduced in Theorem 9.21, $\epsilon \overset{\text{def}}{=} (-1)^{(m+2)d}$, and

$$\alpha \in \mathcal{D}(\mathbb{R}^d), \ \alpha = 1 \text{ on } \omega, \ \operatorname{supp} \alpha \subset \Omega \cap B_{R+1},$$
$$\zeta \in \mathcal{D}(\mathbb{R}^d), \ \zeta = 1 \text{ on } B_{2R+1},$$

where $R \overset{\text{def}}{=} \sup\{|x| : x \in \omega\}$ and $B_r \overset{\text{def}}{=} \{x \in \mathbb{R}^d : |x| \leq r\}$.

1. *Derivative of g.* According to the equality $\partial^\beta(h \diamond \mu) = (-1)^{|\beta|} h \diamond \partial^\beta \mu$ from Theorem 7.17,

$$(\partial_1 \ldots \partial_d)^{m+2} g = \widetilde{\alpha f} \diamond (\partial_1 \ldots \partial_d)^{m+2} (\zeta H_{m+1}). \tag{16.19}$$

Since $(\partial_1 \ldots \partial_d)^{m+2} H_{m+1} = \delta_0$ (Theorem 9.21) and $\zeta = 1$ on B_{2R+1},

$$(\partial_1 \ldots \partial_d)^{m+2} (\zeta H_{m+1}) = (\partial_1 \ldots \partial_d)^{m+2} H_{m+1} = \delta_0 \ \text{ on } B_{2R+1}.$$

Therefore, on $\mathbb{R}^d$,

$$(\partial_1 \ldots \partial_d)^{m+2} (\zeta H_{m+1}) = \delta_0 + \vartheta,$$

where ϑ is a distribution whose support is included in $\{x \in \mathbb{R}^d : |x| \geq 2R + 1\}$. Then, (16.19) gives

$$(\partial_1 \ldots \partial_d)^{m+2} g = \widetilde{\alpha f} \diamond \delta_0 + \widetilde{\alpha f} \diamond \vartheta. \tag{16.20}$$

According to the inclusion of the support of a weighted distribution (Theorem 7.18),

$$\operatorname{supp} \widetilde{\alpha f} \diamond \vartheta \subset \operatorname{supp} \widetilde{\alpha f} - \operatorname{supp} \vartheta \subset \{x : |x| \leq R + 1\} - \{x : |x| \geq 2R + 1\}.$$

Thus, the distribution $\widetilde{\alpha f} \diamond \vartheta$ has its support in $\{x \in \mathbb{R}^d : |x| \geq R\}$, hence

$$\widetilde{\alpha f} \diamond \vartheta = 0 \ \text{ on } \omega.$$

Moreover, the Dirac mass is the neutral element of weighting (Theorem 7.19), $\widetilde{\alpha f} = \alpha f$ on Ω according to Theorem 6.13 and $\alpha = 1$ on ω by hypothesis, therefore equality (16.20) gives

$$(\partial_1 \ldots \partial_d)^{m+2} g = \widetilde{\alpha f} \diamond \delta_0 = \widetilde{\alpha f} = \alpha f = f \ \text{ on } \omega.$$

2. *Continuity of g.* It is sufficient to check that, denoting by Q the support of ζ,

$$\widetilde{\alpha f} \text{ is a distribution of order } m \text{ and } \zeta H_{m+1} \in \mathcal{C}_Q^m(\mathbb{R}^d), \tag{16.21}$$

because, according to Theorem 16.4, this implies

$$\widetilde{\alpha f} \diamond \zeta H_{m+1} \in \mathcal{C}_b(\mathbb{R}^d; E)$$

and hence, after multiplication by the sign ϵ and restriction to Ω,

$$g \in \mathcal{C}_b(\Omega; E). \tag{16.22}$$

2.a. On one hand, f is of order m by hypothesis which, by Definition 16.1 (a), means that, for every semi-norm $\| \ \|_{E;\nu}$ of E, there exists $c_\nu \in \mathbb{R}$ such that, for every $\varphi \in \mathcal{D}(\Omega)$,

$$\|\langle f, \varphi\rangle_\Omega\|_{E;\nu} \leq c_\nu \|\varphi\|_{\mathcal{C}_b^m(\Omega)}.$$

By Definition 6.12 of the localized extension, for every $\phi \in \mathcal{D}(\mathbb{R}^d)$,

$$\langle \widetilde{\alpha f}, \phi\rangle_{\mathbb{R}^d} = \langle f, (\alpha\phi)|_\Omega\rangle_\Omega.$$

Therefore, with the bound of the norms in $\mathcal{C}_b^m(\Omega)$ of a product of functions from Theorem 2.23,

$$\|\langle \widetilde{\alpha f}, \phi\rangle_{\mathbb{R}^d}\|_{E;\nu} \leq c_\nu \|\alpha\phi\|_{\mathcal{C}_b^m(\Omega)} \leq c_\nu 2^m \|\alpha\|_{\mathcal{C}_b^m(\Omega)} \|\phi\|_{\mathcal{C}_b^m(\Omega)} \leq c' \|\phi\|_{\mathcal{C}_b^m(\mathbb{R}^d)},$$

where $c' = c_\nu 2^m \|\alpha\|_{\mathcal{C}_b^m(\Omega)}$. Which proves that

$$\widetilde{\alpha f} \text{ is of order } m.$$

2.b. On the other hand, H_{m+1} belongs to $\mathcal{C}^m(\mathbb{R}^d)$ according to Theorem 9.21, so the product ζH_{m+1} also belongs to it, again according to Theorem 2.23, and its support is included in that of ζ. This finishes the proof of (16.21) and thus of (16.22). □

Global representation. When Ω is "regular enough", the equality $f = (\partial_1 \dots \partial_d)^{m+2} g$ can be obtained on all of Ω, with g not necessarily bounded. □

Support of g**.** In Theorem 16.8, we can impose that g has compact support, i.e.

$$g \in \mathcal{K}(\Omega; E).$$

For this, it suffices to multiply the function obtained above by the localizing function α.

But then, the equality $f = (\partial_1 \dots \partial_d)^{m+2} g$ certainly does not extend to all of Ω if f does not have also a compact support. □

16.5. Distributions in a Banach space as derivatives of functions

We are going to show that every distribution with values in a Banach space is globally a *locally finite sum* of derivatives of uniformly continuous functions with compact support[4]. Recall that we denote

$$\mathcal{K}(\Omega; E) \stackrel{\text{def}}{=} \{\textit{continuous functions with compact support from } \Omega \textit{ into } E\}.$$

4. **History of the decomposition as a locally finite sum of derivatives.** Laurent SCHWARTZ established this decomposition for real-valued distributions in 1950 [69, Chap. III, § 7, Theorem XXX, p. 96]. He proved in 1957 [72, Corollary 2, p. 90] a more general result than Theorem 16.9, namely the said decomposition for any quasi-complete semi-normed space E of type (DF).

Theorem 16.9.– *Let $f \in \mathcal{D}'(\Omega; E)$, where Ω is an open subset of $\mathbb{R}^d$ and*

E is a Banach space.

Then, there exist functions $g_n \in \mathcal{K}(\Omega; E)$ and $\beta_n \in \mathbb{N}^d$ such that

$$f = \sum_{n \in \mathbb{N}} \partial^{\beta_n} g_n \quad on\ \Omega$$

and such that, for every $\omega \Subset \Omega$, all the g_n are zero on ω from a finite rank N_ω, and so

$$f = \sum_{n \leq N_\omega} \partial^{\beta_n} g_n \quad on\ \omega. \tag{16.23}$$

▮

The proof will use the following decomposition into a *finite sum*, when the support of f is compact[5].

Theorem 16.10.– *Let $f \in \mathcal{D}'(\Omega; E)$, where Ω is an open subset of $\mathbb{R}^d$,*

E is a Banach space and the support of f is compact.

Then, there exist functions $g_1, \ldots, g_I$ in $\mathcal{K}(\Omega; E)$ and $\beta_1, \ldots, \beta_I$ in $\mathbb{N}^d$ such that

$$f = \sum_{1 \leq i \leq I} \partial^{\beta_i} g_i. \tag{16.24}$$

For every $r > 0$ and $B \stackrel{\text{def}}{=} \{x \in \mathbb{R}^d : |x| \leq r\}$, the g_i may be choosen so that

$$\operatorname{supp} g_i \subset \operatorname{supp} f + B. \tag{16.25}$$

▮

Proof. The distribution f has finite order due to Theorem 16.3, since its support is compact and E is a Banach space. Theorems 16.6 and 16.7 therefore provide functions

$$g_i \in \mathcal{C}(\Omega; E)$$

satisfying (16.24) and (16.25). By choosing, thanks to the strong inclusion theorem (Theorem A.22), r small enough so that $\operatorname{supp} f + B$ is included in Ω, (16.25) gives

$$g_i \in \mathcal{K}(\Omega; E).$$

Indeed, the support of each g_i is then compact due to Theorem 2.2, since it is included in $\operatorname{supp} f + B$, which is a compact (as is any sum of compact subsets of $\mathbb{R}^d$, see Theorem A.27) subset of Ω. □

5. **History of Theorem 16.10.** Laurent SCHWARTZ proved this result for real-valued distributions in 1950 [69, Chap. III, § 7, Theorem XXVI, p. 91].

We return to the case where the support of f is arbitrary.

Proof of Theorem 16.9. Let us proceed by localization and then gluing.

1. *Localization.* Let $(\kappa_k)_{k\in\mathbb{N}}$ be the cover by crown-shaped sets (Definition 2.18) of Ω, and $(\alpha_k)_{k\in\mathbb{N}}$ a subordinate partition of unity (it exists due to Theorem 2.21).

By Definition 2.20 of a partition of unity, $\alpha_k \in \mathcal{C}^\infty(\Omega)$, so Definition 5.15 of the product with a regular function gives

$$\alpha_k f \in \mathcal{D}'(\Omega; E).$$

The support of such a product being included in the intersection of the supports (Theorem 6.24),

$$\operatorname{supp}(\alpha_k f) \subset \operatorname{supp} \alpha_k \subset \kappa_k \subset \overline{\kappa_k}.$$

Since $\overline{\kappa_k}$ is a compact subset of Ω (Theorem 2.19 (b)), it follows, due to Theorem 6.28, that

the support of $\alpha_k f$ is compact.

Theorem 16.10 therefore gives the existence of a finite number I_k of functions $g_{k,i} \in \mathcal{K}(\Omega; E)$ and of $\beta_{k,i} \in \mathbb{N}^d$ such that

$$\alpha_k f = \sum_{i\leq I_k} \partial^{\beta_{k,i}} g_{k,i}. \tag{16.26}$$

2. *Gluing.* Again by definition of a partition of unity, $\sum_{k\geq 0} \alpha_k = 1$ on Ω, so

$$f = \sum_{k\geq 0} \alpha_k f.$$

Indeed, with again Definition 5.15 of the product by a regular function, for every $\varphi \in \mathcal{D}(\Omega)$,

$$\langle f, \varphi \rangle = \sum_{k\geq 0} \langle f, \alpha_k \varphi \rangle = \sum_{k\geq 0} \langle \alpha_k f, \varphi \rangle.$$

Note that only a finite number of elements in these two sums are not zero. Indeed, the support of φ only intersects a finite number of crown-shaped sets κ_k according to Theorem 2.19 (c) (since $\operatorname{supp} \varphi$ is a compact subset of Ω), thus the functions $\alpha_k \varphi$ related to other κ_k are zero.

With (16.26), it follows

$$f = \sum_{k\geq 0} \sum_{i\leq I_k} \partial^{\beta_{k,i}} g_{k,i}.$$

Each open set $\omega \Subset \Omega$ only intersects a finite number of the κ_k, again according to Theorem 2.19 (c), so only a finite number of the $g_{k,i}$ are not identically zero on ω.

The set of these pairs of indices (k, i) is included in $\mathbb{N}^2$ so, due to Theorem A.2, it is countable, i.e., by Definition A.1, it can be replaced with $\mathbb{N}$. This provides the stated properties. □

Observe that any distribution with values in a Banach space is locally the derivative of a *single* continuous function with compact support[6].

Theorem 16.11.– *Let $f \in \mathcal{D}'(\Omega; E)$, where Ω is an open subset of $\mathbb{R}^d$ and*

E is a Banach space.

Then, for every open set $\omega \Subset \Omega$, there exist $g \in \mathcal{K}(\Omega; E)$ and $\beta \in \mathbb{N}^d$ such that

$$f = \partial^\beta g \ \text{ on } \omega. \ \blacksquare$$

Proof. Let λ be an open set such that

$$\omega \Subset \lambda \Subset \Omega.$$

The distribution f is locally of finite order due to Theorem 16.2, so it is of finite order m on λ. By Theorem 16.8, there exists $h \in \mathcal{C}(\lambda; E)$ such that

$$f = (\partial_1 \dots \partial_d)^{m+2} h \ \text{ on } \omega.$$

Let α be a localizing function given by Uryshon's theorem (Theorem 8.12) such that

$$\alpha \in \mathcal{C}(\mathbb{R}^d), \quad \alpha = 1 \text{ on } \omega, \quad \alpha = 0 \text{ outside of } \lambda.$$

The stated properties are then satisfied by $\beta = (m + 2, \dots, m + 2)$ and

$$g \stackrel{\text{def}}{=} \begin{cases} \alpha h & \text{on } \lambda, \\ 0_E & \text{outside of } \lambda. \ \square \end{cases}$$

6. **History of the representation by a single derivative.** Laurent SCHWARTZ proved in 1950 [69, Chap. III, § 6, Theorem XXI, p. 82] that any distribution of $\mathcal{D}'(\mathbb{R}^d)$ is locally the derivative of a continuous function with compact support. Its proof is based on the Hahn–Banach theorem.
The case of a distribution with values in a Banach space was treated by Jérôme LEMOINE and Jacques SIMON in 1996 [52, Theorem 1, p. 32 and Lemma 2, p. 33], by the method stated here.

16.6. Non-representability of distributions with values in a Fréchet space

Theorem 16.9 on representability by derivatives of continuous functions extends to certain spaces E which are not Banach, see Note 4, p. 335, and it is the same for the other results in § 16.5. Let us show that these results do not extend to every Neumann space, nor even to every Fréchet space.

Theorem 16.12.– *In Theorems 16.9, 16.10 and 16.11, we cannot replace "Banach space" with "Fréchet space" or with "Neumann space".* ▌

Proof. We will do this by considering $E = \mathcal{C}(\mathbb{R})$. It is a Fréchet space according to Theorem A.47, for example for the semi-norms, indexed by $\nu \in \mathbb{N}$,

$$\|h\|_{\mathcal{C}(\mathbb{R});\nu} \stackrel{\text{def}}{=} \sup_{|z|\leq\nu} |h(z)|.$$

It is also a Neumann space, by Definition 1.4 of Neumann and Fréchet spaces.

1. *Construction of f.* We define $h \in \mathcal{C}(\mathbb{R})$ by

$$h(x) \stackrel{\text{def}}{=} \begin{cases} 1 - 2|x| & \text{if } |x| \leq 1/2, \\ 0 & \text{otherwise.} \end{cases}$$

Given Ω an open subset of $\mathbb{R}^d$, $a \in \Omega$ and $\varphi \in \mathcal{D}(\Omega)$, we define $\langle f, \varphi \rangle \in \mathcal{C}(\mathbb{R})$ by: for every $z \in \mathbb{R}$,

$$\langle f, \varphi \rangle(z) \stackrel{\text{def}}{=} \sum_{i\geq 0} \partial_1^i \varphi(a)\, h(z - i). \tag{16.27}$$

Let us show that f satisfies the characterization of distributions from Theorem 3.3. We have

$$\|\langle f, \varphi \rangle\|_{\mathcal{C}(\mathbb{R});\nu} = \sup_{|z|\leq\nu} \left| \sum_{i\geq 0} \partial_1^i \varphi(a)\, h(z - i) \right| \leq \sup_{0\leq i\leq\nu+1} |\partial_1^i \varphi(a)|,$$

since, at a point z, at most one of the $h(z - i)$ is non-zero. Hence, denoting by p_ν the function with constant value $\nu + 1$ on Ω,

$$\|\langle f, \varphi \rangle\|_{\mathcal{C}(\mathbb{R});\nu} \leq \sup_{x\in\Omega} \sup_{0\leq|\beta|\leq p_\nu(x)} p_\nu(x) |\partial^\beta \varphi(x)| = \|\varphi\|_{\mathcal{D}(\Omega);p_\nu}.$$

Since the mapping f is linear, the characterization from Theorem 3.3 therefore gives

$$f \in \mathcal{D}'(\Omega; \mathcal{C}(\mathbb{R})).$$

2. *Impossibility of the representation of Theorem 16.9.* Let us show that decomposition (16.23) in the said theorem is not possible. More precisely, given an open ball ω containing a and such that $\omega \Subset \Omega$, let us show by contradiction that there are no functions $g_n \in \mathcal{C}(\Omega; \mathcal{C}(\mathbb{R}))$ and no $\beta_n \in \mathbb{N}^d$ such that

$$f = \sum_{n \le N} \partial^{\beta_n} g_n \text{ on } \omega. \tag{16.28}$$

If (16.28) was satisfied, for $\varphi \in \mathcal{D}(\omega)$ (again denoting by φ its extension by 0), Definition 6.1 of restriction, the equality $\langle \partial^\beta g, \varphi \rangle = (-1)^{|\beta|} \langle g, \partial^\beta \varphi \rangle$ from Theorem 5.8 and the expression for $\langle g, \phi \rangle$ for a continuous function g from Theorem 3.9 would give

$$\langle f, \varphi \rangle = \langle f, \varphi \rangle_\omega = \sum_{n \le N} (-1)^{|\beta_n|} \int_\omega g_n \partial^{\beta_n} \varphi.$$

For $\nu \in \mathbb{N}$, the bound of the semi-norms of the integral from Theorem 1.23 (b) would give

$$\| \langle f, \varphi \rangle \|_{\mathcal{C}(\mathbb{R});\nu} \le \sum_{n \le N} \Big\| \int_\omega g_n \partial^{\beta_n} \varphi \Big\|_{\mathcal{C}(\mathbb{R});\nu} \le c_\nu \sup_{|\beta \le m} \sup_{x \in \omega} |\partial^\beta \varphi(x)|, \tag{16.29}$$

where $m = \sup_{n \le N} |\beta_n|$ and $c_\nu = N|\omega| \sup_{n \le N} \sup_{x \in \omega} \|g_n(x)\|_{\mathcal{C}(\mathbb{R});\nu}$.

On the other hand, equality (16.27) gives, since $h(\nu - i)$ equals 1 if $i = \nu$ and 0 otherwise,

$$\| \langle f, \varphi \rangle \|_{\mathcal{C}(\mathbb{R});\nu} \ge |\langle f, \varphi \rangle(\nu)| = |\partial_1^\nu \varphi(a)|.$$

For $\nu = m + 1$, we therefore would have, with (16.29),

$$|\partial_1^{m+1} \varphi(a)| \le c_{m+1} \sup_{|\beta| \le m} \sup_{x \in \omega} |\partial^\beta \varphi(x)|. \tag{16.30}$$

But, as we will check in Lemma 16.13, there exists $\varphi \in \mathcal{D}(\omega)$ that does not satisfy this inequality, therefore (16.28) cannot be satisfied.

3. *Impossibility of the representation of Theorem 16.11.* The distribution f cannot locally be the derivative of a single continuous function as in Theorem 16.11 since, by step 2, it cannot even be a sum of such derivatives.

4. *Impossibility of the representation of Theorem 16.10.* The distribution f cannot be a finite sum of derivatives of continuous functions on all of Ω as in Theorem 16.10 since, by step 2, it cannot even be so on ω.

Since Theorem 16.10 assumes that the support of f is compact, we have to check this here. More precisely, let us show that

$$\operatorname{supp} f = \{a\}. \tag{16.31}$$

That is to say, by Definition 6.18 of support, that $\Omega \setminus \{a\}$ is the annihilation domain nihil f of f, namely, by Definition 6.17 of the latter, that

$$\Omega \setminus \{a\} \text{ is the largest open set on which } f = 0.$$

It is an open set, and f indeed is zero on it because, if $\operatorname{supp} \varphi \subset \Omega \setminus \{a\}$, then φ and all its derivatives are zero at a, and thus $\langle f, \varphi \rangle = 0$ according to (16.27). It is indeed the largest of these open sets, since f is not zero on all of Ω. □

It remains to check that inequality (16.30) is not satisfied by all test functions.

Lemma 16.13.– *Let Ω be a non-empty open subset of $\mathbb{R}^d$, $a \in \Omega$, $m \in \mathbb{N}$ and $c \in \mathbb{R}$. Then, there exists $\varphi \in \mathcal{D}(\Omega)$ such that*

$$|\partial_1^{m+1} \varphi(a)| > c \sup_{|\beta| \leq m} \sup_{x \in \Omega} |\partial^\beta \varphi(x)|. \blacksquare$$

Proof. Let $\phi \in \mathcal{D}(\mathbb{R}^d)$ be such that:

$$\partial_1^{m+1} \phi(0) \neq 0, \quad \phi(x) = 0 \text{ if } |x| \geq 1.$$

These properties are, for example, realized by $\phi(x) \stackrel{\text{def}}{=} \rho(2x_1 + 1/2, 2x_2, \ldots, 2x_d)$, where ρ is the function defined by (2.4), p. 23.

Moreover, let $t \geq 1$. We define $\varphi \in \mathcal{D}(\mathbb{R}^d)$ by

$$\varphi(x) \stackrel{\text{def}}{=} \phi(t(x - a)).$$

Then, $\partial^\beta \varphi(x) = t^{|\beta|} \partial^\beta \phi(t(x - a))$, so (since $|t|^{|\beta|} \leq t^m$ when $|\beta| \leq m$)

$$|\partial_1^{m+1} \varphi(a)| - c \sup_{|\beta| \leq m} \sup_{x \in \Omega} |\partial^\beta \varphi(x)| \geq t^m (t \partial_1^{m+1} \phi(0) - c \|\phi\|_{C_b^m(\mathbb{R}^d)}).$$

For t large enough, the right-hand side is > 0 and the stated inequality is satisfied.

Moreover, (the restriction of) φ belongs to $\mathcal{D}(\Omega)$ as soon as t is large enough so that the ball $B(a, 1/t)$ is included in Ω. □

16.7. Extendability of distributions with values in a Banach space

Let us start with the extendability of a distribution on a bounded set. We show that such a distribution with values in a Banach space is extendable if and only if it is the derivative of a uniformly continuous function [7]. For this, we denote

$$\mathbf{C}(\Omega; E) \stackrel{\text{def}}{=} \{\text{uniformly continuous functions from } \Omega \text{ into } E\}.$$

Theorem 16.14.– *Let $f \in \mathcal{D}'(\Omega; E)$, where*

Ω is an open bounded subset of $\mathbb{R}^d$ and E is a Banach space.

Then,

$$f \text{ has an extension } \widetilde{f} \in \mathcal{D}'(\mathbb{R}^d; E)$$

if, and only if, there exist $g \in \mathbf{C}(\Omega; E)$ and $\beta \in \mathbb{N}^d$ such that

$$f = \partial^\beta g \quad \text{on } \Omega. \blacksquare$$

Example of a non-extendable distribution. The distribution associated with the continuous function $f \in \mathcal{C}((0,1))$ defined by

$$f(x) = e^{1/x}$$

does not have an extension $\widetilde{f} \in \mathcal{D}'(\mathbb{R})$. Otherwise, since every real-valued distribution is locally the finite order derivative of a continuous function (Theorem 16.11), the primitives of a certain order m of $\widetilde{f}$ would be continuous, and hence uniformly bounded on every bounded set. By restriction, the primitives of order m of f would be bounded on $(0, 1]$, which is not the case for any m. □

Proof of Theorem 16.14. **1.** *Necessity.* Suppose that

$$f \text{ has an extension } \widetilde{f} \in \mathcal{D}'(\mathbb{R}^d; E).$$

Since E is a Banach space, $\widetilde{f}$ is locally the derivative of a continuous function according to Theorem 16.11. In particular, since $\Omega \Subset \mathbb{R}^d$ (because Ω is bounded), there exist $h \in \mathcal{K}(\mathbb{R}^d; E)$ and $\beta \in \mathbb{N}^d$ such that

$$\widetilde{f} = \partial^\beta h \quad \text{on } \Omega.$$

Then, the restriction $g = h|_\Omega$ satisfies

$$g \in \mathbf{C}(\Omega; E), \quad f = \partial^\beta g \text{ on } \Omega.$$

7. **History of the extendability condition of distributions.** Jérôme LEMOINE and Jacques SIMON established Theorems 16.14 and 16.16 in 1996 [52, Theorem 3, p. 34 and Generalization 4, p. 35].

2. Sufficiency. Conversely, suppose that there are $g \in \mathbf{C}(\Omega; E)$ and $\beta \in \mathbb{N}^d$ such that

$$f = \partial^\beta g \ \text{ on } \Omega.$$

Then, according to the continuous extension theorem (Theorem A.34, since E is sequentially complete), the function g has a continuous extension $\overline{g}$ on $\overline{\Omega}$.

According to Dugandji's extension theorem (Theorem 16.15, afterwards), the function $\overline{g}$ has a continuous extension $\widetilde{g}$ on $\mathbb{R}^d$. Then, $\widetilde{f} = \partial^\beta \widetilde{g}$ is a distribution on $\mathbb{R}^d$ which extends f, since

$$\widetilde{f} = \partial^\beta \widetilde{g} = \partial^\beta g = f \ \text{ on } \Omega. \ \square$$

It remains to establish **Dugundji's extension theorem** [8].

Theorem 16.15.– *Every function $f \in \mathcal{C}(A; E)$, where A is a closed subset of $\mathbb{R}^d$ and E is a separated semi-normed space, has an extension $\widetilde{f} \in \mathcal{C}(\mathbb{R}^d; E)$.* ▮

Proof. **1.** *Construction of $\widetilde{f}$.* Denote $A^c = \mathbb{R}^d \setminus A$ and consider its cover by the open balls $B(z, r_z)$ as z spans it, where

$$r_z \overset{\text{def}}{=} \tfrac{1}{2} d(z, A), \quad d(z, A) \overset{\text{def}}{=} \inf_{y \in A} |z - y|.$$

Let $(\alpha_z)_{z \in A^c}$ be a subordinate partition of unity given by Theorem 2.21, and for each z, let a_z be a point in A such that

$$|a_z - z| = d(z, A) = 2r_z.$$

We define an extension $\widetilde{f}$ of f, for every $x \subset A^c$, by

$$\widetilde{f}(x) \overset{\text{def}}{=} \sum_{z \in A^c} \alpha_z(x) f(a_z),$$

where the sum is taken over the z for which $\alpha_z(x) \neq 0$, which are in finite number by Definition 2.20 of a partition of unity.

2. *Continuity on A^c.* Let $x \in A^c$. By hypothesis, A is closed, so A^c is open and hence contains a closed ball $B(x, r)$. Since this ball is compact, only a finite number of the α_z are not zero on it, again by Definition 2.20, and therefore $\widetilde{f}$ is a finite sum of continuous functions on it. In particular, $\widetilde{f}$ is continuous at the point x.

---

8. **History of Dugundji's extension theorem.** James DUGUNDJI established Theorem 16.14 in 1951 [29, Theorem 4.1, p. 357], in the more general case where A is a closed subset of a metric space. Heinrich TIETZE had proven this result for real functions, for which it is called **Tietze's extension theorem**.

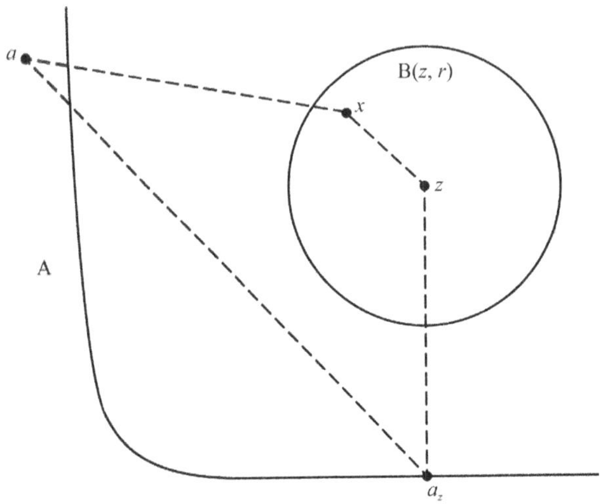

Figure 16.1. *Decomposition of* $a_z - a$

3. Continuity on A. Let $a \in A$, $\| \ \|_{E;\nu}$ be a semi-norm of E and $\epsilon > 0$. The function f is continuous on A by hypothesis, thus there is $\eta > 0$ such that

$$x \in A,\ |x-a| \leq \eta \quad \Rightarrow \quad \|f(x)-f(a)\|_{E;\nu} \leq \epsilon. \tag{16.32}$$

Now consider the case where

$$x \in A^{\mathrm{c}}. \tag{16.33}$$

Then, again by Definition 2.20, $\sum_{z\in A^{\mathrm{c}}} \alpha_z(x) = 1$ and so

$$\tilde{f}(x) - \tilde{f}(a) = \sum_{z\in A^{\mathrm{c}}} \alpha_z(x)(f(a_z) - f(a)).$$

We can restrict ourselves to the z for which $B(z, r_z)$ contains x because otherwise $\alpha_z(x) = 0$ (since, again by Definition 2.20, the support of α_z is included in that ball). Therefore,

$$\|\tilde{f}(x) - \tilde{f}(a)\|_{E;\nu} \leq \sup_{z\,:\,B(z,r_z)\ni x} \|f(a_z) - f(a)\|_{E;\nu}. \tag{16.34}$$

However (see Figure 16.1), if $x \in B(z, r_z)$, we have (since $|z - x| \leq r_z$)

$$2r_z = d(z, A) \leq |z-a| \leq |z-x| + |x-a| \leq r_z + |x-a|,$$

hence $r_z \leq |x - a|$; and thus (since $|a_z - z| = 2r_z$ and $|z - x| \leq r_z$),

$$|a_z - a| \leq |a_z - z| + |z-x| + |x-a| \leq 3r_z + |x-a| \leq 4|x-a|.$$

Thus,

$$x \in B(z, r_z),\ |x-a| \le \eta/4 \quad \Rightarrow \quad |a_z - a| \le \eta. \qquad (16.35)$$

Therefore, inequalities (16.34) and (16.35), along with hypothesis (16.33), give

$$x \in A^c,\ |x-a| \le \eta/4 \quad \Rightarrow \quad \|\widetilde{f}(x) - \widetilde{f}(a)\|_{E;\nu} \le \epsilon.$$

Which, with (16.32), proves that

$$\widetilde{f} \text{ is continuous at the point } a.$$

Hence, with step 2,

$$\widetilde{f} \in \mathcal{C}(\mathbb{R}^d; E). \ \square$$

Now we come to the extension of a distribution on an arbitrary open set. We show that a distribution with values in a Banach space is extendable if and only if, on every bounded set, it is the derivative of a uniformly continuous function.

Theorem 16.16.– *Let $f \in \mathcal{D}'(\Omega; E)$, where Ω is an open subset of $\mathbb{R}^d$ and*

E is a Banach space.

Then,

$$f \text{ has an extension } \widetilde{f} \in \mathcal{D}'(\mathbb{R}^d; E)$$

if and only if, for every bounded open subset ω of Ω, there exist $g \in \mathbf{C}(\omega; E)$ and $\beta \in \mathbb{N}^d$ such that

$$f = \partial^\beta g \ \text{ on } \omega. \ \blacksquare$$

Proof. **1.** *Necessity.* Suppose that

$$f \text{ has an extension } \widetilde{f} \in \mathcal{D}'(\mathbb{R}^d; E).$$

Since E is a Banach space, $\widetilde{f}$ is locally the derivative of a continuous function due to Theorem 16.11. More precisely, for every open bounded subset ω of Ω, since $\omega \Subset \mathbb{R}^d$, there exist $h \in \mathcal{K}(\mathbb{R}^d; E)$ and $\beta \in \mathbb{N}^d$ such that

$$\widetilde{f} = \partial^\beta h \ \text{ on } \omega.$$

Then, $g = h|_\omega$ belongs to $\mathbf{C}(\omega; E)$ and $f = \partial^\beta g$ on ω.

2. *Sufficiency.* Conversely, suppose that for every open bounded subset ω of Ω, there exist $g \in \mathbf{C}(\omega; E)$ and $\beta \in \mathbb{N}^d$ such that

$$f = \partial^\beta g \ \text{ on } \omega. \qquad (16.36)$$

Consider a partition of unity $(\alpha_n)_{n\in\mathbb{N}}$ subordinate to the cover of $\mathbb{R}^d$ by the open crowns $C_n = \{x \in \mathbb{R}^d : n-1 < |x| < n+1\}$. From hypothesis (16.36) applied to $\omega = \Omega \cap C_n$, there exist $g_n \in \mathbf{C}(\Omega \cap C_n; E)$ and $\beta_n \in \mathbb{N}^d$ such that

$$f = \partial^{\beta_n} g_n \quad \text{on } \Omega \cap C_n.$$

By the continuous extension theorem (Theorem A.34), Since E is sequentially complete, the function g_n has an extension

$$\overline{g_n} \in \mathbf{C}(\overline{\Omega \cap C_n}; E).$$

By Dugundji's theorem (Theorem 16.15) the function $\overline{g_n}$ has an extension

$$\widetilde{g_n} \in \mathbf{C}(\mathbb{R}^d; E).$$

Then, $\partial^{\beta_n}\widetilde{g_n} = f$ on $\Omega \cap C_n$. And so,

$$\alpha_n \partial^{\beta_n}\widetilde{g_n} = \alpha_n f \quad \text{on } \Omega \cap C_n.$$

This equality is also true on $\Omega \setminus \operatorname{supp} \alpha_n$, since α_n is zero there by Definition 2.20 of a partition of unity. It is therefore true on their union Ω, due to the gluing theorem for equalities (Theorem 6.10), i.e.

$$\alpha_n \partial^{\beta_n}\widetilde{g_n} = \alpha_n f \quad \text{on } \Omega. \tag{16.37}$$

We define a distribution $\widetilde{f} \in \mathcal{D}'(\mathbb{R}^d; E)$ by: for every $\varphi \in \mathcal{D}(\mathbb{R}^d)$,

$$\langle \widetilde{f}, \varphi \rangle_{\mathbb{R}^d} \overset{\text{def}}{=} \sum_{n\in\mathbb{N}} \langle \alpha_n \partial^{\beta_n}\widetilde{g_n}, \varphi \rangle_{\mathbb{R}^d}.$$

Indeed, this sum only contains a finite number of non-zero terms, since the support of φ only intersects a finite number of the crowns C_n.

It is an extension of f since, again by Definition 2.20 of a partition of unity, $\sum_{n\in\mathbb{N}} \alpha_n = 1$, and hence (16.37) gives

$$\widetilde{f} = \sum_{n\in\mathbb{N}} \alpha_n f = f \quad \text{on } \Omega. \ \square$$

Theorems 16.14 and 16.16 extend to certain spaces E which are not Banach, and more precisely to those for which every distribution is locally a finite sum of derivatives of continuous functions, see Note 4, p. 335.

Let us show that these results do not extend to every Neumann space E, nor even to every Fréchet space. More precisely, there exist Fréchet spaces for which certain distributions are extendable without locally being derivatives of continuous functions.

Theorem 16.17.– *In Theorems 16.14 and 16.16, we cannot replace "Banach space" with "Fréchet space" or with "Neumann space".* ∎

Proof. The distribution f defined in the proof of Theorem 16.12 by (16.27) satisfies

$$f \in \mathcal{D}'(\Omega; \mathcal{C}(\mathbb{R})).$$

Its support reduces to a single point a from (16.31) thus, according to Theorem 6.29,

$$f \text{ has an extension } \widetilde{f} \in \mathcal{D}'(\mathbb{R}^d; \mathcal{C}(\mathbb{R})).$$

However, as shown in (16.28), given an open ball ω containing a included in Ω,

$$\text{there are no } g \in \mathbf{C}_{\mathrm{b}}(\omega; E) \text{ and } \beta \in \mathbb{N}^d \text{ such that } f = \partial^\beta \text{ on } \omega.$$

(In fact, (16.28) states that f is not even in the form $\sum_n \partial^{\beta_n} g_n$ on ω, even if the g_n are not required to be uniformly continuous and bounded.)

This shows that the conditions in Theorems 16.14 et 16.16 for f to be extendable are not necessary when E is $\mathcal{C}(\mathbb{R})$, which is a Fréchet space (Theorem A.47) et thus a Neumann space (by Definition 1.4 of these spaces). □

16.8. Cancellation of distributions with values in a Banach space

Let us show that, for a distribution with values in a Banach space, $\langle f, \varphi\rangle$ cancels[9] as soon as φ and all of its derivatives cancel on the support of f.

Theorem 16.18.– *Let $f \in \mathcal{D}'(\Omega; E)$, where Ω is an open subset of $\mathbb{R}^d$ and*

$$E \text{ is a Banach space,}$$

and let $\varphi \in \mathcal{D}(\Omega)$ be such that, for every $\beta \in \mathbb{N}^d$,

$$\varphi = \partial^\beta \varphi = 0 \text{ on the support of } f. \tag{16.38}$$

Then,

$$\langle f, \varphi\rangle = 0_E. \ \square$$

9. **History of the cancellation condition from Theorem 16.18.** Laurent SCHWARTZ proved in 1950 [69, Chap. III, § 8, Theorem XXXIII, p. 98], for $f \in \mathcal{D}'(\mathbb{R}^d)$, that $\langle f, \varphi\rangle = 0$ as soon as φ and all of its derivatives cancel on the support of f. The present proof is modeled on his.

Proof. Denote S the support of φ and $S_r = S + B(0,r)$, where $r > 0$. Let r_0 be small enough so that the compact set $Q = S_{r_0}$ is included in Ω. Since f is locally of finite order (Theorem 16.2), there exist (Definition 16.1 (b)) $m \in \mathbb{N}$ and $c \in \mathbb{R}$ such that, for every $\phi \in \mathcal{C}_Q^\infty(\Omega)$,

$$\|\langle f, \phi \rangle\|_E \leq c\|\phi\|_{\mathcal{C}_b^m(\Omega)}. \tag{16.39}$$

Let θ_r be a function with values between 0 and 1, equal to 1 on $S_{r/2}$ and to 0 outside of $S_{3r/4}$. Its regular approximation $\alpha_r = \theta_r \diamond \rho_{r/4}$ has values between 0 and 1, equal to 1 on $S_{r/4}$ and to 0 outside of S_r. Then, for every $\beta \in \mathbb{N}^d$ such that $|\beta| \leq m$, we can bound [SCHWARTZ, 69, formula (III,7;18), p. 94], on all of $\mathbb{R}^d$,

$$|\partial^\beta(\theta_r \widetilde{\varphi})| \leq c\eta_r, \quad \text{where } \eta_r = \sup_{|\sigma| \leq m,\ x \in S_r} |\partial^\sigma \varphi(x)|$$

and where c does not depend on r.

When $r \to 0$, $\eta_r \to 0$ since φ and all of its derivatives $\partial^\sigma \varphi$ are zero on S_0, and therefore

$$\|\theta_r \varphi\|_{\mathcal{C}_b^m(\Omega)} \to 0.$$

However, $\theta_r \varphi \in \mathcal{C}_Q^\infty(\Omega)$ as soon as $r < r_0$, since $\alpha_r = 0$ outside of S_r. Hence, according to (16.39),

$$\langle f, \theta_r \varphi \rangle \to 0_E.$$

Finally, $\langle f, \theta_r \varphi \rangle = \langle f, \varphi \rangle$ since $\varphi = 1$ on the neighborhood $S_{r/4}$ of S. Therefore,

$$\langle f, \varphi \rangle = 0_E. \ \square$$

Cancellation of a finite order distribution with values in a Neumann space. When E is a Neumann space, in returning to the proof of Theorem 16.18, we prove that

$$\langle f, \varphi \rangle = 0_E$$

as soon as

$$f \textit{ is locally of finite order, and } \varphi = \partial^\beta \varphi = 0 \textit{ on the support of } f \textit{ for every } \beta \in \mathbb{N}^d, \tag{16.40}$$

or even

$$f \textit{ is of finite order } m, \textit{ and } \varphi = \partial^\beta \varphi = 0 \textit{ on the support of } f \textit{ for every } |\beta| \leq m. \tag{16.41}$$

$\square$

Comparison with the cancellation condition given by the definition of the support. According to Definitions 6.18 of the support and 6.17 of the annihilation domain, $\langle f, \varphi \rangle = 0_E$ as soon as

$$\varphi = 0 \text{ on an open set } \mathcal{O} \text{ containing the support of } f.$$

Which is equivalent, since $\mathcal{O}$ is open, to

$$\varphi = \partial^\beta \varphi = 0 \text{ on } \mathcal{O} \text{ for every } \beta \in \mathbb{N}^d.$$

On the contrary, conditions (16.38), (16.40) and (16.41) only impose that φ and its derivatives are zero on the support of f, not on one of its neighborhoods $\mathcal{O}$. $\square$

Appendix

Reminders

We give here the precise statements of the definitions [1] and results from Volumes 1, *Banach, Fréchet, Hilbert and Neumann spaces* [81], and 2, *Continuous functions* [82], in the series *Analysis for PDEs* that are used in this book, in order to keep it self-contained. Proofs are provided in the said volumes, to which we refer by [Vol. 1, Theorem 12.2] for example. Some statements are limited to simpler cases when that is all we will need, whereas others are complete and proven.

A.1. Notation and numbering

Set theory notation. We use the standard notation (it is presented in § 1.1 of Volume 1). We only recall the difference in notation between the set $\{a, b, \ldots, z\}$ and the ordered set $(a, b, \ldots, z)$. Thus, $(a, b) \neq (b, a)$ if $a \neq b$ whereas we always have $\{a, b\} = \{b, a\}$.

Countability. We assume that the sets $\mathbb{N}$ of natural numbers [2] and $\mathbb{Z}$ of integers, respectively

$$\mathbb{N} \stackrel{\text{def}}{=} \{0, 1, 2, \ldots\}, \quad \mathbb{Z} \stackrel{\text{def}}{=} \{\ldots, -2, -1, 0, 1, 2, \ldots\},$$

and the set $\mathbb{Q}$ of rational numbers are familiar, as well as their addition, subtraction, multiplication, absolute values and orders. We denote $\mathbb{N}^* \stackrel{\text{def}}{=} \mathbb{N} \setminus \{0\}$ and

$$[\![m, n]\!] \stackrel{\text{def}}{=} \{i \in \mathbb{N} : m \leq i \leq n\}, \quad [\![m, \infty]\!] \stackrel{\text{def}}{=} \{i \in \mathbb{N} : i \geq m\} \cup \{\infty\}.$$

A **sequence** in a set U is the data, for every $n \in \mathbb{N}$, of an element $u_n \in U$. It is an ordered set that we denote by $(u_n)_{n \in \mathbb{N}}$.

A **subsequence** of $(u_n)_{n \in \mathbb{N}}$ is any sequence of the form $(u_{\sigma(n)})_{n \in \mathbb{N}}$ such that $\sigma(n) \in \mathbb{N}$ and $\sigma(n+1) > \sigma(n)$ for every $n \in \mathbb{N}$.

Let us define countable sets.

1. **History.** Historical notes for many of the definitions and properties recalled here may be found in Volumes 1 and 2.

2. **CAUTION.** Our notations $\mathbb{N} \stackrel{\text{def}}{=} \{0, 1, 2, \ldots\}$ and $\mathbb{N}^* \stackrel{\text{def}}{=} \{1, 2, \ldots\}$ conform to the standard ISO 80000-2 for mathematical and physics notation (edited in 2009).
Any possible confusion will be of no consequence, apart from surprising readers used to the opposite notation when seeing a term u_0 of a series indexed by $\mathbb{N}$, or the inverse $1/n$ of a number n in $\mathbb{N}^*$.

Definition A.1.– *A set U is **countable** if there exists a bijection from U onto a subset of $\mathbb{N}$.* □

Let us state some properties of countability [SCHWARTZ, 74, p. 104 to 108].

Theorem A.2.– *Any subset of a countable set is countable.*

Any finite product of countable sets is countable.. □

Real numbers. We assume that the set $\mathbb{R}$ of real numbers is familiar, as well as its addition, subtraction, multiplication, absolute value and its order.

A **real interval** is any subset in one of the following forms [3], where a and b are real or infinite:

— open interval $(a,b) \stackrel{\text{def}}{=} \{x \in \mathbb{R} : a < x < b\}$;

— closed interval $[a,b] \stackrel{\text{def}}{=} \{x \in \mathbb{R} : a \leq x \leq b\}$;

— semi-open interval $(a,b] \stackrel{\text{def}}{=} \{x \in \mathbb{R} : a < x \leq b\}$ and $[a,b) \stackrel{\text{def}}{=} \{x \in \mathbb{R} : a \leq x < b\}$.

We say that $m \in \mathbb{R}$ is an **upper bound** of a subset V of $\mathbb{R}$ if every $v \in V$ satisfies $v \leq m$.

We say that m is the **supremum** (or least upper bound) of V if m is an upper bound of V and if every other upper bound v of V satisfies $m \leq v$. If it exists, it is unique and is denoted by $\sup V$.

The **lower bound** and **infimum** are the elements respectively obtained by replacing $\leq$ with $\geq$ in the above. If it exists, the infimum is denoted by $\inf V$.

Let us define convergent or increasing real sequences.

Definition A.3.– *We say that a sequence $(x_n)_{n\in\mathbb{N}}$ in $\mathbb{R}$ **converges** to a **limit** $x \in \mathbb{R}$ if, for every $\epsilon > 0$, there exists $m \in \mathbb{N}$ such that: $n \geq m$ implies $|x_n - x| \leq \epsilon$. If so, we denote $x_n \to x$.*

*We say that $(x_n)_{n\in\mathbb{N}}$ **tends to infinity** if, for every $c \in \mathbb{R}$, there exists $m \in \mathbb{N}$ such that: $n \geq m$ implies $x_n \geq c$. If so, we denote $x_n \to \infty$.*

*We say that $(x_n)_{n\in\mathbb{N}}$ is **increasing** if $x_{n+1} \geq x_n$.* □

A.2. Semi-normed spaces

Vector spaces and semi-norms

Vector spaces. A **vector space** is a set E endowed with an **addition** from $E \times E$ into E, denoted by $+$, with a **multiplication** from $\mathbb{R} \times E$ into E and with an element $0_E \in E$ such that, for every u, v and w in E and s and t in $\mathbb{R}$:

$$(u+v)+w = u+(v+w), \quad u+v = v+u, \quad u+0_E = u,$$
$$t(u+v) = tu+tv, \quad (s+t)u = su+tu, \quad s(tu) = (st)u, \quad 1u = u,$$
$$\text{there exists an element } -u \in E \text{ such that } -u+u = 0_E.$$

A **vector subspace** of E is a subset E which is a vector space for its addition and multiplication.

Norms and semi-norms. A **semi-norm** is a mapping p from a vector space E into $\mathbb{R}$ such that, for every $u \in E$, $v \in E$ and $t \in \mathbb{R}$:

$$p(u) \geq 0, \quad p(tu) = |t|\,p(u), \quad p(u+v) \leq p(u)+p(v).$$

A **norm** is a semi-norm p such that $p(u) > 0$ if $u \neq 0_E$.

---

3. **French notation.** French authors denote the open interval by]a,b[. They denote the closed interval by [a,b], like we do, and furthermore the half-open intervals by]a,b] and [a,b[.

The upper envelope of semi-norms is a semi-norm if it is everywhere finite [Vol. 1, Theorem 12.2]:

Theorem A.4.– *Let $\mathcal{P}$ be a family of semi-norms on a vector space E such that, for every $u \in E$, $\sup_{p\in\mathcal{P}} p(u) < \infty$. Then, the mapping $u \mapsto \sup_{p\in\mathcal{P}} p(u)$ is a semi-norm on E.* □

Sequences, open and closed sets, density, connectedness

Convergent sequences. Let us give some properties of sequences [Vol. 1, Theorems 2.7 and 2.8].

Theorem A.5.– *Every convergent sequence in a separated semi-normed space is bounded, Cauchy and has a unique limit.*

Every Cauchy sequence is bounded.

Every subsequence of a sequence converges, to the same limit. □

Let us enumerate some properties preserved by topological inclusion [Vol. 1, Theorem 3.8].

Theorem A.6.– *Let E and F be two separated semi-normed spaces such that $E \subsetneqq F$, $(u_n)_{n\in\mathbb{N}}$ a sequence in E and $u \in E$. If $(u_n)_{n\in\mathbb{N}}$ has one of the following properties in E, it has it in F too:*

converges, converges to u, is Cauchy, is bounded. □

Open and closed sets. Let us define the open and closed sets.

Definition A.7.– *Let U be a subset of a separated semi-normed space E.*

(a) *We say that U is **open** if, for every $u \in U$, there exist a finite subset N of $\mathcal{N}_E$ and $\eta > 0$ such that: $v \in E$ and $\sup_{\nu\in N} \|v - u\|_{E;\nu} \le \eta$ imply $v \in U$.*

(b) *We say that U is **closed** if $E \setminus U$ is open.*

(c) *We say that U is **sequentially closed** if every sequence in U that converges in E has its limit in U.* □

Let us define the interior and closure [Vol. 1, Definition 2.15 and its justification].

Definition A.8.– *Let U be a subset of a separated semi-normed space.*

*The **interior** of U the largest open set included in U; it is denoted by $\mathring{U}$.*

*The **closure** of U the smallest closed set containing U; it is denoted by $\overline{U}$.*

*The **boundary** of U the set $\overline{U} \setminus \mathring{U}$; we denote it by ∂U.* □

Let us characterize the closure [Vol. 1, Theorem 2.17].

Theorem A.9.– *For every subset U of a separated semi-normed space E,*

$$\overline{U} = \Big\{u \in E : \forall N \subset \mathcal{N}_E \text{ finite}, \forall \eta > 0, \exists v \in U \text{ such that } \sup_{\nu\in N} \|v - u\|_{E;\nu} \le \eta\Big\}. \ \square$$

Closed implies sequentially closed [Vol. 1, Theorem 2.10]:

Theorem A.10.– *Every closed subset of a separated semi-normed space is sequentially closed.* □

Let us state some properties of unions and intersections [Vol. 1, Theorems 2.11 and 2.12].

Theorem A.11.– *In a separated semi-normed space, every (finite or infinite) union and every finite intersection of open sets is open.*

Every finite union and every intersection of closed sets is closed.

Every finite union and every intersection of bounded sets is bounded. □

Density and separability. Let us define the notions of density and separability.

Definition A.12.– *Let U and V be a pair of subsets of the same separated semi-normed space.*

The set V is ***dense*** *in U if $V \subset U \subset \overline{V}$.*

The set V is ***sequentially dense*** *in U if $V \subset U$ and every $u \in U$ is the limit of a sequence in V.*

The set U is ***sequentially separable*** *if it containts a countable subset V that is sequentially dense in U.* □

Let us compare dense with sequentially dense [Vol. 1, Theorem 2.20].

Theorem A.13.– *In a separated semi-normed space, every set that is sequentially dense in another set is dense in it.* □

Connectedness. Let us state some properties of connceted components [Vol. 1, Theorem 2.34].

Theorem A.14.– *In a separated semi-normed space, every set is the union of its connected components.*

The connected components of a set U are connected and pairwise disjoint (or equal), and they are open if U is open. □

Recall that any convex set, and hence any ball, is connected [Vol. 1, Theorem 2.33].

Theorem A.15.– *In a separated semi-normed space, every convex set is connected.* □

Now we come to path connectedness [Vol. 2, Theorem 8.5].

Theorem A.16.– *In a separated semi-normed space, any pair of points of a connected open set U can be connected by a C^1 path in U.* □

Conversely, two points linked by a path belong to the same connected component [Vol. 2, Theorem 8.6]:

Theorem A.17.– *In a separated semi-normed space, two points that are linked by a path in a set U belong to the same connected component of U.* □

Compactness

Definitions. Let us begin by notions of compactness related to covering properties.

A **cover** of a set U is a family of sets whose union contains U. It is called **open** if it is made up of open sets, and **finite** if it has finitely many elements.

A **subcover** is a subfamily which itself constitutes a cover of U.

Definition A.18.– *Let U be a subset of a separated semi-normed space E.*

(a) *We say that U is **compact** if every open cover of U has a finite subcover.*

(b) *We say that U is **relatively compact** if $\overline{U}$ is compact.* □

Now, we come to notions of compactness related to convergence of subsequences.

Definition A.19.– *Let U be a subset of a separated semi-normed space E.*

(a) *We say that U is **sequentially compact** if every sequence in U has a convergent subsequence whose limit belongs to U.*

(b) *We say that U is **relatively sequentially compact** if every sequence in U has a convergent subsequence (in E).* □

Properties. Recall that, in a general separated semi-normed space, sequential compactness is neither stronger nor weaker than compactness [Vol. 1, Properties (2.6) and (2.7), p. 27].

Every relatively sequentially compact set is bounded [Vol. 1, Theorem 2.25], that may be expressed as follows.

Theorem A.20.– *In a separated semi-normed space, every set in which every sequence has a convergent subsequence is bounded.* □

Let us state the **separation theorem** for a closed set and a compact set [Vol. 1, Theorem 3.19].

Theorem A.21.– *Let U be a closed set and K a compact set of a normed space E such that $U \cap K = \emptyset$. Then, there exists a ball $B = \{u \in E : \|u\|_E \leq r\}$, where $r > 0$, such that $(U + B) \cap K = \emptyset$.* □

Let us deduce a **strong inclusion theorem** for a compact subset of an open set.

Theorem A.22.– *Let U be an open subset and K a compact subset of a normed space E such that $K \subset U$. Then, there exists a ball $B = \{u \in E : \|u\|_E \leq r\}$, where $r > 0$, such that $K + B \subset U$.* □

Proof. Since the set $E \setminus U$ is closed, the separation theorem (Theorem A.21) provides $r > 0$ such that $((E \setminus U) + B) \cap K = \emptyset$, which is equivalent to $(E \setminus U) \cap (K + B) = \emptyset$ and thus to $K + B \subset U$. □

Metrizable, Fréchet and Neumann spaces, space $\mathbb{R}^d$

Metrizable spaces. In a metrizable space, sequential properties coincide with topological properties. Among others [Vol. 1, Theorems 4.6 and 4.7]:

Theorem A.23.– *In a metrizable semi-normed space:*

(a) *Closed is equivalent to sequentially closed, and relatively compact is equivalent to relatively sequentially compact.*

(b) *The closure $\overline{U}$ of a set U coincides with the set of limits of its convergent sequences.*

(c) *Every subset of a sequentially separable set is sequentially separable..* □

Fréchet and Neumann spaces. Every closed subspace is sequentially complete [Vol. 1, Theorem 4.12]:

Theorem A.24.– *Every closed, or more generally sequentially closed, topological vector subspace of a Neumann space is a Neumann space.* □

A product of Fréchet spaces is a Fréchet space [Vol. 1, Theorem 6.7]:

Theorem A.25.– *Every product $E_1 \times \cdots \times E_d$ of Fréchet spaces is a Fréchet space.* □

The space $\mathbb{R}^d$. The space

$$\mathbb{R}^d \stackrel{\text{def}}{=} \{x = (x_1, \ldots, x_d) : \forall i,\ x_i \in \mathbb{R}\}$$

is endowed with the **Euclidean norm** and with the scaler product, defined respectively by

$$|x| \stackrel{\text{def}}{=} (x_1^2 + \cdots + x_d^2)^{1/2}, \quad x \cdot y \stackrel{\text{def}}{=} x_1 y_1 + \cdots + x_d y_d.$$

They satisfy the **Cauchy–Schwarz inequality** [Vol. 1, Theorem 5.15]: for every $x \in \mathbb{R}^d$ and $y \in \mathbb{R}^d$,

$$|x \cdot y| \leq |x||y|. \qquad (A.1)$$

Let us give some properties of $\mathbb{R}^d$, and hence of $\mathbb{R}$, respectively established:
— for (a) [Vol. 1, Theorems 4.15 and 5.12];
— for (b), which is the **Borel–Lebesgue theorem**, [Vol. 1, Theorem 5.13];
— for (c), which is the **Bolzano–Weierstrass theorem**, [Vol. 1, Theorems 4.17 and 5.14].

Theorem A.26.– (a) *$\mathbb{R}^d$ is a sequentially separable Banach space.*

(b) *In $\mathbb{R}^d$, a set is compact if and only if it is closed and bounded.*

(c) *Every bounded sequence in $\mathbb{R}^d$ has a convergent subsequence.* □

Observe that the sum of compact sets is compact.

Theorem A.27.– *The sum $U + V$ of two compact subsets of $\mathbb{R}^d$ is compact.* □

Proof. Since they are compact, U and V are bounded (Theorem A.26 (b)), so $U + V$ is bounded too. Since in addition U is closed (Theorem A.26 (b), again), $U + V$ is closed [Vol. 1, Theorem 3.20] and therefore compact (Theorem A.26 (b), once again). □

A.3. Continuous mappings, duality

Continuous mappings

Mappings. A **mapping** from a set X into a set Y is the assignment to each element $u \in X$ of an element $T(u) \in Y$.

We say that $T(u)$ is the **image** of u under T. The **image** of a subset U of X under T is the set

$$T(U) \stackrel{\text{def}}{=} \{T(u) : u \in U\}.$$

The mapping T is **injective** if $T(u) = T(v)$ implies $v = u$; it is **surjective** if $T(X) = Y$, and that it is **bijective** if it is injective and surjective. We also say that T is an **injection**, **surjection** or **bijection**. A bijection is called **bicontinuous** if it is continuous as well as its inverse mapping.

The **preimage** of T of a subset V of Y is the set

$$T^{-1}(V) \stackrel{\text{def}}{=} \{u \in X : T(u) \in V\}.$$

If T is bijective, there exists a unique mapping T^{-1} from Y into X, called the **inverse** of T, such that $T^{-1}(T(u)) = u$ for every $u \in X$. Thus, T is the inverse mapping of T^{-1}, because $T(T^{-1}(v)) = v$.

If T is a mapping from X into Y and S is a mapping from Y into G, we denote by $S \circ T$ or $S(T)$ the **composite mapping** defined from X into G by

$$(S \circ T)(u) \stackrel{\text{def}}{=} S(T(u)).$$

Properties of continuity. Let us give a characterization of continuous mappings [Vol. 1, Theorem 8.2].

Theorem A.28.– *A mapping T from an open subset of a separated semi-normed space E into a separated semi-normed space F is continuous if and only if, for every open subset V of F, the set $T^{-1}(V)$ is open in E.* □

Now let us change the space of values [Vol. 1, Theorem 7.6].

Theorem A.29.– *Let E, F_1 and F_2 be separated semi-normed spaces such that $F_1 \hookrightarrow F_2$, and let $X \subset E$. Then, every continuous or sequentially continuous mapping from X into F_1 has the same property from X into F_2.* □

Let us give properties of continuity of composite mappings [Vol. 1, Theorem 7.9].

Theorem A.30.– *Let T be a mapping from a subset of a separated semi-normed space into a subset Y of a separated semi-normed space, and S a mapping from Y into a separated semi-normed space.*

If S and T are continuous, or sequentially continuous, then so is $S \circ T$. □

Mappings on compact or connected sets. Let us give properties of images [Vol. 1, Theorem 8.7 and 8.6].

Theorem A.31.– *Let T be a continuous mapping from a subset X of a separated semi-normed space into a separated semi-normed space, and $U \subset X$.*

If U is compact, then $T(U)$ is compact.

If U is connected, then $T(U)$ is connected. □

Let us state **Heine's theorem** [Vol. 1, Theorem 8.10].

Theorem A.32.– *Every continuous mapping from a compact subset of a separated semi-normed space into a separated semi-normed space is uniformly continuous and bounded.* □

This theorem implies the following property [Vol. 2, Theorem 1.19].

Theorem A.33.– *Every continuous function with compact support from an open subset of $\mathbb{R}^d$ into a separated semi-normed space is uniformly continuous and bounded.* □

Continuous extension. Let us state the **continuous extension theorem** [Vol. 1, Theorem 10.3].

Theorem A.34.– *Let T be a uniformly continuous mapping from a subset X of a separated semi-normed space E with values in a Neumann space F, and let $V \subset E$ be such that X is sequentially dense in V. Then, T has a unique uniformly continuous extension from V into F.* □

Linear or multilinear mappings

Definitions. Let us define linearity and multilinearity.

Definition A.35.– *A mapping L from a vector space E into another is said to be **linear** if, for every $u \in E$, $v \in E$ and $t \in \mathbb{R}$,*

$$L(u+v) = L(u) + L(v), \quad L(tu) = tL(u).$$

*A mapping T defined on a product of vector spaces is said to be **multilinear** if the partial mappings $u_i \mapsto T(u_1, \dots, u_d)$ are linear. When $d = 2$, we say that T is **bilinear**.* □

An **isomorphism** from a separated semi-normed space onto another is a bicontinuous linear bijection from one onto the other.

Let us define the space of continuous linear mappings.

Definition A.36.– *We denote by $\mathcal{L}(E; F)$ the vector space of continuous linear mappings from a separated semi-normed space E into another F endowed with the semi-norms, indexed by the bounded subsets B of E and by $\mu \in \mathcal{N}_F$,*

$$\|L\|_{\mathcal{L}(E;F);B,\mu} \stackrel{\text{def}}{=} \sup_{u \in B} \|Lu\|_{F;\mu}. \ \square$$

Properties. Let us start with a property of linear mappings [Vol. 1, Theorem 7.15].

Theorem A.37.– *Every continuous or sequentially continuous linear mapping from a separated semi-normed space into another transforms bounded sets into bounded sets.* □

Let us now characterize continuous multilinear mappings [Vol. 1, Theorem 7.20].

Theorem A.38.– *Let T be a multilinear mapping from a product $E_1 \times \cdots \times E_d$ of separated semi-normed spaces into a separated semi-normed space F. Then:*

(a) *T is continuous if and only if, for every $\mu \in \mathcal{N}_F$, there exist finite subsets N_1 of $\mathcal{N}_{E_1}$, ..., N_d of $\mathcal{N}_{E_d}$ and $c \geq 0$ such that: for every $(u_1, \ldots, u_d) \in E_1 \times \cdots \times E_d$,*

$$\|T(u_1, \ldots, u_d)\|_{F;\mu} \leq c \sup_{\nu_1 \in N_1} \|u_1\|_{E_1;\nu_1} \cdots \sup_{\nu_d \in N_d} \|u_d\|_{E_d;\nu_d}.$$

(b) *If the family of semi-norms of each E_i is filtering, T is continuous if and only if, for every $\mu \in \mathcal{N}_F$, there exist $\nu_1 \in \mathcal{N}_{E_1}, \ldots, \nu_d \in \mathcal{N}_{E_d}$ and $c \geq 0$ such that: for every $(u_1, \ldots, u_d) \in E_1 \times \cdots \times E_d$,*

$$\|T(u_1, \ldots, u_d)\|_{F;\mu} \leq c\|u_1\|_{E_1;\nu_1} \cdots \|u_d\|_{E_d;\nu_d}. \ \square$$

Banach–Steinhaus theorem. This theorem says that any simply bounded set of continuous linear mappings is equicontinuous. In terms of semi-norms, it is expressed as follows [Vol. 1, Theorem 10.11].

Theorem A.39.– *Let $\mathcal{T}$ be a set of continuous linear mappings from a Fréchet space F into a separated semi-normed space E such that, for every $u \in F$ and $\nu \in \mathcal{N}_E$,*

$$\sup_{L \in \mathcal{T}} \|L(u)\|_{E;\nu} < \infty.$$

Then, for every $\nu \in \mathcal{N}_E$, there exist a finite subset M of $\mathcal{N}_F$ and $c \in \mathbb{R}$ such that, for every $L \in \mathcal{T}$ and $u \in E$,

$$\|L(u)\|_{E;\nu} \leq c \sup_{\mu \in M} \|u\|_{F;\mu}. \ \square$$

Let us state an analogous result for bilinear mappings [Vol. 1, Theorem 10.12].

Theorem A.40.– *Let $\mathcal{T}$ be a set of continuous bilinear mappings from a product $F_1 \times F_2$ of Fréchet spaces into a separated semi-normed space E be such that, for every $u_1 \in F_1$, $u_2 \in F_2$ and $\nu \in \mathcal{N}_E$,*

$$\sup_{T \in \mathcal{T}} \|T(u_1, u_2)\|_{E;\nu} < \infty.$$

Then, for every $\nu \in \mathcal{N}_E$, there exist $c \in \mathbb{R}$ and finite subsets M_1 of $\mathcal{N}_{F_1}$ and M_2 of $\mathcal{N}_{F_2}$ such that, for every $T \in \mathcal{T}$, $u_1 \in F_1$ and $u_2 \in F_2$,

$$\|T(u_1, u_2)\|_{E;\nu} \leq c \sup_{\mu_1 \in M_1} \|u_1\|_{F_1;\mu_1} \sup_{\mu_2 \in M_2} \|u_2\|_{F_2;\mu_2}. \ \square$$

Weak topology and extractability

Weakly bounded or closed sets. Let us state the **Banach–Mackey** theorem [Vol. 1, Theorem 16.1].

Theorem A.41.– *A set is bounded in a separated semi-normed space E if and only if it is bounded in E-weak.* $\square$

Let us state **Mazur's theorem** [Vol. 1, Theorem 16.4].

Theorem A.42.– *A convex subset of a separated semi-normed space E is closed in E if and only if it is closed in E-weak.* $\square$

Extractability. Let us define extractability.

Definition A.43.– *A separated semi-normed space E is said to be* ***extractable*** *if every bounded sequence in E has a subsequence that converges in E-weak.* $\square$

Any extractable space is "weakly" sequentially complete [Vol. 1, Theorem 18.15].

Theorem A.44.– *If a separated semi-normed space E is extractable, then E-weak is a Neumann space.*

$\square$

A.4. Continuous or differentiable functions

Continuous functions

Cancellation. A continuous function is zero outside of the interior of its support [Vol. 2, Theorem 1.6 (b)]:

Theorem A.45.– *Let f be a continuous function from an open subset Ω of $\mathbb{R}^d$ into a separated semi-normed space E. Then, $f = 0_E$ on $\Omega \setminus \overset{\circ}{\operatorname{supp}} f$.* □

Sequential completion. Let us state properties of sequential completion of spaces of continuous functions [Vol. 2, Theorem 1.12].

Theorem A.46.– *Let $\Omega \subset \mathbb{R}^d$ and E be a Neumann space. Then, $\mathcal{C}(\Omega; E)$, $\mathcal{C}_b(\Omega; E)$ and $\mathbf{C}_b(\Omega; E)$ are Neumann spaces.* □

Observe that $\mathcal{C}(\Omega)$ is a Fréchet space if Ω is open [Vol. 2, Theorem 1.16 (d)].

Theorem A.47.– *Let Ω be an open subset of $\mathbb{R}^d$. Then, $\mathcal{C}(\Omega)$ is a Fréchet space.* □

Image under a linear mapping or a product. Let us start with the image under a linear mapping [Vol. 2, Theorem 3.2].

Theorem A.48.– *Let $f \in \mathcal{C}(\Omega; E)$ and $L \in \mathcal{L}(E; F)$, where Ω is an open subset of $\mathbb{R}^d$ and where E and F are separated semi-normed spaces. Then, $Lf \in \mathcal{C}(\Omega; F)$.* □

Let us give a continuity property of the product by a function.

Theorem A.49.– *Let φ be a continuous real function with compact support on an open subset Ω of $\mathbb{R}^d$, and E a Neumann space. Then, the mapping $f \mapsto f\varphi$ is continuous linear from $\mathcal{C}(\Omega; E)$ into $\mathbf{C}_b(\Omega; E)$.* □

Proof. This mapping is continuous from $\mathcal{C}(\Omega; E)$ into $\mathcal{K}(\Omega; E)$ due to Theorem 3.11 (d) from Volume 2. Since $\mathcal{K}(\Omega; E) \subsetneq \mathbf{C}_b(\Omega; E)$ according to Theorem 1.19 (d) from Volume 2, the stated continuity follows from Theorem A.29 of the current volume. □

Separation of variables. Note that the separation of variables is an isomorphism in the spaces of uniformly continuous functions with bounded supports [Vol. 2, Theorem 1.27 (c)]:

Theorem A.50.– *Let $\Omega_1 \subset \mathbb{R}^{d_1}$, $\Omega_2 \subset \mathbb{R}^{d_2}$, D_1 be a compact subset of $\mathbb{R}^{d_1}$, D_2 a compact subset of $\mathbb{R}^{d_2}$ and E a separated semi-normed space. For every function f defined on $\Omega_1 \times \Omega_2$, let $\underline{f}$ be the function obtained by separating the variables, i.e.*

$$(\underline{f}(x_1))(x_2) \overset{\text{def}}{=} f(x_1, x_2).$$

Then, the mapping $f \mapsto \underline{f}$ is an isomorphism from $\mathbf{C}_{D_1 \times D_2}(\Omega_1 \times \Omega_2; E)$ onto $\mathbf{C}_{D_1}(\Omega_1; \mathbf{C}_{D_2}(\Omega_2; E))$. □

Differentiable functions

Spaces of differentiable functions. Let us compare the topologies of some spaces [Vol. 2, Theorems 2.16 and 2.18].

Theorem A.51.– *Let Ω be an open subset of $\mathbb{R}^d$, E a separated semi-normed space, K a compact subset of Ω and $m \in [\![0, \infty]\!]$. Then:*

(a)
$$\mathcal{C}_K^m(\Omega; E) \subsetneq \mathbf{C}_{\mathrm{b}}^m(\Omega; E) \subsetneq \mathcal{C}_{\mathrm{b}}^m(\Omega; E) \subsetneq \mathcal{C}^m(\Omega; E).$$

(b) *The topologies of $\mathcal{C}^m(\Omega; E)$, $\mathcal{C}_{\mathrm{b}}^m(\Omega; E)$, $\mathcal{C}_K^m(\Omega; E)$ and $\mathbf{C}_{\mathrm{b}}^m(\Omega; E)$ coincide on $\mathcal{C}_K^m(\Omega; E)$.* □

Let us give filtration properties [Vol. 2, Theorem 2.22].

Theorem A.52.– *The families of semi-norms of $\mathcal{C}^m(\Omega)$, $\mathcal{C}_{\mathrm{b}}^m(\Omega)$, $\mathcal{C}_K^m(\Omega)$, where Ω is an open subset of $\mathbb{R}^d$, $0 \leq m \leq \infty$ and K is a compact subset of Ω, are filtering.* □

We give some properties of sequential completion [Vol. 2, Theorem 2.24 (f)].

Theorem A.53.– *Let Ω be an open subset of $\mathbb{R}^d$ and K be a compact subset of Ω. Then, $\mathcal{C}_{\mathrm{b}}^\infty(\Omega)$ and $\mathcal{C}_K^\infty(\Omega)$ are Fréchet spaces.* □

Properties of differentiable functions. Let us start with continuity [Vol. 2, Theorem 2.3].

Theorem A.54.– *Every function from an open subset of $\mathbb{R}^d$ into a separated semi-normed space which is differentiable is continuous.* □

Let us relate differentiability to the existence of partial derivatives [Vol. 2, Theorems 2.10 and 2.13].

Theorem A.55.– *Every function f, from an open subset of $\mathbb{R}^d$ into a separated semi-normed space, which has continuous partial derivatives $\partial_i f$ for every $i \in [\![1, d]\!]$ is continuous and continuously differentiable.*

If f has continuous partial derivatives $\partial^\beta f$ for every $\beta \in \mathbb{N}^d$ such that $|\beta| \leq m$, it is m times continuous differentiable. □

Let us give a differentiability property of a limit [Vol. 2, Theorem 2.23].

Theorem A.56.– *Let f_n, f and g_i be functions in $\mathcal{C}(\Omega; E)$, where Ω is an open subset of $\mathbb{R}^d$ and E is a separated semi-normed space, such that, for every $n \in \mathbb{N}$, $n \to \infty$, and every $i \in [\![1, d]\!]$,*

$$f_n \in \mathcal{C}^1(\Omega; E), \quad f_n \to f \text{ in } \mathcal{C}(\Omega; E), \quad \partial_i f_n \to g_i \text{ in } \mathcal{C}(\Omega; E).$$

Then,

$$f \in \mathcal{C}^1(\Omega; E), \quad \partial_i f = g_i, \quad f_n \to f \text{ in } \mathcal{C}^1(\Omega; E). \ \square$$

Recall that a differentiable function on an interval with a point removed is differentiable on the entire interval if itself and its derivative have continuous extensions [Vol. 2, Theorem 2.28].

Theorem A.57.– *Let f and g be two functions in $\mathcal{C}((a, b); E)$, where (a, b) is an open interval of $\mathbb{R}$ and E is a separated semi-normed space, and let $X = (a, c) \cup (c, b)$, where $c \in (a, b)$. If*

$$f|_X \in \mathcal{C}^1(X; E) \text{ and } (f|_X)' = g|_X,$$

then

$$f \in \mathcal{C}^1((a, b); E) \text{ and } f' = g. \ \square$$

Image under a linear or multilinear mapping

Image under a linear mapping. Let us change the space of values of a differentiable mapping [Vol. 1, Theorem 19.10].

Theorem A.58.– *If a mapping from an open subset of a separated semi-normed space is m times differentiable with values in separated semi-normed space F_1, it is m times differentiable with values in the separated semi-normed space F_2 as soon as $F_1 \subsetneq F_2$.* □

Let us differentiate the image under a continuous linear mapping [Vol. 2, Theorem 3.1].

Theorem A.59.– *Let f be an m times differentiable function from an open subset Ω of $\mathbb{R}^d$ into a separated semi-normed space E, where $1 \leq m \leq \infty$, and L a continuous linear mapping from E into a separated semi-normed space F. Then, Lf is m times differentiable from Ω into F and, for every $i \in [\![1, d]\!]$,*

$$\partial_i(Lf) = L(\partial_i f). \ \square$$

Leibniz's formula. Let us state **Leibniz's formula** for the derivative of a product of functions [Vol. 2, Theorems 3.5 and 3.6].

Theorem A.60.– *Let $f_1 \in \mathcal{C}^m(\Omega; E)$ and $f_2 \in \mathcal{C}^m(\Omega)$, where Ω is an open subset of $\mathbb{R}^d$, E is a separated semi-normed space and $0 \leq m \leq \infty$. Then, $f_1 f_2 \in \mathcal{C}^m(\Omega; E)$ and, if $m \geq 1$, for every $j \in [\![1, d]\!]$,*

$$\partial_j(f_1 f_2) = f_1 \partial_j f_2 + f_2 \partial_j f_1. \ \square$$

Let us state a dual formula to Leibniz's formula [Vol. 2, Theorem 3.8].

Theorem A.61.– *Let $f_1 \in \mathcal{C}^m(\Omega)$ and $f_2 \in \mathcal{C}^m(\Omega)$, where Ω is an open subset of $\mathbb{R}^d$ and $m \in \mathbb{N}$. Then, for every $\beta \in \mathbb{N}^d$ such that $|\beta| \leq m$,*

$$f_1 \partial^\beta f_2 = \sum_{\sigma \in \mathbb{N}^d:\, 0 \leq \sigma \leq \beta} (-1)^{|\beta-\sigma|} C_\beta^\sigma \partial^\sigma (f_2 \partial^{\beta-\sigma} f_1).$$

where $C_\beta^\sigma = \beta!/(\sigma!\,(\beta-\sigma)!)$, $\beta! = \beta_1! \ldots \beta_d!$, $n! = 1 \times 2 \times \cdots \times n$ and $\sigma \leq \beta$ means that $\sigma_i \leq \beta_i$ for every i. □

Image under a multilinear mapping. We denote by $\mathcal{L}^I(E_1 \times \cdots \times E_I; F)$ the vector space of continuous multilinear mapping from a product $E_1 \times \cdots \times E_I$ of separated semi-normed spaces into a separated semi-normed space F.

Let us give a continuity property [Vol. 2, Theorem 3.11 (a)].

Theorem A.62.– *Let $T \in \mathcal{L}^I(E_1 \times \cdots \times E_I; F)$, where $E_1, \ldots, E_I$ and F are separated semi-normed spaces, Ω be an open subset of $\mathbb{R}^d$ and $0 \leq m \leq \infty$. Then, the mapping $(f_1, \ldots, f_I) \mapsto T(f_1, \ldots, f_I)$ is continuous from $\mathcal{C}^m(\Omega; E_1) \times \cdots \times \mathcal{C}^m(\Omega; E_I)$ into $\mathcal{C}^m(\Omega; F)$.* □

Let us bound the image under a multilinear mapping [Vol. 2, Theorem 3.9 (a), for normed spaces].

Theorem A.63.– *Let $f_1 \in \mathcal{C}^m(\Omega; E_1), \ldots, f_I \in \mathcal{C}^m(\Omega; E_I)$ and $T \in \mathcal{L}^I(E_1 \times \cdots \times E_I; F)$, where Ω is an open subset of $\mathbb{R}^d$, $E_1, \ldots, E_I$ and F are normed spaces, and $m \in \mathbb{N}$. Let $c \geq 0$ be such that, for every $u_1 \in E_1, \ldots, u_I \in E_I$,*

$$\|T(u_1, \ldots, u_I)\|_F \leq c \prod_{i=1}^{I} \|u_i\|_{E_i}.$$

Then, for every $\beta \in \mathbb{N}^d$ such that $|\beta| \leq m$ and $x \in \Omega$,

$$\|\partial^\beta (T(f_1, \ldots, f_I))(x)\|_F \leq c I^{|\beta|} \prod_{i=1}^{I} \sup_{|\alpha| \leq |\beta|} \|\partial^\alpha f_i(x)\|_{E_i}. \ \square$$

Derivative of a composite mapping

Partial derivatives. Let us state the expression for the partial derivatives after a change of variable [Vol. 2, Theorems 3.13 (a) and 3.12]. In dimension one, it is the **chain rule theorem**.

Theorem A.64.– *Let $f \in \mathcal{C}^1(\Omega; E)$, where Ω is an open subset of $\mathbb{R}^d$ and E is a separated semi-normed space, and let $T \in \mathcal{C}^1(\Lambda; \Omega)$, where Λ is an open subset of $\mathbb{R}^\ell$. Then:*

(a) *$f \circ T \in \mathcal{C}^1(\Lambda; E)$ and, for every $i \in [\![1, \ell]\!]$,*

$$\partial_i (f \circ T) = \sum_{j=1}^{d} ((\partial_j f) \circ T)\, \partial_i T_j.$$

(b) *In particular, if $d = \ell = 1$,*

$$(f \circ T)' = (f' \circ T)\, T'. \ \square$$

Recall that the composition of regular mappings is regular [Vol. 2, Theorem 3.13].

Theorem A.65.– *Let $f \in \mathcal{C}^m(\Omega; E)$, where Ω is an open subset of $\mathbb{R}^d$, E is a separated semi-normed space and $0 \leq m \leq \infty$, and let $T \in \mathcal{C}^m(\Lambda; \Omega)$, where Λ is an open subset of $\mathbb{R}^\ell$. Then,*

$$f \circ T \in \mathcal{C}^m(\Lambda; E). \ \square$$

Inverse functions. Let us differentiate an inverse function [Vol. 1, Theorem 22.9].

Theorem A.66.– *Let f be a continuous and strictly monotonous real function on a bounded interval (a, b) of $\mathbb{R}$, differentiable at a point $s \in (a, b)$ and such that $f'(s) \neq 0$.*

Then, f is invertible, f^{-1} is continuous, strictly monotonous, differentiable at the $f(s)$, and

$$(f^{-1})'(f(s)) = \frac{1}{f'(s)}. \ \square$$

Some real functions

Powers. Let us state some properties of real powers [Vol. 1, Theorems 22.14 and 22.15].

Theorem A.67.– *For every $s \in \mathbb{R}$, the function $x \mapsto x^s$ is infinitely differentiable from $(0, \infty)$ into $\mathbb{R}$ and, for every $x > 0$,*

$$\frac{\mathrm{d}x^s}{\mathrm{d}x} = s x^{s-1}.$$

For every $x > 0$ and $y > 0$,

$$|x^s - y^s| \leq |s||x - y| \sup\{x^{s-1}, y^{s-1}\}. \ \square$$

Let us bound the sum of a geometric series [Vol. 1, Property (1.10), p. 7].

Theorem A.68.– *Let $x \in \mathbb{R}$, $0 < x < 1$ and consider two integers be such that $n \leq m$. Then:*

$$x^n + x^{n+1} + \cdots + x^m = \frac{x^n - x^m}{1 - x} < \frac{x^n}{1 - x}. \ \square$$

Logarithms. We state some properties of logarithms.

Theorem A.69.– *The function* log *is infinitely differentiable from* $(0, \infty)$ *into* $\mathbb{R}$ *and, for every* $x > 0$,

$$\frac{\mathrm{d} \log x}{\mathrm{d}x} = \frac{1}{x}.$$

For every $x > 0$ *and* $t > 0$,

$$\log\Big(\frac{1}{x}\Big) = -\log x; \quad |\log x| \le \frac{x^{-t}}{t} \textit{ si } x \le 1; \quad |\log x| \le \frac{x^t}{t} \textit{ si } x \ge 1. \ \square$$

Proof. Theorem 22.19 from Volume 1 provides the differentiability properties.

We have $\log(1/x) = -\log x$, since $\log(xy) = \log x + \log y$ [Vol. 1, Theorem 22.18] and $\log 1 = 0$.

Theorem 22.17 from Volume 1 provides the stated upper bound for $x \le 1$ and, for every x, the inequality $(1 - x^{-t})/t \le \log x \le (x^t - 1)/t$. For $x \ge 1$, this gives $0 \le \log x \le (x^t - 1)/t \le x^t/t$. □

Determinant. Let us state some preliminary definitions.

Definition A.70.– *A* ***permutation*** *of* $(1, \dots, d)$ *is any family* $p = (p_1, \dots, p_d)$ *composed of the same elements ordered differently (or identically!). Two integers* i *and* j *such that* $1 \le i < j \le d$ *are said to be* ***inverted*** *by such a permutation if* $p_i > p_j$.

The ***signature*** $\varepsilon(p)$ *of such a permutation is the number* 1 *if it inverts an even number of integers, and* -1 *otherwise. Therefore,*

$$\varepsilon(p) = \prod_{1 \le i < j \le d} \operatorname{sign}(p_j - p_i). \ \square$$

Let us define the determinant of d vectors in $\mathbb{R}^d$.

Definition A.71.– *The* ***determinant*** *of* d *vectors* $v^1, v^2, \dots, v^d$ *in* $\mathbb{R}^d$ *is the real number*

$$\det[v^1, \dots, v^d] \stackrel{\text{def}}{=} \sum_{p \in \mathcal{P}_d} \varepsilon(p)\, v^1_{p_1} \cdots v^d_{p_d}$$

where $\mathcal{P}_d$ *is the set of permutations of* $(1, \dots, d)$, $\varepsilon(p)$ *is the signature of* p *and* v^j_i *is the* i*-th component of* v^j*, namely* $v^j = (v^j_1, \dots, v^j_d)$. □

Let us give some properties of the determinant [Vol. 2, Theorem 5.14 (b) and (f)].

Theorem A.72.– (a) $\det[\mathrm{e}_1, \dots, \mathrm{e}_d] = 1.$

(b) *The absolute value of the determinant of* d *vectors of* $\mathbb{R}^d$ *does not depend on their order.* □

The **product** $[v] = [a]\,[b]$ of two $d \times d$ matrices is defined by

$$v^j_i = \sum_{k=1}^{d} a^k_i b^j_k.$$

Recall that the determinant of a product is the product of the determinants [Vol. 2, Theorem 5.15]:

Theorem A.73.– *If* $[a]$ *and* $[b]$ *are two* $d \times d$ *matrices,*

$$\det([a]\,[b]) = \det[a] \det[b]. \ \square$$

A.5. Integration of uniformly continuous functions

Basic properties

Linearity and image. Let us commence with the linearity of the Cauchy integral [Vol. 2, Theorem 4.11].

Theorem A.74.– *Given an open subset Ω of $\mathbb{R}^d$, ω an open subset of Ω and E a Neumann space, the mapping $f \mapsto \int_\omega f$ is linear from $\mathcal{B}(\Omega; E)$ into E.* □

We arrive now to the commutation of continuous linear mappings [Vol. 2, Theorem 4.13].

Theorem A.75.– *Let $f \in \mathcal{B}(\Omega; E)$, where Ω is an open subset of $\mathbb{R}^d$ and E is a Neumann space, and L be a continuous, or sequentially continuous, linear mapping from E into a Neumann space F. Then, $Lf \in \mathcal{B}(\Omega; F)$ and, for every open subset ω of Ω,*

$$\int_\omega Lf = L \int_\omega f. \ \square$$

Growth. Let us give some growth properties of the real integral [Vol. 2, Theorem 4.14]. Recall that $f \leq g$ means that $f(x) \leq g(x)$ for every x.

Theorem A.76.– *Let $f \in \mathcal{B}(\Omega)$ and $g \in \mathcal{B}(\Omega)$, where Ω is an open subset of $\mathbb{R}^d$, and let ω and λ be open subsets of Ω. Then:*

(a) $$f \leq g \quad \Rightarrow \quad \int_\omega f \leq \int_\omega g.$$

(b) $$\omega \subset \lambda,\ f \geq 0 \quad \Rightarrow \quad \int_\omega f \leq \int_\lambda f.$$

(c) $$f \neq 0,\ f \geq 0 \quad \Rightarrow \quad \int_\Omega f > 0. \ \square$$

Dependence on the domain of integration. Let us observe that the set where the function is zero does not contribute to the integral [Vol. 2, Theorem 4.17].

Theorem A.77.– *Let $f \in \mathcal{B}(\Omega; E)$, where Ω is an open subset of $\mathbb{R}^d$ and E is a Neumann space, be such that $f = 0_E$ on $\Omega \setminus \omega$, where ω is an open subset of Ω. Then,*

$$\int_\Omega f = \int_\omega f.$$

This equality is satisfied by $\omega = \{x \in \Omega : f(x) \neq 0_E\}$ and by $\omega = \mathring{\text{supp}}\, f$. □

Observe that $\{x \in \Omega : f(x) \neq 0_E\} \subset \mathring{\text{supp}}\, f$ and that these two sets can be distinct (for example if $f(x) = |x|$, then $\{x \in \mathbb{R} : f(x) \neq 0\} = \mathbb{R} \setminus \{0\}$ and $\mathring{\text{supp}}\, f = \text{supp}\, f = \mathbb{R}$).

Integration and differentiation in dimension one. For s and t real or infinite, $s \leq t$, we denote, when these quantities are defined,

$$\int_s^t f \overset{\text{def}}{=} \int_{]s,t[} f, \quad \int_t^s f \overset{\text{def}}{=} -\int_{]s,t[} f.$$

Let us state **Chasles' relation**, which is an additivity property [Vol. 2, Theorem 6.2].

Theorem A.78.– *Let $f \in \mathcal{B}((a,b);E)$, where $-\infty \le a \le b \le \infty$ and E is a Neumann space. Then, for every r, s and t in $[a,b]$,*

$$\int_r^t f = \int_r^s f + \int_s^t f. \ \square$$

Let us give the expression for the integral of a derivative, which is the second part of the **fundamental theorem of calculus** [Vol. 2, Theorem 6.4].

Theorem A.79.– *Let $f \in \mathcal{C}^1((a,b);E)$, where $-\infty \le a < b \le \infty$ and E is a Neumann space. Then, for every s and t in (a,b),*

$$\int_s^t f' = f(t) - f(s). \ \square$$

Calculation of some integrals

Additivity. Let us give an additivity property with respect to **crowns** [Vol. 2, Theorem 10.6] which is analogous to Chasles' relation.

Theorem A.80.– *Let $0 \le a \le r \le b \le \infty$ and E be a Neumann space. Then:*

(a) *If $f \in \mathcal{B}(C_{a,b};E)$, where $C_{a,b} \overset{\text{def}}{=} \{x \in \mathbb{R}^d : a < |x| < b\}$,*

$$\int_{C_{a,b}} f = \int_{C_{a,r}} f + \int_{C_{r,b}} f.$$

(b) *If $f \in \mathcal{B}(\mathring{B}_b;E)$, where $\mathring{B}_b \overset{\text{def}}{=} \{x \in \mathbb{R}^d : |x| < b\}$,*

$$\int_{\mathring{B}_b} f = \int_{\mathring{B}_a} f + \int_{C_{a,b}} f. \ \square$$

Integral of powers of $|x|$. Let us integrate $|x|^s$ on a crown [Vol. 2, Theorem 10.6].

Theorem A.81.– *Let $C_{a,b} = \{x \in \mathbb{R}^d : a < |x| < b\}$, where $d \ge 2$ and $0 < a < b < \infty$, $s \in \mathbb{R}$ and $v_d = |\mathring{B}_{\mathbb{R}^d}|$ be the measure of the open unit ball. Then,*

$$\int_{C_{a,b}} |x|^s \, \mathrm{d}x = d v_d \frac{b^{s+d} - a^{s+d}}{s+d} \le \frac{d v_d}{s+d} b^{s+d}. \ \square$$

We give a (non-optimal) upper bound of the integral of the function $x \mapsto (2+|x|)^{-2-d}$ that is independent of the domain of integration [Vol. 2, Lemma 4.23].

Lemma A.82.– *The function $x \mapsto 1/(2+|x|)^{d+2}$ is uniformly continuous on every bounded open subset Ω of $\mathbb{R}^d$ and*

$$\int_\Omega \frac{1}{(2+|x|)^{d+2}} \, \mathrm{d}x \le 2^{d+1}. \ \square$$

Measure of an open set. Let us define the Lebesgue measure of an open subset of $\mathbb{R}^d$.

Definition A.83.– *The **measure** of an open subset ω of $\mathbb{R}^d$ is the nonnegative real or infinite number*

$$|\omega| \overset{\text{def}}{=} \lim_{n \to \infty} 2^{-nd} k_n,$$

where $n \in \mathbb{N}$ and k_n is the number of $s \in \mathbb{Z}^d$ such that the closed cube $\Delta_{s,n}$ of side length 2^{-n} centered at $2^{-n}s$ is included in ω. $\square$

Let us give the measure of a ball [Vol. 2, Theorem 5.4].

Theorem A.84.– *Let* $a \in \mathbb{R}^d$, $0 \le r \le \infty$ *and* $\breve{B}(a,r) = \{x \in \mathbb{R}^d : |x-a| < r\}$. *Then, denoting* $v_d = |\mathring{B}(0,1)|$,

$$|\mathring{B}(a,r)| = v_d r^d. \ \square$$

Change of variables

General change. Let us perform a change of variables in an integral [Vol. 2, Theorem 6.14].

Theorem A.85.– *Let* $f \in \mathcal{B}(\Omega; E)$, *where* Ω *is an open subset of* $\mathbb{R}^d$ *and* E *is a Neumann space, and* T *be a bijection from an open subset* Λ *of* $\mathbb{R}^d$ *onto* Ω *such that* $T \in \mathbf{C}^1_b(\Lambda; \Omega)$ *and* $T^{-1} \in \mathbf{C}^1_b(\Omega; \Lambda)$. *Then:*

(a) $(f \circ T)\,|\det[\nabla T]| \in \mathcal{B}(\Lambda; E)$ *and*

$$\int_{T(\Lambda)} f = \int_\Lambda (f \circ T)\,|\det[\nabla T]|.$$

(b) *Property (a) is also true if, instead of* $T^{-1} \in \mathbf{C}^1_b(\Omega; \Lambda)$, *we suppose that* Λ *is bounded and* $T^{-1} \in \mathcal{C}^1(\Omega; \Lambda)$.

(c) $f\,|\det[\nabla T^{-1}]| \in \mathcal{B}(\Omega; E)$ *and*

$$\int_\Lambda f \circ T = \int_{T(\Lambda)} f\,|\det[\nabla T^{-1}]|. \ \square$$

Particular changes. Let us begin with translation [Vol. 2, Theorem 6.16].

Theorem A.86.– *Let* $f \in \mathcal{B}(\Omega; E)$, *where* Ω *is an open subset of* $\mathbb{R}^d$ *and* E *is a Neumann space. Then, for every* $z \in \mathbb{R}^d$,

$$\int_\Omega f(x)\,\mathrm{d}x = \int_{\Omega - z} f(y+z)\,\mathrm{d}y,$$

where $\Omega - z = \{x - z : x \in \Omega\}$. $\square$

Let us continue with symmetry [Vol. 2, Theorem 6.18].

Theorem A.87.– *Let* $f \in \mathcal{B}(\Omega; E)$, *where* Ω *is an open subset of* $\mathbb{R}^d$ *and* E *is a Neumann space. Then,*

$$\int_\Omega f(x)\,\mathrm{d}x = \int_{-\Omega} f(-y)\,\mathrm{d}y. \ \square$$

Let us now consider a homothety [Vol. 2, Theorem 6.19].

Theorem A.88.– *Let* $f \in \mathcal{B}(\Omega; E)$, *where* Ω *is an open subset of* $\mathbb{R}^d$ *and* E *is a Neumann space. Then, for every* $t > 0$,

$$\int_\Omega f(x)\,\mathrm{d}x = t^d \int_{\Omega/t} f(ty)\,\mathrm{d}y,$$

where $\Omega/t = \{x/t : x \in \Omega\}$. $\square$

Multiple integrals

Separation of variables. Let us recall that it is the same to integrate with respect to several variables simultaneously or successively [Vol. 2, Theorem 6.7].

Theorem A.89.– *Let $f \in \mathbf{C}_{D_1 \times D_2}(\Omega_1 \times \Omega_2; E)$, where Ω_1 is an open subset of $\mathbb{R}^{d_1}$, Ω_2 is an open subset of $\mathbb{R}^{d_2}$, D_1 is a compact subset of $\mathbb{R}^{d_1}$, D_2 is a compact subset of $\mathbb{R}^{d_2}$ and E is a Neumann space.*

Let $\underline{f}$ be the function obtained by separating the variables, namely $(\underline{f}(x_1))(x_2) \stackrel{\text{def}}{=} f(x_1, x_2)$. Then, $\underline{f} \in \mathbf{C}_{D_1}(\Omega_1; \mathbf{C}_{D_2}(\Omega_2; E))$ and, for every open subset ω_1 of Ω_1 and ω_2 of Ω_2,

$$\int_{\omega_1 \times \omega_2} f = \int_{\omega_2} \int_{\omega_1} \underline{f} = \int_{\omega_1} \int_{\omega_2} \underline{f}. \ \square$$

Let us integrate a tensor product [Vol. 2, Theorem 6.9].

Theorem A.90.– *Let $f_1 \in \mathcal{B}(\Omega_1)$ and $f_2 \in \mathcal{B}(\Omega_2)$, where Ω_1 is an open subset of $\mathbb{R}^{d_1}$ and Ω_2 is an open subset of $\mathbb{R}^{d_2}$. Then, $f_1 \otimes f_2 \in \mathcal{B}(\Omega_1 \times \Omega_2)$ and, for every open subset ω_1 of Ω_1 and ω_2 of Ω_2,*

$$\int_{\omega_1 \times \omega_2} f_1 \otimes f_2 = \Big(\int_{\omega_1} f_1\Big)\Big(\int_{\omega_2} f_2\Big). \ \square$$

Permutation of variables. Observe that permutation of variables does not change the integral [Vol. 2, Theorem 6.6].

Theorem A.91.– *Let $f \in \mathbf{C}_{D_1}(\Omega_1; \mathbf{C}_{D_2}(\Omega_2; E))$, where Ω_1 is an open subset of $\mathbb{R}^{d_1}$, D_1 is a compact subset of $\mathbb{R}^{d_1}$, Ω_2 is an open subset of $\mathbb{R}^{d_2}$, D_2 is a compact subset of $\mathbb{R}^{d_2}$ and E is a Neumann space.*

Let $\check{f}$ be the function obtained by permutating the variables, namely $(\check{f}(x_2))(x_1) \stackrel{\text{def}}{=} (f(x_1))(x_2)$. Then, $\check{f} \in \mathbf{C}_{D_2}(\Omega_2; \mathbf{C}_{D_1}(\Omega_1; E))$ and, for every open subset ω_1 of Ω_1 and ω_2 of Ω_2,

$$\int_{\omega_1} \int_{\omega_2} \check{f} = \int_{\omega_2} \int_{\omega_1} f. \ \square$$

Differentiating under the integral sign. Let us differentiate with respect to a variable an integral with respect to another variable [Vol. 2, Theorem 4.27].

Theorem A.92.– *Let $f, g_1, \ldots, g_d$ be functions of $\mathbf{C}_\mathrm{b}(\Omega \times \Lambda; E)$, where Ω is an open subset of $\mathbb{R}^d$, Λ is a bounded open subset of $\mathbb{R}^\ell$ and E is a Neumann space, be such that, for every $y \in \Lambda$, the function $x \mapsto f(x, y)$ is differentiable from Ω into E and its partial differentials are the functions $x \mapsto g_i(x, y)$.*

Then, the function $x \mapsto \int_\Lambda f(x, y)\,\mathrm{d}y$ belongs to $\mathbf{C}^1_\mathrm{b}(\Omega; E)$ and, for every $x \in \Omega$, the function $y \mapsto \partial_i f(x, y)$ belongs to $\mathbf{C}_\mathrm{b}(\Lambda; E)$. Furthermore, $\partial_i f(x, y) = g_i(x, y)$ and

$$\frac{\partial}{\partial x_i} \int_\Lambda f(x, y)\,\mathrm{d}y = \int_\Lambda \frac{\partial f}{\partial x_i}(x, y)\,\mathrm{d}y = \int_\Lambda g_i(x, y)\,\mathrm{d}y. \ \square$$

Bibliography

[1] AGMON, S., DOUGLIS, A. and NIRENBERG, L., Estimates near the boundary for solutions of elliptic partial differential equations satisfying general boundary conditions, I. *Comm. Pure Appl. Math.*, 12 (1959), 623–727.

[2] BERNOULLI, J. (Jacob), Analysis problematis antehac propositi. De Inventione Lineæ descendus a corpore gravi percurrendæ uniformiter. *Opera*, 421–426. Cramer and Philibert, Geneva, 1744.

[3] BERNOULLI, J. (Johann), Lectiones matematicæ de methodo integralium allisque (1691–1692). *Opera omnia* III, 385–559. Original ed. Bousquet, Lausanne and Geneva, 1742 (= Olms, Hildesheim, 1968).

[4] BOCHNER, S., *Vorlesungen über Fouriersche Integrale*, Leipzig, Akad. Verlagsgesellschaft, 1932.

[5] BOCHNER, S., Integration von Funktionen deren Werte die Elemente eines Vektorraumes sind, *Fundamenta Mat.*, 20 (1933), 262–276.

[6] BOLZANO, B., Ein Analytischer Beweis der Lehrsatzes, daß zwischen je zwei Werthen, die ein entgegengesetzes Resultat gewähren, wenigstens eine reelle Wurzel der Gleichung liege, *Abhandlungen K. Böhm. Gesell. Wissen.*, (3), 5 (1817), 1–6 (= in French, *Revue Hist. Sci.*, 17 (1964), 136–164).

[7] BOREL, E., Sur quelques points de la théorie des fonctions, *Ann. Scient. École Normale Supérieure*, ser. 3, 12 (1895), 9–55 (= Thesis, Paris, 1894).

[8] BOURBAKI, N., Sur certains espaces vectoriels topologiques, *Ann. Inst. Fourier Grenoble*, 2 (1950), 5–16.

[9] BOURBAKI, N., *Intégration*, Hermann, 1965.

[10] BOURBAKI, N., *Espaces vectoriels topologiques*, Hermann, new edition, 1967.

[11] BOURBAKI, N., *Fonction d'une variable réelle*, Hermann, new edition, 1976.

[12] BOURBAKI, N., *Éléments d'histoire des mathématiques*, Hermann, new edition, 1974 (= *Elements of History of Mathematics*, Springer, 1998).

[13] BOURBAKI, N., *Variétés différentielles et analytiques : fascicule de résultats*, Hermann, 1971.

[14] CAJORI, F., *A history of mathematical notations*, two volumes, Open Court (Chicago), 1974.

[15] CANTOR, G., *Gesammelte Abhandlungen*, Springer, 1932.

[16] CARTAN, E., *Œuvres complètes*, six volumes, Gauthiers–Villars, 1953–1955.

[17] CARTAN, H., *Cours de calcul différentiel*, Hermann, revised edition, 1977.

[18] CAUCHY, A., *Cours d'Analyse de l'École Royale Polytechnique*, Debure (Paris), 1821.

[19] CAUCHY, A., *Résumé des leçons données à l'École Royale Polytechnique sur le calcul infinitésimal*, Debure (Paris), 1823.

[20] CAUCHY, A., *Mémoire sur les intégrales définies*, Bures frères, Paris, 1825.

[21] CLIFFORD, W. K., *Elements of dynamic*, (1878) (= *Historical Math Monographs*, Cornell University).

[22] CONDORCET, N., *Mémoire sur les équations aux différence partielles*, Mémoires de l'Académie Royale des Sciences, année 1770, Paris, 1773.

[23] CORIOLIS, G.-G., *Du calcul de l'effet des machines, ou considérations sur l'emploi des moteurs et sur leur évaluation, pour servir d'introduction à l'étude spéciale des machines*, Paris, Carilian–Gœury, Libraire des corps royaux des ponts et chaussées et des mines, 1829.

[24] D'ALEMBERT, J., *Réflexions sur la cause générale des vents*, David, Paris, 1747.

[25] DIEUDONNÉ, J., Sur les fonctions continues numériques définies dans un produit de deux espaces compacts, *C. R. Acad. Sci., Paris*, 205, 593-595 (1937).

[26] DIRAC, P.A.M., The physical interpretation of the quantum dynamics, *Proc. Royal. Soc. London*, A, 113 (1926), 621–641.

[27] DREYFUSS, P., Formes différentielles exactes et fermées dans $\mathbb{R}^d$, Université côte d'Azur.

[28] DU BOIS-REYMOND, P., Fortsetzung der Erläuterungen zu den Anfangsgründen der Variationsrechnung, *Mathematische Annalen*, 15, 2 (1879), 564–576.

[29] DUGUNDJI, J., *Topology*, Allyn and Bacon, Inc., Boston, 1966.

[30] EDWARDS, R., *Functional analysis*, Winston, 1965.

[31] EUCLID, *Elementorum Libri* XV, Apud Hieronymum de Marnef & Gulilmum Canellat, Paris, 1573 (= *Les œuvres d'Euclide*, F. Peyrard (1819), A. Blanchard, Paris, 1966).

[32] EULER, L., *Institutiones calculi differentialis*, Acad. Imper. Sci. Petropoli, 1755 (*in Opera omnia, Series prima*, XI, Teubner & O. Füssli, 1911 and 1976).

[33] EULER, L., *Institutionum calculi integralis, Volumen Primum*, Acad. Imper. Sci. Petropoli, 1768 (*in Opera omnia, Series prima*, XI, Teubner & O. Füssli, 1911 and 1976).

[34] EULER, L., De formulis integrabilibus duplicatis, *Novi comentarii academiæ scientiarum Petropolitanæ*, 14 (1770), 72–103 (*in Opera omnia, Series prima*, XVII, Teubner, 1914).

[35] FOURIER, J., *Théorie analytique de la chaleur*, Didot, Paris, 1822 (reed., Gabay, Sceaux, 1988).

[36] GROTHENDIECK, A., Produits tensoriels topologiques et espaces nucéaires, *Memoirs of the American Mathematical Society*, 16, 1955.

[37] HAAR, A., Über die Variation der Doppelintegrale, *J. Reine Angew. Math. (J. de Crelle)*, 149, 1/2 (1926), 1–18.

[38] HAAR, A., Zur Variationsrechnung, *Abhandlungen aus dem Mathematischen Seminar der Universität Hamburg*, 8, 1 (1931/1932), 1–27.

[39] HATCHER, A., *Algebraic topology*, Cambridge University Press, 2002 (free version licensed by the publisher: `http://www.math.cornell.edu/~hatcher`).

[40] HAUSDORFF, F., *Grundzüge der Mengenlehre*, Leipzig, Veit, 1914.

[41] HEAVISIDE, O., On operators in Mathematical Physics, *Proc. Royal Society London*, 52 (1893), 504–529, 54 (1894), 105–143.

[42] HEINE, E., Über trigonometrische Reihen, *J. Reine Angew. Math. (J. de Crelle)*, 71 (1870), 353–365.

[43] HEINE, E., Aus brieflichen Mittelheilungen (namentlich über Variationsrechnung), *Mathematische Annalen*, 2 (1870), 187–191.

[44] HEINE, E., Die Elemente der Functionenlehre, *J. Reine Angew. Math. (J. de Crelle)*, 74 (1872), 172–188.

[45] HORVÁTH, J., *Topological vector spaces and distributions*, Addison-Wesley, 1966.

[46] KOLMOGOROV, A. N., Zur Normierbarkeit eines allgemeinen topologischen linearen Raumes, *Studia Math.*, 5 (1934), 29–33 (*Selected works*, vol. 1, Dordrecht, Kluwer (1991), 183–186).

[47] KRYLOFF, V. I., Sur l'existence des dérivées généralisées des fonctions sommables, *Doklady Akademie Nauk SSSR*, 45 (1947), 375–378.

[48] LADYZHENSKAYA, O. A., *The mathematical theory of viscous incompressible flow*, Gordon and Breach, 1963.

[49] LAGRANGE, J., *Sur une nouvelle espèce de calcul relatif à la différentiation et à l'intégration des quantités variables*, Nouveaux mémoires de l'Académie royale des sciences et belles-lettres de Berlin, année 1772, Berlin, 1774, 185–221.

[50] LEBESGUE, H., Intégrale, longueur, aire, *Ann. di Mat.*, (3), vol. VII (1902), 231–359 (= Thesis, Paris).

[51] LEIBNIZ, W. G., Nova metodus pro maximis et minimis, itemque tangentibus, quae nec fractas nec irrationales quantitates moratur, et singulare pro illis calculi genus, *Acta Eruditorum*, 1884, 467–473 (= *Mathematische Schriften*, vol. I, 220–226, Olms (Hildesheim), 1960).

[52] LEMOINE, J. and SIMON, J., Extension of distributions and representation by derivatives of continuous functions, *Atti. Accad. Naz. Lincei*, ser. 9, 7 (1996), 31–40.

[53] LERAY, J., Sur le mouvement d'un liquide visqueux emplissant l'espace, *Acta Math.*, 63 (1934), 193–248.

[54] LIONS, J.-L., *Quelques méthodes de résolution des problèmes aux limites non linéaires*, Dunod & Gauthier–Villars, 1969.

[55] LÜTZEN, J., *The Prehistory of the Theory of Distributions*, Studies in the History of Mathematics and Physical Sciences, vol. 7, Springer-Verlag, 1982.

[56] NEUMANN, J. VON, On complete topological linear spaces, *Trans. Amer. Math. Soc.*, 37 (1935), 1–20.

[57] NEWTON, I., *The mathematical papers, vol. III, 1670–1673*, Cambridge University Press, 1969.

[58] OSTROGRADSKY, M., Dissertation (in Russian), *Mémoires Acad. Sci. Saint-Pétersbourg, sc. math., phys., nat.*, 1 (1831), 19–53.

[59] PAUMIER, A.-S., Laurent Schwartz (1915-2002) et la vie collective des mathématiques, Thesis, Université Pierre et Marie Curie, Paris, 2014.

[60] PIER, J.-P., *Histoire de l'intégration*, Masson, 1996.

[61] POINCARÉ, H., *Les méthodes nouvelles de la mécanique céleste*, vol. 3, Gauthiers–Villars, 1899.

[62] RADON, J., Theorie und Anwendungen der absolut additiven Mengenfunktionen, *Sitzungsber. Akad. Wissen. Wien.*, 122 (1913), 1295–1438.

[63] DE RHAM, G., *Variétés différentiables*, Hermann, 1955.

[64] RIEMANN, B., *Gesammelte mathematische Werke*, Teubner, second ed., 1892.

[65] RIEMANN, B., Über die Darstellbarkeit einer Function durch eine trigonometrische Reihe, *Abh. K. Gesell. Wiss. Göttingen, Math. Classe*, 3 (1866-1867), 87–132.

[66] RUDIN, W., *Principles of Mathematical Analysis*, McGraw–Hill, 1953.

[67] SCHAEFER, H. H., *Topological vector spaces*, Macmillan, 1966.

[68] SCHWARTZ, L., Généralisation de la notion de fonction, de dérivation, de transformation de Fourier, et applications mathématiques et physiques, *Ann. Univ. Grenoble*, 21 (1945), 57–74.

[69] SCHWARTZ, L., *Théorie des distributions*, Actual. Sci. et Ind., 1091 and 1122, Hermann, 1950–1951 (our historical notes refer to page numbers of the *Nouvelle édition augmentée*, 1973).

[70] SCHWARTZ, L., Sur l'impossibilité de la multiplication des distributions, *C. R. Acad. Sci. Paris*, 239 (1954), 847–848.

[71] SCHWARTZ, L., Espaces de fonctions différentiables à valeurs vectorielles, *J. Analyse Math. Jérusalem*, 4 (1955), 88–148.

[72] SCHWARTZ, L., Distributions à valeurs vectorielles, I and II, *Ann. Inst. Fourier*, 7 (1957) and 8 (1959).

[73] SCHWARTZ, L., *Analyse, Topologie générale et analyse fonctionnelle*, Hermann, 1970.

[74] SCHWARTZ, L., *Analyse I. Théorie des ensembles et topologie*, Hermann, 1991.

[75] SCHWARTZ, L., *A Mathematician Grappling with His Century*, Birkhäuser, 2001 (= *Un mathématicien aux prises avec le siècle*, Odile Jacob, 1997).

[76] SCHWARZ, H. A., Zur integration der partiellen Differentialgleichung, *J. Reine Angew. Math. (J. de Crelle)*, 74 (1872), 218–253

[77] SIMON, J., Démonstration constructive d'un théorème de G. de Rham, *C. R. Acad. Sci. Paris, sér. I*, 316 (1993), 1167–1172.

[78] SIMON, J., Primitives de distributions et applications, Research Report 93-11, Laboratoire de Mathématiques Appliquées, Université Blaise Pascal, 1993.

[79] SIMON, J., Representation of distributions and explicit antiderivatives up to the boundary, in *Progress in partial differential equations: the Metz surveys 2*, M. Chipot ed., Longman, 1993, 201-205.

[80] SIMON, J., Distributions à valeurs dans espace séquentiellement complet, Research Report, Laboratoire de Mathématiques Appliquées, Université Blaise Pascal, 1996.

[81] SIMON, J., *Banach, Fréchet, Hilbert and Neumann Spaces*, Analysis for PDEs set, vol. 1, ISTE Ltd, London, and John Wiley & Sons, New York, 2017.

[82] SIMON, J., *Continuous Functions*, Analysis for PDEs set, vol. 2, ISTE Ltd, London, and John Wiley & Sons, New York, 2020.

[83] SIMON, J., *Espaces semi-normés*, in progress.

[84] SIMON, J., *Plus sur les distributions*, in progress.

[85] SMIRNOV, S. K., Decomposition of solenoidal vector charges into elementary solenoids and the structure of normal one-dimensional currents, *St. Petersburg Math. J.*, 5 (1993), 4, 841–867.

[86] SOBOLEV, S. L., Méthode nouvelle à résoudre le problème de Cauchy pour les équations hyperboliques normales, *Recueil mathématique*, 1 (1936), 39–72.

[87] STEGMANN, F. L., *Lehrbuch der Variationsrechnung und ihrer Anwendung bei Untersuchungen über das Maximum und Minimum*, J.G. Luckhardt, Kassel, 1854.

[88] TARTAR, L., *Topics in nonlinear analysis*, Mathematical Publications, Université d'Orsay, 1978.

[89] THOMAE, J., *Abriss einer Theorie der complexen Funktionen und Thetafunktionen einer Veränderlichen*, Halle (Nebert), 1870.

[90] TREVES, F., *Topological vector spaces, distributions and kernels*, Academic Press, 1967.

[91] VOLTERRA, V., *Leçons sur les fonctions de lignes*, Gauthier-Villars, 1913.

Index

D

E

F

G

H

I

J, K

L

M

N

O

T

U

V, W

Other titles from

iSTE

in

Mathematics and Statistics

2022

DE SAPORTA Benoîte, ZILI Mounir
Martingales and Financial Mathematics in Discrete Time

LESFARI Ahmed
Integrable Systems

2021

KOROLIOUK Dmitri, SAMOILENKO Igor
Random Evolutionary Systems: Asymptotic Properties and Large Deviations

MOKLYACHUK Mikhail
Convex Optimization: Introductory Course

POGORUI Anatoliy, SWISHCHUK Anatoliy, RODRÍGUEZ-DAGNINO Ramón M.
Random Motions in Markov and Semi-Markov Random Environments 1: Homogeneous Random Motions and their Applications
Random Motions in Markov and Semi-Markov Random Environments 2: High-dimensional Random Motions and Financial Applications

PROVENZI Edoardo
From Euclidean to Hilbert Spaces: Introduction to Functional Analysis and its Applications

2020

BARBU Vlad Stefan, VERGNE Nicolas
Statistical Topics and Stochastic Models for Dependent Data with Applications

CHABANYUK Yaroslav, NIKITIN Anatolii, KHIMKA Uliana
Asymptotic Analyses for Complex Evolutionary Systems with Markov and Semi-Markov Switching Using Approximation Schemes

KOROLIOUK Dmitri
Dynamics of Statistical Experiments

MANOU-ABI Solym Mawaki, DABO-NIANG Sophie, SALONE Jean-Jacques
Mathematical Modeling of Random and Deterministic Phenomena

2019

BANNA Oksana, MISHURA Yuliya, RALCHENKO Kostiantyn, SHKLYAR Sergiy
Fractional Brownian Motion: Approximations and Projections

GANA Kamel, BROC Guillaume
Structural Equation Modeling with lavaan

KUKUSH Alexander
Gaussian Measures in Hilbert Space: Construction and Properties

LUZ Maksym, MOKLYACHUK Mikhail
Estimation of Stochastic Processes with Stationary Increments and Cointegrated Sequences

MICHELITSCH Thomas, PÉREZ RIASCOS Alejandro, COLLET Bernard, NOWAKOWSKI Andrzej, NICOLLEAU Franck
Fractional Dynamics on Networks and Lattices

VOTSI Irene, LIMNIOS Nikolaos, PAPADIMITRIOU Eleftheria, TSAKLIDIS George
Earthquake Statistical Analysis through Multi-state Modeling
(Statistical Methods for Earthquakes Set – Volume 2)

2018

AZAÏS Romain, BOUGUET Florian
Statistical Inference for Piecewise-deterministic Markov Processes

IBRAHIMI Mohammed
Mergers & Acquisitions: Theory, Strategy, Finance

PARROCHIA Daniel
Mathematics and Philosophy

2017

CARONI Chysseis
First Hitting Time Regression Models: Lifetime Data Analysis Based on Underlying Stochastic Processes
(Mathematical Models and Methods in Reliability Set – Volume 4)

CELANT Giorgio, BRONIATOWSKI Michel
Interpolation and Extrapolation Optimal Designs 2: Finite Dimensional General Models

CONSOLE Rodolfo, MURRU Maura, FALCONE Giuseppe
Earthquake Occurrence: Short- and Long-term Models and their Validation
(Statistical Methods for Earthquakes Set – Volume 1)

D'AMICO Guglielmo, DI BIASE Giuseppe, JANSSEN Jacques, MANCA Raimondo
Semi-Markov Migration Models for Credit Risk
(Stochastic Models for Insurance Set – Volume 1)

GONZÁLEZ VELASCO Miguel, del PUERTO GARCÍA Inés, YANEV George P.
Controlled Branching Processes
(Branching Processes, Branching Random Walks and Branching Particle Fields Set – Volume 2)

HARLAMOV Boris
Stochastic Analysis of Risk and Management
(Stochastic Models in Survival Analysis and Reliability Set – Volume 2)

KERSTING Götz, VATUTIN Vladimir
Discrete Time Branching Processes in Random Environment
(Branching Processes, Branching Random Walks and Branching Particle Fields Set – Volume 1)

MISHURA YULIYA, SHEVCHENKO Georgiy
Theory and Statistical Applications of Stochastic Processes

NIKULIN Mikhail, CHIMITOVA Ekaterina
Chi-squared Goodness-of-fit Tests for Censored Data
(Stochastic Models in Survival Analysis and Reliability Set – Volume 3)

SIMON Jacques
Banach, Fréchet, Hilbert and Neumann Spaces
(Analysis for PDEs Set – Volume 1)

2016

CELANT Giorgio, BRONIATOWSKI Michel
Interpolation and Extrapolation Optimal Designs 1: Polynomial Regression and Approximation Theory

CHIASSERINI Carla Fabiana, GRIBAUDO Marco, MANINI Daniele
Analytical Modeling of Wireless Communication Systems
(Stochastic Models in Computer Science and Telecommunication Networks Set – Volume 1)

GOUDON Thierry
Mathematics for Modeling and Scientific Computing

KAHLE Waltraud, MERCIER Sophie, PAROISSIN Christian
Degradation Processes in Reliability
(Mathematial Models and Methods in Reliability Set – Volume 3)

KERN Michel
Numerical Methods for Inverse Problems

RYKOV Vladimir
Reliability of Engineering Systems and Technological Risks
(Stochastic Models in Survival Analysis and Reliability Set – Volume 1)

2015

DE SAPORTA Benoîte, DUFOUR François, ZHANG Huilong
Numerical Methods for Simulation and Optimization of Piecewise Deterministic Markov Processes

DEVOLDER Pierre, JANSSEN Jacques, MANCA Raimondo
Basic Stochastic Processes

LE GAT Yves
Recurrent Event Modeling Based on the Yule Process
(Mathematical Models and Methods in Reliability Set – Volume 2)

2014

COOKE Roger M., NIEBOER Daan, MISIEWICZ Jolanta
Fat-tailed Distributions: Data, Diagnostics and Dependence
(Mathematical Models and Methods in Reliability Set – Volume 1)

MACKEVIČIUS Vigirdas
Integral and Measure: From Rather Simple to Rather Complex

PASCHOS Vangelis Th
Combinatorial Optimization – 3-volume series – 2nd edition
Concepts of Combinatorial Optimization / Concepts and Fundamentals – volume 1
Paradigms of Combinatorial Optimization – volume 2
Applications of Combinatorial Optimization – volume 3

2013

COUALLIER Vincent, GERVILLE-RÉACHE Léo, HUBER Catherine, LIMNIOS Nikolaos, MESBAH Mounir
Statistical Models and Methods for Reliability and Survival Analysis

JANSSEN Jacques, MANCA Oronzio, MANCA Raimondo
Applied Diffusion Processes from Engineering to Finance

SERICOLA Bruno
Markov Chains: Theory, Algorithms and Applications

2012

BOSQ Denis
Mathematical Statistics and Stochastic Processes

CHRISTENSEN Karl Bang, KREINER Svend, MESBAH Mounir
Rasch Models in Health

DEVOLDER Pierre, JANSSEN Jacques, MANCA Raimondo
Stochastic Methods for Pension Funds

2011

MACKEVIČIUS Vigirdas
Introduction to Stochastic Analysis: Integrals and Differential Equations

MAHJOUB Ridha

Recent Progress in Combinatorial Optimization – ISCO2010

RAYNAUD Hervé, ARROW Kenneth
Managerial Logic

2010

BAGDONAVIČIUS Vilijandas, KRUOPIS Julius, NIKULIN Mikhail
Nonparametric Tests for Censored Data

BAGDONAVIČIUS Vilijandas, KRUOPIS Julius, NIKULIN Mikhail
Nonparametric Tests for Complete Data

IOSIFESCU Marius *et al.*
Introduction to Stochastic Models

VASSILIOU PCG
Discrete-time Asset Pricing Models in Applied Stochastic Finance

2008

ANISIMOV Vladimir
Switching Processes in Queuing Models

FICHE Georges, HÉBUTERNE Gérard
Mathematics for Engineers

HUBER Catherine, LIMNIOS Nikolaos *et al.*
Mathematical Methods in Survival Analysis, Reliability and Quality of Life

JANSSEN Jacques, MANCA Raimondo, VOLPE Ernesto
Mathematical Finance

2007

HARLAMOV Boris
Continuous Semi-Markov Processes

2006

CLERC Maurice
Particle Swarm Optimization

Printed and bound by CPI Group (UK) Ltd, Croydon, CR0 4YY

13/09/2026

14961039-0008